Environment and Animal Development

genes, life histories and plasticity

EXPERIMENTAL BIOLOGY REVIEWS

Series advisors:

D.W. Lawlor
AFRC Institute of Arable Crops Research, Rothamsted Experimental Station, Harpenden, Hertfordshire AL5 2JQ, UK

M. Thorndyke
School of Biological Sciences, Royal Holloway, University of London, Egham, Surrey TW20 0EX, UK

Environmental Stress and Gene Regulation
Sex Determination in Plants
Plant Carbohydrate Biochemistry
Programmed Cell Death in Animals and Plants
Biomechanics in Animal Behaviour
Cell and Molecular Biology of Wood Formation
Molecular Mechanisms of Metabolic Arrest: Life in Limbo
Environment and Animal Development: Genes, Life Histories and Plasticity

Forthcoming titles include:
Brain Stem Cells
Endocrine Interactions of Insect Parasites and Pathogens

Environment and Animal Development
genes, life histories and plasticity

D. ATKINSON
School of Biological Sciences, University of Liverpool, Liverpool, UK

M. THORNDYKE
School of Biological Sciences, Royal Holloway, University of London, Surrey, UK

Taylor & Francis
Taylor & Francis Group

LONDON AND NEW YORK

First published in 2001

A CIP catalogue record for this book is available from the British Library.

ISBN 1 85996 184 3

Published by Taylor & Francis
2 Park Square, Milton Park, Abingdon, Oxon, OX14 4RN
270 Madison Ave, New York NY 10016

Transferred to Digital Printing 2009

Part of the work described in this book was funded by the NERC, DEMA Thematic Programme.

NATURAL
ENVIRONMENT
RESEARCH COUNCIL

Production Editor: Paul Barlass.
Typeset by Saxon Graphics Ltd, Derby, UK.

Front cover: Larvae of the seastar, *Asterias rubens*, courtesy of M. Thorndyke and C. Moss

Contents

Contributors

Atkinson, D., Population and Evolutionary Research Group, School of Biological Sciences, University of Liverpool, Liverpool L69 3GS, UK

Bale, J.S., School of Biosciences, University of Birmingham, Edgbaston, Birmingham B15 2TT, UK

Beardall, V., Department of Marine Sciences and Coastal Management, Newcastle University, Newcastle Upon Tyne NE1 7RU, UK

Bentley, M.G., Department of Marine Sciences and Coastal Management, Newcastle University, Newcastle Upon Tyne NE1 7RU, UK

Birkbeck, T.H., Division of Infection and Immunity, Institute of Biomedical and Life Sciences, Joseph Black Building, University of Glasgow, Glasgow G12 8QQ, UK

Black, K.D., Dunstaffnage Marine Laboratory, PO Box 3, Oban, Argyll PA34 4AD, UK

Boyle, P.R., Department of Zoology, University of Aberdeen, Tillydrone Avenue, Aberdeen AB24 2TZ, UK

Brown, J.H., Institute of Aquaculture, University of Stirling, Stirling FK9 4LA, UK

Chisholm, J.R.S., Comparative Immunology Group, Gatty Marine Laboratory, University of St Andrews, Fife KY16 8LB, UK

Craig, S., Dunstaffnage Marine Laboratory, PO Box 3, Oban, Argyll PA34 4AD, UK

Crane, M., School of Biological Sciences, Royal Holloway, University of London, Egham, Surrey TW20 0EX, UK

Denlinger, D.L., Department of Entomology, Ohio State University, 1735 Neil Avenue, Columbus, OH 43210, USA

Eads, B.D., Department of Biological Sciences, Louisiana State University Baton Rouge, LA 70803, USA

El Haj, A.J., Keele University, Thornburrow Drive, Hartshill, North Staffordshire ST4 7QB, UK

Emery, A.M., Department of Zoology, University of Aberdeen, Tillydrone Avenue, Aberdeen AB24 2TZ, UK

Frettsome, F., Plymouth Marine Laboratory, Citadel Hill, Plymouth, Devon PL1 2PB, UK

Gotthard, K., University of Neuchatel, Institute of Zoology, Rue Emile-Argand 11, 2007 Neuchatel, Switzerland

Hand, S.C., Department of Biological Sciences, Louisiana State University, Baton Rouge, LA 70803, USA

Holmes, J.M., Keele University, Thornburrow Drive, Hartshill, North Staffordshire ST4 7QB, UK

Johnston, I.A., Gatty Marine Laboratory, School of Biology, Division of Environmental and Evolutionary Biology, University of St Andrews, Fife KY16 8LB, UK

Kingsolver, J.G., Department of Biology, CB-3280, University of North Carolina, Chapel Hill, NC 27599, USA

Last, K., Department of Marine Sciences and Coastal Management, Newcastle University, Newcastle Upon Tyne NE1 7RU, UK

Lewis, C., Department of Marine Sciences and Coastal Management, Newcastle University, Newcastle Upon Tyne NE1 7RU, UK

Lewis, D., IBLS, The Graham Kerr Building, University of Glasgow, Glasgow T12 8QQ, UK

Livingstone, D.R., Plymouth Marine Laboratory, Citadel Hill, Plymouth, Devon PL1 2PB, UK

Logue, J.A., Department of Animal Biology, University of Illinois at Urbana-Champaign, 515 Morrill Hall, 505 South Goodwin Ave., Il 61801, USA

Low, C., Department of Zoology, University of Aberdeen, Tillydrone Avenue, Aberdeen AB24 2TZ, UK

Melvin, W., Department of Molecular and Cell Biology, University of Aberdeen, Aberdeen AB25 2ZD, UK

Morritt, D., School of Biological Sciences, Royal Holloway, University of London, Egham, Surrey TW20 0EX, UK

Neil, D.M., IBLS, The Graham Kerr Building, University of Glasgow, Glasgow T12 8QQ, UK

Nice, H.E., School of Biological Sciences, Royal Holloway, University of London, Egham, Surrey TW20 0EX, UK

Noble, L.R., Department of Zoology, University of Aberdeen, Tillydrone Avenue, Aberdeen AB51 24TZ, UK

O'Hara, S.C.M., Plymouth Marine Laboratory, Citadel Hill, Plymouth, Devon PL1 2PB, UK

Olive, P.J.W. Department of Marine Sciences & Coastal Management, Newcastle University, Newcastle Upon Tyne NE1 7RU, UK

Overnell, J., Dunstaffnage Marine Laboratory, PO Box 3, Oban, Argyll PA34 4AD, UK

Podrabsky, J.E., Section of Integrative Physiology and Neuroscience, Department of EPO Biology, University of Colorado, Boulder, Colorado, 80309–0334, USA

Rees, H.H., School of Biological Sciences, University of Liverpool, Crown Street, Liverpool L69 7ZB, UK

Rogers, A.D., School of Ocean and Earth Science, University of Southampton, Southampton Oceanography Centre, European Way, Southampton SO14 3ZH, UK

Rubio, A., Keele University, Thornburrow Drive, Hartshill, North Staffordshire ST4 7QB, UK

Rundle, J., Plymouth Marine Laboratory, Citadel Hill, Plymouth, Devon PL1 2PB, UK

Secombes, C., Department of Zoology, University of Aberdeen, Tillydrone Avenue, Aberdeen AB24 2TZ, UK

Smith, V.J., Comparative Immunology Group, Gatty Marine Laboratory, University of St Andrews, Fife KY16 8LB, UK

Tatner, M.F., Division of Infection and Immunity, Institute of Biomedical and Life Sciences, Joseph Black Building, University of Glasgow, Glasgow G12 8QQ, UK

Taylor, I., Division of Infection and Immunity, Institute of Biomedical and Life Sciences, Joseph Black Building, University of Glasgow, Glasgow G12 8QQ, UK

Temple, G.K., Gatty Marine Laboratory, School of Biology, Division of Environmental and Evolutionary Biology, University of St Andrews, Fife KY16 8LB, UK

Thorndyke, M., School of Biological Sciences, Royal Holloway, University of London, Egham, Surrey TW20 0EX, UK

Van Breukelen, F., Section of Integrative Physiology and Neuroscience, Department of EPO Biology, University of Colorado, Boulder, Colorado, 80309–0334, USA

Vieira, V.L.A., Gatty Marine Laboratory, School of Biology, Division of Environmental and Evolutionary Biology, University of St Andrews, Fife KY16 8LB, UK

Wainwright, G., School of Biological Sciences, University of Liverpool, Crown Street, Liverpool L69 7ZB, UK

Walters, K.F.A., Central Science Laboratory, Sand Hutton, York YO41 1LZ

Abbreviations

AA	arachidonic acid
AchE	acetyl cholinesterase
AFLP	amplified fragment length polymorphism
AMPK	AMP-activated protein kinase
AP-PCR	arbitrary-primer PCR
APE	alkylphenol polyethoxylates
ARDRA	amplified ribosomal DNA restriction analysis
bHLH	basic helix–loop–helix
C	constant
CD	circular dichroism
CDR	complementarity-determining regions
CHH	crustacean hyperglycaemic hormone
CSA	cross-sectional area
CYP	cytochrome P450
D	diversity
DALP	direct amplification of length polymorphism
DGGE	denaturing gradient gel electrophoresis
DHA	docosahaxaenoic acid
dnMS	dominant negative myostatin transgenic
ECR	ecdysone receptor
EMG	electromyogram
EPA	eicosapentaenoic acid
GIH	gonad-inhibiting hormone
GPX	glutathione peroxidase
GSH	reduced glutathione
HGF	hepatocyte growth factor
IFNγ	interferon gamma
Ig	immunoglobulin
IL-1	interleukin-1
iNOS	inducible NO synthase
ISSR	inter-simple sequence repeat
J	joining
KMBA	2-keto-4-methiolbutyric acid
L	fork length
L-DOPA	L-3,4-dihydroxyphenylalanine
LC-PUFA	long-chain polyunsaturated fatty acids
LDH	lactate dehydrogenase
LPS	lipopolysaccharides
MABs	monoclonal antibodies
MAF	macrophage-activating factor
MDA	malondialdehyde
MF	methyl farnesoate

MHS	multiplex halotype-specific
MIH	moult-inhibiting hormone
MLC1	myosin light chain 1
MO-IH	mandibular organ-inhibiting hormone
mtTfA	mitochondrial transcription factor A
4-NP	4-nonylphenol
1,4-NQ	1,4-naphthaquinone
p.f.	post-fertilization
PAH	polynuclear aromatic hydrocarbon
PCB	polychlorinated biphenyl
PCNA	proliferating cell nuclear antigen
PCR	polymerase chain reaction
PE	phosphatidyl ethanolamine
proPO	prophenoloxidase
PSA	preselective amplification
R.H.	relative humidity
RAG	recombination-activating genes
RAPD	randomly amplified polymorphic DNA
RFLP	restriction fragment length polymorphism
RGR	relative growth rate
ROS	reactive oxygen species
RSS	recombination signal sequences
SA	selective amplification
SCP	supercooling point
smu	slow-muscle-omitted
SOD	superoxide dismutase
SSCP	single-strand conformation polymorphism
SSMC	sea-surface microlayer
SSR	simple sequence repeat
TBA	thiobarbituric acid
TCDD	2,3,7,8-tetrachlorodibenzo-*p*-dioxin
TCR T	cell receptor
TGFβ	transforming growth factor beta
TGGE	temperature-gradient gel electrophoresis
TL	total length
TNFα	tumour necrosis factor alpha
TOSC	total oxidant scavenging capacity
TPC	thermal performance curve
USP	ultraspiracle
V	variable
VHSV	viral haemorrhagic septicaemia virus
VNTR	variable number of tandem repeats

Preface

We believe that the study of interactions between environment and animal development is about to enjoy a resurgence similar to that which has happened to studies that link evolution and development (Evolutionary Developmental Biology) in recent years. A similar sense of expectation is echoed by S.F. Gilbert (2000) in his latest edition of *Developmental Biology* (6th edn, Sinauer):'The exploration of environmental regulation of development is just beginning' (p 674).

We suggest that this predicted growth in interest will result from both increased 'supply' and 'demand'. The 'demand' for research in this area comes from increasing environmental threats and the realization that to understand and successfully predict population responses to changes in weather, climate, pollution, species invasions and introductions usually requires understanding of developmental (including life history) responses to the environment. Animal production and aquaculture also require such understanding, as does medicine and the veterinary profession. One response to this demand was a Natural Environment Research Council (UK) Thematic Programme known as DEMA (Developmental Ecology of Marine Animals). Many of the chapters that contribute to this volume result from research funded under DEMA. DEMA was established because of the realization that most studies of the responses and adaptations of marine species to environmental changes had focused on adult life stages while embryonic and larval stages had been neglected. This was unfortunate because early stages are often more vulnerable to environmental challenge than are adults, Considerable variation in the ability of particular developmental stages to survive a particular environmental challenge has been documented and certain developmental stages and transitions between them have higher survival than others. Consequently, differences in performance and survival among individuals, stages or cohorts may have profound effects on population size, structure and evolution. The research reflected in this volume is concerned with the processes, mechanisms and adaptations involved in environmental effects acting during ontogeny, but especially during early development and its longer term consequences.

We also expect accelerated progress in the 'supply' of scientific solutions to problems. A major influence here is the recent advances in molecular techniques. Understanding should increase as biologists take on the challenge of integrating knowledge of the molecular basis of development with that of developmental and ecological processes at different levels of organisation, and the adaptive significance of phenotypic responses. There are thus many approaches to understanding interactions between environment and development, and this book illustrates this diversity using a wide range of taxa, habitats and techniques. To achieve an integrated understanding will require dialogue between people from different specialisms. A first step towards this is to bring examples of different approaches together into one place, which is what we have done in this book.

The authors in this volume were also contributors to two Society for Experimental Biology symposia held in March 2000 – 'Temperature and Development' and 'Developmental Ecology of Marine Animals' organised by us. We thank the Society for

Experimental Biology, the Company of Biologists, and the Natural Environment Research Council DEMA Programme for financial support. At BIOS, Jonathan Ray showed faith in the concept of the book, and Victoria Oddie and Paul Barlass dealt efficiently with the chapters, and pressed firmly but with understanding when family health problems caused delays in delivery of some material. Acknowledgements are given in individual chapters, but we wish to add the name of Dave Weetman for his helpful comments on Chapter 1. We thank Hayley, Harmony and Patrik (DA) and Lesley (MT) for their tolerance and loving support despite the anti-social hours spent at the computer or with copies of manuscripts and proofs.

D. Atkinson and M. Thorndyke

The nature of animal developmental ecology: integrating the 'how' and the 'why'

D. Atkinson and M. Thorndyke

1. Introduction

Animal development starts with a single cell and progresses through programmed division and differentiation of cells into different functional types, the organization of these into tissues and organs (morphogenesis) and an increase in size (growth) (Gilbert, 2000). On reaching reproductive maturity, the adult can reproduce and continue the developmental cycle. Developmental changes in the broadest sense do not cease at 'maturity' but continue during the adult period as the animal senesces and growth and reproduction often continue. *Figure 1* shows a generalized list of developmental activities that can occur throughout the ontogeny of an animal. Animal developmental biology is concerned mainly with understanding the mechanisms that produce the sequential pattern of changes (or developmental trajectory). Apart from changes along this trajectory (ontogenetic variation, V_o) within an individual, natural developmental variation can be partitioned between individuals into that caused by direct environmental influences (V_e), that due to genetic differences (V_g), and that due to interactions among some or all of these (*Figure 2a,b*). Assuming no observational error, any remaining variation may result from developmental noise arising from instability of processes controlling ontogenetic change, which may itself vary with genotype, environment or stage in ontogeny. The ecological consequences of all these types of variation are within the realm of developmental ecology. In addition, this book examines further how the environment affects development, either alone or interacting with genotype, ontogeny or both (*Figure 2b*). This environmental control of development has been termed 'environmental developmental biology' (Gilbert, 2000). Van der Weele (1999) argues forcefully that environmental influences on development have been much neglected compared to genetic influences.

Environment and Animal Development: Genes, Life Histories and Plasticity, edited by D. Atkinson and M. Thorndyke.

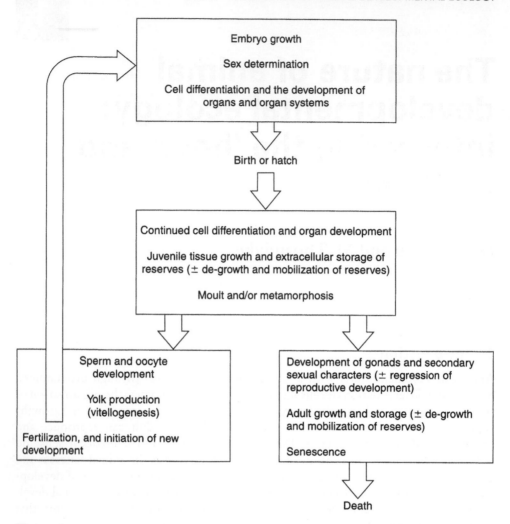

Figure 1. Major developmental activities during animal ontogeny. Not all activities apply to all animals (e.g. moult, metamorphosis).

Examples from this chapter and throughout the book illustrate the ecological importance of the many interactions between environment and developmental processes. Species invasions and introductions, climate change, pollution and natural environmental variation associated with changing seasons, weather and species dynamics can all have profound effects on animal development which lead to major population change. Also, differences in pattern of development between species or genotypes can determine which will thrive or decline under particular environmental conditions. Knowledge of animal development ecology should therefore help in the identification of those species, developmental processes and phases (or 'windows') that will be most susceptible to particular environmental perturbations, and can help us evaluate whether introduced or invading species will be able to establish in a new area. Such information should be of value for the management of biological resources including sustainable harvesting, pest control and species conservation.

(a)

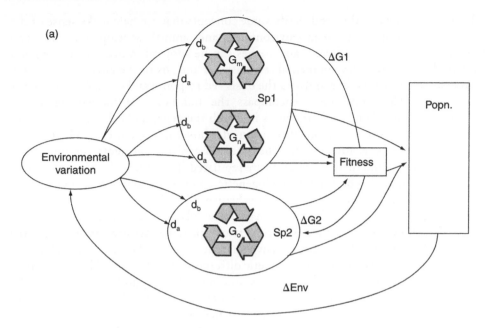

(b)

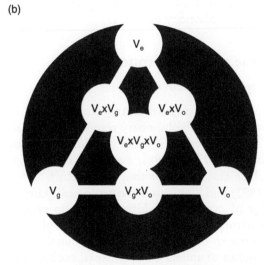

Figure 2. (a) A schematic representation of the main elements of animal developmental ecology. Environmental variation in space and/or time influence one or more developmental processes or events (d_a, d_b ...) (see Figure 1) of one or more genotypes (G_m, G_n, G_o, ...) from one or more species (Sp1, Sp2, ...). Animal developmental ecologists are concerned with those responses that have significant effects on animal performance, hence fitness, and/or consequent changes in populations (Popn.). Effects on fitness will affect the frequency distribution of genetically different developmental trajectories that will be available to respond to environmental change in subsequent generations (evolution; arrows $\Delta G1$ and $\Delta G2$). Population change influences environmental conditions (e.g. crowding), thus affecting subsequent interactions between environment and development (e.g. causing locust phase shifts) (arrow ΔEnv). Variability in developmental trajectories on which environmental change can act also derives from the ecological and evolutionary history of the taxa. (b) A conceptual (not a formal quantitative genetics) representation of variation in development, caused by differences between individuals in environment (V_e) and genotype, including species (V_g), also within an individuals ontogeny (V_o), and interactions among all these. Animal developmental ecology is concerned with the ecological consequences of all of these plus developmental instability (errors). In this book we emphasize environmental effects on development (environmental developmental biology) either alone or interacting with genotype, ontogeny or both (i.e. all types of variation that include the term V_e).

In this chapter we outline and clarify some concepts that can help make sense of the complex relationships between environment and animal development. We also describe how two quite different approaches can be adopted in order to understand ecologically important developmental responses of animals to the environment. One approach attempts to understand *how* the observed response and its ecological impact are achieved. This 'how' must encompass the full range of modern biological approaches to include not only the ways in which the genes and cells respond (functional genomics/proteomics), but also to consider how this is manifest at higher levels of organization where, for example, the whole organism represents the unit upon which populations and their distribution are regulated. The other asks *why* a particular response is seen, and is built upon phylogenetic comparison and adaptive evolution. Therefore, we consider questions about mechanism ('how'; Section 3) and then questions based on evolutionary history – mainly the adaptive significance of developmental responses ('why'; Section 4). These two approaches are each applied to three aspects of developmental responses to environment – effects of environment alone (V_e in *Figure 2b*), how this interacts with genetic differences ($V_e \times V_g$), and how responses to environment vary within an individual's ontogeny ($V_e \times V_o$ and $V_e \times V_o \times V_g$). We contend that, when both mechanistic and adaptationist approaches are adopted and information from one is used to inform the other, overall scientific progress can be accelerated (Section 5). Callahan *et al.* (1997) put forward a similar argument, illustrating it with examples from the developmental plasticity of plants.

2. Types of developmental response to the environment

2.1 *Evolutionary vs developmental responses to the environment*

The environment can affect the evolution of a developmental trajectory, as well as cause developmental responses in a given genotype. Evolutionary responses of developmental trajectories can be studied using approaches from various traditions including evolutionary developmental biology (Hall, 1998), evolutionary ecology, particularly life history evolution (McNamara and Houston, 1996; Roff, 1992; Schlichting and Pigliucci, 1998; Stearns, 1992) and population and ecological genetics (Falconer and Mackay, 1996). Knowledge of evolutionary history can help predict immediate developmental responses to environmental conditions, and their adaptive significance (Section 4) and hence contribute to the understanding of animal development ecology (*Figure 2a*). Moreover, studies of animal developmental ecology can contribute to the understanding of evolution. Those studies that identify the relative success of different genotypes under different environmental conditions can help predict changes in gene frequency (evolution; *Figure 2a*; see also Chapter 16). There is therefore an obvious overlap between the evolution of developmental trajectories and their ecology, and the study of one contributes to the understanding of the other.

2.2 *Environmental vs ontogenetic influences*

Within a genotype, variation in a developmental trait (e.g. growth rate, allometric relationship between limb size and animal weight) has three components (apart from developmental noise throughout ontogeny and between environments, see Section 2.3). These are: (1) caused solely by environmental differences (ecophenotypic or environmentally induced variation = developmental plasticity); (2) due solely to stage in an individual's ontogeny; and (3) due to interactions between environmental variation and ontogenetic stage (*Figure 2b*). When comparing environmental effects on phenotypes, controlling for effects of ontogenetic stage simply involves of comparing phenotypes at a single 'stage', but this is not straightforward if a particular stage is defined by the number of moults or other repeated events, or by the appearance of various organs. The environment can alter both the number of some repeated events (e.g. temperature alters the number of juvenile moults in the grasshopper *Melanoplus femurrubrum;* Bellinger and Pienkowski, 1987) and the order in which particular organs and activities appear, which can be termed 'ecophenotypic (or environmentally induced) heterochrony' (e.g. Gibson and Johnston, 1995; Spicer and El-Gamal, 1999). Such variations should be accounted for when separating environmental effects on a trait from those associated with stage in ontogeny. A solution to the problem of adequately defining stages in ontogeny is to use only developmental events and sequences that are consistently observed in the life cycle of the particular species (Atkinson, 1996).

2.3 Developmental plasticity vs instability

Developmental plasticity is most easily identified among members of a clonal population. This phenotypic variation can be continuous (e.g. mature size affected continuously by temperature) or discrete (polyphenism; e.g. interrupted development; a shift between parthenogenesis and sexual reproduction; environmental sex determination; and non-genetic polymorphisms induced by seasonal, nutritional and predator cues). Responses of a trait to variation in a particular environmental parameter are represented by a reaction norm which can be 'flat' in those cases in which a trait is unaffected by the environment (i.e. fixed, and hence not phenotypically plastic; *Figure 3*). Pigliucci, Schlichting and co-workers (reviewed in Schlichting and Pigliucci, 1998) have used the concept of the *developmental reaction norm* to represent the three-way interaction between environment, ontogeny and allometric relationship between developmental characters (*Figure 3*). Of course there can be interactions between environmental variables; for example, the effect of temperature on development can depend on photoperiod (e.g. Chapters 12 and 15). Moreover, the relationship between two developmental characters may depend on other characters (e.g. Nijhout and Emlen, 1998). Thus, a developmental reaction norm may sometimes require rather more than three dimensions to provide an accurate representation of what may seem at first to be a simple relationship between an environmental factor and the ontogenetic changes in relationships between developmental characters.

Another source of phenotypic variation, which is separate from plasticity, is developmental instability. Instability arises from a failure of developmental buffering or homeostatic mechanisms in a given environment. An important distinction is that plasticity is only observed between environments, whereas instability can also be seen within a given environment. Random errors (developmental noise) generally appear not to be correlated with the degree of plasticity (reviews in Möller and Swaddle, 1997; Schlichting and Pigliucci, 1998).

Random developmental errors have been proposed as indicators of individual performance and of populations under stress. Specifically, traits which have elements that are repeated in the phenotype (e.g. scales of fish, left and right elements of a bilaterally symmetrical trait) may provide useful measures of developmental stability since

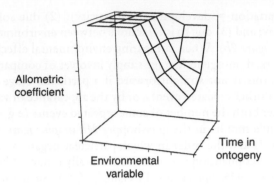

Figure 3. *A three-dimensional representation of the developmental reaction norm. The relationship between two developmental traits is measured as an allometric coefficient. In this example, at low levels of the environmental variable (e.g. temperature), the allometric coefficient is high and unaffected by the environmental variable throughout the whole of ontogeny – hence the left-hand side of the reaction norm is 'flat'. At higher levels of the environmental variable, the allometric coefficient reduces; this is particularly steep early in ontogeny (front of picture). Rather more than three dimensions may be needed to express some relationships adequately because traits can depend on several others, and the response to one environmental variable may depend on the values of others (see text). Developmental 'noise' due to instability of processes controlling ontogeny, would be seen as deviations above and below the plane of the reaction norm (not shown).*

their repeated elements are likely to be under the influence of a single gene complex (Möller and Swaddle, 1997). Fluctuating asymmetry – non-directional deviations from bilateral symmetry – is one such type of trait, but its value as a general indicator of developmental stress is still debated (Bjorksten *et al.*, 2000a,b; Möller, 2000; van Dongen and Lens, 2000). Selection would be expected to favour developmental mechanisms that minimize the instability of those traits whose asymmetry would be particularly costly for the animal. There may also be links between particular stressors and particular traits affected (e.g. a stressor that disrupts calcium metabolism may affect skeletal asymmetry; Eeva *et al.* 2000). However, apart from this, there is little predictive theory to identify which traits will become more asymmetrical in response to particular stressors. Consequently, van Dongen and Lens (2000) and Bjorksten *et al.* (2000b) have emphasized the need for studies of the mechanisms underlying trait development under different levels and types of environmental stressor.

2.4 *Adaptive vs non-adaptive responses*

In a given environment, the adaptive significance of a particular developmental response is the gain in Darwinian fitness by an organism over individuals who differ genetically in that response. Although adaptations are most likely seen in response to long-established environmental conditions (e.g. the seasonal pattern of temperatures and food supply), adaptations can also be observed in response to changes that have occurred in recent decades. For example, dioxin-like compounds are particularly toxic to the early development of fish, and are often abundant in urban estuaries such as New Bedford Harbor, Massachusetts, USA. In laboratory experiments, embryos and larvae of mummichogs, *Fundulus heteroclitus*, from the New Bedford Harbor were much less sensitive to dioxin-like compounds than were fish from reference sites. The responsiveness to these compounds was inherited and was independent of the amount of contaminant contributed by the mother (Nacci et al., 1999).

Not all developmental responses are adaptive. When environmental change is so rapid that there has been insufficient time for an evolutionary response (e.g. when a pollutant is encountered for the first time), a developmental response to it may be maladaptive. Other responses may, in principle at least, be selectively neutral. Thus an adaptationist approach will not always be readily able to explain some cases of developmental plasticity.

The adaptationist approach commonly uses optimality modelling, which can be applied specifically to life history traits (such as age and size at maturity, ontogenetic patterns of reproductive effort and allocation between egg size and number) and may include reaction norms for these traits (Daan and Tinbergen, 1997; McNamara and Houston, 1996; Roff, 1992; Schlichting and Pigliucci, 1998; Sibly and Calow, 1986; Stearns, 1992). The approach has three main stages: (i) identify the range of possible phenotypes that can be produced (or define the phenotypes being compared); (ii) identify the fitness costs and benefits of each under relevant environmental conditions, including any trade-offs between different traits; (iii) based on this information, determine the optimal phenotype in specified environments. Comparisons are then made for observed and predicted associations between developmental traits and environmental conditions. A lack of correspondence could be because errors have occurred during one or more of the stages (i) – (iii). Although optimality modelling may appear largely to ignore genetic and physiological mechanisms, in principle these mechanisms should be included where they define constraints that determine what phenotypes are possible (stage (i)).

Rigorous tests of adaptive significance can be difficult. The phenotypes being compared ideally should represent evolutionary alternatives. However, actual phenotypes will hardly ever differ by just a single trait. Even for very similar phenotypes, an increase in one character (e.g. size at maturity) may involve changes in resource allocation among variable combinations of activities (growth rate, maturation rate, investment in tissue maintenance and other activities that could affect mortality risk; see Chapter 15). Thus to understand the adaptive significance of phenotypic differences between organisms, we may need to examine the full range of traits that differ, including several potential trade-offs between traits.

Particularly convincing tests of adaptive significance of developmental traits involve direct manipulation of the trait on an organism in its natural environment coupled with manipulation of the proposed selective pressure. A critique of alternative approaches to phenotype manipulation (e.g. utilizing genes, hormones, environmental cues or direct physical manipulation) and examples of the more successful studies are provided by Schmitt et al. (1999; and others in the same volume). One such powerful technique is the manipulation of an animal's resource allocation by altering environmental cues (information used to alter developmental decisions). Here, care is required to avoid unwittingly altering environmental favourability in some treatments, or at least to ensure that such effects are accounted for. Otherwise some experimental animals will unintentionally be given additional resources, which may obscure the intended change in resource allocation induced by the cue. Such manipulated phenotypes could no longer be considered plausible evolutionary alternatives. For example, photoperiod is often used by animals as a seasonal cue to adjust developmental decisions and can be used as an experimental tool (e.g. Chapters 12, 15 and 17). But apart from the intended shift in developmental timing, any simultaneous changes in feeding opportunities (hence food uptake) due to extended hours of daylight, especially for visual feeders, should also be accounted for. Adaptive plasticity in animal morphology and life history is discussed further by Gotthard and Nylin (1995) and by Nylin and Gotthard (1998).

Tables 1 and 2 list aspects of variation in development that can be investigated by examining proximate mechanism (Section 3) and by using an adaptationist approach (Section 4).

3. Proximate mechanisms in developmental ecology

When environmental impacts on developmental processes are thought to be the main cause of an animal's population decline, an understanding of mechanisms at different levels of biological organization may be needed in order to establish the chain of causation. For instance, if a pollution spill is associated with a decrease in the population of juvenile alligators in a lake, observations on juveniles that establish the causal links could include: *at the organism level,* elevated circulating levels of oestrogens in females, and depressed circulating concentrations of testosterone in males, coinciding with a decrease in the number of births; *at the tissue/organ level* this decline in birth rate is linked to elevated production of oestrogens from the juvenile testes, reduced penis size and poorly organized seminiferous tubules, and in females changes in enzyme activity in the gonads; *at the cellular level,* ovarian abnormalities that are associated with unusually high oestrogen levels; *at the molecular level,* many of the contaminants in the pollution spill bind to the alligator oestrogen and progesterone receptors leading to overproduction of steroid hormones (Crain and Guillette, 1998; Gilbert, 2000).

To give a flavour of the range of issues raised by a mechanistic approach to animal developmental ecology, we discuss examples that relate to three general questions in Sections 3.1–3.3.

3.1 *How do different environmental conditions affect development?*

Specificity. Some idea of the challenge facing animal developmental ecology as a predictive science can be seen from the sheer diversity of environmental influences and their often complex effects on particular developmental processes or traits. For instance, structurally similar compounds can have rather different effects on a developmental process. Lundholm (1997) demonstrated this in experiments which investigated how pesticide residues caused eggshell thinning and the consequent reproductive failure that was responsible for post-war population crashes in several species of bird (Solomon, 1998; Vos *et al.*, 2000). In the domestic duck, *p,p'*–DDE inhibits prostaglandin synthetase, reducing calcium transport across the mucosa into the eggshell gland lumen. However, the structurally related compounds *o,p'*–DDE, *p,p'*–DDT, *o,p'*–DDT, and *p,p'*–DDD did not cause eggshell thinning or inhibit prostaglandin synthesis in the eggshell gland. By contrast, rather different environmental perturbations can produce similar developmental responses. For example, eggshell thinning in birds can be caused by organochlorines, including the persistent DDT metabolite DDE, but also by methyl mercury (Vos *et al.*, 2000).

The induction of particular responses can also depend upon the summation (or integration) of environmental signals over a particularly sensitive phase of development.

Table 1. Some ways in which proximate mechanisms influence environmentally induced variation in animal development (plasticity, V_e) and its interactions with genotype (V_g) and ontogeny (V_o)

		Sources of variation in development[a]		
			Plasticity varying within ontogeny (± among genotypes) ($V_e \times V_o$ and $V_e \times V_g \times V_o$) (Section 3.3)	
Environmental differences alone (plasticity) (V_e) (Section 3.1)	Plasticity varying among genotypes ($V_e \times V_g$) (Section 3.2)	Within a 'stage'	During ontogeny	Time lags
Specificity *Pollutants causing egg-spell thinning* Summation or integration of environmental signals Spatial distribution *Locust phase shifts* Multiple environmental factors Synergistic effects Experimental separation of effects	Direct genetic differences in development *Continuous vs discontinuous developmental sensitivity* *Use of transgenic strains to investigate resistance to heat stress* *Use of mutant physiology to predict specific environmental effects* *Use of knowledge of genetically based differences in predicting species responses to environmental change* Indirect effects (via niche differences) *Trophic level and biomagnification* *Microhabitat utilisation*	Differential sensitivity of processes *Environmentally induced heterochrony* *Differential responses to hormones* *Hierarchical effects* *'Critical' organs for accumulation of radionuclides* Interdependence of traits *Competition between adjacent imaginal discs in insects*	Sensitive phases *Shift in larval sensitivity to resources* Environmental sex determination *Sensitivity to teratogens* *Age dependence* Commitment of cell fate Set-aside cells *Potential for regeneration after injury* Changes in size *Associated change in physical pressures* *Associated changes in predation risk related to ontogenetic niche shifts*	*Sexual differentiation* *Maternal effects* *Long-lasting effects* *Bioaccumulation*

[a]Apart from variation in mean trait values implied by the column headings, developmental instability ('error') may also vary among environments, genotypes and within ontogeny, and some examples refer to this. Plain typeface refers to subheadings used in the respective section of the chapter, and italics refer to topics discussed under these sub-headings.

Table 2. *Some aspects of adaptive significance of environmentally induced variation in animal development (plasticity, V_e) and its interactions with genotype (V_g) and ontogeny (V_o).*

Sources of variation in development				
Environmental differences alone (plasticity)	Plasticity varying among genotypes	Plasticity varying within ontogeny (± among genotypes) ($V_e \times V_o$ and $V_e \times V_g \times V_o$) (Section 4.3)		
(V_e) (Section 4.1)	($V_e \times V_g$) (Section 4.2)	Within a 'stage'	During ontogeny	Time lags
Timing, spatial distribution and reliability of environmental cues vs speed, amount and reversibility of response, and costs of maintaining it	*Historical/phylogenetic* influences (adaptive or not)	*Usage and differential organ development*	*Links to change in ecological niche*	*Adaptive maternal effects*
Environmental constraints	*Co- and countergradient variation*	Potential structural constraints	Ontogenetic shift in developmental base temperatures	
Stochasticity and bet-hedging			Multiple and state-dependent 'decisions' – *dynamic optimization*	
			Ontogenetic constraints including geometric, biophysical and irreversibility of cell differentiation	

Italics refer to topics discussed under the sub-headings.

An example is the induction of a polyphenic switch (e.g. to a new seasonal morph) in insects (Nijhout, 1999). If the environmental cue (e.g. photoperiod) is experienced for insufficient time during the sensitive period, a polyphenic shift will not be induced. Also, a period of inductive cues can be cancelled by a period of non-inductive environmental cues during the sensitive period.

Spatial distribution of environmental differences. The spatial structure of the environment can alter developmental responses, as shown by the work of Simpson *et al.* (1999) on the desert locust, *Schistocerca gregaria.* Crowding locusts previously kept at low density induces phase changes from a solitary to a gregarious form, involving a complex suite of changes in behaviour, colour, morphology, reproductive development and fecundity. These changes are central to the occasional yet catastrophic impact this species has on humans. The mother is able to alter the phase-state of her offspring by adding a gregarizing chemical to the protective foam in the egg-pods she lays. It appears that the main stimulus experienced prior to egg-laying that elicits this response is tactile contact with conspecifics. Therefore, crowding is perceived on a very localized spatial scale. Consequently, in experiments in which vegetation was clumped (forcing the locusts to aggregate) the induction of gregarization was increased.

Multiple environmental factors. Difficulties in understanding impacts on development are compounded when environmental factors interact so that their combined effects are not simply the summed effects of each factor in isolation (e.g. Hanazato and Dodson, 1995).

Despite these complexities, controlled experiments can identify which of a variety of possible alternative proximate factors cause a particular developmental response. An example is the experiments by Denver *et al.* (1998) to identify the cues used by the western spadefoot toad *Scaphiopus hammondii* as the toads accelerated development towards metamorphosis in response to drying out of their ponds. Although pond desiccation will often correlate with increased temperature and warming normally accelerates development, Denver *et al.* (1998) found a response that did not result from thermal differences between treatments. This developmental response appeared to be due neither to the concentration of compounds in the water nor to chemical or physical interactions among conspecifics. Rather, the response appeared to be related to the reduced swimming volume and perhaps the proximity of the water surface. Moreover, their findings suggest that restriction of swimming volume constrains foraging, and that the cessation of feeding contributes to accelerated development.

3.2 *How are different responses produced among genotypes?*

Developmental plasticity may differ among genotypes, including species, because of direct genetic differences in developmental physiology. In addition, genetically determined differences in ecological niche will cause developmental differences at least partly because the genotypes or species experience different environments.

Direct effects. Direct physiological differences appear to explain variation in the induction of predator-resistant forms within a population of the intertidal barnacle *Chthamalus anisopoma* when exposed to a gastropod predator (Lively *et al.*, 2000). One genetic explanation for this variation – the 'continuous-sensitivity' model –

assumes that there was normally distributed genetic variation for sensitivity to the cue and predicts that given sufficient exposure to the predator, all individuals would develop into the predator-resistant form. A second model – the 'discontinuous sensitivity' model – asserts a genetic polymorphism for inducibility such that some individuals are not able to respond to the cue while others can. This predicts that with repeated exposures to the predator, any dose–response curve would reach an asymptote at less than 100% of the population. Lively et al. (2000) produced a dose–response curve that showed a significant asymptote at about 22% induction which supported the discontinuous sensitivity model, and indicates the possibility of a developmental polymorphism in this barnacle.

A powerful technique for examining the direct genetic basis of developmental responses is to produce transgenic strains containing single genetic differences that have major developmental consequences, and examine environmental effects in controlled experiments. Roberts and Feder (1999) exposed larvae of two strains of *Drosophila melanogaster* to natural heat stress. One strain had the wild-type number of genes encoding the major inducible heat shock protein Hsp70 ($n=10$), whereas the other was a transgenic sister strain with 22 copies of the *hsp70* gene. Under these conditions adults with the wild-type number of copies developed abnormal wings significantly more frequently than the strain with additional copies of the *hsp70* gene.

When the physiological effect of a genetic mutation is well understood and involves a clear link to the external environment it may be possible to suggest what environmental perturbation might induce this developmental defect. An example is found in the study by Zinke et al. (1999) of the *Drosophila* mutant *pumpless* (*ppl*), which is defective in food intake and growth at the larval stage. *Pumpless* larvae can feed normally when they first hatch but during late first-instar they fail to pump the food from the pharynx into the oesophagus. At the same time they begin moving away from the food source. Although *ppl* larvae do not feed, neither do they upregulate genes involved in gluconeogenesis or lipid breakdown which would typically be found in starving animals. When developmental anomalies caused by environmental stressors mimic the effects of known mutations they are called phenocopies. The data of Zinke et al. (1999) suggest the existence of a signal arising from the fat body that is dependent on amino acids and induces cessation of feeding in the larva. Therefore, Zinke et al. (1999) suggest that feeding wild-type larvae high levels of amino acids could phenocopy the feeding and growth defects of *ppl* mutants. They also propose that the signalling system may also mediate growth transition from larval to the pupal stage during *Drosophila* development.

Clearly, knowledge of genetically based developmental differences can be utilized to predict environmental impacts on different populations or species. For instance, we could start to produce predictions about the relative impacts of an extreme hot spell of weather on each member of a group of species, if we were armed with knowledge of their thermal optima and maxima for performance measures (e.g. growth and reproductive rate) and their ability to aestivate (see also Chapter 18 for an examination of overwintering capacity).

Indirect effects (via niche differences). There are many ways in which genetically determined differences in ecological niche will affect environmental impacts on the ontogeny of different species. An obvious example is trophic level: pollutants such as organochlorines that accumulate in tissues can be biomagnified through the food chain so that top predators will accumulate particularly high concentrations of pollutants in their tissues compared with species lower in the food chain. Another basic niche difference is between planktonic and benthic lifestyles, and this probably partly explains why benthic crustacea are prone to accumulating greater quantities of radionuclides near to localized outfalls of radioactive waste than are planktonic species. Sedimentation processes ensure that there is accumulation of radionuclides on the seabed with corresponding contamination of the infauna (Clark *et al.*, 1997). Thus exposure to radionuclides will be greater for benthic species.

3.3 *How are different responses produced within ontogeny?*

Understanding how developmental responses to environment vary during ontogeny (and also why; Section 4.3) can help predict population impacts of a short-lived environmental perturbation such as a pollution spillage or heat-wave and may even help conservation managers target sensitive individuals or age-classes for protection from such a perturbation.

Here, we discuss (i) variation that can occur at a given ontogenetic stage, (ii) temporal changes in response to environment during ontogeny, and (iii) the occurrence of time lags between initial exposure to particular environmental conditions and the subsequent effects on the animal's performance.

(i) Variation among developmental responses that can occur at a given ontogenetic stage.

Differential sensitivity of processes and traits – Exposure to certain environmental challenges early on can accelerate the development of particular regulatory mechanisms relative to the appearance of some organs, resulting in ecophenotypic heterochrony. For example, in the brine shrimp, *Artemia franciscana*, the ability to regulate oxygen uptake during exposure to acutely declining oxygen tensions developed early, co-occurring with the appearance of a functional heart and gills. However, culture under chronic hypoxia caused this regulation to be brought forward both in development and in time, occurring before the heart and gills formed (Spicer and El-Gamal, 1999).

A specific developmental response can occur when some tissues or at least some of the component target cells which comprise the tissue either have lower response thresholds to a regulatory factor (such as a hormone) or require shorter periods of exposure in order to achieve the normal response (Nijhout, 1999). The ability to respond to environmental variables (for example via hormone secretion) and the characteristics of that response (via hormone receptors on target cells) are controlled by different mechanisms. In insects, for example, hormone secretion is controlled by the central nervous system whereas target cells can regulate their receptor levels either entirely autonomously or in response to growth factors and inhibitors produced by nearby cells (Nijhout, 1999).

The environment can cause developmental responses by a hierarchy of mechanisms. First, it can cause direct but differential effects on rates or timing of particular biochemical processes so that there is a change in relative rates (e.g. changes of enzyme activity with temperature). Conditions that slow development (e.g. low temperature) are associated with the alteration of colour patterns in *Bicyclus* butterflies (Roskam and Brakefield, 1996) and this may be an example of such direct effects (Nijhout, 1999). Second, the environment can alter the level of expression of certain genes so that they produce more (or less) gene product (e.g. hormone), which leads to a change in phenotype. These two mechanisms come under the heading 'allelic sensitivity' (Schlichting and Pigliucci, 1998) which covers direct environmental effects that produce quantitative changes in development. However, Nijhout (1999) pointed out many instances in which artificial elevation of a hormone titre appears to be sufficient to cause a polyphenic shift in phenotype which can often be perceived as qualitative. In a third mechanism, the environment brings about the expression of new genes (and/or repression of others) hence leading to different developmental programs in different environments. This developmental switch has been called 'regulatory plasticity' (Schlichting and Pigliucci, 1998), and is most likely to lead to strong qualitative shifts in development as are seen in polyphenism. Environmental effects on development can be considered hierarchical because each of the alternative developmental pathways produced by regulatory plasticity can themselves exhibit allelic sensitivity (Nijhout, 1999; *Figure 4*).

Differential effects on particular organs can also be observed when particular substances, including pollutants, accumulate only in certain 'critical' organs. The existence of these organs is particularly important for radionuclides that emit α- and β-

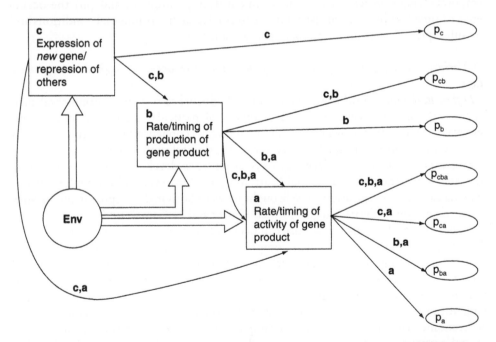

Figure 4. Schematic representation of the hierarchical effect of environment on development. Environment (Env) can alter phenotypes by directly affecting rates of activity of gene products ((a); e.g. temperature affecting enzyme activity), here producing phenotype p_a. Environment can also affect the level of gene expression, hence the amount (or timing) of gene product (b), producing phenotype p_b. These responses have been called 'allelic sensitivity' (see text). Phenotypic change resulting from altered rates of gene expression (b) can be further modified by the environment-sensitivity of their products (represented as b, a) producing phenotype p_{ba}. In addition, the environment can cause the expression of new genes (or repression of others) (c), producing phenotype p_c ('regulatory plasticity'). However, each alternative developmental pathway produced by regulatory plasticity can exhibit either or both forms of allelic sensitivity, hence modifying phenotypes further (p_{cb}, p_{cba}, p_{ca}).

radiation because the intensely ionizing effect is damaging only over a short distance. The critical organ for the radioisotope iodine-131 is the thyroid because the thyroid accumulates 600 times the iodine concentration of other tissues in the body. The critical organ for strontium-90, which behaves chemically like calcium, is bone (Clark *et al.*, 1997).

Interdependence of traits – Holometabolous insects such as beetles and butterflies undergo a drastic change when the larva pupates, and then metamorphoses from a non-feeding pupa into an adult. In these species, many adult structures develop in the pupa from groups of cells called imaginal disks, relying on resources accumulated during larval life. Emlen (1997) has shown experimentally that when larval male dung beetles *Onthophagus acuminatus* are fed a poor quality diet the resulting adults have smaller horns relative to body size. However, this reduction in horn size is associated with a relative increase in the size of the eyes. This inverse relationship between relative horn and eye sizes within dung beetle species can also be achieved by application of a juvenile hormone mimic and by artificial selection (Nijhout and Emlen, 1998). This implies that a change in allocation of resources to one trait (horns) appears to be traded-off against allocation to other traits (eyes). This is consistent with the idea that the imaginal disks are competing for limited resources in the haemolymph. Notably, traits in close physical proximity interact and/or compete with each other more directly than traits that are further apart – an observation that may help in the prediction of trade-offs (reviewed in Emlen, 2000).

(ii) Temporal changes during ontogeny.

Sensitive phases – One problem is to understand how certain developmental phases and processes respond to the environment, and how this varies during ontogeny. For example, in taxa as diverse as amphibians and Crustacea, age at metamorphosis can be modified early in larval life by resource availability, but later the timing becomes fixed (independent of resources). Thus, only during late larval phase is the plasticity lost (Hentschel, 1999).

Critical developmental phases (or windows) when the environment can alter particular traits are ecologically important in species with environmental sex determination. In these, sex is determined well after fertilization, during organogenesis. In sea turtles, for example, eggs in nests are not actively incubated by females but are subject to environmental temperatures which determine the sex of the hatchlings. Davenport (1997) describes how global warming could have major consequences for population survival. Sea turtles usually return to lay eggs on the beach where they themselves had hatched. High temperatures are likely to produce only females, which could subsequently affect the number of eggs that will be fertilized. Moreover, the long period before sea turtles become sexually mature is likely to prevent a rapid evolutionary response to the unprecedented rate of warming. Unless the turtles choose to use cooler parts of the breeding season or cooler parts of the beach, the population consequences could be very severe.

The sensitive period for sex determination is also particularly susceptible to a variety of other environmental influences (Vos *et al.*, 2000). For instance, in turtles, lizards and alligators, whose sex is determined by environmental temperature during egg incubation, sex is also sensitive during this period to exposure to exogenous hormones and to polychlorinated biphenyls (PCBs). In a variety of species, when 17β-estradiol is administered to eggs during the temperature-sensitive period, phenotypic females are produced at male-producing temperatures (Crain and Guillette, 1998). Bergeron *et al.*

(1994) found that exposure of red-eared slider turtle (*Trachemys scripta elegans*) eggs to artificial hydroxy-PCBs incubated at a temperature known to produce 100% males significantly altered the sex ratio, producing more females.

It appears that the enzyme aromatase, which can convert testosterone to oestrogen, is important in temperature-dependent sex determination. For instance, in diamond-backed terrapins, *Malaclemys terrapin*, aromatase activity is temperature dependent, and when it is inhibited the gonads become masculinized (Jeyasuria *et al.*, 1994). Moreover, if an aromatase inhibitor is injected into eggs of an all-female partheno-genetic species of lizard, males are formed (Wibbels and Crews, 1994).

The critical period during which many, though not all, external disruptors of devel-opment (teratogens) induce morphological defects in mammals is a relatively short period of embryogenesis between differentiation of the germ layer (gastrulation) and completion of major organ formation (Gilbert, 2000). For example, the development of external genitalia in humans is particularly sensitive to chemically induced perturba-tions in weeks 8 and 9 of gestation, which can lead to a variety of abnormalities in the adult. While humans are sensitive to various endocrine-disrupting and toxic chemicals during foetal life, rodents are also susceptible as neonates. By contrast, similarly exposed young adults in both groups are only transiently affected (Gray, 1998). The concept of *critical windows* for exposure is explored in more detail in Chapter 10 where both short-term and long-term (transgenerational) effects are considered. One of the main points to emerge is that not only are there critical windows for exposure to contaminants or for environmental influences but that the quality and sensitivity of critical windows can vary temporally throughout the developmental period, including during periods of adult gametogenesis.

Sensitivity to environmental conditions can depend on another major feature of ontogeny–age. When the colonial bryozoan *Membranipora membranacea* is exposed to chemical cues from predators, defensive spines can be induced in those modules around the periphery of the colony. To determine whether this spatial pattern of induction was a direct consequence of the peripheral position of these modules or because they were all the same age, Harvell (1991) experimentally altered the normal centre-to-edge gradient. This revealed that age was the more important – spines were induced by local action on each module of the appropriate age group, irrespective of its position within the colony.

Commitment of cell fate – An important characteristic of animal ontogeny is that as cells differentiate they lose the potential to differentiate again into another cell type. This commitment of cell type and function means that cell differentiation and devel-opmental consequences of this are irreversible. Thus, for new tissues to develop during ontogeny, there must be a supply of cells that are set aside for future differentiation. Such 'set-aside' cells have a relatively unlimited capacity to divide and retain the ability to differentiate into different types (i.e. are pluripotent; Peterson *et al.*, 1997). These cells include the stem cells of vertebrates, the imaginal disc of insect larvae, as well as many examples of pluripotent stem cells in other invertebrate phyla including the neoblasts of platyhelminthes and the set-aside cells of echinoderm larvae. It is perhaps not surprising that developmental stages retain the capacity for plasticity and this is borne out especially by recent studies on echinoderm larvae which show remarkable potential for regeneration (Vickery *et al.*, 2001). One important area that until recently has been largely overlooked is that of adult regeneration. In effect this is an *adult* developmental phenomenon and so is open to the same stresses and pressures as is normal ontogeny. Moreover, it is important to emphasize that regeneration is a quite normal part of these animals' physiology. For example, brittlestars and featherstars are commonly preyed upon by flatfish and grazers such as wrasse, and their regenerative potential is thus clearly a feature that has proved to be of considerable adaptive value and sensitive to environmental pressures (Patruno *et al.*, 2001; Thorndyke and Candia Carnevali, 2001). A good example of this is seen in a recent study of arm regeneration in the crinoid *Antedon mediterranea* in response to PCB exposure (Candia Carnevali *et al.*, 2001). Here, the presence of contaminant had a significant impact on regeneration, considerably enhancing the normal process. Of particular interest was the recruitment of muscle cells into the cell cycle suggesting that the PCB congeners were able to interact with endogenous receptors and promote cell plasticity (Candia Carnevali *et al.*, 2001). It is clear then that certain, perhaps all, animals retain some capacity for plasticity at the cellular level as adults. Once more, this lends support to the idea that development is a process which continues thoroughout the lifespan of all animals.

Size – Another major feature of ontogeny is that size typically increases by several orders of magnitude. Consequently, crucial physical parameters affecting the developing animal are also likely to change. Viscosity, surface tension and diffusional processes are likely to dominate when the animal is small, whereas inertia, streamlining, gravity and convectional processes are likely to become more important when it becomes large (Horn *et al.*, 1982).

Size can also strongly influence the ability to avoid predators, to forage in a particular habitat, or forage on a particular prey type. Therefore, biological pressures also change as an animal grows, and can cause ontogenetic niche shifts (see examples in Sillett and Foster, 2000). Olson (1996) showed that variation in early growth of largemouth bass, *Micropterus salmoides*, caused by differences in temperature and other environmental factors could have particularly large effects on later growth and probably subsequent survival and population dynamics. In this species fast growth during the early, invertebrate-feeding phase can produce an early size advantage over young-of-the-year bluegills, *Lepomis macrochirus*, their primary fish prey, facilitating a rapid diet shift to piscivory. By contrast, slow early growth reduced the size advantage over bluegills and delayed or even prevented the shift to piscivory in the first year. Since a shift to piscivory led to increased growth, and since survival of young-of-the-year bass is strongly size-dependent, especially through the first winter, Olson (1996) suggested that the timing of the shift to piscivory may be important in determining survival and recruitment rates.

An ontogenetic niche shift can differ between populations of a species that have different degrees of development of defensive structures against predators, as has been shown for three-spine sticklebacks, *Gasterosteus aculeatus*, from two Alaskan lakes (Sillett and Foster, 2000). Adults from a lake that lacked other fish species had reduced pelvic girdles and generally lacked pelvic spines, whereas those from a lake that contained other fish species had more developed pelvic girdles and spines. Also, in predation experiments, large size and the possession of pelvic spines gave juvenile sticklebacks some protection from predation, including cannibalism by adults. Consistent with smaller individuals responding to their greater predation risk, small juveniles were more abundant among vegetation (containing more refuges from predators) than in open water in the littoral zone of each lake. However, in the population with pelvic spines, a greater proportion of large juveniles was observed in open water than among vegetation. This was in contrast to the population without pelvic spines whose large juveniles were spread more evenly between microhabitats. Further, a habitat choice experiment for juveniles in which the presence of adult conspecifics (risk of cannibalism) was manipulated showed that large juveniles of either population showed no microhabitat preference in the absence of adults, but that a higher proportion of large juveniles from the population of fish with spines was observed in open water than large juveniles from the population without spines. These results support the idea that habitat preference is affected by a combination of body size, the possession of defensive structures, and the immediate presence of predators.

(iii) Time lags between environmental exposure and effect on performance. While some exposures to particular environmental conditions have immediate effects on an animal's performance (e.g. as lowered temperature often immediately slows growth of ectothermic species), others do not show obvious ecologically relevant effects until later in ontogeny. An example is the delay between foetal or perinatal exposure to endocrine-disrupting chemicals and its subsequent effects on adult reproductive performance in humans and rodents (Gray, 1998). A lag also occurs between maternal exposure during gametogenesis and performance of progeny after birth or hatch, which can be thought of as 'transgenerational phenotypic plasticity' (Mousseau and Fox, 1998a). For example, in field studies of a tropical damselfish, *Pomacentrus amboinensis*, McCormick (1998) showed that effects of the maternal environment dramatically influenced the size of larval progeny at hatching, 4 days after laying. Much of the variability in progeny size could be explained by levels of the stress-associated steroid hormone, cortisol, in the mother. A field experiment manipulating maternal levels of cortisol found that cortisol concentrations strongly influence the morphology and yolk size of larval progeny at hatching. Moreover, 38% of the observed variance in maternal cortisol levels could be explained by variation in the density of egg predators and competitors. Similar findings are now being reported for marine invertebrates (see Chapter 10). Here, exposure to the endocrine disruptor 4-nonylphenol not only has a direct toxic effect on oyster larvae but also a latent influence on surviving larvae and even a transgenerational effect on subsequent adult performance and fecundity.

Progeny size or other small changes in early development can permanently and dramatically alter developmental trajectories and fitness (Bernardo, 1996; Fox and Savalli, 2000; Henry and Ulijaszek, 1996; Lindström, 1999). In the case of snapping turtles, *Chelydra serpentina*, intermediate egg incubation temperatures (23–27°C) produce males (the larger sex) while extreme temperatures produce females. O'Steen (1998) found that experimentally manipulated egg temperature was linearly and negatively correlated with growth rate of both male and female juveniles, and with water temperatures chosen by juveniles. Moreover, these temperature choices were highly repeatable and persistent, even after the turtles had hibernated for 6 months at 7°C.

Other delayed effects on performance arise through bioaccumulation of heavy metals and halogenated hydrocarbons (e.g. the insecticides DDT and dieldrin; industrial chemicals in the group of PCBs). Those that cannot be fully excreted remain in the body and are continually added as the animal grows. Chlorinated hydrocarbons are lipid-soluble and so occur in much higher concentrations in fatty tissues than in other tissues. Consequently, in times when food uptake is low, animals mobilize fat reserves and hence increase the concentration of chlorinated hydrocarbons circulating in the body, possibly to a dangerous level (Clark *et al.*, 1997).

3.4 Environmental genomics and proteomics

Clearly interpretation of any of the responses noted in the previous sections needs to take into account their complexity. That is, for example, responses to heat stress or radionuclide exposure will involve changes (up or down) to many gene expression patterns and hence downstream protein biosynthetic pathways. The powerful techniques of genomics and proteomics are now beginning to be used to address these problems (Ferea and Brown, 1999). For example, gene expression profiling and differential protein profiling are emerging as instructive tools with which to explore such important biological issues as regeneration in both vertebrates and invertebrates (Li *et al.*, 2000; Vickery *et al.*, 2001). The important point here is that it has now become possible to monitor global changes in gene expression and protein biosynthetic patterns in response to environmental variation and change. In other words we are in a position to identify all those genes and their protein products which are influenced by an event with adaptive potential. Thus, we can begin to understand the interrelationship between genes and the way in which gene clusters respond both spatially (tissue specific) and temporally to the environment and so derive an integrated picture of the whole organism response. A good example of this is the recent study of hypoxia-induced gene expression in the long-jaw mudsucker, *Gillichthys mirabilis*, a burrow-dwelling fish that is frequently exposed to periods of low oxygen (Gracey *et al.*, 2001). In this study, fish were exposed to decreasing oxygen tensions and tissues sampled temporally. Gene expression profiles were taken from brain, liver and both skeletal and cardiac muscle. As well as the identification of genes that might be predicted to change, such as those involved in glycolysis, other cognate genes were seen to be differentially regulated in a co-ordinated fashion. These included genes expressed in the liver and involved in haemoglobin metabolism (Gracey *et al.*, 2001). Of particular interest was the possible indication that hypoxia induced a gluconeogenic mechanism to maintain blood glucose levels during bouts of anoxic metabolism (Gracey *et al.*, 2001).

It should be clear then that this type of approach, together with that of proteomics, when applied at the population and generational level should provide a powerful means by which to assess changes in gene expression that have adaptive significance, and their recruitment into successive and successful generations.

4. Adaptive significance of developmental responses

4.1 Why does the environment induce particular developmental responses?

Before considering adaptive explanations, one should identify which environmental features are likely to evoke particular types of adaptive response. Adaptive developmental plasticity is expected to evolve when an environmental cue varies between or within generations and/or between areas occupied by a population, and when the developmental response to this cue improves fitness. When environmental changes are very rapid in relation to the speed of the developmental response, adaptive plasticity may not be favoured by natural selection (Padilla and Adolph, 1996).

For an adaptive response to become advantageous and therefore incorporated into the genome (evolve) it is important that the cue is a good predictor of conditions expected throughout the functional life of the response (e.g. Chapter 14). Thus, if an induced morphological response to predator cues, for example, is irreversible, the cues initiating the response should correlate with increased predation risk throughout the rest of the animal's life so that the overall fitness benefits of inducing the response outweigh the costs. It may be this requirement for reliability of environmental cues that has led to the evolution in insects of the requirement for signals to be summed throughout the environment-sensitive developmental phase before a polyphenic switch can be induced (Section 3.1).

Irreversible responses are likely to be advantageous compared with reversible ones if the environmental variance between generations is large compared to the variation experienced within a generation (Gabriel, 1999). Thus for irreversible inducible defences against predators to evolve, for instance, a predator would normally have to appear with low probability but for a long period. Reversible plastic responses, by contrast, are expected to evolve if the trait is advantageous for a short time period compared with the entire life span (Gabriel, 1999). The adaptive value of discarding a morphological response when it is not beneficial is particularly clear when there are costs of maintaining the structure. The suitability of different types of cue for inducing defences against enemies, and the various types of cost associated with these defences, are discussed by Tollrian and Harvell (1999).

Sometimes an environmental variable (e.g. food availability) may act simultaneously both as a cue for future conditions (sometimes called a 'proximate factor') and also as a constraint which directly affects fitness ('ultimate factor'). Thus, if one observes a single developmental response in isolation, it can sometimes be difficult to determine whether the organism has altered a developmental decision (which may at least potentially be adaptive) or whether activities have been reduced merely because lack of resources didn't allow an increase. For instance, Gage (1995) found that adult male Indian meal moths, *Plodia interpunctella*, developed a smaller thorax (hence possibly less investment in flight) and survived for shorter periods when reared as larvae at high density. This might initially suggest that the animals incurred direct costs of crowding which reduced resources available for investment in thorax and somatic maintenance. However, these males also developed relatively larger testes and produced larger ejaculates. Thus, crowding appears to have evoked a shift in resource allocation between different activities, and early death may have been partly a cost of increased investment in reproduction. Whether or not this response is adaptive is not proven, although intriguingly, increased expenditure in ejaculates is predicted to be favoured at high density (hence high female encounter rates) because of the high expected frequency of copulations and/or elevated risk of sperm competition (Parker *et al.*, 1997).

Another type of adaptive plastic response is 'diversified risk spreading' (a form of bet-hedging) in which a single genotype produces multiple phenotypes to hedge against the chance of environmental change. This type of response is predicted to occur especially in response to increased environmental uncertainty. An example would be when a mother produces a variety of offspring sizes in response to conditions that indicate that the environment is becoming more unpredictable. Another is when individuals of the same genotype may or may not diapause under identical cues (e.g. daylength at a particular stage in development). In this case these large phenotypic differences may arise from small differences in conditions experienced during early life, possibly exaggerated by developmental instability (Simons and Johnston, 1997). Some evidence for risk spreading in insects does exist, especially in facultative diapause (Hopper, 1999). Yet Hopper (1999) was unable to find unequivocal results that risk spreading has been a major factor in the evolution of insect life histories.

4.2 Why do particular developmental responses differ among genotypes?

Phylogenetically related species may share particular developmental responses. Differences between taxonomic groups may be historical legacies that are no longer adaptive. On the other hand, such differences could still be adaptive if the taxonomically related species have similar niches and selection pressures (Westoby *et al.*, 1995). After eliminating the possibility that a difference is the result of a time lag between environmental change and evolutionary response, a genetic difference may be due to a difference in niche and selection pressure.

It is sometimes assumed that the evolutionary response to an environmental gradient will be the same as the developmental response (termed 'co-gradient variation'). In *Drosophila* for example, laboratory evolution at different temperatures produces an inverse genetic relationship between temperature and adult size which is mirrored in the direct developmental response of body size to temperature (Partridge *et al.*, 1994; see also Chapter 14). The evolutionary response to temperature is also reflected in the genetic increase in body size observed with increasing latitude (hence reduced mean temperature; James *et al.*, 1995). Yet an evolutionary response may sometimes be opposite to the plastic response ('counter-gradient variation'). For example, pumpkinseed sunfish, *Lepomis gibbosus*, when small, can experience competition from bluegill sunfish, *L. macrochirus*, so that they grow more slowly than when bluegills are absent. To test the idea that rapid growth has evolved in juvenile pumpkinseed sunfish to allow passage through the competing size classes, Arendt and Wilson (1999) performed an experiment using juvenile pumpkinseed derived from lakes with bluegill and from lakes without bluegill. Pumpkinseed were raised in cages without bluegill, with a low density of bluegill, or with a high density of bluegill. Bluegill decreased growth for all pumpkinseed, following the predicted competition gradient. However, pumpkinseed derived from lakes with bluegill grew faster than those from lakes without bluegill under all treatments, thereby indicating counter-gradient evolution.

4.3 Why particular differences occur within ontogeny

(i) Variation among responses that can occur at a given ontogenetic stage. It is well known that an organ will often enlarge in response to increasing demands placed on it, so that the overall allocation between organs changes in what appears to be an adaptive manner. The effect of exercise on muscle development is one such example; another is how neuronal activity influences which nervous connections (synapses) are retained by the adult animal (Gilbert, 2000). Sizes of organs and aspects of metabolic physiology in vertebrates may show great 'flexibility' (a term implying reversible changes; Piersma and Lindström, 1997) over timescales of weeks and even days in response to changing demands imposed by the environment (Piersma and Lindström, 1997). Again, the principles outlined in Section 4.1 should apply when identifying which environmental stimuli are likely to evoke particular types of adaptive response: frequency, predictability and duration of the environmental change, and the capacity to generate and reverse a given size and speed of response, will all affect the evolution of such adaptive plasticity.

Food shortage will stimulate the growth of longer ciliated arms in the pluteus larva of some echinoderms, which increases the ability of the larva to obtain food. This presumably adaptive response appears to incur a cost to the adult rudiment growing within the larva since even when they can acquire more food it takes longer for them to metamorphose (Hart and Strathmann, 1994).

Several features of development may limit the amount of phenotypic variation that is potentially available from which an optimal response is chosen by natural selection. One example of such a developmental constraint from holometabolous insects is the possibility that competition between imaginal disks for limiting resources in the haemolymph is necessarily more intense the closer together in location they are (Emlen, 2000). Therefore, trade-offs between the sizes of structures are more likely to be seen between spatially adjacent characters.

(ii) Temporal changes during ontogeny. Where developmental responses to environmental conditions change during ontogeny to reflect changes in niche, the adaptive significance may appear obvious. An example is the much-reduced phenotypic sensitivity to many environmental conditions once animals have entered diapause, which is associated with increased environmental stress (Chapters 11 and 12). However, the adaptive significance of other ontogenetic changes may not be so obvious, and many require testing. An example is the increase in the developmental threshold or base temperature (below which development ceases) with each successive developmental phase in several insect species (e.g. Hart *et al.*, 1997; Rodriguez-Saona and Miller, 1999) which corresponds with warming normally experienced during the period of juvenile development. Whether or not this represents adaptive variation is not proven, but in an interesting parallel from plants, populations of wheat from environments that get warmer during the juvenile phase also had increasing base temperatures (Slafer and Savin, 1991) whereas two species from tropical habitats in which temperature did not show a trend during ontogeny showed no changes in developmental base temperature (Atkinson and Porter, 1996). Moreover, predicted advantages of this ontogenetic shift during periods of warming include an improved ability to maintain a developmental trajectory that is adapted to seasonal changes, including the ability to 'catch up' if early development has been delayed (Atkinson and Porter, 1996).

Ontogeny can be thought of as a period in which many developmental 'decisions' are made. For predicting sequential decisions, 'dynamic optimization' models are particularly appropriate (McNamara and Houston, 1996). Animals with very highly developed sensory mechanisms and regulatory control of development are more likely to be able to utilize complex arrays of environmental and physiological cues (which collectively represent the 'state' of the animal) to adaptively adjust development (for examples, see Marrow *et al.*, 1996; Newman, 1992).

Direct adaptive significance of a change in trait should not be assumed. Some changes in plasticity during ontogeny are likely to reflect changes in constraint. These are likely to include the irreversible commitment of cell fate once a cell has differentiated, and geometrical or biophysical limitations such as surface area: volume limitations that may operate at large sizes to limit growth capacity even when environmental conditions appear favourable (Atkinson and Sibly, 1996).

(iii) Time lags. The evolution of a response that improves performance at some future time after the response is induced will require that the animal utilizes cues that are good predictors of future conditions. Thus in terrestrial and shallow-water environments at

temperate latitudes, temperature alone – which often varies stochastically from week to week – will often be a poor predictor of time of year. Therefore, a more predictable cue such as daylength is usually used to evoke an adaptive seasonal developmental response (e.g. Chapters 12 and 15).

Environmental conditions experienced by mothers during gametogenesis can affect the size of eggs she produces, and hence the success of her offspring, although the adaptive significance of these changes is sometimes debatable (Chapter 14; Bernardo, 1996; Mousseau and Fox, 1998a,b). The argument for adaptive significance is strengthened when the fitness benefits can be demonstrated, and when the response runs counter to (or is independent of) constraints on egg size imposed by maternal size, age or nutrient availability. An example is the seed beetle, *Stator limbatus*, whose females respond to the host plant on which they lay eggs by laying substantially larger eggs on poorer quality plants (Fox and Mousseau, 1996). However, this size increase incurs a cost, as mothers laying large eggs also lay fewer because of the environmentally induced trade-off between size and number of progeny (Fox and Mousseau, 1996).

5. Integrating the 'how' and the 'why' – synergy from synthesis

To what extent are studies of proximate mechanism and of evolutionary influences (or adaptation) on developmental response already integrated? In reality, neither approach totally ignores the other. For instance, studies of developmental mechanism do not ignore the likely adaptive function of a particular trait (e.g. Brakefield and French, 1999). However, in such studies the analysis of adaptive design is typically not so detailed as would be obtained from a formal optimality model. A similar point can be made for studies of adaptive significance which may to some extent identify constraints on what developmental responses are possible and identify how the organisms interact with their environment, both of which require a consideration of mechanisms. Indeed the best experimental studies of adaptation (e.g. Kingsolver, 1995, 1996) specifically elucidate the mechanisms by which environment interacts with phenotype. However, we consider that there is often a need, especially in theoretical studies (which are often then tested empirically), for a more detailed mechanistic understanding of environmental developmental biology in order to identify realistic constraints on adaptive responses (see examples in Section 4).

When studying adaptive significance of developmental interactions, specific benefits that can be gained from a good knowledge of proximate mechanism include:

(i) Avoiding wasting time on those ('black box') models of optimal phenotypes that are incompatible with underlying mechanisms.

(ii) Identifying means of manipulating phenotypes or environments in ways that will allow rigorous tests of optimality models. Thus an understanding of mechanism should help us to avoid producing confounding phenotypic changes when attempting to manipulate a single trait or resource allocation 'decision'. In the case of genetic manipulation, pleiotropic effects may confuse the interpretation of the adaptive significance of a trait. In the case of manipulation of environmental cues, additional resources may produce phenotypes that are not plausible evolutionary alternatives; Section 2.4).

Specific benefits for animal developmental ecology that can arise from knowledge of evolutionary influences include:

(i) Predicting mechanisms that are taxon-specific (from phylogenetic comparisons).
(ii) Predicting similar responses to the environment in different species because of similar selection pressures, even though they may not be closely related taxonomically (i.e. cases of convergent evolution) and may even have quite different developmental mechanisms (e.g. similar effects of photoperiod on reproductive activity in many mid-latitude vertebrates, invertebrates and plants).

We therefore advocate a close dialogue between those who specialize in answering 'how' questions and those who focus mainly on the 'why'. A first step towards achieving this is to bring together the variety of approaches into one place, which is what we have done in this book.

References

Arendt, J.D. and Wilson, D.S. (1999) Countergradient selection for rapid growth in pump-kinseed sunfish: Disentangling ecological and evolutionary effects. *Ecology* 80: 2793–2798.

Atkinson, D. (1996) Ectotherm life-history responses to developmental temperature. In: *Animals and Temperature: Phenotypic and Evolutionary Adaptation* (eds I.A. Johnston and A.F. Bennett). Cambridge University Press, Cambridge, pp. 183–204.

Atkinson, D. and Porter, J.R. (1996) Temperature, plant development and crop yields. *Trends Plant Sci.* 1: 119–124.

Atkinson, D. and Sibly, R.M. (1996) On the solutions to a major life-history puzzle. *Oikos* 77: 359–365.

Bellinger, R.G. and Pienkowski, R.L. (1987) Developmental polymorphism in the red-legged grasshopper *Melanoplus femurrubrum* (De Geer) (Orthoptera: Acrididae). *Environ. Entomol.* 16: 120–125.

Bergeron, J.M., Crews, D. and McLachlan, J.A. (1994) PCBs as environmental estrogens – turtle sex determination as a biomarker of environmental contamination. *Environ. Health Persp.* 102: 780–781.

Bernardo, J. (1996) Maternal effects in animal ecology. *Am. Zool.* 36: 83–105.

Bjorksten, T.A., Fowler, K. and Pomiankowski, A. (2000a) What does sexual trait FA tell us about stress? *Trends Ecol. Evol.* 15: 163–166.

Bjorksten, T.A., Fowler, K. and Pomiankowski, A. (2000b) Symmetry, size and stress: reply. *Trends Ecol. Evol.* 15: 331.

Brakefield, P.M. and French, V. (1999) Butterfly wings: the evolution of development of colour patterns. *BioEssays* 21: 391–401.

Callahan, H.S., Pigliucci, M. and Schlichting, C.D. (1997) Developmental phenotypic plasticity: where ecology and evolution meet molecular biology. *BioEssays* 19: 519–525.

Candia Carnevali, M.D., Galassi, S., Bonasoro, F., Patruno, M. and Thorndyke M.C. (2001) Regenerative response and endocrine disrupters in crinoid echinoderms: arm regeneration in *Antedon mediterranea* after experimental exposure to polychlorinated biphenyls. *J. Exp. Biol.* 204: 835–842.

Clark, R.B., Frid, C. and Attrill, M. (1997) *Marine Pollution*, 4th edn. Oxford University Press, Oxford.

Crain, D.A. and Guillette L.L. Jr (1998) Reptiles as models of contaminant-induced endocrine disruption. *Animal Reprod. Sci.* 53: 77–86.

Daan, S. and Tinbergen, J.M. (1997) Adaptation of life histories. In: *Behavioural Ecology: an Evolutionary Approach*, 4th edn (eds J.R. Krebs and N.B. Davies). Blackwell Science, Oxford, pp. 311–333.

Davenport, J. (1997) Temperature and the life-history strategies of sea-turtles. *J. Therm. Biol.* 22: 479–488.

Denver, R.J., Mirhadi, N. and Phillips, M. (1998) Adaptive plasticity in amphibian metamorphosis: response of *Scaphiopus hammondii* tadpoles to habitat desiccation. *Ecology* 79: 1859–1872.

Eeva, T., Tanhuanpaa, S., Rabergh, C., Airaksinen, S., Nikinmaa, M. and Leikoinen, E. (2000) Biomarkers and fluctuating asymmetry as indicators of pollution-induced stress in two hole-nesting passerines. *Funct. Ecol.* 14: 235–243.

Emlen, D.J. (1997) Diet alters male horn allometry in the dung beetle *Onthophagus acuminatus*. *Proc. R. Soc. Lond. B* 264: 567–574.

Emlen, D.J. (2000) Integrating development with evolution: a case study with beetle horns. *BioScience* 50: 403–418.

Falconer, D.S. and Mackay, T.F.C. (1996) *Introduction to Quantitative Genetics*, 4th edn. Longman, Harlow, Essex.

Ferea, T.L. and Brown, P.O. (1999) Observing the living genome. *Curr. Opin. Genet. Devl.* 9: 715–722.

Fox, C.W. and Mousseau, T.A. (1996) Larval host plant affects the fitness consequences of egg size in the seed beetle *Stator limbatus*. *Oecologia* 107: 541–548.

Fox, C.W. and Savalli, U.M. (2000) Maternal effects mediate host expansion in a seed-feeding beetle. *Ecology* 81: 3–7.

Gabriel, W. (1999) Evolution of reversible plastic responses: inducible defenses and environmental tolerance. In: *The Ecology and Evolution of Inducible Defenses* (eds R. Tollrian and C.D. Harvell). Princeton University Press, Princeton, NJ, pp. 286–305.

Gage, M.J.G. (1995) Continuous variation in reproductive strategy as an adaptive response to population density in the moth *Plodia interpunctella*. *Proc. R. Soc. Lond. B* 261: 25–30.

Gibson, S. and Johnston, I.A. (1995) Temperature and development in larvae of the turbot *Scophthalmus maximus*. *Mar. Biol.* 124: 17–25.

Gilbert, S.F. (2000) *Developmental Biology*, 6th edn. Sinauer, Sunderland, MA.

Gotthard, K. and Nylin, S. (1995) Adaptive plasticity and plasticity as an adaptation: a selective review of plasticity in animal morphology and life history. *Oikos* 74: 3–17.

Gracey, A.Y., Troll, J.V. and Somero, G. (2001) Hypoxia-induced gene expression profiling in the euryoxic fish *Gillichthys mirabilis*. *Proc. Natl Acad. Sci. USA* 98: 1993–1998.

Gray, L.E. Jr (1998) Chemical-induced alterations of sexual differentiation: a review of effects in humans and rodents. *J. Clean Technol., Environ. Toxicol., Occup. Med.* 7: 121–145.

Hall, B.K. (1998) *Evolutionary Developmental Biology*, 2nd edn. Kluwer Academic, Dordrecht.

Hanazato, T. and Dodson, S.I. (1995) Synergistic effects of low oxygen concentration, predator kairomone and a pesticide on the cladoceran *Daphnia pulex*. *Limnol. Oceanogr.* 40: 700–709.

Hart, A.J., Bale, J.S. and Fenlon, J.S. (1997) Developmental threshold, day-degree requirements and voltinism of the aphid predator *Episyrphus balteatus* (Diptera: Syrphidae). *Ann. Appl. Biol.* 130: 427–437.

Hart, M.W. and Strathmann, R.R. (1994) Functional consequences of phenotypic plasticity in echinoid larvae. *Biol. Bull.* 186: 291–299.

Harvell, C.D. (1991) Coloniality and inducible polymorphism. *Am. Nat* 138: 1–14.

Henry, C.J.K. and Ulijaszek, S.J. (1996) *Long-term Consequences of Early Environment: Growth, Development and the Lifespan Developmental Perspective*. Cambridge University Press, Cambridge.

Hentschel, B.T. (1999) Complex life cycles in a variable environment: predicting when the timing of metamorphosis shifts from resource dependent to developmentally fixed. *Am. Nature* 154: 549–558.

Hopper, K.R. (1999) Risk-spreading and bet-hedging in insect population biology. *Annu. Rev. Entomol.* 44: 535–560.

Horn, H.S., Bonner, J.T., Dohle, W., Katz, M.J., Koehl, M.A.R., Meinhardt, H., Raff, R.A., Reif, W.-E., Stearns, S.C. and Strathmann, R. (1982) Adaptive aspects of development. In: *Evolution and Development* (ed J.T. Bonner). Springer, Berlin, pp. 215–235.

James, A.C., Azevedo, R.B.R. and Partridge, L. (1995) Cellular basis and developmental timing in a size cline of *Drosophila melanogaster*. *Genetics* 140: 659–666.

Jeyasuria, P., Roosenburg, W.M. and Place, A.R. (1994) Role of P-450 aromatase in sex determination of the diamondback terrapin, *Malaclemys terrapin*. *J. Exp. Zool.* 270: 95–111.

Kingsolver, J.G. (1995) Fitness consequences of seasonal polyphenism in western white butterflies. *Evolution* 49: 942–952.

Kingsolver, J.G. (1996) Experimental manipulation of wing pigment pattern and survival in western white butterflies. *Am. Nat* 147: 296–306.

Li, X., Mohan, S., Gu,W., Miyakoshi, N. and Baylink, D.J. (2000) Differential protein profile in the ear-punched tissue of regeneration and non-regeneration strains of mice: a novel approach to explore the candidate genes for soft-tissue regeneration. *Biochem. Biophys. Acta* 1524: 102–109.

Lindström, J. (1999) Early development and fitness in birds and mammals. *Trends Ecol. Evol.* 14: 343–348.

Lively, C.M., Hazel, W.N., Schellenberger, M.J. and Michelson, K.S. (2000) Predator-induced defense: Variation for inducibility in an intertidal barnacle. *Ecology* 81: 1240–1247.

Lundholm, C-E. (1997) DDE-induced eggshell thinning in birds: effects of p,p′-DDE on the calcium and prostaglandin metabolism of the eggshell gland. *Comp. Biochem. Physiol. C* 118: 113–128.

Marrow, P., McNamara, J.M., Houston, A.I., Stevenson, I.R. and Clutton-Brock T.H. (1996) State-dependent life history evolution in Soay sheep: dynamic modelling of reproductive scheduling. *Phil. Trans. R. Soc. Lond. B* 351: 17–32.

McCormick, M.I. (1998) Behaviorally induced maternal stress in a fish influences progeny quality by a hormonal mechanism. *Ecology* 79: 1873–1883.

McNamara, J.M. and Houston, A.I. (1996) State-dependent life histories. *Nature* 380: 215–221.

Möller, A.P. (2000) Symmetry, size and stress. *Trends Ecol. Evol.* 15: 330.

Möller, A.P. and Swaddle, J.P. (1997) *Asymmetry and Developmental Stability*. Oxford University Press, Oxford.

Mousseau, T.A. and Fox, C.W. (1998a) The adaptive significance of maternal effects. *Trends Ecol. Evol.* 13: 403–407.

Mousseau, T.A. and Fox, C.W. (eds) (1998b) *Maternal Effects as Adaptations*. Oxford University Press, Oxford.

Nacci, D., Coiro, L., Champlin, D., Jayaraman, S., McKinney, R., Gleason, T.R., Munns, W.R., Specker, J.L. and Cooper, K.R. (1999) Adaptations of wild populations of the estuarine fish *Fundulus heteroclitus* to persistent environmental contaminants. *Mar. Biol.* 134: 9–17.

Newman, R.A. (1992) Adaptive plasticity in amphibian metamorphosis. *BioScience* 42: 671–678.

Nijhout, H.F. (1999) Control mechanisms of polyphenic development in insects. *BioScience* 49: 181–192.

Nijhout, H.F.and Emlen, D.J. (1998) Competition among body parts in the development and evolution on insect morphology. *Proc. Natl Acad. Sci. USA* 95: 3685–3689.

Nylin, S. and Gotthard, K. (1998) Plasticity in life history traits. *Annu. Rev. Entomol.* 43: 63–83.

Olson, M.H. (1996) Ontogenetic niche shifts in largemouth bass: variability and consequences for first-year growth. *Ecology* 77: 179–190.

O'Steen, S. (1998) Embryonic temperature influences juvenile temperature choice and growth rate in snapping turtles *Chelydra serpentina*. *J. Exp. Biol.* 201: 439–449.

Padilla, D.K. and Adolph, S.C. (1996) Plastic inducible morphologies are not always adaptive: the importance of time delays in a stochastic environment. *Evol. Ecol.* 10: 105–117.

Patruno, M., Thorndyke, M.C., Candia Carnevali, M.D., Bonasoro, F. and Beesley, P.W. (2001) Growth factors, heat shock proteins and regeneration in echinoderms. *J. Exp. Biol.* 204: 843–848.

Parker, G.A., Ball, M.A., Stockley, P. and Gage, M.J.G. (1997) Sperm competition games: a prospective analysis of risk assessment. *Proc. R. Soc. Lond. B* **264**: 1793–1802.

Partridge, L., Barrie, B., Fowler, K. and French, V. (1994) Evolution and development of body size and cell size in *Drosophila melanogaster* in response to temperature. *Evolution* **48**: 1269–1276.

Peterson, K.J., Cameron, R.A. and Davidson, E.H. (1997) Set-aside cells in maximal indirect development: evolutionary and developmental significance. *BioEssays* **19**: 623–631.

Piersma, T. and Lindström, A. (1997) Rapid reversible changes in organ size as a component of adaptive behaviour. *Trends Ecol. Evol.* **12**: 134–138.

Roberts, S.P. and Feder, M.E. (1999) Natural hyperthermia and expression of the heat shock protein Hsp70 affect developmental abnormalities in *Drosophila melanogaster*. *Oecologia* **121**: 323–329.

Rodriguez-Saona, C. and Miller, J.C. (1999) Temperature-dependent effects on development, mortality and growth of *Hippodamia convergens* (Coleoptera: Coccinellidae). *Environ. Entomol.* **28**: 518–522.

Roff, D.A. (1992) *The Evolution of Life Histories: Theory and Analysis*. Chapman and Hall, New York.

Roskam, J.C. and Brakefield, P.M. (1996) Comparison of temperature-induced polyphenism in African *Bicyclus* butterflies from a seasonal savannah-rainforest ecotone. *Evolution* **50**: 2360–2372.

Schlichting, C.D. and Pigliucci, M. (1998) *Phenotypic Evolution: a Reaction Norm Perspective*. Sinauer, Sunderland, MA.

Schmitt, J., Dudley, S.D. and Pigliucci, M. (1999) Manipulative approaches to testing adaptive plasticity: phytochrome-mediated shade-avoidance responses in plants. *Am. Nat* **154**: S43–S54.

Sibly, R.M. and Calow, P. (1986) *Physiological Ecology of Animals: An Evolutionary Approach*. Blackwell Scientific, Oxford.

Sillett, K.B. and Foster, S.A. (2000) Ontogenetic niche shifts in two populations of juvenile threespine stickleback, *Gasterosteus aculeatus*, that differ in pelvic spine morphology. *Oikos* **91**: 468–476.

Simons, A.M. and Johnston, M.O. (1997) Developmental instability as a bet-hedging strategy. *Oikos* **80**: 401–406.

Simpson, S.J., McCaffery, A.E. and Hägele, B.F. (1999) A behavioural analysis of phase change in the desert locust. *Biol. Rev.* **74**: 461–480.

Slafer, G.A. and Savin, R. (1991) Developmental base temperature in different phenological phases of wheat (*Triticum aestivum*). *J. Exp. Bot.* **42**: 1077–1082.

Solomon, K.R. (1998) Endocrine-modulating substances in the environment: the wildlife connection. *Int. J. Toxicol.* **17**: 159–172.

Spicer, J.I. and El-Gamal, M.M. (1999) Hypoxia accelerates the development of respiratory regulation in brine shrimp–but at a cost. *J. Exp. Biol.* **202**: 3637–3646.

Stearns, S.C. (1992) *The Evolution of Life Histories*. Oxford University Press, Oxford.

Thorndyke, M.C. and Candia Carnevali M.D. (2001) Growth factors and regeneration in Echinoderms. *Can. J. Zool.* (in press).

Tollrian, R. and Harvell, C.D. (1999) The evolution of inducible defenses: current ideas. In: *The Ecology and Evolution of Inducible Defenses* (eds R. Tollrian and C.D. Harvell). Princeton University Press, Princeton, NJ, pp. 306–321.

van der Weele, C. (1999) *Images of Development: Environmental Causes in Ontogeny*. SUNY Press, Albany, NY.

van Dongen, S. and Lens, L. (2000) Symmetry, size and stress. *Trends Ecol. Evol.* **15**: 330–331.

Vickery, M.C.L., Vickery, M.S., McClintock, J.B. and Amsler, C.D. (2001) Utilization of a novel deuterostome model for the study of regeneration genetics: molecular cloning of genes that are differentially expressed during the early stages of larval seastar regeneration. *Gene* **262**: 73–80.

Vos, J.G., Dybing, E., Greim, H.A., Ladefoged, O., Lambre, C., Tarazona, J.V., Brandt, I. and Vethaak, A.D. (2000) Health effects of endocrine-disrupting chemicals on wildlife, with special reference to European situation. *Crit. Rev. Toxicol.* **30**: 71–133.

Westoby, M., Leishman, M.R. and Lord, J.M. (1995) On misinterpreting the phylogenetic correction. *J. Ecol.* **83**: 531–534.

Wibbels, T. and Crews, D. (1994) Putative aromatase inhibitor induces male sex determination in a female unisexual lizard and in a turtle with temperature-dependent sex determination. *J. Endocrinol.* **141**: 295–299.

Zinke, I., Kirchner, C., Chao, L.C., Tetzlaff, M.T. and Pankratz, M.J. (1999) Suppression of food intake and growth by amino acids in *Drosophila*: the role of *pumpless,* a fat body expressed gene with homology to vertebrate glycine cleavage system. *Development* **126**: 5275–5284.

2

Molecular ecology and identification of marine invertebrate larvae

A.D. Rogers

1. Introduction

The majority of marine animal species have evolved multiphasic life cycles, with the exception of the nematodes and chaetognaths (Wray, 1995). These life histories involve indirect development through an intermediate life history phase, which primitively is a pelagic larva (Havenhand, 1995). Pelagic larvae, which are usually planktotrophic, are released and disperse away from adult populations to feed and grow in the water column. When the larvae have grown sufficiently, they become competent to metamorphose and settle in a suitable habitat that they will occupy as juveniles and adults (Shanks, 1995).

For a long period of time the importance of the larval stage was not fully appreciated by ecologists studying populations of marine animals. For example, during the 1960s and 1970s, studies on intertidal ecology attributed the distribution of species to a variety of physical and biological factors. Increasing exposure to desiccation and extremes of temperature with increasing height on the shore were often cited as limiting the upper distribution of intertidal organisms (Connell, 1972, 1975; Paine, 1974). In contrast, predation and competition were considered as important in setting lower levels of distribution (Connell, 1961a,b, 1970; Paine, 1974). During the late 1970s and 1980s it was recognized that the distribution of intertidal animals and plants was more complicated than previously considered. This was because of the role of larval recruitment in determining species distribution (Hawkins and Hartnoll, 1982; Connell, 1985; reviewed by Underwood and Denley, 1984; Underwood and Fairweather, 1989). Recruitment to intertidal habitats was found to be partially determined by larval dispersal, influenced in turn by a variety of physical and biological factors in the pelagic environment (Hawkins and Hartnoll, 1982; Gaines and Bertness, 1992; Le Fèvre and Bourget, 1992). Fisheries biologists have related physical and biological influences to larval survival and recruitment in fish populations for some time (Carruthers, 1938; Walford, 1946).

Environment and Animal Development: Genes, Life Histories and Plasticity, edited by D. Atkinson and M. Thorndyke.
© 2001 BIOS Scientific Publishers Ltd, Oxford.

Larval dispersal has been recognized as crucial in shaping the range and spatial genetic structure of marine populations. The extent of larval dispersal is directly related to the length of time spent in the water column and to advection. This is a function of the amount of growth larvae must complete prior to metamorphosis or settlement. A long-lived, planktotrophic larva will tend to disperse over long distances. In such a situation, high levels of migration may be expected between populations, leading to widespread panmixia and genetic homogeneity over large geographic areas (e.g. the urchin *Strongylocentrotus purpuratus*, Palumbi and Wilson, 1990). Marine species in which the larval phase has been abbreviated to a short period of time, or lost altogether, may be expected to show high levels of genetic differentiation between allopatric populations. Because populations of species with different life histories are likely to show differing spatial genetic structures, it is likely that they will speciate at different rates or even through different mechanisms (Palumbi, 1992). The fossil record has been cited as providing evidence that planktotrophic species have broader ranges, lower rates of extinction and lower speciation rates than non-planktotrophic species (Jablonski, 1986).

However, the relationship between species life history and spatial genetic structure is not as simple as first appears. Several studies have demonstrated that potential larval dispersal in planktotrophic species may not be recognized because of a variety of mechanisms. Limited advection of larvae, through poor water movement in the proximity of a spawning population, may lead to high levels of self-recruitment and a high degree of genetic structure at a local scale, despite planktotrophy (Parsons, 1996). In a similar way, currents forming gyres or eddies may act to concentrate larvae over spawning populations preventing dispersal (Hill *et al.*, 1996; Wolanski and Sarsenski, 1997). Alternatively species may show behavioural mechanisms to avoid dispersal and thus may exhibit a higher degree of spatial genetic structure than expected from experimental observations of the length of time at the larval stage (Knowlton and Keller, 1986; Todd *et al.*, 1998).

Highly fecund marine animals with planktotrophic larvae have a very high mortality in the early life stages that can have a profound effect on the genetic structure of populations. Mortality rates of up to 20% a day have been recorded in natural populations of larvae (Wray, 1995). Larvae and juveniles may be carried away from adult populations by unfavourable currents (Le Fèvre and Bourget, 1992), or they may succumb to unfavourable environmental conditions, such as starvation (Fenaux *et al.*, 1994), disease or predation. As a consequence, variation in recruitment is very common in such populations (Underwood and Fairweather, 1989 for review). These factors are at least partially responsible for a substantial variation in the reproductive success of individuals within a population (Hedgecock, 1994). Many individuals, under such conditions, fail to match reproductive activity with favourable oceanographic conditions. However, a small number, because of the high individual fecundity of such species, may replace the entire population by fortunate matching of reproductive activity with the most favourable environmental conditions (e.g. the match–mismatch hypothesis; Cushing, 1959). Other factors, important in determination of individual reproductive success, may include a high variance in individual fecundity, gamete viability, fertilization of eggs during gamete release (e.g. sperm limitation) and sex ratio (McEdward and Coulter, 1987; Havenhand, 1995; Levitan and Petersen, 1995; variance in egg size see Levitan, 1996). If variance in reproductive success is high then the effective population size (i.e. the number of individuals contributing their genes to the next generation) may be orders of magnitude smaller than the actual numbers of individuals in a population (Knowlton and Jackson, 1993; Hedgecock, 1994).

The lowest effective population sizes of all animals have been recorded for marine invertebrates with planktonic larvae (Frankham, 1995). Extremely low effective population sizes may lead to decreases in heterozygosity within populations and higher than expected levels of genetic differentiation between populations as a result of random genetic drift. Temporal variance in allele frequencies may also account for contradictory observations of spatial genetic structure in some populations of marine invertebrates (e.g. krill; see Fevolden and Ayala, 1981; Fevolden and Schneppenheim, 1989).

Larval dispersal is clearly important in shaping marine communities and influencing the genetic structure and evolution of marine populations. However, studies of the distribution of planktonic stages of animals have been extremely difficult for a variety of reasons. The larvae of many species, even from well-known groups, such as fish, have not been identified (Hare et al., 1994). For many groups of invertebrates, larval identification has relied on time-consuming examination of species-specific characters using microscopy (Hu et al., 1993). In such cases, a high degree of skill and experience is often required and if specimens are damaged or have not reached a certain stage in development, they may not be identifiable. Larvae may also show significant phenotypic plasticity leading to a large degree of error in species identification or they may be indistinguishable because of conservative morphology within a group (Medeiros-Bergen et al., 1995). It is only recently, with the advent of molecular techniques, particularly those with a genetic basis, that reliable methods for larval identification have become available. Furthermore, this means that studies of the genetics of larval populations are now possible. In the present paper, molecular methods for larval identification and genotyping are reviewed. The chemical or genetic bases of these methods are considered, along with practical aspects of application and the types of studies to which particular techniques are suited.

2. Biochemical methods for larval identification

2.1 Allozyme electrophoresis

Biochemical genetics is based on enzymes that usually catalyse important reactions in major biochemical pathways, such as those involved in the tricarboxylic acid cycle. All enzymes are proteins with an amino acid sequence that is coded for by specific genes. Alterations in the DNA sequence coding for a protein, through mutation, can result in a change in the amino acid sequence and an alteration in the charge/mass ratio of a protein. Different forms of a specific protein arising from such mutational processes are termed isozymes, isoenzymes, allozymes or alloenzymes and represent different molecular forms arising from different alleles at the same genetic locus (IUPAC-IUB Commission on Biochemical Nomenclature, 1977; Manchenko, 1994). As such, differences in the occurrence of homologous proteins in related populations or species can provide an indication of the levels of genetic difference between them (Thorpe, 1982).

Different allozymes are detected using electrophoresis. This is based on the differential migration of water soluble protein molecules, extracted from tissues, through a gel, across which is an electrical potential (Murphy et al., 1990; Manchenko, 1994; see Figure 1). The gel medium is usually made of starch, cellulose acetate or polyacrylamide

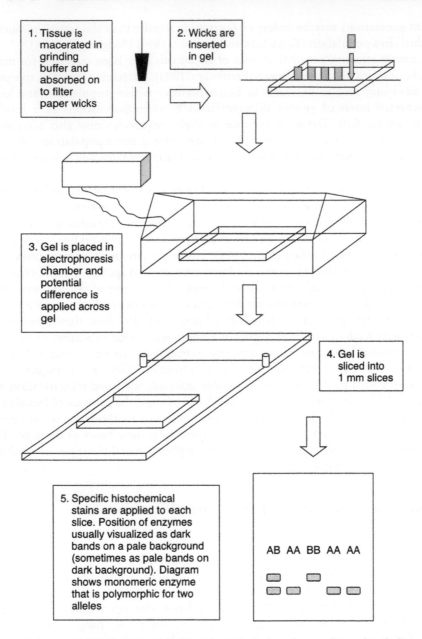

Figure 1. *Starch gel electrophoresis of allozymes. Other gel media may be used including polyacrylamide and cellulose acetate. Note the latter does not require slicing.*

(Smithies, 1955; Kohn, 1957; Raymond and Weintraub, 1959). The rate of migration of a protein depends on its net charge, size and shape and the viscosity of the electrophoretic medium and the strength of the electric field across the gel (Alberts *et al.*, 1983; Murphy *et al.*, 1990). Specific enzymes are marked by applying specific histochemical stains to the electrophoretic medium (Shaw and Prasad, 1970; Harris and Hopkinson, 1976; Murphy *et al.*, 1990). Details of the methods employed for allozyme electrophoresis are provided by a number of authors (Harris and Hopkinson, 1976; Ferguson, 1980; Murphy *et al.*, 1990).

Allozymes are usually coded for by unlinked genetic loci scattered throughout the genome. The loci have codominant alleles that are usually inherited in a Mendelian fashion. This means that allozyme data can be used for a variety of analyses at the intraspecific level including estimates of genetic variation, congruence of genotype frequencies to Hardy–Weinberg expectations, genetic distance, spatial genetic structure and effective population size. Allozymes can also be used to differentiate between closely related species and can provide systematic information up to the level of confamilial genera (see Thorpe, 1982; Murphy *et al.*, 1990).

Allozyme electrophoresis is an older technique but it has some advantages when compared to other molecular genetic markers. A number of loci can be scored for a large number of specimens relatively cheaply. However, because the technique is decreasing in use it is often the case that electrophoresis equipment has to be specially built for this method, especially when using thick gels that are sliced so that a set of samples can be scored for multiple enzymes. Another major advantage is that there is a very large number of published allozyme studies. Because of this large body of data, measures of genetic identity and distance (Nei, 1978) can often provide an accurate guide to the taxonomic relationships of sample populations (Thorpe, 1982, 1983). However, there are exceptions to this in certain taxa, notably birds and other groups where there is less than expected genetic differentiation associated with speciation (Thorpe, 1982; D'Incao *et al.*, 1998).

One of the disadvantages of allozyme electrophoresis, especially with the identification of larvae, is that material must be fresh or frozen. This makes conducting an extensive larval sampling programme difficult, especially as the larvae of interest must be sorted from the sample prior to genetic analysis. In addition to this a substantial quantity of tissue is often required for allozyme electrophoresis. One way of avoiding this problem is to actually capture the larvae of interest alive and allow them to grow to a size where they can be subsequently electrophoresed. Such an approach has been used in studies of larval distribution in anemones (Orr *et al.*, 1982), sipunculids (Staton and Rice, 1999) and bivalves (Crisp *et al.*, 1983; Tyler-Walters and Crisp, 1989; Hu *et al.*, 1992). Alternatively, cellulose acetate gels require a much smaller amount of tissue than starch or polyacrylamide slab gels and the crude protein extract from a single small individual is sufficient to run several gels (Yund, 1995).

2.2 *Application of allozyme electrophoresis in larval ecology and the identification of marine larvae*

There are few examples of the use of allozymes to identify larvae in plankton samples probably because of the limitations described above. Staton and Rice (1999) used allozymes to identify larvae from two genetically distinct forms (sibling species) of the sipunculid *Apionsoma misakianum* from off the coast of the southeastern United States. Differences in allele frequencies across several enzyme loci were used to identify individual larvae after they had been reared for 16 months to adult size. The study examined the occurrence of the distinct adult types in southern, tropical localities (Florida Keys, Bahamas) and a northern temperate locality (Central Florida) and the larval types in the Gulf Stream off Florida. It was found that, despite the existence of large numbers of larvae of the southern genetic form (species) of *A. misakianum* in the waters overlying the northern site, most adults inhabiting the site were of the northern genetic form. In other words large-scale dispersal of larvae of the southern type was not realized in large-scale recruitment and survival at the northern site (Staton and Rice, 1999).

Other studies have analysed larval dispersal by comparing allele frequencies in adult and larval populations using allozyme electrophoresis. Kordos and Burton (1993) analysed genetic differentiation in populations of the estuarine crab, *Callinectes sapidus*, along the Gulf coast of America. This species has a planktotrophic megalopa larva that was thought to develop outside the estuarine environment over a period of 4–8 weeks. It was hypothesized that geographically separated populations of *C. sapidus* would show little genetic structure and larval populations would have similar genotype frequencies to adult populations. The study found that significant genetic differentiation occurred between adult populations of *C. sapidus*. There was also significant genetic differentiation between adult and adjacent larval populations, between larvae taken from different localities and between larvae sampled at different times. Interpretation of this data set is difficult but it would seem that populations along the Texas coast are not genetically homogenous and this is reflected in both adult and larval populations. It was suggested that recruitment in blue crab populations was probably episodic. If this is the case, the variation in allele frequencies of larvae, on time scales from months to years, would suggest that low effective population size is probably a strong factor influencing the genetic structure of populations of this crab.

Studies on another decapod, the western rock lobster, *Panulirus cygnus*, indicate that temporal genetic variation in larval recruits to populations in western Australia is also an important factor in determination of the genetic variation in populations (Johnson and Wernham, 1999). Studies of the larvae at two sites over a period of 3 years indicated that allele frequencies varied significantly on a monthly and annual basis. This was reflected by temporally unstable genetic differences between adult populations. Such temporal instability of spatial genetic structure, at a local scale, has been termed chaotic genetic patchiness (Johnson and Black, 1982). This phenomenon has been detected by the same group of researchers for siphonarian limpets and the urchin *Echinometra mathaei* (Johnson and Black, 1982, 1984; Watts *et al.*, 1990) in Australia and by other scientists for the bivalve *Spisula ovalis* (David *et al.*, 1997). These studies emphasize the need for careful interpretation of data on spatial genetic structure of marine species when temporal variation has not been considered and when studies only include a few sites.

Allozyme electrophoresis has also been used to study the asexual production of larvae in a number of species, mostly cnidarians. Ottaway and Kirby (1975) compared the catalase genotypes of adult and brooded juveniles of the anemone *Actinia tenebrosa* (*Figure 2*) from southern Australia. They found that the genotypes of all the juveniles they examined were identical to those of the brooding adults. Because *A. tenebrosa* develops male and female gonads it was thought that clonal reproduction in this species was unlikely, despite the obvious conclusion that a large portion of the offspring were produced by budding or by parthenogenesis. Alternative explanations of the results included an underestimate of genetic diversity of larvae because of the use of only a single enzyme locus or that this species ejected juveniles that were of a different genotype to the adult (Ottaway and Kirby, 1975).

A later study, using three enzyme loci, confirmed that the production of larvae in *A. tenebrosa* was clonal (Black and Johnson, 1979). This study also investigated the

Figure 2. Actinia tenebrosa. *Adult individual, Lyall Bay, Wellington, New Zealand.*

genetic structure of *A. tenebrosa* populations and found that they exhibited many features associated with asexual reproduction (large genetic distance between populations, significant departures from Hardy–Weinberg equilibrium, local dominance of a few genotypes). They also found that over the entire study area (Rottnest Island, western Australia) there were a large number of genotypes, almost in expected frequencies according to a random mating model. This suggested that, while the predominant mode of reproduction in *A. tenebrosa* was asexual, sexual reproduction probably occurred and resulted in planktonic larvae (Black and Johnson, 1979). It is interesting to note that many invertebrates that show mixed reproductive strategies have high genetic variability (Solé-Cava and Thorpe, 1991).

Studies on four separate enzyme loci in *Actinia equina* from the United Kingdom (*Figure 3*) also showed that juveniles were of an identical genotype to the adults in which they were brooded (Orr *et al.*, 1982). Similar results have also been obtained for the species *Actinia bermudensis* collected from the coast of Brazil (Monteiro *et al.*, 1998). Observations that *A. equina* maintained in isolation, in filtered or artificial seawater continued to produce broods of young over long periods of time supported evidence from genetic studies that demonstrated widespread asexual reproduction in this species and probably in other members of the genus.

Figure 3. Actinia equina. *Adult individual, Hope Cove, UK.*

Several species of the genus *Epiactis* that occur on the western coast of North America have also been subject to studies of the occurrence of asexual *vs* sexual reproduction. In this case brooded juveniles were compared to adults using both allozyme electrophoresis and multilocus DNA fingerprinting (Edmands, 1995; see below). In *Epiactis prolifera*, a gynodioecious hermaphrodite, allozyme mobility and DNA fingerprints were identical between brooding adults and their associated juveniles (Edmands, 1995). This indicated that this species reproduced asexually. For the dioecious (gonochoric) species *Epiactis lisbethae*, parent individuals were found to have identical allozyme mobility, across 4–5 loci, to brooded offspring. However, some variation was found in DNA fingerprints and it was suggested that juveniles were produced sexually (Edmands, 1995). There was the possibility that some externally produced juveniles could have been 'adopted' by females (Edmands, 1995) or they were produced by self-fertilization or meiotic parthenogenesis, as banding patterns were only slightly different. This study shows the low resolution, for the purposes of assessing the occurrence of clones, provided by using a few allozyme loci, in some cases.

Asexual production of larvae has been detected in other cnidarians using allozymes. The comparison of larval and parental genotypes across five polymorphic enzyme loci in the scleractinian coral *Pocillopora damicornis* from western Australia indicated that genotypes were identical in all cases (Stoddart, 1983). As expected, populations of *P. damicornis* showed significant departures from Hardy–Weinberg equilibrium and linkage disequilibria for the enzyme loci investigated. However, this species is a simultaneous hermaphrodite and when data from the three investigated populations was pooled, deviations from Hardy–Weinberg expectations and linkage disequilibria were greatly reduced, suggesting the occurrence of some sexual reproduction (Stoddart, 1983).

Other studies on corals have revealed contrasting results. The zooxanthellate corals *Acropora palifera* and *Seriatopa hystrix*, from western Australia, were found to have larval genotypes that were different to parent individuals, consistent with at least some cross-fertilization during sexual reproduction (Ayre and Resing, 1986). However, planulae from two azooxanthellate species, *Tubastrea diaphana* and *T. coccinea*, were identical to parent individuals (Ayre and Resing, 1986). These studies not only confirmed the existence of both modes of planula production in corals but also demonstrated that problems associated with the presence of symbionts in host tissues could be avoided when using allozyme electrophoresis.

There are few studies on the genetic relationships of larvae and adults in other phyla. The ophiuroid, *Ophiomyxa brevirima*, from New Zealand, is dioecious and its larvae are brooded. Multilocus genotype comparisons indicated that in this species nearly half the breeding females produced larvae asexually (Garrett *et al.*, 1997). Remaining individuals showed evidence of crossing with at least one male and, in some cases, evidence of multiple paternity. There was no evidence of significant deviations of genotype frequencies from Hardy–Weinberg expectations in adult populations, although the sex ratio was skewed with a significant excess of females (Garrett *et al.*, 1997).

Ascidians have been the subject of studies on sperm dispersal using allozyme analysis of brooded larvae as a measure of fertilization success (Grosberg, 1991; Yund and McCartney, 1994; Yund, 1995). Grosberg (1991), analysed sperm dispersal in the colonial species *Botryllus schlosseri* (*Figure 4*). Colonies homozygous for rare alleles were placed into dense conspecific populations and fertilization success was assessed at different distances for the presence of rare alleles in brooded embryos. It was found that there was a significant negative relationship between the numbers of embryos

Figure 4. Botryllus schlosseri. *Adult colony from Elk Reef off Plymouth, UK.*

fertilized by the inserted colony and the distance away from it (Grosberg, 1991). Subsequent investigations indicated that fertilization can occur at considerable distances (>10 m) away from acting male colonies at lower colony densities (Yund and McCartney, 1994). It has also been found that the density of colonies has a strong influence on fertilization success in *B. schlosseri* (Yund, 1995).

3. Molecular methods

3.1 *Extracting DNA from marine larvae*

Molecular genetic methods of larval identification all rely on assessing variation in genomic DNA. They have the advantage that they usually utilize polymerase chain reaction (PCR) amplification of the region of interest and therefore require a very small amount of tissue for analysis. This tissue can be fresh, stored at low temperature, or preserved in ethanol or buffer. However, DNA can also be recovered from formalin-fixed material or ancient tissue samples recovered from the environment (e.g. resting stages). This provides enormous advantages over allozyme-based methods for identifying larvae, especially when carrying out extensive sampling programmes where full laboratory facilities may not be available in the field.

There is an increasing variety of molecular genetic methods used to study marine larvae. The resolution of these methods depends on the intrinsic variation of the region(s) of the genome being studied and on the method used to detect this variation. All methods require some form of DNA extraction. Conventional phenol/chloroform or high salt extraction methods (e.g. Sambrook *et al.*, 1989; Palumbi *et al.*, 1991) have been used in some larval studies, especially when subsequent protocols have required high quality DNA (Bilodeau *et al.*, 1999; Makinster *et al.*, 1999; Barki *et al.*, 2000). Alternatively, a number of simpler methods have been utilized to release DNA from marine larvae. At the simplest these techniques have involved rehydrating individual larvae preserved in absolute ethanol, followed by immersion in PCR buffer for several hours. After this, the remaining constituents of a PCR reaction can be added to the tube and the amplification reaction carried out (Bucklin *et al.*, 1995; used in Lindeque *et al.*, 1999). Alternatively, ethanol-preserved larvae have been rehydrated and then frozen to rupture cells. An aliquot of the subsequent solution, containing DNA, was then added to a conventional PCR reaction as template (Evans *et al.*, 1998). Commercially available solutions may also be used to release template DNA from whole larvae. Individual larvae have been immersed in Chelex solution, heated and then the resultant solution used directly or centrifuged to pellet template DNA (Medeiros-Bergen *et al.*, 1995; Rocha-Olivares, 1998; Bishop *et al.*, 2000). Larvae have also been immersed in Microlysis® (Microzone Ltd, Lewes, East Sussex, UK) solution followed by cycles of heating and cooling on a PCR machine. The resulting solution has been directly used in PCR (Morgan and Rogers, 2001).

3.2 *Analysis of sequence variation*

The most common region of DNA that has been used to differentiate between species has been the 16S mitochondrial rRNA gene (e.g. Olsen *et al.*, 1991; Evans *et al.*, 1998; Lindeque *et al.*, 1999). This mitochondrial region has also been often used in identifying species of holoplankton (e.g. Bucklin *et al.*, 1995) and in the assessment of interspecific relationships in numerous other species (e.g. Maggioni *et al.*, 2001). Other regions used to identify larvae have included the mitochondrial cytochrome-*b* gene (Rocha-Olivares, 1998), and the nuclear internal transcribed spacers (ITS1 and ITS2) that form part of the rRNA multigene family that also includes the 18S, 5.8S and 28S rRNA genes (Dixon *et al.*, 1995a; Dixon and Dixon, 1996; Toro, 1998). However, any region that shows significant interspecific variation for the taxa of interest is useful for species identification.

Conceptually, the simplest approach to larval identification is direct sequencing of a specific region of the genome that shows variation between all potential species in a sample (*Figure 5*). This method has been used to differentiate between 16S sequences of morphologically similar larvae of the holothurians, *Psolus fabricii* and *Cucumaria frondosa* off the coast of New Hampshire in the northwestern Atlantic (Olsen *et al.*, 1991). However, such methods are generally time-consuming and expensive. An alternative approach has been to develop an alternative means to detect sequence variation, in the DNA region of interest, that provides a simple and less expensive means to screen large numbers of individuals.

Oligonucleotide probes. One method to do this has been to develop specific oligonucleotide probes for the species of interest. This approach was used by Medeiros-Bergen *et al.* (1995) to differentiate between the larvae of three species of sea cucumbers, *Cucumaria frondosa*, *Psolus fabricii* and *Chiridota laevis*, from the eastern coast of the United States (*Figure 6*). The 16S rRNA gene was amplified for individual larvae and aliquots of the PCR reaction were dotted onto a nylon membrane. The membrane-bound DNA was exposed to isotopically labelled or biotinylated oligonucleotide probes complementary to a 15 bp region of the 16S rDNA known from previous sequencing to show substantial differences between all three species. The method was cross-checked by sequencing the larvae and probes were found to be highly species specific. Only a small number of individuals failed to hybridize with probes because of intraspecific variation in a single nucleotide position complementary to the probe (Medeiros-Bergen *et al.*, 1995).

An approach used to identify the eggs of sturgeons (caviar) has been to sequence several regions of DNA and then to design species specific primers targeted at parts of the sequence that show interspecific differences (DeSalle and Birstein, 1996;

Figure 7). Primers produced amplification only if they were perfectly complementary to template DNA. Template DNA was subject to PCR with a series of primers and then an identification of the species was made on the basis of positive amplifications.

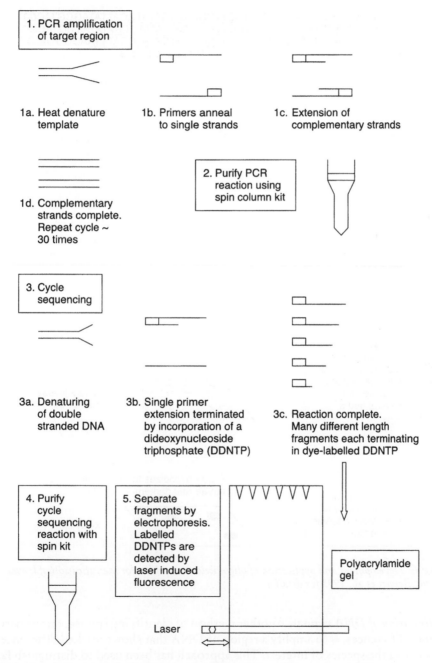

1. PCR amplification of target region

1a. Heat denature template

1b. Primers anneal to single strands

1c. Extension of complementary strands

1d. Complementary strands complete. Repeat cycle ~ 30 times

2. Purify PCR reaction using spin column kit

3. Cycle sequencing

3a. Denaturing of double stranded DNA

3b. Single primer extension terminated by incorporation of a dideoxynucleoside triphosphate (DDNTP)

3c. Reaction complete. Many different length fragments each terminating in dye-labelled DDNTP

4. Purify cycle sequencing reaction with spin kit

5. Separate fragments by electrophoresis. Labelled DDNTPs are detected by laser induced fluorescence

Polyacrylamide gel

Laser

Figure 5. Current method of DNA sequencing from the PCR stage. Note sequence detection is achieved with an automated sequencer (e.g. Applied Biosystems 377).

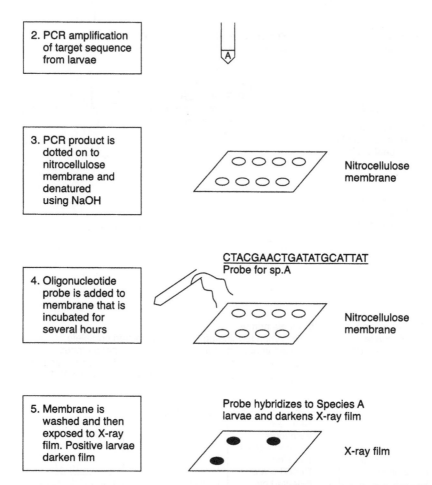

1. Sequence region of interest. Choose site that shows specific differences in DNA sequence
Probes complementary to site are designed

Species A GACTACCTATAGGTATGGGCTACGAACTGATATGCATTATAGCGTACGT
Species B GACTACCTATAGGTATGGGCTTGGAAG-GATATGCAATATAGCGTACGT
Species C GACTACCTATAGGTATGGGCTACGTAGTGATATGCATTTAAGCGTACGT
Probe sequences

2. PCR amplification of target sequence from larvae

A

3. PCR product is dotted on to nitrocellulose membrane and denatured using NaOH

Nitrocellulose membrane

CTACGAACTGATATGCATTAT
Probe for sp.A

4. Oligonucleotide probe is added to membrane that is incubated for several hours

Nitrocellulose membrane

5. Membrane is washed and then exposed to X-ray film. Positive larvae darken film

Probe hybridizes to Species A larvae and darkens X-ray film

X-ray film

Figure 6. *Development and application of oligonucleotide probes for identification of larvae (see Medeiros-Bergen* et al., *1995 for details).*

Amplification of DNA regions. Another method of identifying species, on the basis of sequence differences, is to amplify a region of DNA that shows marked differences in size between the species of interest. This approach has been used to distinguish larval mussels of three different species (Dixon *et al.*, 1995b; *Figure 8*). This study relied on amplification of the ITS2 region of nuclear rDNA. Species were distinguished by the

1. Design primers for several regions
   of DNA targeted at part of
   sequence that shows interspecific
   variation

Species A GACTACCTATAGGTATGGGCTACGAACTGATATGCATTATAGCGTACGT
Forward primer (e.g. 16SrDNA)   GATGCTTGACTATACGTAATATCG

Species B GACTACCTATAGGTATGGGCTTGGAAG-GATATGCAATATAGCGTACGT

Species C GACTACCTATAGGTATGGGCTACGTAGTGATATGCATTTAAGCGTACGT

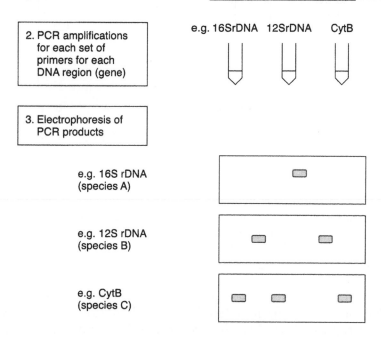

2. PCR amplifications
   for each set of
   primers for each
   DNA region (gene)

e.g. 16SrDNA   12SrDNA   CytB

3. Electrophoresis of
   PCR products

e.g. 16S rDNA
(species A)

e.g. 12S rDNA
(species B)

e.g. CytB
(species C)

Figure 7. Development and application of species specific amplification of mitochondrial DNA sequences (see DeSalle and Birstein, 1996 for details).

size of the ITS2 amplification product assayed through agarose gel electrophoresis (Dixon *et al.*, 1995b). ITS sequences often show marked differences in size between closely related species because of the presence of microsatellite regions within the sequence (Harris and Crandall, 2000). Size variation has also been used to distinguish between alleles at a single genetic locus. This has been used to distinguish between alleles at a neutral locus in an intron of the calmodulin-1 gene in *Mytilus edulis* (Côrte-Real *et al.*, 1994). It has also been used to distinguish between alleles in anonymous single copy nuclear genes in *Crassostrea virginica* (Hu and Foltz, 1996).

Different sized amplification products can be utilized to distinguish species through allele-specific PCR. The first step in this method is sequencing of a specific PCR product for the species of interest. Sequences are used to identify a common priming site for all species and then species-specific reverse priming sites. Primers may be used in multiplexed PCRs giving size-specific products to each of the species being studied (Bucklin, 2000). This method has been used to differentiate between species of holoplankton (Bucklin *et al.*,

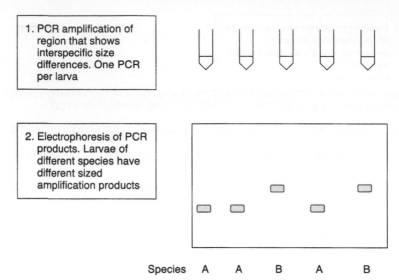

Figure 8. *Method for identification of larvae based on size-specific differences in a specific region of DNA (see Dixon* et al., *1995a,b for details).*

1998). A variant of this method termed multiplex haplotype-specific (MHS; *Figure 9*) PCR involves using combinations of forward and reverse primers aimed at taxon-specific point mutations in amplified mitochondrial sequences. This method provides an inexpensive means to differentiate between a large number of taxa and has been used to identify larvae of *Sebastes* species from the eastern Pacific (Rocha-Olivares, 1998).

RFLP. A more popular approach to identification of marine larvae has been the use of restriction analysis or restriction fragment length polymorphism (RFLP) to assess variation of amplified regions (*Figure 10*). This entails digesting DNA with restriction enzymes that have specific sequence recognition sites, usually of 4–8 bp in length. The DNA, usually an amplification product (Evans *et al.*, 1998), but sometimes the entire mitochondrial genome (Silberman *et al.*, 1994a), is cut at the restriction sites if they are present in the sequence. This produces fragments of DNA of different lengths depending on whether restriction sites are present or absent and depending on length variation of the sequence between restriction sites. Fragments are separated according to size by electrophoresis. Species that show interspecific variation in a sequence will produce different banding patterns if they show differences in the presence or absence of restriction sites.

RFLP of the entire mitochondrial genome has been used to analyse intraspecific variation in decapod larvae (Silberman *et al.*, 1994a). Isolation of mitochondrial DNA, however, is technically difficult and usually requires substantial amounts of tissue. It is more common use RFLP to analyse variation in PCR amplified sequences, especially in regions coding for ribosomal RNA, in which case it may be known as amplified ribosomal DNA restriction analysis (ARDRA; Dixon and Dixon, 1996; Evans *et al.*, 1998; Lindeque *et al.*, 1999).

Other methods. Other methods are available to provide indirect assessment of DNA sequence variation but they have not been frequently used for studies of marine larvae.

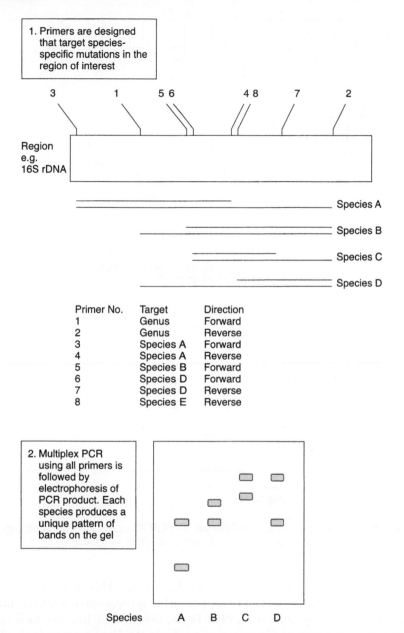

Figure 9. *Development and application of method for identification of larvae through multiplex haplotype specific PCR (see Rocha-Olivares, 1998 for details).*

Such methods include single-strand conformation polymorphism (SSCP), denaturing gradient gel electrophoresis (DGGE) and heteroduplex analysis. SSCP begins with double-stranded PCR product that is denatured by heating and then rapidly cooled by plunging into ice (*Figure 11*). Under such conditions the denatured strands do not reanneal but fold up on themselves to form a secondary structure that is dependent on sequence. Single strands with different sequences have different conformations and

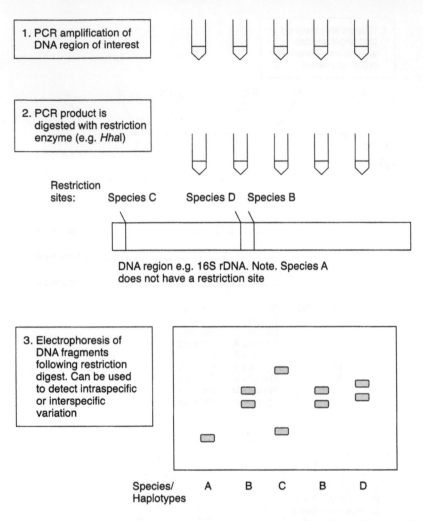

Figure 10. *Restriction fragment length polymorphism (RFLP) analysis as used to identify species or to assess intraspecific sequence variation (see Evans et al., 1998 for details).*

therefore migrate at different rates in an acrylamide gel. This method can separate DNA molecules of between 100 and 1000 bp long but is reported to work optimally on sequences of 200 bp long (Orita *et al.*, 1989; Potts, 1996). SSCP has been used to study genetic variation in larvae of *Crassostrea gigas* from the Pacific coast of the United States (Li and Hedgecock, 1998).

DGGE relies on the fact that the point at which DNA denatures, because of heating or the presence of a chemical denaturant, depends on its sequence (*Figure 12*). As DNA denatures, its electrophoretic mobility is retarded. Thus two DNA molecules of the same size but with different sequences will denature at different temperatures or concentrations of chemical denaturant and will have different mobilities on a gel (Myers *et al.*, 1987; Potts, 1996). This method is reported to be optimal for studies of sequences between 200 and 1000 bp long. At present this technique would not seem to have been used in studies of marine larvae.

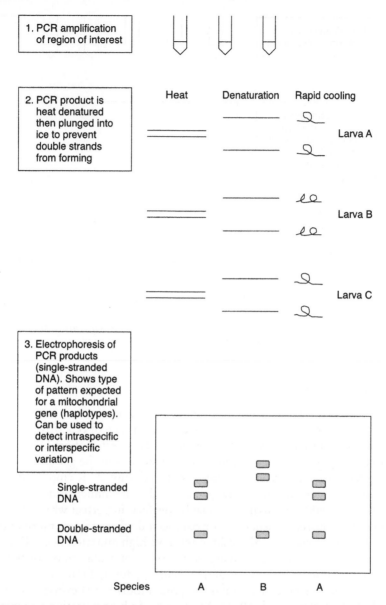

Figure 11. Detection of sequence variation through SSCP analysis (see Li and Hedgecock, 1998 for details).

Heteroduplex analysis involves heating DNA fragments until they denature, followed by slow cooling (*Figure 13*). Single-stranded DNA molecules reanneal to complementary strands, forming homoduplexes. However, they also reanneal to non-complementary strands that have mismatches, forming heteroduplexes. Heteroduplexes have a retarded electrophoretic mobility because of changes in secondary structure and thus DNA fragments that show differences in DNA sequence (i.e. multiple alleles) can be separated (White *et al.*, 1992; Potts, 1996). Heteroduplex analysis has not been used to study marine larvae to date.

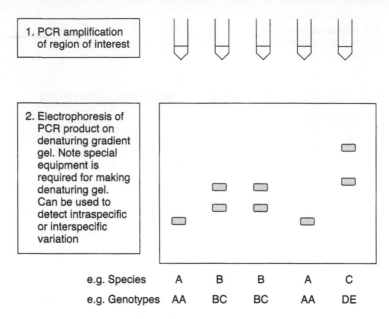

Figure 12. *Detection of sequence variation through DGGE. Note denaturing may be achieved through a gradient of chemical denaturant in gel or through a temperature gradient (TGGE).*

3.3 *Repetitive sequences*

There are several types of repetitive DNA sequence, some of which have been used to identify or study marine larvae. The most widely used type of repetitive sequences to date, are microsatellites. Microsatellites are tandemly repeated short DNA sequences between 1 and 6 bp long. The commonest microsatellites in the genome are poly (A) or poly (T) regions. However, these are unstable during PCR amplification so the most commonly used microsatellites in genetic studies are dinucleotide repeats such as $(GT)_n$ (Hancock, 1999). Microsatellites maybe perfect, imperfect where the sequence is interrupted by non-repeat sequence, or compound where two or more microsatellites lie side by side (*Figure 14*). Microsatellites have high mutation rates (Dallas, 1992; Weber and Wong, 1993). The underlying mechanism for mutation is not fully understood but is thought to involve slip-strand mispairing, during DNA replication and/or unequal cross-over or gene-conversion during recombination (Levinson and Gutman, 1987; Smith, 1976; Jeffreys *et al.*, 1994). As a result of a high mutation rate microsatellites show polymorphism in the number of repeats. This polymorphism gives rise to the use of microsatellites as codominant, single-locus genetic markers. As microsatellites are generally non-coding, these markers are usually thought to be selectively neutral.

Because microsatellites have high mutation rates they are generally unconserved between all but the most closely related species (Primmer *et al.*, 1996; D'Amato *et al.*, 1999). This usually means that new microsatellites must be found prior to studying a new organism, although new loci are being reported all the time for a variety of species. There are several methods of detecting microsatellites in the genome. One technically easy method is to amplify DNA using arbitrary primers (random amplified polymorphic DNA; see below and *Figure 15*). This produces multiple amplified fragments

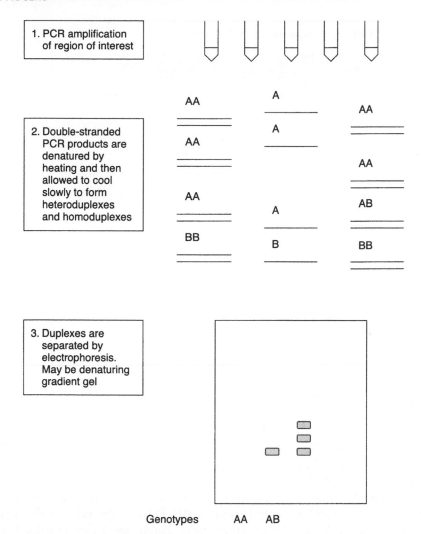

1. PCR amplification of region of interest

2. Double-stranded PCR products are denatured by heating and then allowed to cool slowly to form heteroduplexes and homoduplexes

AA

AA

AA

BB

A

A

A

A

B

AA

AA

AA

AB

BB

3. Duplexes are separated by electrophoresis. May be denaturing gradient gel

Genotypes   AA   AB

Figure 13. *Detection of sequence variation through heteroduplex analysis.*

of DNA that are separated on an agarose gel. This gel is blotted on to a nitrocellulose membrane to bind and denature the DNA. This membrane is then incubated with an isotopic or chemically labelled microsatellite probe (e.g. $(CA)_{15}$). The probe hybridizes to bands containing microsatellites. Electrophoresis of amplified fragments can be repeated and microsatellite-containing fragments can be excised from the gel and sequenced.

Alternatively, genomic DNA can be digested using a restriction enzyme with a four-base recognition site (*Figure 16*). The resulting fragments can be separated by elec-trophoresis and those of a length between 200 and 500 bp long can be excised and purified from the gel. These fragments are then directly used to construct a DNA library or they go through a microsatellite enrichment step prior to library construction (e.g. Kandpal *et al.*, 1994; Morgan *et al.*, 2000). Duplicate DNA libraries are produced and one set of colonies is blotted on to nitrocellulose membranes and incubated with microsatellite

Perfect microsatellite: *Lophelia pertusa*

GCTCCCCC<u>CACACACACACACACACACACACACACA</u>
<u>CACACACACACACACACACACACACACACA</u>CCCCT

Imperfect microsatellite: *Euphausia superba*

GATAC<u>CACACACACACACACACACACACACACACACACA</u>
<u>CACACACACACACACACACAC</u>ATACACACACACATACGCAC
ACACACACACACACACACAC</u>ATCCGCACACACACATACAC
GC

Compound microsatellite: *Penaeus (Farfantepenaeus)
paulensis*

TATGATA<u>ATTATTATTATTATTATT</u>ATTA<u>CACACA</u>
<u>CACACA</u>CGC<u>ACACACACACACACACAC</u>ATTGA

Figure 14. Examples of the three different types of microsatellites identified from three different species (A.D. Rogers, unpublished data).

probes. Positive colonies contain microsatellites and these are picked from the duplicate plate for growing overnight prior to plasmid purification and sequencing of microsatellites. Whilst these methods are technically more complex than RAPD microsatellite isolation, they tend to produce more microsatellites over a given period of time, especially enrichment protocols (see Morgan and Rogers, 2001). Polymorphic microsatellites can be scored using polyacrylamide gel electrophoresis. Bands are either visualized using ethidium bromide or silver staining or by end labelling a primer with ^{32}P or ^{33}P and exposing the gel to X-ray film or a phosphor imaging system. Automated sequencing machines can now detect flourescently labelled DNA fragments including microsatellites.

As with other PCR-based methods, microsatellites can be amplified from very small quantities of DNA. They have been used to study genetic variation in cohorts of larval cod, *Gadus morhua*, in the western Atlantic (Ruzzante et al., 1996; Herbinger et al., 1997). Another study has exploited the taxon-specific amplification of microsatellite loci to identify larvae of the native oyster, *Ostrea edulis*, off the south coast of the UK (Morgan and Rogers, 2001). In this case, a single PCR followed by visualization of amplification product was all that was required to identify an oyster larva. This method also allowed the identification of the presence of oyster larvae in DNA from mixed samples of zooplankton. Extensive trials were required with template DNA from other common bivalves in the study area as some microsatellite loci amplified across several species. The frequency of cross-specific amplification was correlated with genetic relatedness of taxa. The quantity of amplification product could also be roughly equated to the number of larvae in a DNA extraction.

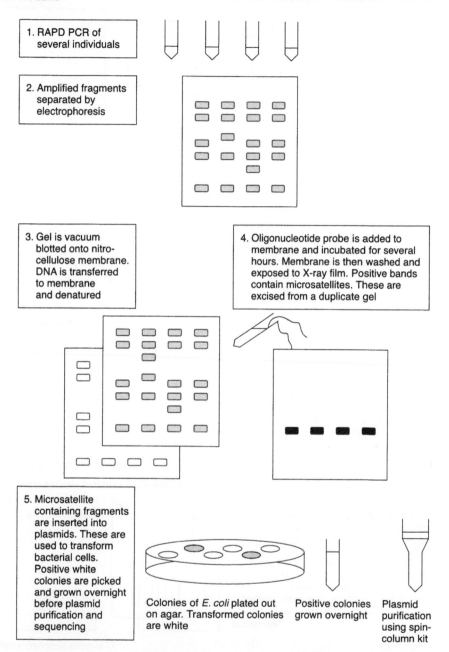

Figure 15. The RAPD method for identification of microsatellite loci. Note that once microsatellite sequences are detected, primers must be designed for amplification from genomic DNA. The microsatellites can then be assessed for variation (see Ender et al., 1996 for details).

Minisatellites are tandem repeats of up to 65 bp in length. This type of marker has not been used in larval studies to date. Middle repetitive elements are sequences present in 10^3–10^5 copies per genome (Bilodeau *et al*., 1999). They are present in many species, dispersed throughout the genome and often consist of transposable elements

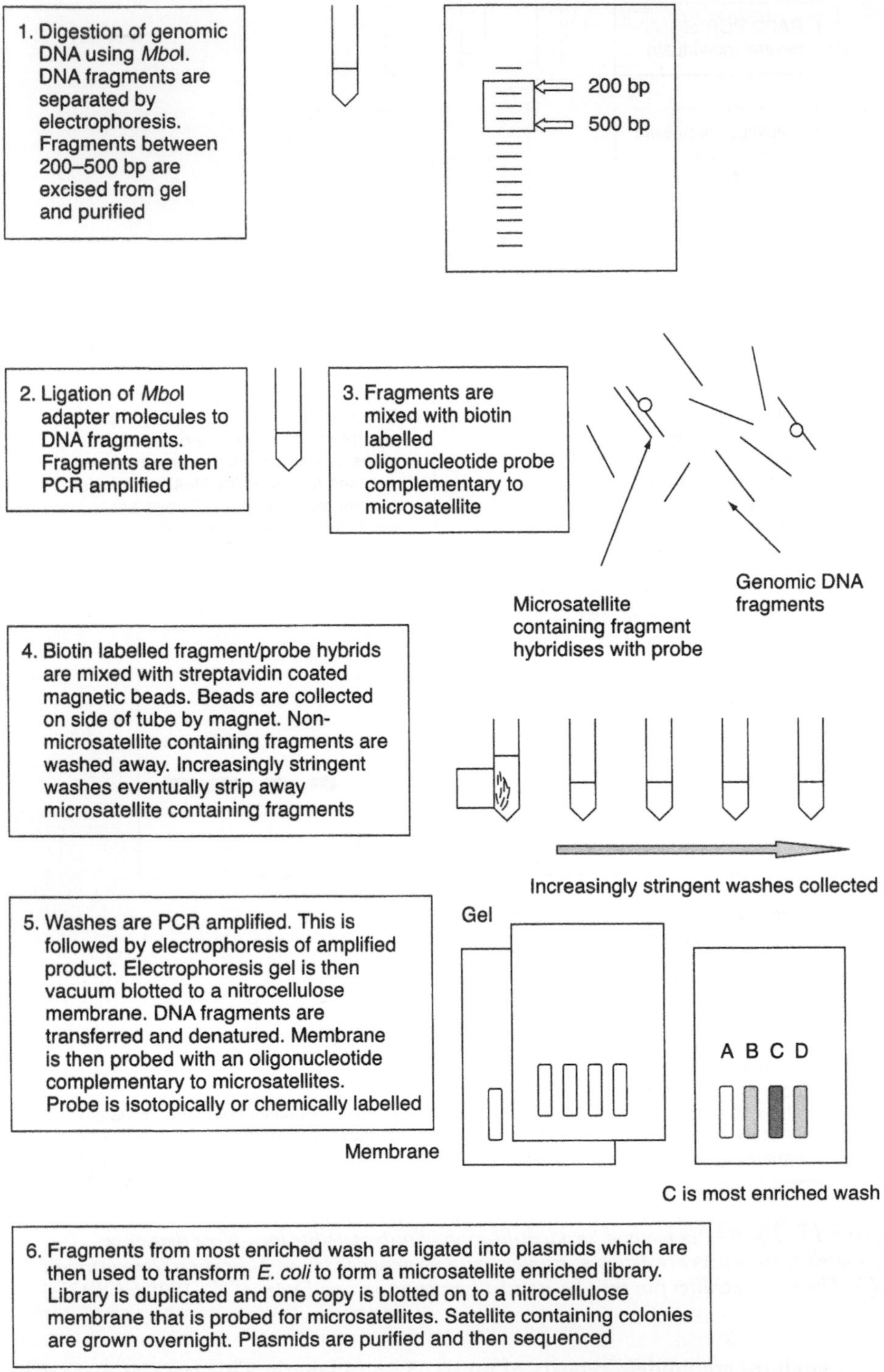

1. Digestion of genomic DNA using *Mbo*I. DNA fragments are separated by electrophoresis. Fragments between 200–500 bp are excised from gel and purified

200 bp

500 bp

2. Ligation of *Mbo*I adapter molecules to DNA fragments. Fragments are then PCR amplified

3. Fragments are mixed with biotin labelled oligonucleotide probe complementary to microsatellite

Microsatellite containing fragment hybridises with probe

Genomic DNA fragments

4. Biotin labelled fragment/probe hybrids are mixed with streptavidin coated magnetic beads. Beads are collected on side of tube by magnet. Non-microsatellite containing fragments are washed away. Increasingly stringent washes eventually strip away microsatellite containing fragments

Increasingly stringent washes collected

5. Washes are PCR amplified. This is followed by electrophoresis of amplified product. Electrophoresis gel is then vacuum blotted to a nitrocellulose membrane. DNA fragments are transferred and denatured. Membrane is then probed with an oligonucleotide complementary to microsatellites. Probe is isotopically or chemically labelled

Gel

Membrane

A B C D

C is most enriched wash

6. Fragments from most enriched wash are ligated into plasmids which are then used to transform *E. coli* to form a microsatellite enriched library. Library is duplicated and one copy is blotted on to a nitrocellulose membrane that is probed for microsatellites. Satellite containing colonies are grown overnight. Plasmids are purified and then sequenced

Figure 16. The enrichment protocol for identification of microsatellites. Note that once microsatellite sequences are detected, primers must be designed for amplification from genomic DNA. The microsatellites can then be assessed for variation (see Kandpal et al., 1994 for details).

(Spradling and Rubin, 1981). These regions evolve at high rates and therefore only tend to be conserved amongst closely related species (Makinster *et al.*, 1999). Such regions may be located by creating a genomic library for the species of interest and analysing the sequences in the library for variation either through direct sequencing or by restriction analysis (Bilodeau *et al.*, 1999; *Figure 17*). The presence of a large number of similar or overlapping sequences may indicate the presence of a middle repetitive element. Because middle repetitive elements are not conserved across taxa they can make useful species or genus specific markers for larvae (Bilodeau *et al.*, 1999; Makinster *et al.*, 1999).

3.4 *Fingerprinting*

Fingerprinting includes several techniques that have different genetic characteristics, all of which produce highly polymorphic banding patterns after gel electrophoresis. These techniques are generally applied to studies of intraspecific variation, particularly in relation to reproductive behaviour and parentage. The oldest method is multilocus DNA fingerprinting (Jeffreys *et al.*, 1985; Fleischer, 1996; *Figure 18*). Genomic DNA is digested using a restriction enzyme with a four-base recognition site. The resultant DNA fragments are then separated by electrophoresis. The electrophoretic gel is then blotted onto a nitrocellulose membrane, transferring the denatured DNA. The membrane is then incubated with an isotopic or chemically labelled minisatellite or microsatellite probe that hybridizes to complementary fragments of the digested DNA. The gel is visualized by exposure to X-ray film for both isotopic or chemiluminescent detection methods (Fleischer, 1996).

The image obtained from multilocus fingerprinting consists of multiple bands usually corresponding to independently assorting, Mendelian inherited loci (Jeffreys *et al.*, 1985, 1986, 1987). The advantage of this method is that a single probe can detect 10–25 highly variable loci and a membrane can usually be stripped and reprobed with several independent oligonucleotides (Fleischer, 1996). However, multilocus DNA fingerprinting requires substantial amounts of DNA, presenting a problem for very small larvae. This has probably restricted the use of this technique to only a few studies on marine larvae (Coyer *et al.*, 1995 (algal spores rather than larvae); Edmands, 1995). In these cases the larvae used had already grown to a juvenile stage or they were on grown in the laboratory.

Randomly amplified polymorphic DNA (RAPD) markers, also known as arbitrary primed PCR (AP-PCR) is a PCR-based fingerprinting technique (*Figure 19*). PCR is carried out with small (usually 10 bp) arbitrary primers, usually with a G+C content of 50% (Grosberg *et al.*, 1996). A single primer is used and amplification usually only takes place in parts of the genome where there are priming sites in opposite orientations, close enough for amplification to take place. The resultant PCR products, fragments of DNA usually 200–2000 bp long, are then separated by electrophoresis (Grosberg *et al.*, 1996). RAPD bands are generally considered as dominant markers with two alleles, one of which is amplifiable, and which are inherited in a Mendelian fashion. In such a case, allele and genotypic frequencies can be estimated for RAPD loci if it was assumed that populations are in Hardy–Weinberg equilibrium.

Experimental data suggest that the majority of RAPD loci are inherited in a Mendelian fashion (Levitan and Grosberg, 1993), but there is some evidence that this is not always the case (Riedy *et al.*, 1992). There is also evidence that RAPD loci often have more than two alleles (Levitan and Grosberg, 1993; Smith *et al.*, 1994). However, the main problems with RAPDs appear to arise from practical difficulties in applying the method. Bands may comigrate during electrophoresis leading to incorrect estimates of allele frequency. Also template quality and other PCR conditions can have

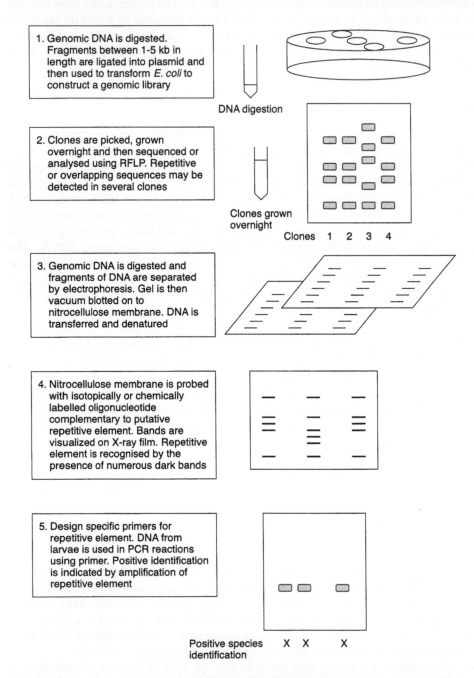

Figure 17. Development and application of middle repetitive elements for identification of larvae (see Bilodeau et al., 1999 for details).

dramatic effects on RAPD amplification leading to difficulties in repeatability of experiments (Grosberg *et al.*, 1996 and references therein). Some of this can arise from interactions between DNA strands during PCR or even from amplification of DNA from commensal or parasitic organisms in the tissue of the study species (Rabouam *et al.*, 1999). Optimization of RAPD protocols can be a lengthy procedure and for some marine taxa can be particularly problematic. Because of fundamental and practical

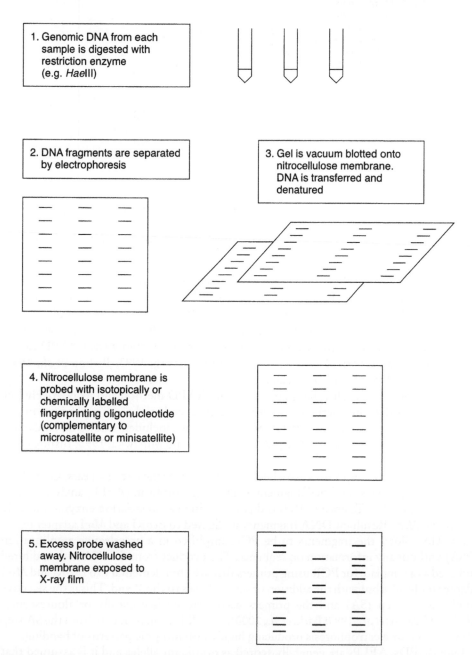

1. Genomic DNA from each sample is digested with restriction enzyme (e.g. *Hae*III)

2. DNA fragments are separated by electrophoresis

3. Gel is vacuum blotted onto nitrocellulose membrane. DNA is transferred and denatured

4. Nitrocellulose membrane is probed with isotopically or chemically labelled fingerprinting oligonucleotide (complementary to microsatellite or minisatellite)

5. Excess probe washed away. Nitrocellulose membrane exposed to X-ray film

Figure 18. Multilocus DNA fingerprinting (see Edmands, 1995 for details).

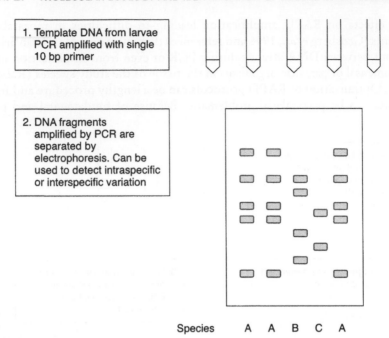

1. Template DNA from larvae
PCR amplified with single
10 bp primer

2. DNA fragments
amplified by PCR are
separated by
electrophoresis. Can be
used to detect intraspecific
or interspecific variation

Species   A   A   B   C   A

Figure 19. RAPD analysis (see Levitan and Grosberg, 1993; Bishop et al., 1996 for details)

problems with RAPD PCR it is generally the practice to analyse RAPD data using cluster analysis based on band presence/absence. This has the advantage of pooling data across all loci and reducing the possibility that a few spurious bands lead to incorrect interpretation of results (Grosberg *et al.*, 1996). Only where extensive mating studies have been carried out is it possible to comfortably treat RAPD loci as having Mendelian inherited alleles (Levitan and Grosberg, 1993; Bishop *et al.*, 1996; Grosberg *et al.*, 1996; Bishop *et al.*, 2000).

Despite potential pitfalls, the simplicity of the RAPD method and its relatively low cost has meant that this method has been used in several studies on reproduction and larval identification in marine invertebrates. These have included studies on cnidarians and ascidians (Levitan and Grosberg, 1993; Coffroth and Mulawka, 1995; Bishop *et al.*, 1996, 2000).

Other types of fingerprinting have been developed over the last few years. One of the most widely applied is amplified fragment length polymorphism (AFLP) analysis (Vos *et al.*, 1995; *Figure 20*). Template DNA is digested using two restriction enzymes, usually *Eco*RI and *Mse*I. Resultant DNA fragments are ligated to *Eco*RI and *Mse*I adapter molecules. This allows the fragments to be PCR amplified in a preselective amplification (PSA) without any information on sequence. The product from the PSA is then diluted and used as a template for PCR using primers that are complementary to *Eco*RI and *Mse*I adapter molecules but with the addition of several bases to the 3' end. This is the selective amplification step (SA) and the primers used may be isotopically or flourescently labelled (Palacios *et al.*, 1999; Barki *et al.*, 2000). Amplification products from the SA step are separated by electrophoresis producing highly polymorphic patterns of banding.

Like RAPDs, AFLPs are generally scored as dominant alleles and it is assumed that these consist of biallelic loci inherited in a Mendelian fashion (Yan *et al.*, 1999).

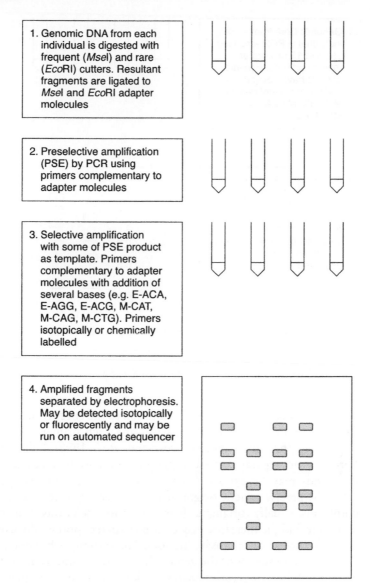

1. Genomic DNA from each individual is digested with frequent (*Mse*I) and rare (*Eco*RI) cutters. Resultant fragments are ligated to *Mse*I and *Eco*RI adapter molecules

2. Preselective amplification (PSE) by PCR using primers complementary to adapter molecules

3. Selective amplification with some of PSE product as template. Primers complementary to adapter molecules with addition of several bases (e.g. E-ACA, E-AGG, E-ACG, M-CAT, M-CAG, M-CTG). Primers isotopically or chemically labelled

4. Amplified fragments separated by electrophoresis. May be detected isotopically or fluorescently and may be run on automated sequencer

Figure 20. AFLP analysis (see Barki et al., 2000 for details).

However, some AFLP loci are reported to behave as codominant loci and some loci probably consist of more than two alleles. Confirmation of mode of inheritance in these markers requires breeding experiments. The main advantage of AFLPs over RAPDs appears to be a high level of reproducibility. To date little work has been carried out on marine larvae using this technique. There is one study on the larvae of the soft coral *Parerythropodium fulvum fulvum* from the Red Sea (Barki *et al.*, 2000).

Two other fingerprinting techniques are yet to be employed in studies of marine larvae. The first of these is known as direct amplification of length polymorphism (DALP; *Figure 21*). This method involves PCR amplification of nuclear DNA with a set of primers, the core sequence of which correspond to M-13 forward and reverse

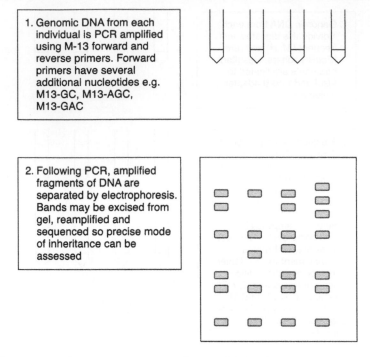

Figure 21. DALP analysis (see Desmarais et al., 1998 for details).

sequencing primers (Desmarais *et al.*, 1998). In addition to the core sequence, different sets of primers have a number of additional bases and each primer set generates a different set of products for a given species (Desmarais *et al.*, 1998). Following amplification, PCR products are separated by electrophoresis and produce a pattern of polymorphic bands. Bands may be excised from sequencing gels and reamplified and sequenced so that the precise genetic origin of band polymorphism can be detected. These polymorphisms usually originate from insertion/deletions but some are microsatellites. This technique therefore may be used to detect potentially useful polymorphic anonymous nuclear or microsatellite loci. The technique has been found to work successfully on adult mussels of the genus *Mytilus* (Desmarais *et al.*, 1998).

The final method of DNA fingerprint analysis is known as simple sequence repeat, or inter-simple sequence repeat analysis (SSR or ISSR; *Figure 22*). This method is a variant of RAPD analysis but primers are targeted at microsatellite regions. Template DNA is subject to PCR using primers that consist of a repeat section and then an anchoring sequence of two or more bp (e.g. $(CA)_6RY$; Esselman *et al.*, 1999). PCR products are separated by electrophoresis to produce highly polymorphic banding patterns. ISSR can produce more polymorphic loci and a higher reproducibility than RAPD (Yang *et al.*, 1996; Wolfe *et al.*, 1998; Esselman *et al.*, 1999). The method is also very inexpensive (Yang *et al.*, 1996).

3.5 *Molecular case-studies of marine larvae*

As with allozymes there is a relatively small number of studies on marine larvae using molecular techniques. Most publications in this area have simply dealt with methods to

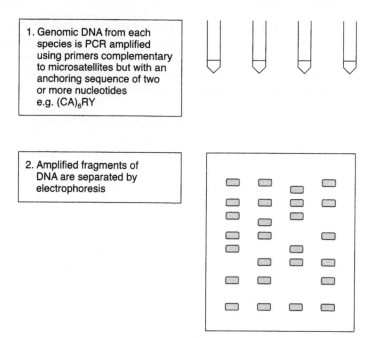

1. Genomic DNA from each species is PCR amplified using primers complementary to microsatellites but with an anchoring sequence of two or more nucleotides e.g. $(CA)_6RY$

2. Amplified fragments of DNA are separated by electrophoresis

Figure 22. SSR method for fingerprint analysis (see Esselman et al., 1999 for details).

identify larvae rather than the actual application of these methods to study ecological questions. This lack of application of molecular methods probably partially reflects the infancy of many of the molecular techniques now available to study marine larvae. However, it also probably reflects the inherent difficulties in designing sampling regimes to test questions about the ecological and genetic relationships of larval and adult populations in the marine environment. Such questions require extremely large sample sets, integrated over several seasons and several years. It is likely that the application of modelling to describe the population dynamics and genetics of marine populations against a background of carefully observed hydrographic data will be of great benefit to marine biologists in the design of rigorous sampling programmes. Such an approach is likely to provide the data required to understand the relationships between adults and larvae both in terms of population dynamics and in terms of microscale evolutionary processes that influence speciation in the marine environment.

In terms of relating genetic constitution of larval and adult populations the first study was that of Silberman *et al.* (1994a) on the spiny lobster, *Panulirus argus*, from the Florida Keys. Samples of larvae were collected over a period of three years and genetic variability was estimated by RFLP analysis of the mitochondrial genome. Haplotype frequencies were found to be very similar throughout the study period, although it was found that larvae came from two distinct haplotype lineages, also detected in adult populations (Silberman *et al.*, 1994b). It was concluded that effective population size was relatively high, compared to census population size in this species, and that larval populations were well mixed, reflecting a homogenous spatial genetic structure in adult populations (Silberman *et al.*, 1994b). The separate mitochondrial lineages may have represented the existence of two separate species or subspecies within *P. argus* or a historical event separating populations that subsequently came into secondary contact (Silberman *et al.*, 1994a). Homogeneity of haplotype frequencies contrasted with previous allozyme studies that suggested differentiation between lobster populations (Menzies and Kerrigan, 1979; Menzies, 1980). However, subsequent allozyme studies also reported genetic homogeneity of populations (Hately and Sleeter, 1993). Contrasting results from biochemical studies may reflect systematic problems within the species.

Recent studies on the larvae of cod (*Gadus morhua*) in the Northwestern Atlantic (Ruzzante *et al.*, 1996; Herbinger *et al.*, 1997) have focused on questions relating to temporal variance in allele frequencies and low effective population size (Hedgecock, 1994). Variation in six microsatellite loci was used to analyse the genetic structure of an aggregation of cod larvae over the Scotian Shelf off Canada. It was found that genotype frequencies of larvae, sampled over three weeks, showed significant deviations from Hardy–Weinberg expectations and strong heterozygote deficiencies. These results were consistent with the existence of several cohorts of larvae, differentiated by size and probably originating from separate spawning events (Ruzzante *et al.*, 1996). Within a single cohort, observed genotype frequencies did not deviate significantly from expected values. Evidence suggested that the larvae from different spawning events during the year were genetically differentiated, although adults formed part of the same population (Ruzzante *et al.*, 1996). Each spawning event involved a large number of adults and there was no evidence that larvae within a cohort were siblings (Herbinger *et al.*, 1997). As cohorts appear at different times of the year they are subject to different environmental conditions (match–mismatch hypothesis) and survival and growth is likely to be different. This provides a mechanism to explain the high temporal variance in allele frequencies in new recruits to populations of some marine animals (low effective population size; Hedgecock and Sly, 1990; Hedgecock *et al.*, 1992; Hedgecock, 1994).

Similar results have been obtained with marine invertebrates. Temporal variance in allele frequencies has been demonstrated in new recruits to the Pacific oyster, *Crassostrea gigas*, population from Dabob Bay in Washington (Hedgecock, 1994). A survey on genetic variation of larvae collected over a breeding season, using SSCP analysis of part of the mitochondrial 16S rDNA region, detected significant differentiation between samples collected at different times, particularly at the beginning and end of the sampling period (Li and Hedgecock, 1998). These results are consistent with reproduction by a finite number of adults at any given time, resulting in significant temporal genetic variation in larvae and therefore, probably, in new recruits.

Studies of the distribution of morphologically similar larvae of holothurians off the coast of Maine have been carried out using oligonucleotide probes to identify three species (Medeiros-Bergen *et al.*, 1995). This study demonstrated that such a method could be used to track the spatial distribution of marine larvae. All three species investigated (*Cucumaria frondosa*, *Psolus fabricii* and *Chiridota laevis*) were found in near-surface waters, close to the coast in proximity to spawning populations. As time progressed larvae were advected further offshore and to the south. In a similar way to sipunculid larvae off the coast of Florida (Staton and Rice, 1999) the distribution of larvae did not completely reflect the distribution of adults. Adult populations of *C. frondosa* did not occur in the southern part of the sampling area, so larvae advected to the south presumably died at a pre- or post-settlement stage (Medeiros-Bergen *et al.*, 1995). These studies indicate that populations near the limits of the geographic distribution of a species may suffer considerable losses of larvae as they are advected away from areas of favourable biological and environmental conditions.

Mitochondrial DNA sequencing has also been used to study fish larvae from the eastern coast of the United States. In this case two morphotypes of larvae of the tropical labrid genus *Xyrichtys*, were sampled in the Middle Atlantic Bight (Hare *et al.*, 1994). The morphotypes differed particularly in eye shape but analysis of a partial sequence of the cytochrome-*b* gene indicated that they belonged to the same species, *Xyrichtys novacula*. The two morphotypes showed significantly different length frequencies and clearly represented two separate cohorts (Hare *et al.*, 1994). They were also found in different areas above the continental slope. It would seem likely that the larvae were transported in two separate events, probably associated with the Gulf Stream. The question as to whether differing eye shape between the two cohorts resulted from the cohorts growing in different environmental conditions or whether they represented larvae from genetically distinct populations remained unresolved. Maintenance of separate populations would be surprising given the dispersal potential of these larvae (1000 km or more; Hare *et al.*, 1994). However, analysis of a less conserved region of DNA may reveal the occurrence of sibling species (i.e. *X. novacula* is a species complex). It is possible that differences in physical and biological parameters may be experienced by different cohorts of larvae, leading to differential survival (Ruzzante *et al.*, 1996).

Along the Mid-Atlantic Ridge, one of the conspicuous species found on deep-sea hydrothermal vents is the shrimp, *Rimicaris exoculata* (Van Dover, 1995). Allozyme data has indicated that genetic differentiation between populations of *R. exoculata*, located at vents separated by distances of over 650 km, is extremely low (Creasey *et al.*, 1996; Shank *et al.*, 1998; Creasey and Rogers, 1999). The question arose as to whether panmixia between populations was maintained by a planktonic larval stage. Adult shrimps do not have conventional crustacean eyes and appear to be at least partially dependent on vents for nutrition (by grazing sulphur bacteria or through sulphur symbionts) judging by the presence of lipid of bacterial origin in body tissues (Dixon and Dixon, 1996).

Trawl surveys of the water column above the Mid-Atlantic Ridge recovered shrimp post-larvae with three distinct morphologies. One of these resembled adults of the vent shrimp genus *Alvinocaris* and one resembled adults of the genus *Chorocaris* (Dixon and Dixon, 1996). Comparison of restriction profiles of the ITS2 region of nuclear rDNA of both adult and larval shrimps indicated that *Chorocaris*-like larvae consisted of two species, *Chorocaris chacei* and *R. exoculata*. Larvae of the latter species had not been separated from larvae of *C. chacei* as they undergo considerable morphological change in moults to the adult stage, involving the loss of eyes (Dixon and Dixon, 1996). Work on the lipid composition of juvenile *R. exoculata* indicated that they had substantial reserves of wax-esters with characteristics typical of an origin in the euphotic zone (Pond *et al.*, 1997). Genetic and lipid composition data together indicated that *R. exoculata* has a dispersive planktotrophic 'eyed' larval stage that feeds on material originating in surface waters (Dixon and Dixon, 1996). Such methods should prove useful for identification of other hydrothermal vent larvae. Preliminary analysis of sequence from the nuclear 28S rDNA region indicated that sufficient interspecific variability existed to distinguish various vent-endemic invertebrates including polychaetes and vestimentiferans (Williams *et al.*, 1993; Dixon *et al.*, 1995a).

One area in which molecular methods have been successfully used is in the identification of threatened species or products from them (Baker *et al.*, 2000). Commercial products from animals are often impossible to identify to species level using conventional methods. Highly specific primers have been used to identify the origin of caviar for sale in the United States and Russia (DeSalle and Birstein, 1996). In this case, nearly one-quarter of the tested samples were labelled incorrectly and several of these originated from extremely rare or threatened species of sturgeons. Such molecular tools are extremely valuable in efforts to conserve stocks of threatened species and to enforce international regulations on export and trade in threatened wildlife (DeSalle and Birstein, 1996; Baker *et al.*, 2000).

As with allozymes, the other major area in which molecular methods have been used with larvae is in reproductive studies. Some of these studies have specifically investigated the inheritance of genetic markers. The inheritance of the calmodulin-1 intron 3 locus in *Mytilus edulis* was investigated by controlled crosses (Côrte-Real *et al.*, 1994). This single-copy nuclear locus has two alleles that are amplified by PCR and separated on the basis of size. It was found that the majority of crosses produced larvae with genotypes that fitted Hardy–Weinberg expectations. However, a small number of larvae failed to produce expected genotypes (Côrte-Real *et al.*, 1994). Explanations for this ranged from artefacts arising from experimental procedure (contamination of samples with sperm, contamination of PCR reactions) to chromosomal loss in some larvae (aneuploidy). A similar, but more marked deviation from expected genotype frequencies was detected in scnDNA polymorphism in *Crassostrea virginica* (Hu and Foltz, 1996). Artefacts and aneuploidy were also discussed as possible origins of these unexpected genotype frequencies but null alleles were also considered as being potentially responsible (Hu and Foltz, 1996).

Most studies on reproduction in marine organisms have been carried out using fingerprinting techniques, particularly RAPDs. In a study of the free spawning hydroid *Hydractinia symbiolongicarpus* 13 RAPD primers gave 133 putative polymorphic loci of which all but four appeared to be inherited in a Mendelian fashion (Levitan and Grosberg, 1993). Cluster analysis across these loci allowed the correct assignment of genealogical relationships among individuals of this species.

RAPD loci that show Mendelian patterns of inheritance have been used to study reproduction in the colonial ascidian *Diplosoma listerianum* (Bishop *et al.*, 1996, 2000; Bishop, 1998). This species is a hermaphrodite and eggs are retained by individuals, whilst sperm is released into the water column and fertilization takes place internally. The number of gametes released by each individual zooid is small, compared to many free-spawning marine invertebrates. This gives rise to many questions regarding prevention of self-fertilization in this species and the mechanisms that allow fertilization between colonies, given problems of sperm dispersal (Levitan, 1993; Levitan and Peterson, 1995). Experimental crosses between clones of *D. listerianum* confirmed findings from studies of radio-labelled sperm that cross-fertilization was prevented between identical clones. This occurred through blockage and destruction of sperm in part of the oviduct leading to the ovary. Furthermore, a weak negative correlation was found between fertilization success and relatedness of colonies as estimated by RAPD analysis (Bishop *et al.*, 1996). However, a direct relationship between relatedness and mating success was not supported by experimental observations and it was suggested that an incompatibility mechanism, possibly controlled by a few genetic loci, was in operation (Bishop *et. al.*, 1996). Further experiments indicated that the sperm of *D. listerianum* last for at least 24 h. Sperm are efficiently filtered from the water by this ascidian and can be stored for up to 55 days prior to fertilization (Bishop, 1998). It was also found, through experimental manipulation of mating between different clones, that there was strong precedence in fertilizations by the first sperm to arrive at a functioning female (Bishop *et al.*, 2000). Sperm precedence declined with time, possibly as sperm was used up but also possibly because of ageing effects in sperm from the first mating.

Overall, mating experiments with *Diplosoma listerianum* show that reproduction in a colonial marine invertebrate can be highly complex. Genetic mechanisms exist to prevent self-fertilization and sperm limitation may be prevented by sperm longevity coupled with sperm storage and a highly efficient mechanism for accumulating very dilute sperm from the water column, possibly aided by the presence of the benthic boundary layer. These are important adaptations for a sessile marine invertebrate in which colonies may be separated by a considerable distance on the seabed or on the shore. Similar studies on outcrossing in sessile freshwater invertebrates have also been carried out using RAPD (e.g. the freshwater bryozoan, *Cristatella mucedo*; Jones *et al.*, 1994).

Multilocus DNA fingerprinting has rarely been employed in studies of marine larvae. However, it has been used to confirm clonal reproduction in the anemone *Epiactis prolifera* from the Pacific coast of the United States (Edmands, 1995). As part of the same study, *E. lisbethae* was also investigated using allozymes and multilocus DNA fingerprinting. Allozymes failed to detect any differences between brooded juveniles and parent anemones but multilocus fingerprinting did reveal variation, demonstrating the higher resolution of this highly polymorphic marker (Edmands,1995). It was thought that this variation arose through either outcrossing or through capture of some larvae that were produced externally.

The genetic relationships between larvae and parents have also been investigated in the soft coral *Parerythropodium fulvum fulvum* from the Red Sea. The planula larvae of this species are brooded on the surface of the colonies for up to a week before crawling away to found new colonies (Barki *et al.*, 2000). An investigation of fingerprint patterns of four parental colonies and 10 larvae from each was carried out using the AFLP method. This revealed that all the larvae showed different banding patterns from each other, demonstrating that the major mode of reproduction in this species was sexual (Barki *et al.*, 2000).

4. Alternative methods for identifying marine larvae

Other methods exist for studying the larvae of marine invertebrates. One of the most obvious is the direct observation of larvae in the environment. This has only generally been possible where larvae are very large, so to date most investigations of this type have been restricted to ascidians (Young, 1986; Davis and Butler, 1989). An alternative approach has been to 'tag' larvae with a chemical tracer. In this context, rare earth elements have been of particular interest to biologists. These elements may be taken up by organisms and become radioactive when they are irradiated with neutrons, a process known as neutron activation (Levin *et al.*, 1993). Trials with such methods have indicated that some elements may be useful as tags (e.g. lutetium or selenium; Levin *et al.*, 1993; Anastasia *et al.*, 1998) but the majority are either not taken up in sufficient quantities by larvae to be detectable above background levels, or they are often toxic. The use of these techniques still remains to be proven in studies of larval ecology. Other methods include release of dyes or drogues with larvae or even the release of larval mimics such as microspheres. However, all these methods suffer the drawback that real larvae may not act as passive particles.

5. Conclusions

Both biochemical and molecular genetic techniques have proved useful in studying marine larvae. These methods allow unambiguous identification of larvae, assignment of larval parentage and comparisons of genotype frequencies in larval and adult populations. Significant discoveries from these investigations include widespread temporal genetic variation in larvae and recruits to marine populations resulting from low effective population size. This has a direct influence on the spatial genetic structure of marine populations and gives an insight into mortality processes that play an important role in regulating recruitment. It has also been found that larval dispersal is not necessarily reflected by distribution of adult populations. Larvae may be advected into waters in which adults are not found, especially if they are from populations close to the limit of the geographic distribution of a species. This can lead to a large wastage of larvae. Molecular and biochemical studies have also been very important in elucidating the reproductive processes in sessile marine invertebrates. This has led to a resolution of the relative roles of sexual and asexual reproduction in population maintenance for several species. It has also revealed specific physiological adaptations of reproductive systems to the sessile mode of existence that may be compared to mechanisms in plants.

The numbers of molecular genetic studies on marine larvae, however, remain low. This is probably related to the extreme technical difficulties in designing statistically meaningful sampling programmes for very large populations, over significant periods of time. Only by addressing such problems will a full understanding of the contribution of larval processes to population dynamics be reached.

References

Alberts, B., Bray, D., Lewis, J., Raff, M., Roberts, K. and Watson, J.D. (1983) *Molecular Biology of the Cell.* Garland Publishing, New York.

Anastasia, J.R., Morgan, S.R. and Fisher, N.S. (1998) Tagging crustacean larvae: assimilation and retention of trace elements. *Limnol. Oceanogr.* 43(2): 362–368.

Ayre, D.J. and Resing, J.M. (1986) Sexual and asexual production of planulae in reef corals. *Mar. Biol.* 90: 187–190.

Baker, C.S., Lento, G.M., Cipriano, F. and Palumbi, S.R. (2000) Predicted decline of protected whales based on molecular genetic monitoring of Japanese and Korean markets. *Proc. R. Soc. Lond. B.* 267: 1191–1199.

Barki, Y., Douek, J., Graur, D., Gateño, D. and Rinkevich, B. (2000) Polymorphism in soft coral larvae revealed by amplified fragment-length polymorphism (AFLP) markers. *Mar. Biol.* 136: 37–41.

Bilodeau A.L., Lankford, W.S., Kim, T.J., Felder, D.L. and Neigel, J.E. (1999) An ultrasensitive method for detection of single crab larvae (*Sesarma reticulatum*) by PCR amplification of a highly repetitive DNA sequence. *Mol. Ecol.* 8: 683–684.

Bishop, J.D.D. (1998) Fertilization in the sea: are the hazards of broadcast spawning avoided when free-spawned sperm fertilize retained eggs? *Proc. R. Soc. Lond. B.* 265: 725–731.

Bishop, J.D.D., Jones, C.S. and Noble, L.R. (1996) Female control of paternity in the internally fertilizing compound ascidian *Diplosoma listerianum*. II. Investigation of male mating success using RAPD markers. *Proc. R. Soc. Lond. B.* 263: 401–407.

Bishop, J.D.D., Pemberton, A.J. and Noble, L.R. (2000) Sperm precedence in a novel context: mating in a sessile marine invertebrate with dispersing sperm. *Proc. R. Soc. Lond. B.* 267: 1107–1113.

Black, R. and Johnson, M.S. (1979) Asexual viviparity and population genetics of *Actinia tenebrosa*. *Mar. Biol.* **53**: 27–31.

Bucklin, A. (2000) Methods for population genetic analysis of zooplankton. In: *ICES Zooplankton Methodology Manual*. Academic Press, London, pp. 533–570.

Bucklin, A., Frost, B.W. and Kocher, T.D. (1995) Molecular systematics of six *Calanus* and three *Metridia* species (Calanoida: Copepoda). *Mar. Biol.* **121**: 655–664.

Bucklin, A., Bentley, A.M. and Franzen, S.P. (1998) Distribution and relative abundance of *Pseudocalanus moultoni* and *P. newmani* (Copepoda: Calanoida) on Georges Bank using molecular identification of sibling species. *Mar. Biol.* **132**: 97–106.

Carruthers, J.N. (1938) Fluctuations in the herrings of the East Anglian Autumn fishery, the yield of the Ostend spent herring fishery and the haddock of the North Sea in the light of relevant wind conditions. *Rapp. Procès-Verb. Cons. int. Explor. Mer* **107**: 1–15.

Coffroth, M.A. and Mulawka, J.M. (1995) Identification of marine invertebrate larvae by means of PCR-RAPD species-specific markers. *Limnol. Oceanogr.* **40**(1): 181–189.

Connell, J.H. (1961a) Effects of competition, predation by *Thais lapillus*, and other factors on natural populations of the barnacle *Balanus balanoides*. *Ecol. Monogr.* **31**: 61–104.

Connell, J.H. (1961b) The influence of interspecific competition and other factors on the distribution of the barnacle *Cthalamus stellatus*. *Ecology* **42**: 710–723.

Connell, J.H. (1970) A predator-prey system in the marine intertidal region. I. *Balanus glandula* and several species of *Thais*. *Ecol. Monogr.* **40**: 49–78.

Connell, J.H. (1972) Community interactions on marine rocky intertidal shores. *A. Rev. Ecol. Syst.* **3**: 169–192.

Connell, J.H. (1975) Some mechanisms producing structure in natural communities: a model and evidence from field experiments. In: *Ecology and Evolution of Communities* (eds M.S. Cody and J.M. Diamond). Harvard University Press, Cambridge, MA, pp. 460–490.

Connell, J.H. (1985) The consequences of variation in initial settlement vs. post-settlement mortality in rocky intertidal communities. *J. Exp. Mar. Biol. Ecol.* **93**: 11–45.

Côrte-Real, H.B.S.M, Dixon, D.R. and Holland, P.W.H. (1994) Intron-targeted PCR: a new approach to survey neutral DNA polymorphism in bivalve populations. *Mar. Biol.* **120**: 407–413.

Coyer, J.A., Robertson, D.L. and Alberte, R.S. (1995) Genetic variability and parentage in *Macrocystis pyrifera* (Phaeophyceae) using multi-locus DNA fingerprinting. *J. Phycol.* **31**: 819–823.

Creasey, S. and Rogers, A.D. (1999) Population genetics of bathyal and abyssal organisms. *Adv. Mar. Biol.* **35**: 1–151.

Creasey, S., Rogers, A.D. and Tyler P.A. (1996) A genetic comparison of two populations of the deep-sea vent shrimp *Rimicaris exoculata* (Decapoda: Caridea: Bresiliidae) from the Mid-Atlantic Ridge. *Mar. Biol.* **125**(3): 473–482.

Crisp, D.J., Burfitt, A., Rodrigues, K. and Budd, M.D. (1983) *Lasaea rubra*: an apomictic bivalve. *Mar. Biol. Lett.* **4**: 127–136.

Cushing, D.H. (1959) On the nature of production in the sea. *Fish. Invest. Lond. Ser. 2.* **22**(6): 40pp.

Dallas, J.F. (1992) Estimation of microsatellite mutation rates in recombinant inbred strains of mouse. *Mammalian Genome* **3**: 452–456.

D'Amato, M.E., Lunt, D.H. and Carvalho, G.R. (1999) Microsatellite markers for the hake *Macruronus magellanicus* amplify other gadoid fish. *Mol. Ecol.* **8**: 1086–1088.

David, P., Perdieu, M.-A., Pernot, A.-F. and Jarne, P. (1997) Fine-grained spatial and temporal genetic structure in the marine bivalve *Spisula ovalis*. *Evolution* **51**: 1318–1322.

Davis, A.R. and Butler, A.J. (1989) Direct observations of larval dispersal in the colonial ascidian *Podoclavella moluccensis* Sluiter: evidence for closed populations. *J. Exp. Mar. Biol. Ecol.* **127**: 189–203.

DeSalle, R. and Birstein, V.J. (1996) PCR identification of black caviar. *Nature (Lond.)* **381**: 197–198.

Desmarais, E., Lanneluc, I. and Lagne, J. (1998) Direct amplification of length polymorphisms (DALP) or how to get and characterize new genetic markers in many species. *Nucleic Acids Res.* 26(6): 1458–1465.

D'Incao, F., Delevedove, G., Maggioni, D.E. and Maggioni, R. (1998) Evidência genética presença de *Farfantepenaeus paulensis* (Pérez Farfante, 1967) no litoral nordeste do Brasil (Decapoda: Penaeidae). *Nauplius Rio Grande* 6: 129–137.

Dixon, D.R. and Dixon, L.R.J. (1996) Results of DNA analyses conducted on vent-shrimp postlarvae collected above the Broken Spur vent field during the CD95 cruise, August 1995. *BRIDGE Newsletter* 11: 9–15.

Dixon, D.R., Jollivet, D.A.S.B., Dixon, L.R.J., Nott, J.A. and Holland, P.W.H. (1995a) The molecular identification of early life-history stages of hydrothermal vent organisms. In: *Hydrothermal Vents and Processes* (eds L.M. Parson, C.L. Walker and D.R. Dixon), Geological Society Special Publication no. 87, The Geological Society of London, London, pp. 343–350.

Dixon, D.R., Solé-Cava, A.M., Pascoe, P.L. and Holland, P.W.H. (1995b) Periostracal adventitious hairs on spat of the mussel *Mytilus edulis. J. Mar. Biol. Assoc. UK* 75: 363–372.

Edmands, S. (1995) Mating systems in the sea anemone genus *Epiactis. Mar. Biol.* 123: 723–733.

Ender, A., Schwenk, K., Stadler, T., Streit, B. and Schierwater, B. (1996) RAPD identification of microsatellites in *Daphnia. Mol. Ecol.* 5(3): 437–441.

Esselman, E.J., Jianqiang, L., Crawford, D.J., Windus, J.L. and Wolfe, A.D. (1999) Clonal diversity in the rare *Calamagrostis porteri* ssp. *insperata* (Poaceae): comparative results for allozymes and random amplified polymorphic DNA (RAPD) and intersimple sequence repeat (ISSR) markers. *Mol. Ecol.* 8: 443–451.

Evans, B.S., White, R.W.G. and Ward, R.D. (1998) Genetic identification of asteroid larvae from Tasmania, Australia, by PCR-RFLP. *Mol. Ecol.* 7: 1077–1082.

Fenaux, L., Strathmann, M.F. and Strathmann, R.R. (1994) Five tests of food-limited growth of larvae in coastal waters by comparisons of rates of development and form of echinoplutei. *Limnol. Oceanogr.* 39: 84–98.

Ferguson, A. (1980) *Biochemical Systematics and Evolution.* Blackie, Glasgow.

Fevolden, S.E. and Ayala, F.J. (1981) Enzyme polymorphisms in Antarctic krill (Euphausiacea); genetic variation between populations and species. *Sarsia* 66: 167–181.

Fevolden, S.E. and Schneppenheim, R. (1989) Genetic homogeneity of krill (*Euphausia superba* Dana) in the Southern Ocean. *Polar Biol.* 9: 533–539.

Fleischer, R.C. (1996) Application of molecular methods to the assessment of genetic mating systems in vertebrates. In: *Molecular Zoology: Advances, Strategies, and Protocols* (eds J.D. Ferraris and S.R. Palumbi). John Wiley, New York, pp. 133–161.

Frankham, R. (1995) Effective population size/adult population size ratios in wildlife: a review. *Genet. Res. Camb.* 66: 95–107.

Gaines, S.D. and Bertness, M.D. (1992) Dispersal of juveniles and variable recruitment in sessile marine species. *Nature (Lond.)* 360: 579–580.

Garrett, F.K., Mladenov, P.V. and Wallis, G.P. (1997) Evidence of amictic reproduction in the brittle-star *Ophiomyxa brevirima. Mar. Biol.* 129: 169–174.

Grosberg, R.K. (1991) Sperm-mediated gene flow and the genetic structure of a population of the colonial ascidian *Botryllus schlosseri. Evolution* 45: 130–142.

Grosberg, R.K., Levitan, D.R. and Cameron, B.B. (1996) Characterization of genetic structure and genealogies using RAPD-PCR markers: a random primer for the novice and nervous. In: *Molecular Zoology: Advances, Strategies, and Protocols* (eds J.D. Ferraris and S.R. Palumbi). John Wiley, New York, pp. 67–100.

Hancock, J.M. (1999) Microsatellites and other simple sequences: genomic context and mutational mechanisms. In: *Microsatellites Evolution and Applications* (eds D.B. Goldstein and C. Schlötterer). Oxford University Press, Oxford, pp. 1–9.

Hare, J.A., Cowen, R.K., Zehr, J.P., Juanes, F. and Day, K.H. (1994) Biological and oceanographic insights from larval labrid (Pisces: Labridae) identification using mtDNA sequences. *Mar. Biol.* 118: 17–24.

Harris, D.J. and Crandall, K.A. (2000) Intragenomic variation within ITS1 and ITS2 of fresh-water crayfishes (Decapoda: Cambaridae): implications for phylogenetic and microsatellite studies. *Mol. Biol. Evol.* **17**(2): 284–291.

Harris, H. and Hopkinson, D.A. (1976) *Handbook of Enzyme Electrophoresis in Human Genetics.* North Holland, Amsterdam.

Hately, J.G. and Sleeter, T.D. (1993) A biochemical genetic investigation of the spiny lobster (*Panulirus argus*) stock replenishment in Bermuda. *Bull. Mar. Sci.* **53**: 993–1008.

Havenhand, J.N. (1995) Evolutionary ecology of larval types. In: *Ecology of Marine Invertebrate Larvae.* CRC Press, Boca Raton, FL, pp. 79–122.

Hawkins, S.J. and Hartnoll, R.G. (1982) Settlement patterns of *Semibalanus balanoides* (L.) in the Isle of Man (1977–1981). *J. Exp. Mar. Biol. Ecol.* **62**: 271–283.

Hedgecock, D. (1994) Does variance in reproductive success limit effective population sizes of marine organisms. In: *Genetic and Evolution of Aquatic Organisms* (ed A.R. Beaumont). Chapman & Hall, London, pp. 122–134.

Hedgecock, D. and Sly, F.L. (1990) Genetic drift and effective population sizes of hatchery-propagated stocks of the Pacific oyster *Crassostrea gigas. Aquaculture* **88**: 21–38.

Hedgecock, D., Chow, V. and Waples, R.E. (1992) Effective population numbers of shellfish broodstocks estimated from temporal variance in allele frequencies. *Aquaculture* **108**: 215–232.

Herbinger, C.M., Doyle, R.W., Taggart, C.T., Lochmann, S.E., Brooker, A.L., Wright, J.M. and Cook, D. (1997) Family relationships and effective population size in a natural cohort of Atlantic cod (*Gadus morhua*) larvae. *Can. J. Fish. Aquat. Sci.* **54** (Suppl. 1): 11–18.

Hill, A.E., Brown, J. and Fernand, L. (1996) The western Irish Sea gyre: a retention system for Norway lobster (*Nephrops norvegicus*)? *Ocean. Acta.* **19** (3–4): 357–368.

Hu, Y.P. and Foltz, D.W. (1996) Genetics of scnDNA polymorphisms in juvenile oysters (*Crassostrea virginica*) Part I: Characterising the inheritance of polymorphisms in controlled crosses. *Mol. Mar. Biol. Biotech.* **5**(2): 123–129.

Hu, Y.P., Lutz, R.A. and Vrijenhoek, R.C. (1992) Electrophoretic identification and genetic analysis of bivalve larvae. *Mar. Biol.* **113**: 227–230.

Hu, Y.P., Fuller, S.C., Castagna, M., Vrijenhoek, R.C. and Lutz, R.A. (1993) Shell morphology and identification of early life history stages of congeneric species of *Crassostrea* and *Ostrea. J. Mar. Biol. Assoc. UK* **73**: 471–496.

IUPAC-IUB Commission on Biochemical Nomenclature (1977) Nomenclature of multiple forms of enzymes: recommendations (1976). *J. Biol. Chem.* **252**: 5939.

Jablonski, D. (1986) Larval ecology and macroevolution in marine invertebrates. *Bull. Mar. Sci.* **39**(2): 565–587.

Jeffreys, A.J., Wilson, V. and Thein, S.L. (1985) Individual specific "fingerprints" of human DNA. *Nature (Lond.)* **316**: 76–79.

Jeffreys, A.J., Wilson, V., Thein, S.L., Weatherall, D.J. and Ponder, B.A.J. (1986) DNA "fingerprints": and segregation analysis of multiple markers in human pedigrees. *Am. J. Hum. Genet.* **39**: 11–24.

Jeffreys, A.J., Wilson, V., Kelly, R., Taylor, B.A. and Bulfield, G. (1987) Mouse DNA "finger-prints": analysis of chromosome location and germ-line stability of hypervariable loci in recombinant inbred strains. *Nucleic Acids Res.* **15**: 2823–2836.

Jeffreys, A.J., Tamaki, K., MacLeod, A., Monckton, D.G., Neil, D.L. and Armour, J.A.L. (1994) Complex gene conversion events in germline mutation at human minisatellites. *Nature Genet.* **6**: 136–145.

Johnson, M.S. and Black, R. (1982) Chaotic genetic patchiness in an intertidal limpet, *Siphonaria* sp. *Mar. Biol.* **70**: 157–164.

Johnson, M.S. and Black, R. (1984) Pattern beneath the chaos: the effect of recruitment on genetic patchiness in an intertidal limpet. *Evolution* **38**: 1371–1393.

Johnson, M.S. and Wernham, J. (1999) Temporal variation of recruits as a basis of ephemeral genetic heterogeneity in the western rock lobster *Panulirus cygnus. Mar. Biol.* **135**: 133–139.

Jones, C.S., Okamura, B. and Noble, L.R. (1994) Parent and larval RAPD fingerprints reveal outcrossing in freshwater bryozoans. *Mol. Ecol.* 3: 193–199.

Kandpal, R.P., Kandpal, G. and Weissman, S.M. (1994) Construction of libraries enriched for sequence repeats and jumping clones, and hybridization selection for region-specific markers. *Proc. Natl Acad. Sci. USA* 91: 88–92.

Knowlton, N. and Jackson, J.J. (1993) Inbreeding and outbreeding in marine invertebrates. In: *The Natural History of Inbreeding and Outbreeding* (ed N.W. Thornhill). University of Chicago Press, Chicago, IL, pp. 200–249.

Knowlton, N. and Keller, B.D. (1986) Larvae which fall short of their potential: highly localized recruitment in an alpheid shrimp with extended larval development. *Bull. Mar. Sci.* 39(2): 213–223.

Kohn, J. (1957) A cellulose acetate supporting medium for zone electrophoresis. *Clin. Chim. Acta* 2: 297.

Kordos, L.M. and Burton, R.S. (1993) Genetic differentiation of Texas Gulf Coast populations of the blue crab *Callinectes sapidus*. *Mar. Biol.* 117: 227–233.

Le Fèvre, J. and Bourget, E. (1992) Hydrodynamics and behaviour: transport processes in marine invertebrate larvae. *Trends Ecol. Evol.* 7(9): 227–233.

Levin, L.A., Huggett, D., Myers, P., Bridges, T. and Weaver, J. (1993) Rare-earth tagging methods for the study of larval dispersal by marine invertebrates. *Limnol. Oceanogr.* 38(2): 346–360.

Levinson, G. and Gutman, G.A. (1987) Slipped-strand mispairing: a major mechanism for DNA sequence evolution. *Mol. Biol. Evol.* 4: 203–221.

Levitan, D.R. (1993) The importance of sperm limitation to the evolution of egg size in marine invertebrates. *Am. Nature* 141: 517–536.

Levitan, D.R. (1996) Effects of gamete traits on fertilization in the sea and the evolution of sexual dimorphism. *Nature (Lond.)* 382: 153–155.

Levitan, D.R. and Grosberg, R.K. (1993) The analysis of paternity and maternity in the marine hydrozoan *Hydractinia symbiolongicarpus* using randomly amplified polymorphic DNA (RAPD) markers. *Mol. Ecol.* 2: 315–326.

Levitan, D.R. and Petersen, C. (1995) Sperm limitation in the sea. *Tr. Evol. Ecol.* 10(6): 228–231.

Li, G. and Hedgecock, D. (1998) Genetic heterogeneity, detected by PCR-SSCP, among samples of larval Pacific oysters (*Crassostrea gigas*) supports the hypothesis of large variance in reproductive success. *Can. J. Fish. Aquat. Sci.* 55: 1025–1033.

Lindeque, P.K., Harris, R.P., Jones, M.B. and Smerdon, G.R. (1999) Simple molecular method to distinguish the identity of *Calanus* species (Copepoda: Calanoida) at any developmental stage. *Mar. Biol.* 133: 91–96.

Maggioni, R., Rogers, A.D., Maclean, N. and D'Incao, F. (2001) Molecular phylogeny of western Atlantic *Farfantepenaeus* and *Litopenaeus* shrimp based on mitochondrial 16S partial sequences. *Mol. Phylog. Evol.* 18(1): 66–73.

Makinster, J.G., Roberts, J.E., Felder, D.L., Chlan, C.A., Boudreaux, M., Bilodeau, A.L. and Neigel, J.E. (1999) PCR amplification of a middle repetitive element detects larval stone crabs (Crustacea: Decapoda: Menippidae) in estuarine plankton samples. *Mar. Ecol. Prog. Ser.* 188: 161–168.

Manchenko, G.P. (1994) *Handbook of Detection of Enzymes on Electrophoretic Gels*. CRC Press, London.

McEdward, L.R. and Coulter, L.K. (1987) Egg volume and energetic content are not correlated among sibling offspring of starfish: implications for life-history theory. *Evolution* 41(4): 914–917.

Medeiros-Bergen, D.E., Olsen, R.R., Conroy, J.A. and Kocher, T.D. (1995) Distribution of holothurian larvae determined with species-specific genetic probes. *Limnol. Oceanogr.* 40: 1225–1235.

Menzies, R.A. (1980) Biochemical population genetics and the spiny lobster larval recruitment problem: an update. *Proc. Gulf Caribb. Fish. Inst.* 33: 230–243.

Menzies, R.A. and Kerrigan, J.M. (1979) Implications of spiny lobster recruitment patterns of the Caribbean – a biochemical genetic approach. *Proc. Gulf Caribb. Fish. Inst.* **31**: 164–178.

Monteiro, F.A., Russo, C.A.M. and Solé-Cava, A.M. (1998) Genetic evidence for the asexual origin of small individuals found in the coelenteron of the sea anemone *Actinia bermudensis* McMurrich. *Bull. Mar. Sci.* **63**(2): 257–264.

Morgan, T.S. and Rogers, A.D. (2001) Specificity and sensitivity of microsatellite markers for the identification of larvae. *Mar. Biol.* (in press).

Morgan, T.S., Rogers, A.D. and Iyengar, A. (2000) Novel microsatellite markers for the European oyster *Ostrea edulis*. *Mol. Ecol.* **9**: 489–504.

Murphy, R.W., Sites, J.W., Buth, D.G. and Haufler, C.H. (1990) Proteins I: Isozyme electrophoresis. In: *Molecular Systematics* (eds D.M. Hillis and C. Moritz). Sinauer Associates, New York, pp. 45–146.

Myers, R.M., Maniatis, T. and Lerman, L. (1987) Detection and localization of single base changes by denaturing gradient gel electrophoresis. *Methods Enzymol.* **155**: 501–527.

Nei, M. (1978) Estimation of average heterozygosity and genetic distance from a small number of individuals. *Genetics* **89**: 583–590.

Olsen, R.R., Runstadler, J.A. and Kocher, T.D. (1991) Whose larvae? *Nature (Lond.)* **351**: 357–358.

Orita, M., Iwahana, H., Kanazawa, H. and Sekiya, T. (1989) Detection of polymorphism of human DNA by gel electrophoresis as single-strand conformation polymorphisms. *Proc. Natl Acad. Sci. USA* **86**: 2766–2770.

Orr, J., Thorpe, J.P. and Carter, M.A. (1982) Biochemical genetic confirmation of the asexual reproduction of brooded offspring in the sea anemone *Actinia equina*. *Mar. Biol.* **7**: 227–229

Ottaway, J.R. and Kirby, G.C. (1975) Genetic relationships between brooding and brooded *Actinia tenebrosa*. *Nature (Lond.)* **255**: 221–223.

Paine, R.T. (1974) Intertidal community structure: experimental community studies on the relationship between a dominant competitor and its principle predator. *Oecologia (Berlin)* **15**: 93–120.

Palacios, C., Kresovich, S. and González-Candelas (1999) A population genetic study of the endangered plant species *Limonium dufourii* (Plumbaginaceae) based on amplified fragment length polymorphism (AFLP). *Mol. Ecol.* **8**: 645–657.

Palumbi, S.R. (1992) Marine speciation on a small planet. *Tr. Evol. Ecol.* **7**: 114–118.

Palumbi, S.R. and Wilson, A.C. (1990) Mitochondrial DNA diversity in the sea urchins *Strongylocentrotus purpuratus* and *S. droebachiensis*. *Evolution* **44**: 403–415.

Palumbi, S.R., Martin, A., Romano, S., McMillan, W.O., Stice, L. and Grabowski, G. (1991) *The Simple Fool's Guide to PCR*. University of Hawaii, Honolulu.

Parsons, K.E. (1996) The genetic effects of larval dispersal depend on spatial scale and habitat characteristics. *Mar. Biol.* **126**: 403–414.

Pond, D., Dixon, D. and Sargent, J. (1997) Wax-ester reserves facilitate dispersal of hydrothermal vent shrimps. *Mar. Ecol. Prog. Ser.* **146**: 289–290.

Potts, W.K. (1996) PCR-based cloning across large taxonomic distances and polymorphism detection: MHC as a case study. In: *Molecular Zoology: Advances, Strategies, and Protocols* (eds J.D. Ferraris and S.R. Palumbi). John Wiley, New York, pp. 181–194.

Primmer, C.R., Moller, A.P. and Ellegren, H. (1996) A wide-range survey of cross-species microsatellite amplification in birds. *Mol. Ecol.* **5**: 365–378.

Rabouam, C., Comes, A.M., Bretagnolle, V., Humbert, J.-F., Periquet, G. and Bigot, Y. (1999) Features of DNA fragments obtained by random amplified polymorphic DNA (RAPD) assays. *Mol. Ecol.* **8**: 493–503.

Raymond, S. and Weintraub, L. (1959) Acrylamide gel as a supporting medium for zone electrophoresis. *Science* **130**: 711.

Riedy, M.F., Hamilton, W.J. and Aquadro, C.F. (1992) Excess non-parental bands in offspring from known pedigrees assayed using RAPD PCR. *Nucleic Acids Res.* **20**: 918.

Rocha-Olivares, A. (1998) Multiplex haplotype-specific PCR: a new approach for species identification of the early life stages of rockfishes of the species-rich genus *Sebastes* Cuvier. *J. Exp. Mar. Biol. Ecol.* **231**: 279–290.

Ruzzante, D.E., Taggart, C.T. and Cook, D. (1996) Spatial and temporal; variation in the genetic composition of a larval cod (*Gadus morhua*) aggregation: cohort contribution and genetic stability. *Can. J. Fish. Aquat. Sci.* **53**: 2695–2705.

Sambrook, J., Fritsch, E. and Maniatis, T. (1989) *Molecular Cloning: a Laboratory Manual.* Cold Spring Harbor Laboratory Press, New York.

Shank, T.M., Lutz, R.A. and Vrijenhoek, R.C. (1998) Molecular systematics of shrimp (Decapoda: Bresiliidae) from deep-sea hydrothermal vents, I: Enigmatic "small orange" shrimp from the Mid-Atlantic Ridge are juvenile *Rimicaris exoculata*. *Mol. Mar. Biol. Biotechnol.* **7**(2): 88–96.

Shanks, A.L. (1995) Mechanisms of cross-shelf dispersal of larval invertebrates and fish. In: *Ecology of Marine Invertebrate Larvae*. CRC Press, Boca Raton, FL, pp. 323–367.

Shaw, C.R. and Prasad, R. (1970) Starch gel electrophoresis of enzymes: a compilation of recipes. *Biochem. Genet.* **4**: 297–320.

Silberman, J.D., Sarver, S.K. and Walsh, P.J. (1994a) Mitochondrial DNA in seasonal cohorts of spiny lobster (*Panulirus argus*) postlarvae. *Mol. Mar. Biol. Biotechnol.* **3**(3): 165–170.

Silberman, J.D., Sarver, S.K. and Walsh, P.J. (1994b) Mitochondrial DNA variation and population structure in the spiny lobster *Panulirus argus*. *Mar. Biol.* **120**: 601–608.

Smith, G.P. (1976) Evolution of repeated DNA sequences by unequal crossover. *Science* **191**: 528–535.

Smith, J.L., Scott-Craig, J.S., Leadbetter, J.R., Bush, G.L., Roberts, D.L. and Fulbright, D.W. (1994) Characterization of random amplified polymorphic DNA (RAPD) products from *Xanthomonas campestris* and some comments on the use of RAPD products in phylogenetic analysis. *Mol. Phylogenet. Evol.* **3**: 135–145.

Smithies, O. (1955) Zone electrophoresis in starch gels: group variations in the serum proteins of normal human adults. *Biochem. J.* **61**: 629–641.

Solé-Cava, A.M. and Thorpe, J.P. (1991) High levels of genetic variation in natural populations of marine lower invertebrates. *Biol. J. Linn Soc.* **44**: 65–80.

Spradling, A.C. and Rubin, G.M. (1981) *Drosophila* genome organization: conserved and dynamic aspects. *Annu. Rev. Genet.* **15**: 219–264.

Staton, J.L. and Rice, M.E. (1999) Genetic differentiation despite teleplanic larval dispersal: allozyme variation in sipunculans of the *Apionsoma misakianum* species-complex. *Bull. Mar. Sci.* **65**(2): 467–480.

Stoddart, J.A. (1983) Asexual production of planulae in the coral *Pocillopora damicornis*. *Mar. Biol.* **76**: 279–284.

Thorpe, J.P. (1982). The molecular clock hypothesis: biochemical evolution, genetic differentiation and systematics. *Annu. Rev. Ecol. Syst.* **13**: 139–168.

Thorpe, J.P. (1983) Enzyme variation, genetic distance and evolutionary divergence in relation to levels of taxonomic separation. In: *Protein Polymorphism: Adaptive and Taxonomic Significance* (eds G.S. Oxford and D. Rollinson). Academic Press, London, pp. 131–152.

Todd, C.D., Lambert, W.J. and Thorpe, J.P. (1998) The genetic structure of intertidal populations of two species of nudibranch molluscs with planktotrophic and pelagic lecithotrophic larval stages: are pelagic larvae "for" dispersal? *J. Exp. Mar. Biol. Ecol.* **228**(1): 1–28.

Toro, J.E. (1998) Molecular variation of four species of mussels from southern Chile by PCR-based nuclear markers: the potential use in studies involving planktonic surveys. *J. Shellfish Res.* **17**(4): 1203–1205.

Tyler-Walters, H. and Crisp, D.J. (1989) The modes of reproduction in *Lasaea rubra* (Montagu) and *L. australis* (Lamarck): (Erycinidae; Bivalvia). In: *Reproduction, Genetics and Distributions of Marine Organisms, 23rd European Marine Biology Symposium, 1988* (eds J.S. Ryland and P.A. Tyler). Olsen and Olsen, Denmark, pp. 189–200.

Underwood, A.J. and Denley, E.J. (1984) Marine Community Paradigms. In: *Ecological Communities: Conceptual Issues and the Evidence* (eds L.G. Arele and A.B. Thistle). Princeton University Press, Princeton, NJ, pp. 151–180.

Underwood, A.J. and Fairweather, P.G. (1989) Supply-side ecology and marine benthic assemblages. *Trends Ecol. Evol.* 4: 16–19.

Van Dover, C.L. (1995) Ecology of mid-Atlantic ridge hydrothermal vents. In: *Hydrothermal Vents and Processes* (eds L.M. Parson, C.L. Walker and D.R. Dixon), Geological Society Special Publication no. 87. The Geological Society of London, London, pp. 257–294.

Vos, P., Hogers, R., Bleeker, M., Reijans, M., van de Lee, T., Hornes, M., Frijters, A., Pot, J., Peleman, J., Kuiper, M. and Zabeau, M. (1995) AFLP: a new technique for DNA fingerprinting. *Nucleic Acids Res.* 23: 4407–4414.

Walford, L.A. (1946) Correlation between fluctuations in abundance of the Pacific sardine (*Sardinops caerulea*) and salinity of sea waters. *J. Mar. Res.* 6(1): 48–53.

Watts, R.J., Johnson, M.S. and Black, R. (1990) Effects of recruitment on genetic patchiness in the urchin *Echinometra mathaei* in Western Australia. *Mar. Biol.* 105: 145–151.

Weber, J.L. and Wong, C. (1993) Mutation of human short tandem repeats. *Hum. Mol. Genet.* 2: 1123–1128.

White, M.B., Carvalho, M., Derse, D., O'Brien, S.J. and Dean, M. (1992) Detecting single base substitutions as heteroduplex polymorphisms. *Genomics* 12: 301–306.

Williams, N.A., Dixon, D.R., Southward, E.C. and Holland, P.W.H. (1993) Molecular evolution and diversification of the vestimentiferan tube worms. *J. Mar. Biol. Assoc. UK* 73: 437–452.

Wolanski, E. and Sarsenski, J. (1997) Larvae dispersion in coral reefs and mangroves. *Am. Sci.* 85: 236–243.

Wolfe, A.D., Xiang, Q.-Y. and Kephart, S. (1998) Assessing hybridization in natural populations of *Penstemon* (Scrophulariaceae) using hypervariable inter-simple sequence repeat markers. *Mol. Ecol.* 7: 1107–1125.

Wray, G.A. (1995) Causes and consequences of heterochrony in early echinoderm development. In: *Evolutionary Change and Heterochrony* (ed K.J. McNamara). John Wiley, Chichester, pp. 197–223.

Yan, G., Romero-Severson, J., Walton, M., Chadee, D.D. and Severson, D.W. (1999) Population genetics of the yellow fever mosquito in Trinidad: comparisons of amplified length polymorphism (AFLP) and restriction fragment length polymorphism (RFLP) markers. *Mol. Ecol.* 8: 951–963.

Yang, W., de Oliveira, A.C., Godwin, I., Schertz, K. and Bennetzen, J.L. (1996) Comparison of DNA marker technologies in characterizing plant genome diversity: variability in Chinese sorghums. *Crop Sci.* 36: 1669–1676.

Young, C.M. (1986) Direct observations of field swimming behaviour in larvae of the colonial ascidian, *Ecteinascidia turbinata*. *Bull. Mar. Sci.* 39: 279–289.

Yund, P.O. (1995) Gene flow via the dispersal of fertilizing sperm in a colonial ascidian (*Botryllus schlosseri*): the effect of male density. *Mar. Biol.* 122: 649–654.

Yund, P.O. and McCartney, M.A. (1994) Male reproductive success in sessile invertebrates: competition for fertilizations. *Ecology* 75: 2152–2167.

3

Hormonal regulation of reproductive development in crustaceans

G. Wainwright and H.H. Rees

1. Introduction

Crustaceans undergo punctuated development throughout their life-time, characterized by a series of moults. Moulting occurs in cycles and involves the shedding of the old exoskeleton, uptake of water resulting in increased size, and the final hardening of the new exoskeleton (for reviews, see Warner, 1977; Aiken, 1980; Skinner, 1985). Within this cyclical growth, reproductive development occurs.

Reproductive development can be sub-divided into two main phases, pre-vitellogenesis and vitellogenesis. Pre-vitellogenesis is, in a sense, a premature state, when the ovary tissue is not developing into a reproductively competent organ. At certain times of the year (dependent upon species and environmental conditions), ovaries exit this reproductively dormant pre-vitellogenic state and vitellogenesis begins. This is characterized most readily in early stages of development by a change in colour of the haemolymph, indicating the presence of yolk protein. Crustacean yolk protein (vitellin) is a phospho-lipo-glyco-carotenoprotein. As vitellogenesis progresses, vitellin accumulates in the oocytes and the ovary tissue becomes pigmented (usually either orange or green). It is noteworthy that in some crustacean species, vitellin is synthesized in more than one tissue, the hepatopancreas and ovary tissues, whilst in others it is synthesized in the hepatopancreas and transported to the ovary via the circulating haemolymph or synthesized solely in the ovary tissue (Adiyodi and Adiyodi, 1970). In the ovary tissue, the follicle cells surrounding developing oocytes are the site of synthesis of vitellin; thus, at some stage, vitellin must be actively taken up by developing oocytes. At the ultrastructural level, pre-vitellogenesis is characterized by formation of rough endoplasmic reticulum and accumulation of glycoproteins in the cytosol of the developing oocytes. At the end of the pre-vitellogenic period, meiosis in the developing oocytes arrests at prophase-I, thus synchronizing oocytes to the same stage of development. During vitellogenesis, oocytes begin to grow synchronously in preparation for eventual spawning (for a review of ovarian development in crustaceans, see Charniaux-Cotton and Payen, 1988).

Environment and Animal Development: Genes, Life Histories and Plasticity, edited by D. Atkinson and M. Thorndyke.
© 2001 BIOS Scientific Publishers Ltd, Oxford.

These two key facets of crustacean life, namely reproduction and growth, occur is cycles related to seasonal changes in each year.

2. Effects of temperature and photoperiod

The seasonal reproductive cycle of crustaceans is clearly affected by environmental factors such as temperature and photoperiod. For example, it has been reported that lobster populations in different fisheries spawn at different times of the year, but that the temperatures in the individual fisheries at the time of spawning are comparable (Aiken and Waddy, 1986). This indicates that specific temperatures are a pre-requisite in determining the time of spawning. Additionally, as in ectotherms generally, temperature has been shown to affect the rate of larval development of crustaceans (Templeman, 1936) and, in adults, increased summer temperatures can induce early reproductive maturation of male and female lobsters (Aiken and Waddy, 1986). Interestingly, with the lobster, *Homarus americanus*, it has been shown that the decrease in water temperature during winter months is crucial for the correct development of gametes later in the year (Waddy and Aiken, 1992). These reports serve to highlight the importance of seasonal changes in temperature for correct reproductive development in crustaceans. However, it must be noted that, at a population level in the wild, the effects of temperature have far more complex interactions with lobster life cycles, such as the availability of food at different temperatures and the susceptibility to disease and stress. However, these complex interactions are beyond the scope of this review.

Photoperiod is another seasonal input that can affect the reproductive development cycle of crustaceans, and it has been suggested that it is absolutely crucial to the completion of oocyte maturation (Waddy and Aiken, 1992). The mode by which changes in photoperiod can affect the reproductive system of crustaceans is uncertain at present, but it is clear that eyes contain factors that regulate the processes of moulting for growth and reproductive development. Within the eyestalk of crustaceans, there is a neurohaemal body, the X-organ–sinus gland complex, that is a site of synthesis and release of neuropeptide hormones (for reviews, see Keller, 1992; Van Herp, 1998; Webster, 1998). Thus, a hypothesis might be that photoperiod may ultimately regulate release of key neuropeptide hormones, and other factors, from the X-organ–sinus gland that regulate reproduction and development in crustaceans. This hypothesis has begun to be investigated with work reporting the effects of neurotransmitters and D-glucose on release of crustacean hyperglycaemic hormone (CHH), from single *in vitro* cultured X-organ cells (Glowik *et al.*, 1997). CHH is the hormone that regulates glucose levels in the haemolymph of crustaceans (for reviews see Keller, 1992; Van Herp, 1998). This work clearly demonstrated that the neurotransmitters, γ-aminobutyric acid and serotonin, as well as D-glucose, caused hyperpolarization of CHH X-organ cells. Therefore, it is not inconceivable that changes in photoperiod could effect neurotransmitter-mediated synthesis/release of peptides from the X-organ–sinus gland complex of the eyestalk, thereby affecting the levels of circulating peptides that regulate key metabolic processes involved in growth and reproduction of crustaceans. The effect of D-glucose suggests a feedback regulation mechanism, whereby release of CHH from these cells is switched off in the presence of high levels of D-glucose in the haemolymph of the crustacean.

3. The eyestalk

The eyestalk has long been known as a site of production of factors that regulate moulting and reproductive development. Since the early experiments of Zeleny (1905) and Panouse (1943), the characterization of such moult- and reproduction-regulating factors has been of prime interest to crustacean biologists. Early work demonstrated that bilateral ablation of eyestalks induced the onset of a premature moult, thereby stimulating the growth process (Zeleny, 1905). Later, experiments using female crab, *Leander serratus*, showed that, depending on the time of eyestalk removal in the developmental cycle of the crustacean, ablation resulted in either a premature moult or precocious development of ovaries (Panouse, 1943). Thus, eyestalk-derived factors responsible for regulating moulting and reproduction were postulated, placing the eyestalk itself in a central role in the regulation of development of crustaceans. Indeed, the practice of unilateral eyestalk ablation of prawns in commercial farms to improve fecundity and growth rates, highlights the central importance of the eyestalk with respect to regulation of growth and reproduction in crustaceans (Browdy, 1992).

4. Moulting hormones

In most arthropod species, the major active moulting hormone (ecdysteroid) is 20-hydroxyecdysone, which is generally accompanied by much lower concentrations of ecdysone (*Figure 1*; Rees, 1989; Lafont, 2000). In fact, 20-hydroxyecdysone was first isolated from a crustacean species (Hampshire and Horn, 1966). A source of ecdysteroid production in crustaceans is the Y-organs, which consist of a single cell type of epithelial origin and were first described by Gabe (1953). The role of Y-organs in control of moulting in the shore crab, *Carcinus maenas*, was established by Echalier (1954, 1955) and ecdysone was shown to be a secretory product of the Y-organs of the crabs, *Pachygrapsus crassipes* and *Cancer antennarius* (Chang and O'Connor, 1977) and the crayfish, *Orconectes limosus* (Willig and Keller, 1976). However, in Y-organs of *C. maenas*, 25-deoxyecdysone is a major ecdysteroid secretory product (Lachaise *et al.*, 1989), whereas in *C. antennarius* (Spaziani *et al.*, 1989) and a crayfish, *Procambarus clarkii*, Y-organs (Sonobe *et al.*, 1991), 3-dehydroecdysone is secreted.

Since it was known that eyestalks of crustaceans contain factor(s) that negatively regulate moulting (Zeleny, 1905), and that Y-organs are a source of moulting hormone, an assay was established to measure the effects of eyestalk extracts on ecdysteroid production by Y-organs. Using this assay, the first moult-inhibiting hormone (MIH) from crustacean eyestalks was isolated and characterized (Webster, 1998). This MIH was a 78 amino acid neuropeptide isolated from the X-organ–sinus gland complex of

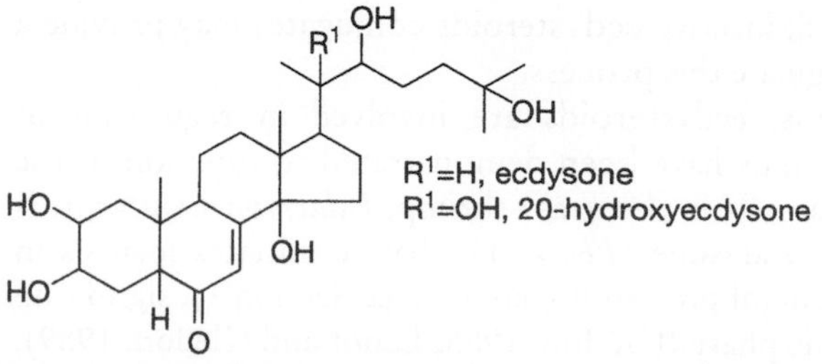

R^1=H, ecdysone
R^1=OH, 20-hydroxyecdysone

Figure 1. Ecdysteroid structures.

the eyestalk of the shore crab, *C. maenas*, and exhibited sequence similarities with the previously isolated CHH from the same species (Kegel *et al.*, 1989).

In related work, Chang *et al.* (1990), isolated an MIH from the sinus glands of the American lobster, *H. americanus*. In this case, the peptide was isolated using an *in vivo* bioassay and resulted in the isolation of a 72 amino acid neuropeptide that was almost identical to the CHH previously isolated from *C. maenas* (Kegel *et al.*, 1989). Since CHH peptides from crustacean sinus glands regulate glucose levels in the circulating haemolymph, it appeared that the *H. americanus* MIH could also act as a CHH. This pleiotropy of effects, e.g. MIH acting as CHHs and vice versa, suggested that CHH peptides, besides acting to regulate glucose levels in haemolymph, could also modulate levels of moulting hormone. This does not seem surprising when it is considered that the process of moulting is a very energy-demanding process and that glucose is a primary metabolic 'fuel' source. These peptides were the first few members of the CHH/MIH/vitellogenesis-inhibiting hormone (VIH) family of crustacean neuropeptides, which to date comprise of more than 32 peptides isolated from a variety of species (for reviews, see Keller, 1992; Van Herp, 1998; Webster, 1998). The family is subdivided into two main groupings, the CHH-like peptides and the MIH-like peptides (Lacombe, *et al.*, 1999). Presently there are two members of this neuropeptide family that are not from crustaceans, the locust ion transport peptide (Meredith *et al.*, 1996) and a CHH-like peptide from the silkmoth, *Bombyx mori* (Endo *et al.*, 2000).

During the reproductive maturation of oocytes in insects and crustaceans, maternally derived ecdysteroids are deposited within the oocytes, in many cases, as conjugates, the nature of which depends upon species (for reviews, see Isaac and Slinger, 1989; Rees, 1995). For instance, in the locust, *Schistocerca gregaria*, the major storage forms of ecdysteroids are phosphate esters (Dinan and Rees, 1981; Rees and Isaac, 1984), whilst in crickets and cockroaches (Whiting and Dinan, 1989; Slinger *et al.*, 1986) and in ticks (Crosby *et al.*, 1986) they are fatty acid esters, and in prawns the major forms can be phosphorylated, acetylated or fatty acylated depending on the species of crustacean (Young *et al.*, 1993a,b). These conjugated ecdysteroids are thought to be deposited within the developing oocytes as maternal storage forms of hormone for utilization by embryos during the early stages of development, prior to the formation of ecdysteroid-synthesizing machinery (e.g. Y-organs in crustaceans or prothoracic glands in insects; Lanot *et al.*, 1989). During embryogenesis in crustaceans, a number of successive moults occur prior to hatching of free-swimming nauplii. It has been demonstrated that increases in ecdysteroid titre within the eggs coincide with the deposition of new cuticle in the developing crustacean embryo (McCarthy and Skinner, 1979; Lachaise and Hoffmann, 1982; Chaix and De Reggi, 1982; Spindler *et al.*, 1987; Goudeau *et al.*, 1990; Young *et al.*, 1991). Thus, as well as regulating moulting and reproduction in adult crustaceans, it appears that ecdysteroids also play a key role in the control of embryonic moulting. At least in some species, hydrolysis of the maternally derived, inactive ecdysteroids conjugates may provide a source of moulting hormones to regulate this process.

In larval insects and crustaceans, ecdysteroids are involved in regulation of moulting, whilst in adults, ecdysteroids have been demonstrated to function in the regulation of reproduction. For example, in the grass shrimp, *Palaemon serratus*, it is the moulting hormone, 20-hydroxyecdysone (*Figure* 1), that re-initiates meiosis in prophase I-arrested oocytes at the end of pre-vitellogenesis (see Section 1), signifying entry of oocytes into the vitellogenic phase (Clédon, 1985; Lanot and Clédon, 1989).

The arrest of meiosis is important for synchronizing developing oocytes to the same stage prior to fertilization and spawning. In adult male crustaceans, ecdysteroids have been suggested to function in androgenic gland development and their titre is correlated with reproductive competence (Laufer et al., 1993).

Thus, ecdysteroids and MIH (or CHH) play key roles in the control of growth and of reproductive and embryonic development in crustaceans.

5. Juvenile hormones

5.1 *Juvenile hormones in insects*

In insects, whereas ecdysteroid triggers moulting, the nature of the moult is determined by the presence of juvenile hormone (*Figure 2*). In larval stages, juvenile hormone acts to maintain the larval character of the moulting insect, whereas in holometabolous insects, the pupal moult occurs in the presence of low concentration of juvenile hormone, with the hormone being virtually absent during the adult moult (for reviews, see Riddiford, 1994; Truman and Riddiford, 1999). Because of this ability to prevent metamorphosis, juvenile hormone has been referred to as a *status quo* hormone (Riddiford, 1994).

In adult insects, production of juvenile hormone by the corpora allata resumes and the hormone regulates aspects of reproductive development, including yolk protein synthesis in fat body of insects (for a review, see Bellés, 1998). However, another role of juvenile hormone in insects is the stimulation of patency of ovaries for uptake of yolk proteins in *Rhodnius prolixus* (Davey, 1996; Wyatt and Davey, 1996). In this, it appears that juvenile hormone interacts with receptors on cell membranes of the follicle cells that surround developing oocytes in ovary tissue. When juvenile hormone binds to these receptors, protein kinase C activity is stimulated that ultimately stimulates Na^+/K^+-ATPase activity in the follicle cell membranes and gaps begin to appear between the follicle cells. This gap allows extracellular fluid (that will contain proteins such as vitellin) to directly contact the cell surface of the developing oocytes, providing a possible mechanism of uptake of vitellin into oocytes. This would provide a mechanism of synchronising vitellin production by one tissue and facilitating uptake by another.

5.2 *Methyl farnesoate in crustaceans*

The mandibular organs of crustaceans were first described by Le Roux (1968) and, based on morphological features, were thought to resemble tissues that produced steroids. Experiments by Yudin et al. (1980) showed that injecting homogenates of mandibular

(a)   (b)

Methyl farnesoate   Juvenile hormone-III

Figure 2. Structures of methyl farnesoate and juvenile hormone-III.

organ into the blue crab, *Callinectes sapidus*, accelerated progression towards moulting. The authors tested the extracts for the presence of moulting hormone (ecdysteroids), but were unsuccessful. It was also noted around the same time that, as crustaceans progressed throughout the moult cycle, the mandibular organs underwent hypertrophy (Hinsch, 1977). These two reports suggested a link between mandibular organ function and the moult cycle. It was not until 1987 that methyl farnesoate (MF) was first isolated from the mandibular organs of crustaceans and a correlation between haemolymph levels of MF and ovarian development was demonstrated (Laufer *et al.*, 1987a,b). MF) is structurally similar to insect juvenile hormone (*Figure 2*) and was originally isolated from the spider crab, *Libinia emarginata* (Laufer *et al.*, 1987a). It lacks the epoxide ring found in juvenile hormones. Although the reason for this difference is unclear, it is interesting that a recent report shows that a terrestrial isopod crustacean, *Porcellio scaber*, contains juvenile hormone III (Trieblemayr and Lackner, 1999).

5.3 *Methyl farnesoate and reproductive development*

The levels of circulating MF have been correlated with ovarian maturation in a number of species of crustaceans, including the spider crab, *L. emarginata* (Laufer *et al.*, 1987a), the lobster, *H. americanus* (Borst *et al.*, 1987), the prawn, *Macrobrachium rosenbergii* (Wilder and Aida, 1995; Wilder *et al.*, 1995) and the edible crab, *Cancer pagurus* (Wainwright *et al.*, 1996a). The foregoing reports strongly suggest a role for MF in regulation of ovarian maturation in crustaceans. Indeed, studies have demonstrated in the white shrimp, *Penaeus vannamei*, that treatment of oocytes cultured *in vitro* with physiological doses of MF, stimulated an increase in size compared with untreated oocytes (Tsukimura and Kamemoto, 1991). This observation is pertinent when taken together with separate reports which demonstrate that injections of MF into female crustaceans can stimulate ovarian growth and development (Jo *et al.*, 1999; Laufer *et al.*, 1998; Reddy and Ramamurthi, 1998). These findings become even more relevant when taking into account a recent report indicating that MF can stimulate protein kinase C activity in ovaries of a crayfish, *Cherax quadricarinatus* (Soroka *et al.*, 2000). This appears to be akin to the system described earlier (see Section 5.1) whereby juvenile hormone in insects stimulates patency of ovary for uptake of vitellin by activating a signalling cascade involving a protein kinase C (Davey, 1996). This stimulation of ovarian maturation by MF has formed the basis of a patent application covering the use of MF in aquaculture feed to stimulate fecundity of crustaceans (Laufer, 1992). However, in a non-decapod crustacean species, *Triops longicaudatus*, MF does not seem to stimulate ovarian maturation or vitellin synthesis (Riley and Tsukimura, 1998).

In male crustaceans, evidence suggests that MF is involved in regulation of reproductively active male morphotype development of *L. emarginata* (Laufer *et al.*, 1993) and testicular growth in the freshwater crab, *Oziotelphusa senex senex fabricius* (Klavarthy *et al.*, 1999).

Thus, a potential role for MF in adult decapod crustaceans is in the regulation of reproductive maturation, and it may be considered as a gonadotropin.

5.4 *Methyl farnesoate in regulation of metamorphosis*

Evidence for a possible role for MF as a classical juvenile, or status quo, hormone has been difficult to obtain. This may be in part due to the difficulties involved in culturing crustaceans under laboratory conditions, particularly in maintaining synchronous populations of larval crustaceans for such studies. However, despite this, evidence is now available to suggest such a function for MF. Following the initial report of the discovery of MF in crustaceans (Laufer *et al.*, 1987a), it was shown that MF could cause a small, but significant, delay in moulting between stages in juvenile lobsters, *Homarus americanus* (Borst *et al.*, 1987). Supporting evidence for this demonstrated that MF delayed the onset of larval *M. rosenbergii* metamorphosis (Abdu *et al.*, 1998). This study used a novel method of delivery of MF that involved feeding an MF-rich diet to *Artemia salina*, which were then used as food for larval *M. rosenbergii*. However, despite these studies on decapod crustaceans, the most striking evidence for a role for MF as a juvenile hormone has come from work on barnacles. It has been suggested that MF, acting through activation of protein kinase C and ion channels, can induce settlement of larval barnacles (*Balanus amphitrite*; Yamamoto *et al.*, 1997) and annelids (*Capitella*; Biggers and Laufer, 1996, 1999). However, these studies also noted that the settlement and metamorphosis (akin to the larval-to-pupal moult in insects) was abnormal. This strongly suggested that these effects might be due to the toxicity of the comparatively high doses of compounds used. A recent study in *B. amphitrite* demonstrated a physiologically relevant effect of MF for the first time (Smith *et al.*, 2000). This latter work demonstrated that, at physiological concentrations, MF inhibited settlement of the cypris larvae of *B. amphitrite*, thus exhibiting juvenile hormone-like effect. However, at higher concentrations, similar to those used in the foregoing reports (Biggers and Laufer, 1996; Yamamoto *et al.*, 1997; Biggers and Laufer, 1999), the cypris larvae settled but underwent an abnormal and often fatal metamorphosis. The role of ion channels and protein kinase C in the normal process of settlement and metamorphosis has yet to be established.

5.5 *Regulation of reproduction by sinus gland neuropeptides*

Since the early experiments of Panouse (1943), the existence of a VIH secreted from the crustacean eyestalk was postulated. The first VIH was isolated from sinus glands of the European lobster, *Homarus americanus*, using an *in vivo* bioassay (Soyez *et al.*, 1991). In this assay, extracts of *H. americanus* sinus gland were injected into grass shrimp, *Palaeamonetes varians*, and the effect on vitellogenesis monitored. *P. varians* was used, as opposed to *H. americanus*, because *H. americanus* undergoes a vitellogenic cycle approximately once every 2 years, making it unsuitable for such experiments. In subsequent experiments, cDNA encoding VIH was isolated and found to be present in male lobsters (DeKleijn *et al.*, 1992). As a result, this peptide has been renamed as a gonad-inhibiting hormone (GIH). It appears from work using the lobster, *H. americanus*, that GIH action inhibits uptake of vitellin into oocytes (Van Herp and Payen, 1991). In other species, it has been suggested that GIH may act to regulate the production of vitellin in ovarian tissues (Lee and Watson, 1995). A second GIH, Arv-GIH, was recently isolated from isopod, *Armadillidium vulgare*, using both *in vivo* and *in vitro* homologous bioassays (Greve *et al.*, 1999). From the *in vivo* assay, Arv-GIH appears to work by inhibiting the onset of vitellogenesis in females, whilst the *in vitro* assays suggested that Arv-GIH regulated vitellogenin synthesis in the fat body (hepatopancreas) of *A. vulgare*. Thus, a direct link between an eyestalk factor, viz. Arv-GIH, and a key process in reproductive development of crustaceans (vitellogenesis) was established.

5.6 Mandibular organ-inhibiting hormones

It is germane to consider that the peptides that regulate production of MF in crustaceans, may, through their activity, also be considered to regulate ovarian maturation since the latter hormone has a putative role in regulation of reproductive development. To date, such peptides have been isolated and characterized from only two species (Liu and Laufer, 1996; Wainwright *et al.*, 1996b). Peptides from both species negatively regulate the production of MF by mandibular organs and have been termed mandibular organ-inhibiting hormones (MO-IHs).

In the spider crab, *L. emarginata*, three isoforms of MO-IH have been isolated, with one exhibiting structural similarities to CHH peptides (e.g. 72 amino acids in length with blocked N- and C-termini). Indeed, *in vivo* these peptides were shown to have hyperglycaemic activity, whilst *in vitro*, they behaved as MO-IHs (Liu and Laufer, 1996).

In the edible crab, *C. pagurus*, there are two isoforms of MO-IH isolated thus far, MO-IH-1 and -2. These MO-IHs are structurally more similar to the previously isolated MIH peptides (e.g. 78 amino acids long with unblocked N- and C-termini). The isoforms of MO-IH in *C. pagurus* differ from each other by a single amino acid (a lysine for glutamine substitution at position 33 in MO-IH; Wainwright *et al.*, 1996b) that is encoded by a single nucleotide change (A to C) in the cDNA sequence that encodes these two peptides (Tang *et al.*, 1999). During the isolation of MO-IH-1 and -2, it is apparent from HPLC chromatograms that the amount of MO-IH-1 is approximately twice that of MO-IH-2 (Wainwright *et al.*, 1996b). This suggested that there may be twice as many copies of *MO-IH-1* genes as there are of *MO-IH-2*, but still did not explain why there are two rather similar peptides (both structurally and functionally) in a single species. When the genes for the MO-IHs of *C. pagurus* were isolated from a genomic library an interesting observation was made (Lu *et al.*, 2000). Clones were identified that contained single copies of the *MIH* gene, single copies of the *MO-IH-2* gene and clones where the *MO-IH-1* gene was clustered within a 6.5 kbp DNA fragment with an *MIH* gene. Computer analysis for putative promoters indicated that the *MO-IH-1* and *MIH* are convergently transcribed. Between the *MIH* and *MO-IH-1* genes, GC and AC repeat elements were present, whereas upstream of the *MIH* gene, GT and CT repeats occurred. One could envisage a possible mechanism whereby *MO-IH* originally arose through some mechanism of gene duplication of the *MIH* gene. Additionally, there were approximately twice as many copies of the *MO-IH-1* gene present in the library as there were *MO-IH-2* genes. When other species of *Cancer* were tested for the presence of MO-IH activity, only one isoform was found (S.G. Webster, unpublished results) suggesting that, during the speciation of *Cancer* to *C. pagurus*, ca 12 million years ago (Harrison and Crespi, 1999), some gene duplication had occurred giving rise to *MO-IH-2* from *MO-IH-1*. Indeed, the similarity of the intron-exon structure of these *MO-IH/MIH* genes and *CHH*-like genes previously isolated (Chan *et al.*, 1998; Gu and Chan, 1998) clearly suggests that *CHH* is the ancestral gene from which the *MIH* and *MO-IH* genes have evolved through a mechanism involving gene duplication.

5.7 Regulation of methyl farnesoate biosynthesis

Methyl farnesoate is an isoprenoid synthesized using acetate as a metabolic precursor (Wainwright *et al.*, 1998). In experiments where mandibular organs cultured *in vitro* in

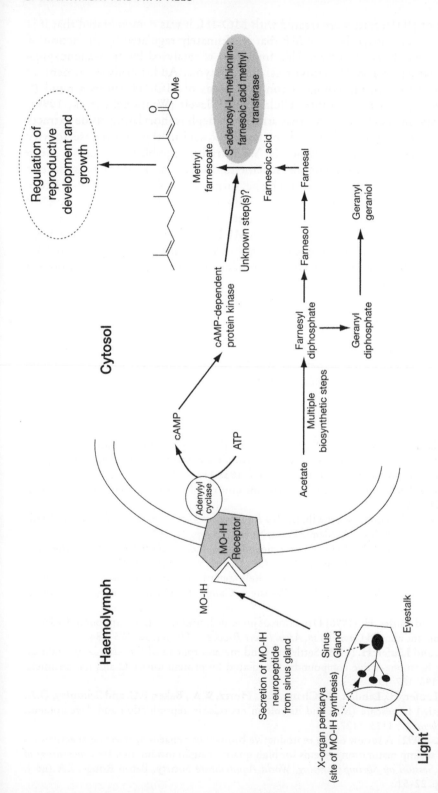

Figure 3. Possible outline mechanism by which mandibular organ-inhibiting hormone (MO-IH) acts to regulate the biosynthesis of methyl farnesoate in mandibular organs. Upon release of MO-IH from the X-organ–sinus gland of the eyestalk, under the control of environmental cues (eg. light), MO-IH interacts with mandibular organ cell surface receptor, stimulating an increase in intracellular cAMP, which ultimately leads to inhibition of farnesoic acid methyl transferase activity and reduced production of methyl farnesoate.

the presence of [³H]acetate were treated with MO-IH, it was demonstrated that it is the final step in the biosynthesis of MF that is ultimately regulated by the action of MO-IH (Wainwright et al., 1998). This final step is catalysed by an S-adenosyl-L-methionine: farnesoic acid O-methyl trasferase enzyme. Additionally, treatment of mandibular organs with physiological concentrations of MO-IH causes a rapid (5 min) and reversible increase in intracellular cAMP levels (Wainwright et al., 1999). Additional experiments have shown that artificially dephosphorylating crude extracts of mandibular organ cytosol that contains farnesoic acid methyl transferase activity stimulates the activity of the enzyme (G. Wainwright, unpublished results). From these foregoing observations, a potential scheme for the regulation of MF biosynthesis in mandibular organs of C. pagurus might be formulated. Binding of MO-IH to a receptor on the cell surface of the mandibular organs stimulates an increase in cAMP production, with the increase in cAMP ultimately leading to a decrease in farnesoic acid methyl transferase activity, shutting down MF biosynthesis (Figure 3). The release of MO-IH from X-organ–sinus glands in the eyestalk may well be regulated by photoperiod (see Section 2), thus environmental change may regulate production of a gonadotropic hormone, MF.

Acknowledgements

We thank NERC, BBSRC and The Leverhulme Trust for their generous support. We are also indebted to Drs P. C. Turner and M.R.H. White (University of Liverpool) and Dr S.G. Webster (University of Wales, Bangor) for their fruitful collaboration.

References

Abdu, U., Takac, P., Laufer, H. and Sagi, A. (1998) Effect of methyl farnesoate on late larval development and metamorphosis in the prawn, Macrobrachium rosenbergii (Decapoda, Palaemonidae): A juvenoid-like effect? Biol. Bull. 195: 112–119.

Adiyodi, K.G. and Adiyodi, R.G. (1970) Endocrine control of reproduction in Crustacea. Biol. Rev. 45: 121–165.

Aiken, D.E. (1980) Molting and growth. In: The Biology and Management of Lobsters, vol. 1 (eds J.S. Cobb and B.F. Phillips). Academic Press, New York, pp. 91–162.

Aiken, D.E. and Waddy, S.L. (1986) Environmental influence on recruitment of the American Lobster, Homarus americanus: a perspective. Can. J. Fish. Aquat. Sci. 43: 2258–2270.

Bellés, X. (1998) Endocrine effectors in insect vitellogenesis. In: Recent Advances in Arthropod Endocrinology (eds G.M. Coast and S.G. Webster). Cambridge University Press, Cambridge, pp. 71–90.

Biggers, W. J. and Laufer, H. (1996) Detection of juvenile hormone-active compounds by larvae of the the marine annelid Capitella sp. Arch. Insect Biochem. Physiol. 32: 475–484.

Biggers, W. J. and Laufer, H. (1999) Settlement and metamorphosis of Capitella larvae induced by juvenile hormone-active compounds is mediated by protein kinase C and ion channels. Biol. Bull. 196: 187–198.

Borst, D.W., Laufer, H., Landau, M., Chang, E.S., Hertz, W.A., Baker, F.C. and Schooley, D.A. (1987) Methyl farnesoate (MF) and its role in crustacean reproduction and development. Insect Biochem. 17: 1123–1127.

Browdy, C.L. (1992) A review of the reproductive biology of penaeus species: perspectives on controlled shrimp maturation systems for high quality nauplii production. In: Proceedings of the Special Session on Shrimp Farming, World Aquaculture Society, Baton Rouge, LA (ed J. Wyban), pp. 22–51.

Chaix, J.-C. and De Reggi, M. (1982) Ecdysteroid levels during ovarian development and embryogenesis in the spider crab, *Acanthonyx lunulatus*. *Gen. Comp. Endocrinol.* **47**: 7–14.

Chan, S.-M., Chen, X.-G. and Gu, P.-L. (1998) PCR cloning and expression of the molt-inhibiting hormone gene for the crab (*Charybdis feriatus*). *Gene* **224**: 23–33.

Chang, E.S. and O'Connor, J.D. (1977) Secretion of α-ecdysone by crab Y-organs *in vitro*. *Proc. Natl Acad. Sci. USA* **74**: 615–618.

Chang, E.S., Prestwich, G.D. and Bruce, M.J. (1990) Amino acid sequence of a peptide with both molt-inhibiting and hyperglycemic activities in the lobster, *Homarus americanus*. *Biochem. Biophys. Res. Commun.* **171**: 818–826.

Charniaux-Cotton, H. and Payen, G. (1988) Crustacean reproduction. In: *Endocrinology of Selected Invertebrate Types* (eds H. Laufer and R.G.H. Downer). Alan R. Liss, New York, pp. 279–303.

Clédon, P. (1985) Cytological and experimental-analysis of oocyte maturation and activation in the prawn, *Palaemon serratus* (Crustace Decapode Natantia). *C. R. Acad. Sci., Paris Sér. D* **301**: 317–322.

Crosby, T., Evershed, R.P., Lewis, D., Wigglesworth, K.P. and Rees, H.H. (1986) Identification of ecdysone 22-long-chain fatty acyl esters in newly laid eggs of the cattle tick, *Boophilus microplus*. *Biochem. J.* **240**: 131–138.

Davey, K.G. (1996) Hormonal control of the follicular epithelium during vitellogenin uptake. *Invert. Reprod. Devl.* **30**: 249–254.

DeKleijn, D.P.V., Coenen, T., Laverdure, A.M., Tensen, C.P. and Van Herp, F. (1992) Localization of mRNAs encoding the crustacean hyperglycaemic hormone (CHH) and gonad-inhibiting hormone (GIH) in the X-organ sinus gland complex of the lobster, *Homarus americanus*. *Neuroscience* **51**: 121–128.

Dinan, L.N. and Rees, H.H. (1981) The identification and titres of conjugated and free ecdysteroids in developing ovaries and newly-laid eggs of *Schistocerca gregaria*. *J. Insect Physiol.* **27**: 51–58.

Echalier, G. (1954) Recherches expérimentales sur le rôle de "l'organe Y" dans la mue de *Carcinus maenas* (L.) Crustacés Décapodes. *C. R. Acad. Sci. Paris Sér. D* **238**: 523–525.

Echalier, G. (1955) Rôle de l'organe Y dans la determinisme de la mue de *Carcinus maenas* (L.) Expériences d'implantation. *C. R. Acad. Sci. Paris Sér. D* **240**: 1581–1583.

Endo, H., Nagasawa, H. and Watanabe, T. (2000) Isolation of a cDNA encoding a CHH-family peptide from the silkworm, *Bombyx mori*. *Insect Biochem. Mol. Biol.* **30**: 355–361.

Gabe, M. (1953) Sur l'existence, chez quelque Crustacés Malocostracés d'un organe comparable á la glande de la mue des insects. *C. R. Acad. Sci. Paris Sér. D* **237**: 1111–1113.

Glowik, R.G., Golowasch, J., Keller, R. and Marder, E. (1997) D-Glucose-sensitive neurose-cretory cells of the crab *Cancer borealis* and negative feedback regulation of blood glucose level. *J. Exp. Biol.* **200**: 1421–1431.

Goudeau, M., Lachaise, F., Carpentier, G. and Goxe, B. (1990) High titres of ecdysteroids are associated with the secretory process of embryonic envelopes in the European lobster. *Tiss. Cell* **22**: 269–281.

Greve, P., Sorokine, O., Berges, T., Lacombe, C., Van Dorsselaer, A. and Martin, G. (1999) Isolation and amino acid sequence of a peptide with vitellogenesis inhibiting activity from the terrestrial isopod, *Armadillidium vulgare* (*Crustacea*). *Gen. Comp. Endocrinol.* **115**: 406–414.

Gu, P.-L. and Chan, S.-M. (1998) The shrimp hyperglycaemic hormone-like neuropeptide is encoded by multiple copies of genes arranged in a cluster. *FEBS Lett.* **441**: 397–403.

Hampshire, F. and Horn, D.H.S. (1966) Structure of crustecdysone, a crustacean moulting hormone. *J. Chem. Soc. Chem. Commun.* 37–38.

Harrison, M.K. and Crespi, B.J. (1999) Phylogenetics of *Cancer* crabs (Crustacea: Decapoda: Brachyura). *Mol. Phylogenet. Evol.* **12**: 186–199.

Hinsch, G.W. (1977) Fine structural changes in the mandibular gland of the male spider crab, *Libinia emarginata* (L.) following eyestalk ablation. *J. Morphol.* **154**: 307–316.

Isaac, R.E. and Slinger, A.J. (1989) Storage and excretion of ecdysteroids. In *Ecdysone* (ed J. Koolman). Georg Thieme, Stuttgart, pp. 250–259.

Jo, Q.T., Laufer, H., Biggers, W.J. and Kang, H.S. (1999) Methyl farnesoate induced ovarian maturation in the spider crab, *Libinia emarginata*. *Invert. Reprod. Dev.* **36**: 79–85.

Kegel, G., Reichwein, B., Wesse, S., Gaus, G., Peter-Katalinic, J. and Keller, R. (1989) Amino acid sequence of the crustacean hyperglycemic hormone (CHH) from the shore crab, *Carcinus maenas*. *FEBS Lett.* **255**: 10–14.

Keller, R. (1992) Crustacean neuropeptides: Structures, functions and comparative aspects. *Experientia* **48**: 439–448.

Klavarthy, Y., Mamatha, P. and Reddy, P.S. (1999) Methyl farnesoate stimulates testicular growth in the freshwater crab, *Oziotelphusa senex senex fabricus*. *Naturwissenschaften* **86**: 394–395.

Lachaise, F. and Hoffmann, J. A. (1982) Ecdysteroids and embryonic development in the shore crab, *Carcinus maenas*. *Hoppe Seylers Zool. Physiol. Chem.* **363**: 1059–1067.

Lachaise, F., Carpentier, G., Sommé, G., Colardeau, J. and Beydon, P. (1989) Ecdysteroid synthesis by crab Y-organs. *J. Exp. Zool.* **252**: 283–292.

Lacombe, C., Greve, P. and Martin, G. (1999) Overview on the subgrouping of the crustacean hyperglycemic hormone family. *Neuropeptides* **33**: 71–80.

Lafont, R. (2000) Understanding insect endocrine systems: molecular approaches. *Entemol. Exp. Appl.* **97**: 123–136.

Lanot, R. and Clédon, P. (1989) Ecdysteroids and meiotic reinitiation in *Palaemon* serratus (Crustacea Decapoda Natantia) and in *Locusta migratoria* (Insecta Orthopetra): a comparative study. *Invert. Reprod. Dev.* **16**: 169–175.

Lanot, R., Dorn, A., Gunster, B., Thiebold, J., Lagueux, M. and Hoffmann, J.A. (1989) Functions of ecdysteroids in oocyte maturation and embryonic development of insects. In: *Ecdysone*. (ed J. Koolman). Georg Thieme, Stuttgart, pp. 262–278.

Laufer, H. (1992) Method for increasing crustacean larval production. United States Patent 5,161,481.

Laufer, H., Borst, D.W., Baker, F.C., Carrasco, C., Sinkus, M., Reuter, C.C., Tsai, L.W. and Schooley, D.A. (1987a) Identification of a juvenile hormone-like compound in a crustacean. *Science* **235**: 202–205.

Laufer, H., Landau, M., Homola, E. and Borst, D.W. (1987b) Methyl farnesoate: its site of synthesis and regulation of secretion in a juvenile crustacean. *Insect Biochem.* **17**, 1129–1131.

Laufer, H., Biggers, W.J. and Ahl, J.S.B. (1998) Stimulation of ovarian maturation in the crayfish, *Procambarus clarkii*, by methyl farnesoate. *Gen. Comp. Endocrinol.* **111**: 113–118.

Laufer, H., Wainwright, G., Young, N.J., Sagi, A., Ahl, J.B.S. and Rees, H.H. (1993) Ecdysteroids and juvenoids in two male morphotypes of *Libinia emarginata*. *Insect Biochem. Mol. Biol.* **23**: 171–179.

Le Roux, A. (1968) Description d'organes mandibulaires nouveaux chez les Crustacés Décapodes. *C. R. Acad. Sci. Paris Sér. D* **226**: 1414–1417.

Lee, C.Y. and Watson, R.D. (1995) *In vitro* study of vitellogenesis of the blue crab (*Callinectes sapidus*): site and control of vitellin synthesis. *J. Exp. Zool.* **271**: 364–372.

Liu, L. and Laufer, H. (1996) Isolation and characterisation of sinus gland neuropeptides with both mandibular organ inhibiting and hyperglycaemic effects from the spider crab, *Libinia emarginata*. *Arch. Insect Biochem. Physiol.* **32**: 375–385.

Lu, W., Wainwright, G., Webster, S.G., Rees, H.H. and Turner, P.C. (2000) Clustering of mandibular organ-inhibiting hormone and moult-inhibiting hormone genes in the crab, *Cancer pagurus*, and implications for regulation of expression. *Gene* **253**: 197–207.

McCarthy, J.F. and Skinner, D.M. (1979) Changes in ecdysteroids during embryogenesis of the blue crab, *Callinectes sapidus Rathbun*. *Devl. Biol.* **69**: 627–633.

Meredith, J., Ring, M., Macins, A., Marschall, J., Cheng, N.N., Theilmann, D., Brock, H.W. and Philips, J.E. (1996) Locust ion transport peptide (ITP): primary structure, cDNA and expression in a baculovirus system. *J. Exp. Biol.* **199**: 1053–1061.

Panouse, J.B. (1943) Influence de l'ablation du pédoncule oculaire sur la croissance de l'ovaire chez la crevette *Leander serratus*. *C. R. Acad. Sci. Paris Sér. D* **217**: 553–555.

Reddy, P.S. and Ramamurthi, P. (1998) Methyl farnesoate stimulates ovarian maturation in the freshwater crab, *Oziotelphusa senex senex fabricus*. *Curr. Sci.* **74**: 68–70.

Rees, H.H. (1989) Zooecdysteroids: structures and occurrence. In *Ecdysone*. (ed J. Koolman). Georg Thieme, Stuttgart, pp. 28–28.

Rees, H.H. (1995) Ecdysteroid biosynthesis and inactivation in relation to function. *Eur. J. Entomol.* **92**: 9–39.

Rees, H.H. and Isaac, R.E. (1984) Biosynthesis of ovarian ecdysteroid phosphates and their metabolic fate during embryogenesis in *Schisticerca gregaria*. In: *Biosynthesis, Metabolism and Mode of Action of Invertebrate Hormones* (eds J.A. Hoffmann and M. Porchet). Springer, Berlin, pp.181–195.

Riddiford, L.M. (1994) Cellular and molecular actions of juvevile hormone. I. General considerations and premetamorphic actions. *Adv. Insect Physiol.* **24**: 213–274.

Riley, L.G. and Tsukimura, B. (1998) Yolk protein synthesis in the Riceland tadpole shrimp, *Triops longicaudatus*, measured by *in vitro* incorporation of [^{3}H]leucine. *J. Exp. Zool.* **281**: 238–247.

Skinner, D.M. (1985) Molting and regeneration. In: *The Biology of Crustacea*, Vol 9. Academic Press, New York, pp. 43–146.

Slinger, A.J., Dinan, L.N. and Isaac, R.E. (1986) Isolation of apolar ecdysteroid conjugates from newly-laid oothecae of *Periplaneta americana*. *Insect Biochem.* **16**: 115–119.

Smith, P.A., Clare, A.S., Rees, H.H., Prescott, M.C., Wainwright, G. and Thorndyke, M.C. (2000) Identification of methyl farnesoate in the cypris larvae of the barnacle, *Balanus amphitrite* and its role as a juvenile hormone. *Insect Biochem. Mol. Biol.* **30**: 885–890.

Sonobe, H., Kamba, M., Ohta, K., Ikeda, M. and Naya, Y. (1991) *In vitro* secretion of ecdysteroids by Y-organs of the crayfish, *Procambarus clarkii*. *Experientia* **47**: 948–952.

Soroka, Y., Sagi, A., Khalaila, I., Abdu, U. and Milner, Y. (2000) Changes in protein kinase C during vitellogenesis in the crayfish, *Cherax quadricarinatus* – possible activation by methyl farnesoate. *Gen. Comp. Endocrinol.* **118**: 200–208.

Soyez, D., Le Caer, J.P., Noel, P. Y. and Rossier, J. (1991) Primary structure of two isoforms of the vitellogenesis-inhibiting hormone of the lobster, *Homarus americanus*. *Neuropeptides* **20**: 25–32.

Spaziani, E., Rees, H.H., Wang, W.L. and Watson, R.D. (1989) Evidence that Y-organs of the crab, *Cancer antennarius*, secrete 3-dehydroecdysone. *Mol. Cell. Endocrinol.* **66**: 17–25.

Spindler, K.-D., Van Wormhoudt, A., Sellos, D. and Spindler-Barth, M. (1987) Ecdysteroid levels during embryongenesis in the shrimp, *Palaemon serratus* (Crustacea Decapoda): Qualitative and quantitative changes. *Gen. Comp. Endocrinol.* **66**: 116–122.

Tang, C., Lu, W., Wainwright, G., Webster, S.G., Rees, H.H. and Turner, P. C. (1999) Molecular characterization and expression of mandibular organ-inhibiting hormone, a recently discovered neuropeptide involved in the regulation of growth and reproduction in the crab, *Cancer pagurus*. *Biochem. J.* **343**: 355–360.

Templeman, W. (1936) The influence of temperature, salinity, light and food conditions on the survival and growth of the larvae of the lobster (*Homarus americanus*). *J. Biol. Board Can.* **2**: 485–497.

Trieblemayr, K. and Lackner, B. (1999) Juvenile hormone III in the tegumental glands of *Porcellio scaber* (Crustacea, Isopoda). *Seventh International Conference of the Juvenile Hormones*, Jerusalem, Israel.

Truman, J.W. and Riddiford, L.M. (1999) The origins of insect metamorphosis. *Nature* **40**: 447–452.

Tsukimura, B. and Kamemoto, F.I. (1991) *In vitro* stimulation of oocytes by presumptive mandibular organ secretions in the shrimp, *Penaeus vannamei*. *Aquaculture* **92**: 59–62.

Van Herp, F. (1998) Molecular, cytological and physiological aspects of the crustacean hyperglycaemic hormone family. In: *Recent Advances in Arthropod Endocrinology* (eds G.M. Coast and S.G. Webster). Cambridge University Press, Cambridge, pp.53–70.

Van Herp, F. and Payen G.G. (1991) Crustacean neuroendocrinology: perspectives for the control of reproduction in aquacultural systems. *Bull. Inst. Zool., Academica Sinica* 16: 513–539.

Waddy, S.L. and Aiken, D.E. (1992) Seasonal variation in spawning by preovigerous American lobster (*Homarus americanus*) in response to temperature and photoperiod manipulation. *Can. J. Fish. Aquat. Sci.* 49: 1114–1117.

Wainwright, G., Prescott, M.C., Webster, S.G. and Rees, H.H. (1996a) Mass spectrometric determination of methyl farnesoate profiles and correlation with ovarian development in the edible crab, *Cancer pagurus. J. Mass Spectrom.* 31: 1338–1344.

Wainwright, G., Webster, S.G., Wilkinson, M.C., Chung, J.S. and Rees, H.H. (1996b) Structure and significance of mandibular organ-inhibiting hormone in the crab, *Cancer pagurus*; involvement in multi-hormonal regulation of growth and reproduction. *J. Biol. Chem.* 271: 12749–12754.

Wainwright, G., Webster, S.G. and Rees, H.H. (1998) Neuropeptide regulation of biosynthesis of the juvenoid, methyl farnesoate, in the edible crab, *Cancer pagurus. Biochem. J.* 334: 651–657.

Wainwright, G., Webster, S.G. and Rees, H.H. (1999) Involvement of second messenger systems in mandibular organ-inhibiting hormone-mediated inhibition of methyl farnesoate biosynthesis in mandibular organs of the edible crab, *Cancer pagurus. Mol. Cell. Endocrinol.* 154: 55–62.

Warner, G.F. (1977) Life histories. In: *The Biology of Crabs*. Paul Elek, London, pp. 119–140.

Webster, S.G. (1998) Neuropeptides inhibiting growth and reproduction in crustaceans. In: *Recent Advances in Arthropod Endocrinology* (eds G.M. Coast and S.G. Webster). Cambridge University Press, Cambridge, pp. 33–52.

Whiting, P. and Dinan, L.N. (1989) Indentification of the endogenous apolar ecdysteroid conjugates present in newly-laid eggs of the house cricket (*Acheta domesticus*) as 22-long-chain fatty acyl esters of ecdysone. *Insect Biochem.* 19: 759–765.

Wilder. M.N. and Aida, K. (1995) Crustacean ecdysteroids and juvenoids: chemistry and physiological roles in two species of prawn, *Macrobrachium rosenbergii* and *Penaeus japonicus. Israeli J. Aquaculture-Bamidgeh* 47: 129–136.

Wilder, M.N., Okada, S., Fusetani, N. and Aida, K. (1995) Hemolymph profiles of juvenoid substances in the giant fresh-water prawn, *Macrobrachium rosenbergii* in relation to reproduction and molting. *Fish. Sci.* 61: 175–176.

Willig, A. and Keller, R. (1976) Biosynthesis of α- and β-ecdysone by the crayfish, *Orconectes limosus, in vivo* and by its Y-organs *in vitro. Experientia* 32: 936–937.

Wyatt, G.R. and Davey, K.G. (1996) Cellular and molecular actions of juvenile hormone. 2. Roles of juvenile hormone in adult insects. *Adv. Insect Physiol.* 26: 1–155.

Yamamoto, H., Okino, T., Yoshimura, E., Tachibana, A., Shimizu, K. and Fusetani, N. (1997) Methyl farnesoate induces larval metamorphosis of the barnacle, *Balanus amphitrite*, via protein kinase C activation. *J. Exp. Zool.* 278: 349–355.

Young, N.J., Webster, S.G., Jones, D.A. and Rees, H.H. (1991) Profile of embryonic ecdysteroids in the decapod crustacean, *Macrobrachium rosenbergii. Invert. Reprod. Devl.* 20: 201–212.

Young, N.J., Webster, S.G. and Rees, H.H. (1993a) Ecdysteroid profiles and vitellogenesis in *Penaeus monodon* (Crustacea: Decapoda). *Invert. Reprod. Devl.* 24: 107–118.

Young, N.J., Webster, S.G. and Rees, H.H. (1993b) Ovarian and haemolymph titers during vitellogenesis in *Macrobrachium rosenbergii. Gen. Comp. Endocrinol.* 90: 183–191.

Yudin, A.I., Diener, R.A., Clark, W.H. and Chang, E.S. (1980) Mandibular gland of the blue crab, *Callinectes sapidus. Biol. Bull.* 159: 760–772.

Zeleny, C. (1905) Compensatory regulation. *J. Exp. Zool.* 2: 1–102.

4

Rearing lobsters at different temperatures; effects on muscle phenotype and molecular expression

Janet M. Holmes, Ana Rubio, Dominic Lewis, Douglas M. Neil and Alicia J. El Haj

1. Introduction

1.1 *The life cycle of a lobster*

The life cycle of a clawed lobster such as *Homarus gammarus*, the European lobster, or *Homarus americanus*, the American lobster comprises pre-larval, larval, post-larval (early benthic), young juvenile and adult phases (*Figure 1*; see Aiken and Waddy, 1980; Ennis, 1995 for reviews). After the eggs are spawned they are retained on the abdominal appendages of the female for up to one year, during which time the embryos pass through a series of pre-larval stages (nauplius, prezoea). The pre-larval hatchlings emerge from the egg envelopes and moult into stage 1 larvae, which are planktonic, migrating vertically in the water column in response to changes in light intensity, and feeding on the rich supplies of food within the plankton. They also undergo three moults during this planktonic phase, gradually increasing in size (from 8 mm at stage 1 to 11 mm at stage 3), and then transforming more extensively through a metamorphosis to stage 4 post-larvae, which have a morphology very similar to the adult lobster. These post-larvae swim actively to the bottom to begin a benthic existence. The duration of the larval and post-larval stages can vary from 11 days at 10°C to 60 days at 22°C, dependent upon food availability and particularly upon temperature (Aiken and Waddy, 1986; MacKenzie, 1988), and thus development times are influenced by seasonal changes in the weather and the flow of oceanic currents. Furthermore, post-larvae can delay settlement for up to 5 weeks, until they find a suitable benthic environment for further development.

During the early benthic phase, the post-larvae are still extremely vulnerable, and in order to survive they either excavate burrows within the sediment or find small

Environment and Animal Development: Genes, Life Histories and Plasticity, edited by D. Atkinson and M. Thorndyke.
© 2001 BIOS Scientific Publishers Ltd, Oxford.

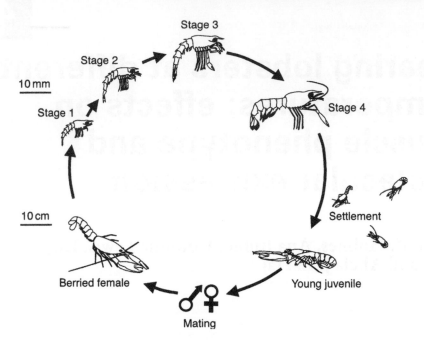

Figure 1. *Life cycle of* Homarus gammarus. *Larval stages 1, 2, and 3 are planktonic, settlement occurs at post-larval stage 4 when the larvae undergoes metamorphosis, taking on the adult form. Early benthic post-larvae and young juveniles dwell in burrows, until large enough to defend themselves. Sexual maturity and mating occurs between 5 and 8 years.*

crevices within rocky substrata, and feed by creating a current of water through the burrow, to provide a supply of small planktonic prey items (Wahle and Steneck, 1991). Young juvenile lobsters may spend between 3 and 4 years within their burrows, but as they grow and become less vulnerable they begin to undertake longer foraging forays. Sexual maturity is reached between 5 and 8 years of age, depending on water temperature and food availability (Waddy *et al.*, 1995).

1.2 *Lobster aquaculture and ranching*

The commercial importance of lobsters as a luxury food item has resulted in the implementation of many aquaculture programmes (Aiken and Waddy, 1995; Bannister and Howard, 1991). The main obstacle faced by these programmes, however, is the length of time (up to 8 years) that is required for the lobsters to develop into mature adults, and consequently the space, maintenance and cost that this requires. The aquaculture of lobsters begins with collecting hatchlings as they emerge from gravid females, and holding the free-swimming larvae (stages 1–3) in rapidly circulating seawater bins. Under these conditions cannibalism is minimized, and survival rates of 70–80% can be achieved. Moreover, the planktonic phase of development can be reduced to less than 30 days by elevating water temperature to 18°C. At settlement, the post-larvae have to be distributed into individual trays in order to prevent antagonistic interactions between these highly aggressive animals. Keeping the lobsters under these conditions for periods of years is extremely labour-intensive and costly, and the exercise is also prone to failure if the rearing conditions are not strictly maintained. A less expensive alternative to aquaculture is a lobster ranching programme, which is undertaken in order to re-stock natural lobster populations after they have been depleted by over-fishing, or to seed artificial reefs. Ranching involves rearing the planktonic larvae in bins, as described above, but then ongrowing the post-larvae for only a short time (around 6 months, until they are stage 9), before releasing them back into the sea. The most expensive and protracted phase of the rearing programme is thus eliminated.

1.3 *The rationale for our experimental approach*

In order to maximize growth of farmed lobsters for either aquaculture or ranching, it is standard practice to elevate rearing temperatures. Elevated temperatures accelerate metabolic processes, and growth rate is proportional to temperature between 8°C and 24°C (Waddy *et al.*, 1995). However, the consequences of raising water temperature on the muscle physiology, performance and behaviour of these lobsters once released back into the marine environment is not known. What is apparent, however, is that modulation of the thermal environment during the early stages of development, when moulting frequency is naturally at its highest level, may have a greater impact and effect on muscle growth and plasticity. We have therefore performed a study in order to answer the question: do lobster ranching programmes rear animals that are suitably equipped to compete against their wild counterparts for food and space once released? In this study, groups of post-larval European lobsters, *Homarus gammarus* were raised at 11, 15 and 19°C from post-larval stage 6 (approximately 3 months old) to 18 months, and, as well as keeping full records of their times of moulting, regular morphometric and physiological measurements were taken throughout this period of development. Our aim was to identify any changes in muscle growth and phenotype that may affect how these hatchery-reared lobsters behave once released into the marine environment. We have concentrated on muscle systems of particular importance for lobster behaviour: the abdominal muscles, which produce swimming, and the muscles of the cutter and crusher claws, which are important for feeding and for antagonistic behaviour. These muscles also represent the commercially important body tissues.

2. Muscles and moulting

2.1 *The muscle systems studied*

The abdominal muscles form the major muscle system in the body, and their arrangement within the abdomen is shown in transverse section in *Figure 2b*. The organization and segregation of the abdominal musculature is established early on in development, and these muscles are functional when the larvae first hatch (Cole and Lang, 1980). The deep fast flexor and extensor muscles represent the majority of the abdominal muscle mass and control the powerful flexions and extensions of the abdomen, which propel the lobster through the water during the escape tail-flip response (Kennedy and Takeda, 1965; Wine and Krasne, 1982). Thin sheets of slow superficial extensor and flexor muscles surround the deep fast muscle, and control the posture of the abdomen (Kennedy and Takeda, 1965; Pilgrim and Wiersma, 1963). Unlike the deep muscles, which are homogeneous populations of fast type fibres, the superficial muscles can be further sub-divided into two slow types, S1 and S2, which

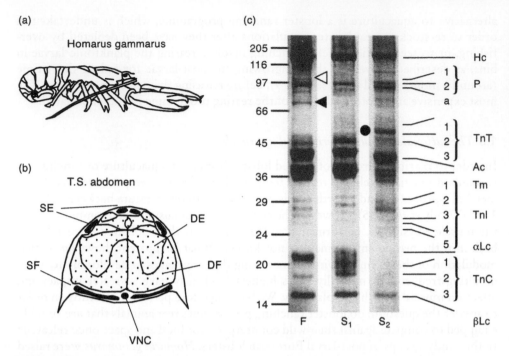

Figure 2. Morphology of the abdomen of Homarus gammarus. *The animal (a) has been sectioned transversely in the third segment of the abdomen to show the arrangement of the deep and superficial extensor and flexor muscles. Each SFM comprises a medial and lateral muscle bundle and shows regional segregation into fibre subtypes. The medial bundle comprises fibres of the S2 phenotype, whilst the lateral bundle comprises approximately 90% S1 fibres and 10% S2 fibres (b). Discontinuous SDS–PAGE 12% gel of fibres from the fast abdominal flexor muscle and the S1 and S2 fibres from the SFM stained with Coomassie blue (c). SE, superficial extensor; SF, superficial flexor; DE, deep extensor; DF, deep flexor; Hc, myosin heavy chain; P, paramyosin; TnT, troponin T (three isoforms); Ac, Actin; Tm, tropomyosin; TnI, troponin I (five isoforms); αLc, alpha light chain; TnC (three isoforms), troponin C; βLc, beta light chain. A filled arrowhead indicates the* P_1 *isoform of paramyosin, an open arrow-head indicates the* 75×10^3 M_r *protein and a filled circle indicates the* T_1 *isoform of troponin T.*

play different roles in postural control (Fowler and Neil, 1992; Holmes *et al.*, 1999). The fast, S1 and S2 fibre phenotypes, which are found within the skeletal muscles of decapod crustaceans, are well characterized and can be identified by their morphological (Fowler and Neil, 1992; Govind *et al.*, 1987), mechanical (Galler and Neil, 1994; Holmes *et al.*, 1999), biochemical (Mykles, 1988) and histological (Li and Mykles, 1990; Mykles, 1985a,b; Neil *et al.*, 1993) properties. As in other crustaceans, fibres persist throughout development, and attain their different contractile properties through the expression of fibre-type specific isoforms, which can be visualized using one-dimensional SDS–PAGE gel electrophoresis (*Figures 2c* and *3b*). Thus at least seven isoforms are unique to the fast fibres (fMHC, fTm, paramyosin$_1$, P75, troponin I$_1$, I$_3$ and I$_5$), five isoforms are unique to the S1 slow fibres (sMHC, sTm$_1$, sTm$_2$, troponin T$_3$ and troponin I$_4$), and the T$_1$ isoform of troponin is unique to S2 slow fibres (Mykles, 1985a,b; Neil *et al.*, 1993).

Much of the research into growth of lobster muscle has centred on the limbs, such as the leg and claws. Quite contrasting growth patterns exist for these two muscle

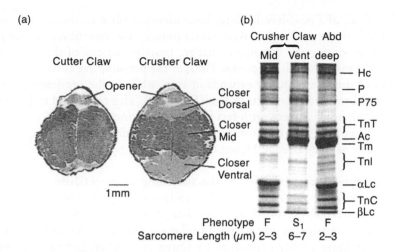

(a)

Cutter Claw   Crusher Claw

Opener

Closer
Dorsal

Closer
Mid

Closer
Ventral

1mm

(b)

Crusher Claw Abd

Mid Vent deep

— Hc
— P
— P75
}— TnT
— Ac
— Tm
}— TnI
— αLc
}— TnC
}— βLc

Phenotype   F   S₁   F
Sarcomere Length (μm) 2–3   6–7   2–3

Figure 3. *Relationships between morphological, histochemical and biochemical properties of the claw muscles of the lobster* Homarus gammarus. *(a) Cross sections through the middle of the claws of a juvenile (stage 15) lobster, raised for 12 months (from stage 6) at 19°C. The sections have been stained histochemically for myofibrillar ATPase, and show that the majority of fibres in the cutter claw closer muscle have high ATPase activity (dark staining), whereas in the crusher claw the high ATPase fibres are limited to a central band. In a fully differentiated crusher muscle this dark band is normally narrower (or totally absent), but in this case the claw is a 'false crusher', having been deprived of the stimulation of manipulating hard material during rearing.*
(b) SDS–PAGE gel lanes showing the myofibrillar proteins present in different regions of the crusher claw muscle, and for comparison in the deep abdominal flexor muscle of the same lobster. Particular proteins (e.g. P75) and isoforms (e.g. of TnI) identify fibres with a fast (F) phenotype in the deep abdominal muscle and in the central region of the crusher claw closer muscle, while the ventral region of this claw muscle has fibres of the S1 phenotype, with no P75 band and a different combination of TnI isoforms. The sarcomere lengths of fibres of these two phenotypes also show a consistent difference: 2–3 μm for F fibres and 6–7 μm for S1 fibres. Abbreviations are listed in the legend to Figure 2.

groupings. In the leg muscles the extensor and flexor muscle is organized as two discrete pennate muscles which span from central apodemes. The fibre types present within the extensor muscle have been well characterized, being predominantly fast with a smaller proportion of slow type fibres (Houlihan and El Haj, 1985). Early studies demonstrated that with growth of the muscle the ratio of slow and fast muscle remained constant, which may reflect on the constant functional requirements for walking throughout post-larval, juvenile and adult life.

The claw muscles also comprise fibres of either the fast or slow (predominantly S1) phenotypes, according to the various myofibrillar protein isoforms that they express (Govind et al., 1987; Mykles, 1997). *Figure 3* shows the expression of myofibrillar protein isoforms for individual S1 and fast fibres taken from the mid and ventral region of a *H. gammarus* crusher claw. However, in contrast to the abdominal muscles, the claw muscles undergo considerable transformation during the post-larval stages (Govind and Lang, 1978; Ogonowski et al., 1980). This occurs in association with changes in the claws themselves, which are isomorphic in early post-larval lobsters, but over the early benthic and juvenile phases differentiate into the characteristic cutter and crusher claws of the adult lobster (Govind, 1992). The closer muscles of the isomorphic claws of a post-larval lobster have identical fibre arrangements, with fast fibres located in the central region of the closer muscle, and slow fibres located peripherally. However, during the dimorphic change, transformations of the closer muscle fibre types take place in both directions. Thus, in the presumptive cutter claw the fast-fibre region of the closer muscle expands in area, while in the presumptive crusher-claw this fast-fibre region contracts in area (*Figure 3a*). This process of transformation to the dimorphic adult pattern, with fast fibres comprising at least 80% of the muscle bulk in the cutter claw and the crusher claw composed almost entirely of slow fibres, can be complete as early as stage 13, but in some cases is not attained until the lobster is 1–2 years old.

Fibre type transformation is due to the switching of pre-existing, fully-differentiated fibres from one type to the other (slow to fast in the presumptive cutter and fast to slow in the presumptive crusher). This change in fibre phenotype is due to the changes in the expression of certain myofibrillar protein isoforms, rather than to the replacement of fibres of one phenotype by fibres of another, generated *de novo* by hyperplasia. This process of fibre transformation occurs along a distinct boundary between the regions of fast and slow fibres, and takes place over a relatively short period of the moult cycle, i.e. immediately post moult, when protein synthesis is elevated and haemolymph ecdysteroid concentrations are low. It is only at this time that fibres at the boundary between the fast and slow regions have been found to display intermediate myofibrillar ATPase activities, or to co-express the messenger RNAs for fast and slow myosin heavy chains (Mykles, 1997).

2.2 *The moult cycle*

Unlike vertebrates, such as fish, which grow continuously throughout the year, the indeterminate growth of lobsters is coupled to the moult cycle (El Haj, 1996; El Haj and Whiteley, 1997; Whiteley and El Haj, 1997). In brief, the moult cycle can be divided into a number of phases: premoult, when the animal is preparing to moult; moult or ecdysis, when the old exoskeleton is shed; postmoult, when the new external cuticle is hardening; and intermoult. Prior to moulting, the lobster absorbs minerals from the old exoskeleton, causing it to become soft, and by taking in water, blood volume is increased, swelling the new exoskeleton and causing the old one to split. Longitudinal growth in crustacean muscle fibres, which occurs by either the addition of new sarcomeres or by sarcomere lengthening, has been shown to occur in abdominal and leg muscles over ecdysis and the immediate postmoult period (El Haj and Houlihan, 1987; El Haj *et al.*, 1984; Houlihan and El Haj, 1985). Similarly, myofibrillar splitting, which indicates an increased number of fibres per cross sectional area, has been suggested to occur over the moult (El Haj *et al.*, 1984). In contrast, the claw muscle undergoes atrophy during the premoult to enable the muscles to pass through the narrow basi-ischum joints during ecdysis (Mykles and Skinner, 1982). Claw muscle atrophy is associated with elevated levels of protein synthesis pre- and post-moult (El Haj *et al.*, 1996; Mykles and Skinner, 1985a,b; Skinner, 1965), whereas synthesis rates are only slightly raised during late premoult in the muscles of the abdomen and legs (El Haj *et al.*, 1996). The complex patterns of cellular and physiological processes associated with the moult cycle (El Haj, 1996; El Haj *et al.*, 1992; El Haj and Whiteley, 1997) make it extremely difficult to establish how muscle growth and phenotype are modified in response to other variables such as environmental temperature. For example, elevated protein synthesis rates may simply reflect the early onset of moult and not an effect of temperature. Moreover, regulation of different sets of muscles occurs independently.

3. Temperature effects on lobster growth, muscle phenotype and protein expression

3.1 *Morphometric growth of lobsters*

In our study, the rate of growth, as measured by increases in carapace length, increased with temperature, so that at the end of the acclimation period the carapace length of lobsters raised at 19°C was 1.5 times greater than that of lobsters raised at 15°C, and 2.5 times greater than that of those raised at 11°C (*Figure 4*). The different rates of growth at varying temperatures were due primarily to different rates of moulting, so that over the acclimation period the lobsters raised at 19°C had reached stage 16 (the median value), while those raised at 15°C had reached stage 14, and those raised at 11°C had reached only stage 12 (*Figure 4*). It can also be seen that the lobsters raised at the highest temperature reached stage 12 in only 9 months, compared with 18 months for the lobsters at the lowest temperature. Plotting these growth data in relation to moult stage (*Figure 5*), rather than age, illustrates that the lobsters raised at the different temperatures attain relatively similar sizes at a given stage, which reflects the fact that the growth increment over a moult shows no systematic trend with temperature (*Figure 5*).

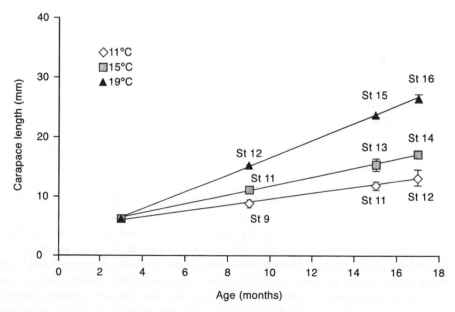

Figure 4. The rate of growth, measured as increasing carapace length, of groups of 250 post-larval Homarus gammarus *lobsters, reared from 3 months of age (stage 6) up to 18 months of age at each of three temperatures (11, 15 and 19°C). Data (mean ± SE) at the three further sampling times (9, 15 and 17 months of age) represent values for the lobsters in the median moult stage for each temperature group, as indicated.*

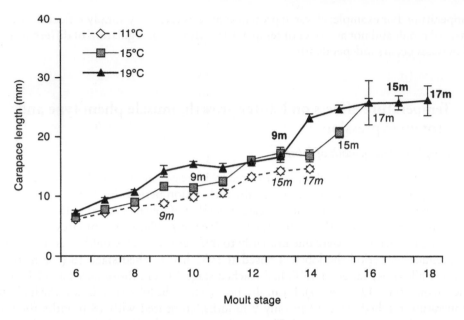

Figure 5. The rate of growth of post-larval Homarus gammarus *lobsters, reared from 3 months of age (stage 6) up to 18 months of age at each of three temperatures (11, 15 and 19°C) expressed in relation to moult stage. The ages of the lobsters at the last three sampling times are indicated for each temperature group.*

As growth is dependent on overall tissue turnover, which is a function of tissue synthesis and degradation, this would imply that temperature affects both rate processes in a similar fashion, resulting in no elevated growth at comparable stages.

3.2 *Protein synthesis rates*

Our previous work on intermoult animals has demonstrated that the effect of temperature on protein synthesis rates (k_s) is tissue- and species-dependent according to the normal thermal regime (Whiteley *et al.*, 1997). In general, k_s increases with temperature and this is accompanied by an elevation in RNA activity at a constant RNA:protein ratio (Whiteley and El Haj, 1997; Whiteley *et al.*, 1992, 1997). This occurs in crustaceans with different thermal tolerances, as demonstrated by the temperate eurythermal isopod *Idotea rescata* (2–24°C; Whiteley *et al.*, 1996), the stenothermal Antarctic isopod, *Glypotonotus antarcticus* (0–5°C; Whiteley *et al.*, 2001) and the tropical tiger prawn, *Penaeus esculentus* (Hewitt, 1992). Little is known, however, of the effects of rearing temperature on tissue degradation rates and, in many cases, growth measurements. In our study, we have demonstrated how moulting frequency and corresponding whole body protein synthesis rates increased with rearing temperature (Intanai and Taylor, personal communication), but that this was not associated with an increase in size for a given moult stage. This would support our hypothesis above that elevated temperatures not only result in elevated ribosomal activity and rate of protein synthesis in the muscles, but must also increase protein degradation rates, resulting in an unchanged overall protein turnover. Hence, although the developmental processes are accelerated, the growth at any stage point is constant between rearing temperatures.

3.3 Claw muscle transformation

It is well documented that the claws of lobsters grow allometrically in the juvenile and adult (Waddy et al., 1995). However, over the period of our study on Homarus gammarus (from 3 to 18 months of age) the claws grew isometrically to the rest of the body, and we found no evidence that elevated temperatures increased their relative size at a given moult stage. We also mapped the time course of dimorphic development in the claws of lobsters by applying myofibrillar ATPase histochemistry to the claw muscles of representative animals from the three temperature cohorts at regular intervals during development. We found that fibres of the closer muscle of the presumptive cutter claw transformed more rapidly with stage from the slow to fast phenotype at elevated temperatures, although ultimately the adult fully dimorphic condition was reached by animals at all temperatures. Thus, in the cutter claw at least, a higher rearing temperature can increase the rate of transformation to the fully dimorphic condition, over and above its effect on the rate of moulting.

3.4 Muscle gene expression over the moult cycle

Cloning of the major sarcomeric protein isoforms (actin (Harrison and El Haj, 1994), fast myosin (Cotton and Mykles, 1993), slow myosin (Mykles and Cotton, 1994) and fast and slow tropomyosin (Mykles et al., 1998)) has enabled the molecular regulation of the protein isoforms to be studied in relation to muscle growth, to the moult cycle and to the thermal environment. In a number of Crustacean species, the expression of the major sarcomeric proteins have been assessed over the moult cycle and shown to be moult stage-dependant and temperature-dependent (El Haj, 1996; El Haj and Whiteley, 1997; El Haj et al., 1992; El Haj et al., 1996; Harrison and El Haj, 1994; Whiteley and El Haj, 1997; Varadaraj et al., 1996). Current evidence suggests that the variation in protein gene expression may be regulated by different processes, with the sarcomeric protein actin being regulated by post-translatory processes (El Haj et al., 1992; Varadaraj et al., 1996), whereas the myosin isoforms are transcriptionally regulated (El Haj, 1999). Slot blot analysis of total RNA extracted from the muscles of the crusher claw of lobsters maintained at a constant temperature of 18°C hybridized with a cDNA probe for lobster slow myosin (Mykles and Cotton, 1994) demonstrates elevated levels of slow myosin gene expression in premoult lobsters (Figure 6).

3.5 Muscle gene expression following thermal acclimation

In this study, changes in muscle gene expression have been measured using total RNA extracted from the fast and slow muscles of European lobsters, acclimated to 9, 11 and 19°C as outlined above. In general, messenger RNA levels for actin, slow myosin and fast myosin are up-regulated with increased water temperature, but show similar trends over the moult at the three temperatures to patterns we have observed previously in adult lobsters (Figure 6). We have also begun to characterize the transitions in muscle sarcomeric protein isoforms combined with an analysis of the regulatory regions of the myosin heavy chain gene found within the abdominal muscles of developing Homarus gammarus from larvae through to adult. The stage of larval development where muscle gene expression is determined is not yet known or indeed which is the critical stage where temperature may influence muscle development and differentiation. Based on

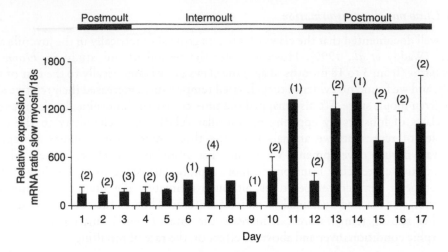

Figure 6. *Levels of slow myosin mRNA over the moult cycle in* Homarus gammarus. *Total RNA was extracted daily from the slow muscles of the crusher claw from different animals over a 17 day moult cycle. Total RNA was separated on a standard 1% agarose denaturing gel, transferred to a nylon membrane and hybridized with a cDNA probe for* Homarus americanus *slow myosin (Mykles and Cotton, 1994), labelled with [α]³²P and exposed to autoradiographic film at –70°C. Bars represent mean value ± SE,* n *values are indicated in parentheses.*

our present knowledge of the life cycle of the lobster (see Aiken and Waddy, 1980; Ennis, 1995 for reviews), it is likely that the developmental myosin isoforms, which are present in the free-swimming pelagic larvae, are switched off prior to or soon after settlement and that the myosin isoforms characteristic of the adult lobster are expressed from stage 5 through young juvenile and adult. Our results show by combined northern and sequence analysis that developmental isoforms are expressed from egg through to early larval free swimming stages. During the early larval stages 1–3, the adult sarcomeric isoforms for actin fast myosin and tropomyosin are switched on. Expression of the sarcomeric protein actin and tropomyosin is also observed in the developing embryo. Expression is up-regulated during post-larval development and with temperature.

3.6 *Sequence characterization of the myosin heavy chain gene – potential changes with temperature*

Studies on fish have shown that temperature can alter the contractile characteristics of muscle by determining the type of myosin crossbridge and the molecular motors that produce force. For example, in eurythermal fish, swimming speed and levels of myofibrillar ATPase activity are modulated by thermal acclimation (Johnston *et al.*, 1985), associated with expression of different isoforms of myosin (Imai *et al.*, 1997). Different molecular motors are coded for by different myosin heavy chain genes, with isoform expression, and hence muscle function determined by changes in temperature (Goldspink, 1995, 1998; Watabe *et al.*, 1998). Our present work has been focusing on the properties of two hypervariable regions in the globular head, surface loops 1 and 2, which are associated with the hydrolysis site for myofibrillar ATPase and the actin binding pocket, respectively. The structure of the loop 1 region has been proposed as the primary determinant of the rate of ADP release from myosin, which in turn controls the actin filament sliding rate (Sweeney et al., 1998). The rate of ADP release also appears to be a function of loop size, with larger, more flexible loops giving faster rates of release (Sweeney et al., 1998). Loop 2 interacts with the negatively charged amino-terminal part of actin in the highly bound state of the cross-bridge cycle. Since actin binding is necessary for the activation of myosin ATPase and for ADP release (Spudich, 1994) the structural properties of loops 1 and 2 modulate motor function, determining speed of contraction (Spudich, 1994; Uyeda et al., 1994) and how economical a muscle is for developing and maintaining force (Gauvry et al., 1997). Sequence variation in loops 1 and 2 has been correlated with muscle phenotype and to environmental temperature in fish (Imai et al., 1997). In our studies, we have designed two sets of degenerate primers, which amplify specific regions of the crustacean myosin genes of 216 and 663 bp in length, spanning loops 1 and 2, respectively. Our initial studies have investigated changes in myosin sequence structure in the claw muscles of adult American lobster acclimated to an extremely cold environment, close to the limits of their thermal tolerance. Figure 7 shows the loop 1 sequence from Homarus americanus held at two temperatures, 0 and 13°C. A marked shift in sequence structure was identified by comparing sequences for the hypervariable region with identified S2 slow type fibres. This may not reflect a switch to a low temperature-specific isoform, but a shift away from a slow fibre type to a fast type phenotype following acclimation for a period of 6 weeks. We are now investigating variation in the loop 1 and 2 regions from the fast and slow fibre types from the European lobster (Homarus gammarus) held at different rearing temperatures in order to determine whether temperature-dependent transitions occur during development.

4. Conclusion

In conclusion, our results suggest that rearing lobsters at higher temperatures in order to maximize growth for aquaculture simply accelerates the pre-programmed series of events, which are associated with the moult cycle and would take place naturally at ambient water temperature. Intermoult period is reduced, moulting is more frequent and hence growth is more rapid. In addition to this our results indicate that temperature does not appear to increase animal size for any given moult stage. If anything, moult increment is smaller in animals raised at higher temperatures. At the molecular level, temperature may, however, have an effect on muscle plasticity during development and also fibre type transitions in adults. These changes may be measured at the level of protein isoform expression or by changes directly on the sarcomeric genes at key regions of the gene which influence function, such as the ATPase binding region or

Muscle	Temp (°C)													
Slow S2 fibres	10°C	S	T	K	K	A	G	E	D	T	K	P	N	L	E	D
Crusher claw	0°C	S	T	K	K	V	A	E	D	T	K	P	N	L	E	D
Crusher claw	13°C	T	T	K	K	R	G	E	E	T	K	Q	N	L	E	D

Figure 7. Deduced amino-acid sequences for myosin heavy chain loop 1 from slow type 2 (S2) fibres from the European lobster, Homarus gammarus held at 10°C and the slow muscles of the crusher claws from the American lobster, Homarus americanus held at 0 and 13°C.

the actin binding region. The changes in muscle phenotype will be reflected in muscle performance and hence behaviour, which could potentially have implications for the success of lobster ranching programmes where animals are released into the wild as young juveniles.

References

Aiken, D.E. and Waddy, S.L. (1980) Reproductive biology. In: *The Biology and Management of Lobsters* (eds J.S. Cobb and B.F. Phillips), Vol. 1. Academic Press, New York, pp. 275–276.

Aiken, D.E. and Waddy, S.L. (1986) Environmental influences on recruitment of the American lobster, (*Homarus americanus*): A perspective. *Can. J. Fish. Aquat. Sci.* **43**: 2258–2270.

Aiken, D.E. and Waddy, S.L. (1995) Aquaculture. In: *Biology of the Lobster* Homarus americanus (ed J.R. Factor). Academic Press, New York, pp. 153–176.

Bannister, R.C.A. and Howard, A.E. (1991) A large-scale experiment to enhance a stock of lobster (*Homarus gammarus* L.) on the English east coast. *ICES Mar. Sci. Symp.* **192**: 99–107.

Cole, J.L. and Lang, F. (1980) Spontaneous and evoked postsynaptic potentials in an embryonic neuromuscular system of the lobster *Homarus americanus*. *J. Neurobiol.* **11**: 459–470.

Cotton, J.L. and Mykles, D.L. (1993) Cloning of a crustacean myosin heavy chain isoform: exclusive expression in fast muscle. *J. Exp. Zool.* **267**: 578–586.

El Haj, A.J. (1996) Crustacean genes involved in growth. In: *Gene Regulation in Aquatic Organisms* (eds S. Ennion and G. Goldspink), Society for Experimental Biology Seminar Series no. 46. Cambridge University Press, Cambridge, pp. 94–112.

El Haj, A.J. (1999) Regulation of muscle growth and sarcomeric protein gene expression over the intermoult cycle. *Am. Zool.* **39**(3): 570–579.

El Haj, A.J. and Houlihan, D.F. (1987) In vitro and in vivo protein synthesis rates in a crustacean muscle during the moult cycle. *J. Exp. Biol.* **127**: 413–426.

El Haj, A. J. and Whiteley, N. M. (1997) Molecular regulation of muscle growth in crustacea. *J. Mar. Biol. Assoc.* **77**: 1–15.

El Haj, A.J., Govind, C.K. and Houlihan, D.F. (1984) Growth of lobster leg muscle fibres over intermoult and moult. *J. Crust. Biol.* **4**(4): 536–545.

El Haj, A.J., Whiteley, N.M. and Harrison, P. (1992) Molecular regulation of muscle growth over the crustacean moult cycle. In: *Molecular Biology of Muscle* (ed A.J. El Haj), Society for Experimental Biology Seminar Series no. 58. Cambridge University Press, Cambridge, pp. 151–165.

El Haj, A.J., Clarke, S.R., Harrison, P. and Chang, E.S. (1996) In vivo muscle protein synthesis rates in the American lobster *Homarus americanus* during the moult cycle and in response to 20-hydroxyexdysone. *J. Exp. Biol.* **199**: 579–585.

Ennis, G.P. (1995) Larval and postlarval ecology. In: *Biology of the Lobster* Homarus americanus (ed J.R. Factor). Academic Press, New York, pp. 23–46.

Fowler, W.S. and Neil, D.M. (1992) Histochemical heterogeneity of fibres in the abdominal superficial flexor muscles of the Norway lobster, *Nephrops norvegicus* (L.). *J. Exp. Zool.* **264**: 406–418.

Galler, S. and Neil, D.M. (1994) Unloaded shortening of skinned mammalian skeletal muscle fibres: effects of the experimental approach and passive force. *J. Muscle Res. Cell Motil.* **15**: 400–412.

Gauvry, L., Mohan-Ram, V., Ettelaie, C., Ennion, S. and Goldspink, G. (1997) Molecular motors designed for different tasks and to operate at different temperatures. *J. Therm. Biol.* **6**: 367–373.

Goldspink, G. (1995) Adaptation of fish to different environmental temperature by qualitative and quantitative changes in gene expression. *J. Therm. Biol.* **1–2**: 167–174.

Goldspink, G. (1998) Selective gene expression during adaptation of muscle in response to different physiological demands. *Comp. Biochem. Physiol.* **120**: 5–15.

Govind C.K. (1992) Claw asymmetry in lobsters: case study in developmental neuroethology. *J. Neurobiol.* **23**: 1423–1445.

Govind, C.K. and Lang, F. (1978) Development of the dimorphic claw closer muscles of the lobster Homarus americanus. III. Transformation to dimorphic muscles in juveniles. *Biol. Bull.* **154**: 55–67.

Govind, C.K., Mellon, De F. and Quigley, M.M. (1987) Muscle and muscle fiber type transformation in clawed crustaceans. *Am. Zool.* **27**: 1079–1098.

Harrison, P. and El Haj, A.J. (1994) Actin mRNA levels and myofibrillar growth in leg muscles of the European lobster (*Homarus gammarus*) in response to passive stretch. *Mol. Mar. Biol. Biotechnol.* **3**: 35–41.

Hewitt, D. R. (1992) Response of protein turnover in the brown tiger prawn Penaeus esculentus to variation in dietary protein content. *Comp. Biochem. Physiol.* **103A**: 183–187.

Holmes, J.M., Hilber, K., Galler, S. and Neil, D.M. (1999) Shortening properties of two biochemically defined fibre types from the Norway lobster Nephrops norvegicus L. *J. Muscle Res. Cell Motil.* **20**: 265–278.

Houlihan, P.F. and El Haj., A.J. (1985) An analysis of muscle growth. In: *Factors in Adult Growth* (ed A.M. Wenner). A.A. Blaskema Press, Amsterdam.

Imai, J., Hirayama, Y., Kikuchi, K., Kakinuma, M. and Watabe, S. (1997) cDNA cloning of myosin heavy chain isoforms from carp fast skeletal muscle and their gene expression associated with temperature acclimation. *J. Exp. Biol.* **200**: 27–34.

Johnston, I.A., Sidell, B.D. and Driedzic, W.R. (1985) Force-velocity characteristics and metabolism of carp muscle fibres following temperature acclimation. *J. Exp. Biol.* **119**: 239–249.

Kennedy, D. and Takeda, K. (1965) Reflex control of abdominal flexor muscle in the crayfish. I. The phasic system. *J. Exp. Biol.* **43**: 211–227.

Li, Y. and Mykles, D.L. (1990) Analysis of myosins from lobster muscles: fast and slow isozymes differ in heavy-chain composition. *J. Exp. Zool.* **255**: 163–170.

MacKenzie, B.R. (1988) Assessment of temperature effects on interrelationships between stage durations mortality, and growth in laboratory-reared *Homarus americanus* Milne Edwards larvae. *J. Exp. Biol. Ecol.* **116**: 87–98.

Mykles, D.L. (1985a) Heterogeneity of myofibrillar proteins in lobster fast and slow muscles: variants of troponin, paramyosin and myosin light chains comprise four distinct protein assemblages. *J. Exp. Zool.* **234**: 23–32.

Mykles, D.L. (1985b) Multiple variants of myofibrillar proteins in single fibres of lobster claw muscles. Evidence for two types of fibres in the cutter claw. *Biol. Bull.* **169**: 476–483.

Mykles, D.L. (1988) Histochemical and biochemical characterisation of two slow fibre subtypes in decapod crustacean muscle. *J. Exp. Zool.* **245**: 232–243.

Mykles, D.L. (1997) Crustacean muscle plasticity: Molecular mechanisms determining mass and contractile properties. *Comp. Biochem. Physiol. B* **117**: 367–378.

Mykles, D.L. and Cotton, J.L.S. (1994) Isolation of cDNAs encoding fast and slow isoforms of lobster myosin heavy chain. *J. Cell Biochem.* **18D** (Suppl.): 520.

Mykles, D.L. and Skinner, D.M. (1982) Crustacean muscle: atrophy and regeneration during moulting. In: *Basic Biology of Muscles: A Comparative Approach* (eds B.M. Twarog, R.J.C. Levine and M.M. Dewey). Raven Press, New York, pp. 37–357.

Mykles, D.L. and Skinner, D.M. (1985a) Muscle atrophy and restoration during moulting. In: *Crustacean Growth* (ed A. Wenner). A.A. Balkema, Rotterdam, pp. 31–46.

Mykles, D.L. and Skinner, D.M. (1985b) The role of calcium dependent proteinases in moult induced claw muscle atrophy. In: *Intracellular Protein Catabolism* (eds E. Khairallah, J.S. Bow, and J.W. Bird). Alan R. Liss, New York, pp. 141–150.

Mykles, D.L., Cotton J.L.S., Tangiguchi, H., Sano, K. and Maeda, Y. (1998). Cloning of tropomyosins from lobster (*Homarus americanus*) striated muscles: fast and slow isoforms may be generated from the same transcript. *J. Muscle Res. Cell Motil.* **19**: 105–115.

Neil, D.M., Fowler, W.S. and Tobasnick, G. (1993) Myofibrillar protein composition correlates with histochemistry in fibres of the abdominal flexor muscles of the Norway lobster *Nephrops norvegicus. J. Exp. Biol.* 183: 185–201.

Ogonowski, M.M., Lang, F. and Govind, C.K. (1980) Histochemistry of lobster claw-muscles during development. *J. Exp. Zool.* 213: 359–367.

Pilgrim, R.L.C. and Wiersma, C.A.G. (1963) Observations on the skeletal and somatic muscu-lature of the abdomen and thorax of *Procambarus clarkii* (Girard), with notes on the thorax of *Pandulirus interruptus* (Randall) and *Astacus. J. Morph.* 113: 453–487.

Skinner, D.M. (1965) Amino acid incorporation into protein during the moult cycle of the land crab, *Gecarcinus lateralis. J. Exp. Zool.* 161: 222–234.

Spudich, J.A. (1994) How molecular motors work. *Nature* 372: 515–518.

Sweeney, H.L, Rosenfeld, S.S., Brown, F., Faust, L., Smith J., Xing, J., Stein, L.A. and Sellers, J.R. (1998) Kinetic tuning of myosin via a flexible loop adjacent to the nucleotide binding pocket. *J. Biol. Chem.* 273: 6262–6270.

Uyeda, T.Q.P., Ruppel, K.M. and Spudich, J.A. (1994) Enzymatic activities correlate with chimaeric substitutions at the actin-binding face of myosin. *Nature* 368: 567–569.

Varadaraj, K., Kumari, S.S. and Skinner, D.M. (1996) Actin-encoding cDNAs and gene expression during the intermolt cycle of the Bermuda land crab *Gecarcinus lateralis. Gene* 171: 177–184.

Waddy, S.L., Aiken, D.E. and de Kleijn, D.P.V. (1995) Control of Growth and Reproduction. In: *Biology of the Lobster* Homarus americanus (ed J.R. Factor). Academic Press, New York, pp. 217–266.

Wahle, R.A. and Steneck, R.S. (1991) Recruitment habitats and nursery grounds of the American lobster, *Homarus americanus*: a demographic bottleneck. *Mar. Ecol. Prog. Ser.* 69: 231–243.

Watabe, S., Hirayama, Y., Nakaya, M., Kakinuma, M., Kikuchi, K., Guo, X-F., Kanoh, S., Chaen, S. and Ooi, T. (1998) Carp expresses fast skeletal myosin isoforms with altered motor functions and structural stabilities to compensate for changes in environmental temperature. *J. Therm. Biol.* 22: 375–390.

Whiteley, N.M. and El Haj, A.J. (1997) Regulation of muscle gene expression over the moult cycle. *Comp. Biochem. Physiol.* 117B(2): 30–32.

Whiteley, N.M., Taylor, E.W. and El Haj, A.J. (1992) Actin gene expression during muscle growth in *Carcinus maenas. J. Exp. Biol.* 167: 277–284.

Whiteley, N.M., Taylor, E.W. and El Haj, A.J. (1996) A comparison of the metabolic cost of protein synthesis in stenothermal and eurythermal isopod crustaceans. *Am. J. Physiol. Integr. Comp. Physiol.* 40(5): R1295–R1303.

Whiteley, N.M., Taylor, E.W. and El Haj, A.J. (1997) Seasonal and latitudinal adaptation to temperature in crustaceans. *J. Therm. Biol.* 22: 419–427.

Whiteley, N.M., Robertson, R.F., Meagor, J., El Haj, A.J. and Taylor, E.W. (2001) Protein synthesis and specific dynamic action in crustaceans: effects of temperature. *Comp. Biochem. Physiol.* 128(3): 595–606.

5

Impact of temperature on the growth and differentiation of muscle in herring larvae

Ian A. Johnston, Genevieve K. Temple and Vera L.A. Vieira

1. Introduction

For several centuries the Atlantic herring (*Clupea harengus* L.) has been one of the most important commercially exploited fish species in Northern Europe. Distinct populations of herring can be identified from the Baltic, Iceland, Irish Sea, North Sea and Norwegian Sea (Parrish and Saville, 1967). Sexually mature fish show a high fidelity to their spawning grounds (Wheeler and Winters, 1984), although there is some evidence for migration between stocks (Anokhina, 1971). Herring from different stocks vary with respect to their size at maturity and numerous morphological characters including otolith fine structure and numbers of vertebrae, fin rays, keeled scales and gill rakers (Parrish and Saville, 1967; Rosenberg and Palmén, 1982). Genetic studies based on allozyme frequencies, mitochondrial DNA and microsatellites, however, indicate considerable homogeneity of populations over large geographical distances (Ryman *et al.*, 1984; Turan *et al.*, 1998). Exceptions are some isolated deep water resident populations in Norwegian fjords, which can be genetically differentiated from the coastal stocks (Jørstad *et al.*, 1994; Turan *et al.*, 1998). Herring populations spawn in almost every month at sea temperatures ranging from 3°C to 16°C (Parrish and Saville, 1967). In this chapter we discuss the extent to which temperature during early development can influence morphology, physiology and behaviour, giving rise to phenotypic differences both within and between populations.

2. Larval morphology

Herring is a shoaling species that deposits eggs on the substrate in relatively shallow water to form dense mats several inches deep. Although the surrounding water is likely to be well saturated with oxygen, diffusion gradients within the egg mats may well

---

Environment and Animal Development: Genes, Life Histories and Plasticity, edited by D. Atkinson and M. Thorndyke.
© 2001 BIOS Scientific Publishers Ltd, Oxford.

affect development. Hatching success in artificially fertilized Pacific herring (*Clupea pallasii*), a beach spawner, decreased sharply with egg mass thickness (Taylor, 1971). Stratoudakis *et al.* (1998) found that embryos from the bottom layer of egg mats sampled from the Ballantrae Bank, Firth of Clyde, had a lower developmental age than those at the top. They suggested that variability in larval length at hatch might be related to the position of the embryos in multi-layered mats of eggs. In laboratory rearing experiments, a single layer of eggs is usually attached to glass plates to correspond to the well-oxygenated conditions found on the surface of egg mats in the wild.

Herring larvae spend up to 6 months in the plankton depending on the time of spawning (Doyle, 1977). At hatch, the transparent yolk-sac larvae (*Figure 1a*) are around 7–10 mm in length depending on population origin and rearing temperature (Hempel and Blaxter, 1961). Changes in body morphology in laboratory reared Clyde herring larvae at 8°C are illustrated in *Figure 1a–f*. The dorsal fin begins to differentiate from the dorsal primordial fin fold at 12–13 mm total length (TL; *Figure 1b*). There was a progressive increase in the number and length of dorsal fin rays and associated erector and depressor muscles as body length increased (*Figure 1b–e*). Flexion of the notochord and caudal fin rays was apparent in larvae at 16.5–18 mm TL (*Figure 1d*). The anal fin developed from the ventral primordial fin at 16 mm TL (*Figure 1c*), and the final number of fin rays and associated muscles were present by 22 mm TL (*Figure 1e*). Pelvic fin rudiments were apparent in a proportion of larvae at 20–22 mm TL, and there was a progressive reduction in the size of the primordial fin (Johnston *et al.*, 1998). Respiration is entirely cutaneous during the first part of planktonic life until around 25 mm TL (Batty, 1984). Metamorphosis to the juvenile stages takes place between around 22 (*Figure 1e*) and 35 mm TL (*Figure 1f*). The

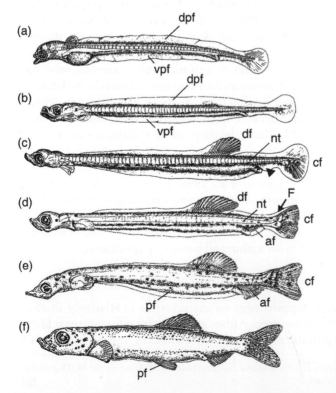

Figure 1. *Drawings to show the morphology of Clyde herring* (Clupea harengus) *larvae from hatch to completion of metamorphosis. Fish were reared in the laboratory at 8°C. (a) Yolk-sac larvae; (b) 12 mm TL larva; (c) 16 mm TL larva (arrowhead shows the site of development of the anal fin); (d) 18 mm TL larva; (e) 22 mm TL and (f) 37 mm TL fish (metamorphosis to the juvenile stage was judged to be largely complete at this stage. vpf, ventral primordial fin-fold; dpf, dorsal primordial fin-fold; df, dorsal fin; nt, notochord; F, notochord flexion; af, anal fin; cf, caudal fin; pv, pelvic fin. Reproduced from Johnston* et al. *(1998). J. Exp. Biol. 201: 623–646, with permission of the Company of Biologists Ltd.*

juveniles possess haemoglobin, well-developed gill filaments, scales and show a characteristic silvering of the body (DeSilva, 1974).

3. Swimming behaviour and muscle fibre types

3.1 Juvenile and adult stages

The muscle fibre types of adult teleosts can be characterized with respect to their colour, innervation patterns, contraction speed and metabolic characteristics. The relative proportion of each fibre type varies along the trunk and between species in relation to their swimming style and behaviour (see Bone, 1978; Johnston and Altringhham, 1991; Altringham and Ellerby, 1999 for reviews). The herrings (Clupea sp.) represent a relatively simple condition in that only two muscle fibre types, red and white, can be identified (see Figure 1b in Crockford and Johnston, 1993). Red muscle fibres form a triangular strip of muscle at the level of the major horizontal septum. White fibres constitute the majority of the myotomal muscle. Red and white muscles in herring are composed of multi-terminally innervated slow fibres and focally innervated fast fibres, respectively (Bone, 1964). Aerobic and anaerobic metabolic pathways provide the main energy supply for contraction in adult slow and fast muscle fibres, respectively (Bone, 1978; Johnston, 1981). In Pacific herring (Clupea harengus pallasi) of 15–17.5 cm fork length (L) electromyograms (EMGS) were recorded from the red muscle layer alone at sustained swimming speeds of up to 4–5 L s^{-1} (Bone et al., 1978). Fish of this size were able to maintain these speeds for at least 5 h. At higher speeds, steady swimming was interspersed with larger amplitude tail-beats and EMGs were also recorded from the white muscle. Above 5 L s^{-1} endurance was limited to less than 2 min, reflecting the transition from aerobic to anaerobic energy sources (Bone et al., 1978).

3.2 Larval stages

The myotomes of yolk-sac herring larvae contain two muscle fibre types in common with the adults (Batty, 1984; Vieira and Johnston, 1992; Johnston, 1993). The general arrangement of the muscle fibre types in the yolk-sac stage of a spring-spawning Clyde herring larva is illustrated in Figure 2a. The lateral surfaces of the myotomes are covered with a single superficial layer of small diameter fibres, which stain intensely for mitochondrial enzymes (Batty, 1984). Each superficial muscle fibre contains two to four myofibrils and a dense aggregation of mitochondria towards the fibre surface facing the skin (Figure 2b). The more numerous inner muscle fibres have larger diameters and contain a higher proportion of myofibrils (Figure 2b).

The muscle fibres in yolk-sac larvae correspond to embryonic types, which are gradually transformed to the adult fibre types. The inner muscle fibres of larvae are much more aerobic in character than adult fast muscle. Thus, up to 26% of the volume of inner muscle fibres in larvae is comprised of mitochondria, compared to only 1–3% in adult fast muscle fibres (Vieira and Johnston, 1992; Johnston, 1993). There is evidence that the swimming activity of early fish larvae is largely fuelled by aerobic metabolism (Wieser, 1995). Studies with cyprinid species have shown that the activities of glycolytic enzymes are initially very low and the aerobic H$_4$-form of lactate dehydrogenase (LDH) is the predominant isoenzyme (El-Fiky et al., 1987; Hinterleitner et al., 1987). In the roach Rutilus rutilus the anaerobic M$_4$ isoenzyme of LDH only becomes

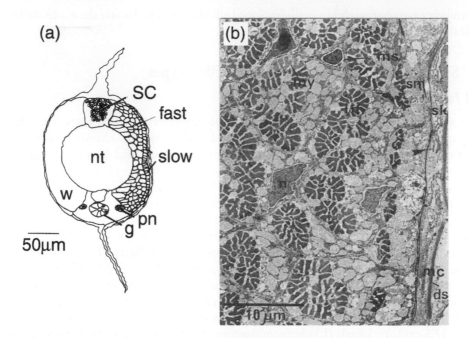

Figure 2. *Muscle fibre types in yolk-sac larva of Clyde herring* (Clupea harengus L.). *(a) Camera lucida drawing of the muscle fibres at a point immediately posterior to the yolk-sac. g, gut; pn, pronephros; nt, notochord; sc, spinal chord. (b) Electron micrograph of myotomal muscle from a 1 day-old herring larva reared at 12°C. ds, dermal scale; im, fast muscle fibre; mc, mucocyte; ms, persistent myoblast; mt, mitochondria; my, myofibril; n, nucleus; sk, skin; sm, slow muscle fibre.*

the major form 2 weeks after hatching (El-Fiky *et al.*, 1987). Consistent with these findings, yolk-sac herring larvae are able to make repeated fast starts and are hard to fatigue. The lactic acid that does accumulate is cleared from the body an order of magnitude faster than in adult fish at the same temperature, presumably through oxidative metabolism (Franklin *et al.*, 1996).

The main components of the myofibril are the contractile proteins actin and myosin and the Ca^{2+} regulatory proteins of the tropomyosin–troponin complex. All myofibrillar proteins except actin exist as multiple isoforms, each with a slightly different structure and distinct functional properties (Schiaffino and Reggiani, 1996). Different isoforms arise both from different gene families and from alternate splicing of the same gene and other post-translational modifications. The composition of myosin heavy and light chains determines many aspects of contractile function including shortening speed.

It is possible to isolate pure samples of larval fast muscle for the analysis of myofibrillar proteins using electrophoretic techniques (Crockford and Johnston, 1993; Johnston *et al.*, 1997, 1998). Peptide mapping studies have indicated distinct isoforms of the myosin heavy chains in embryonic fast muscle, adult fast muscle and adult slow muscle (Crockford and Johnston, 1993). However, myosin alkali light chain 1 (MLC1), MLC3 and tropomyosin from larval inner and adult fast muscle had the same relative mobility on two-dimensional gels (Crockford and Johnston, 1993). There was evidence for an embryonic isoform of myosin light chain 2 (MLC2) in larval fast muscle, which was also present as a minor component in adult fast muscle fibres (Crockford and

Johnston, 1993). Adult slow and fast, and larval inner muscle also contained unique isoforms of troponin T and troponin I (Crockford and Johnston, 1993).

In the fast muscle of larvae there was a progressive replacement of embryonic with larval and adult isoforms of the myofibrillar proteins (Crockford and Johnston, 1993; Johnston et al., 1997). The length at which different components of the myofibril assumed their adult characteristics was not the same for all the myofibrillar proteins. For example, the transition from the embryonic to the adult pattern of myosin heavy chain isoform expression in fast muscle occurred between 20 and 25 mm TL (Johnston et al., 1998), whereas the adult MLC2 composition was attained at 11–15 mm TL depending on the temperature (Johnston et al., 1997). The existence of developmental-stage-specific isoforms enables the contractile properties of the muscle to be adjusted as the fish increases in length.

Scaling considerations would dictate that the superficial muscle fibres of larvae should have a relatively fast contraction speed compared to adult slow muscle. A monoclonal antibody to herring fast muscle MLC3 strongly stained the superficial muscle layer in yolk-sac larvae (Johnston and Horne, 1994). Expression of MLC3 was progressively switched off, and slow muscle fibres that were negative for MLC3 and presumably expressed slow myosin light chain isoforms were added externally to the larval superficial muscle layer in the region of the major horizontal septum starting at 22–25 mm TL (Johnston et al., 1998). In larvae of 31–33 mm TL the slow muscle at the horizontal septum consisted of a layer five or six fibres thick, increasing to 12–15 fibres thick towards the end of metamorphosis (Johnston and Horne, 1994).

Both the superficial and inner muscle fibres of Atlantic herring are focally inner-vated at their myoseptal ends (Johnston et al., 1997). The endplates of a sub-set of the superficial muscle fibres present at the level of the major horizontal septum have a distinct morphology, and correspond to the muscle pioneer cells. Muscle pioneers are the first fibres to form and become innervated in the embryo. The multi-terminal pattern of slow muscle innervation has been found to develop progressively between 12 and 22 mm TL (Johnston et al., 1997).

Ontogenetic changes in muscle characteristics, fins, body shape and length have a profound impact on the swimming behaviour of larval stages (Batty, 1984; Bailey and Batty, 1984). In early post-yolk-sac larvae (11 mm TL) the amplitude of swimming movements increases linearly towards the tail, and the wave speed is constant along the body (Batty, 1984). This so-called anguilliform style of swimming results in a constant lateral acceleration of water along the body, indicating resistive forces are important for generating thrust. In 22 mm TL larvae, the first third of the trunk is almost rigid and there is a non-linear increase in body amplitude towards the tail (Batty, 1984). This so-called subcarangiform style of swimming involves a more rapid acceleration of water by the fish body towards the tail, indicating the predominance of reactive (inertial) forces for generating thrust. The transition between the anguilliform and subcarangiform styles between 15 and 20 mm coincides with the appearance of caudal and dorsal fins and the development of the adult patterns of slow muscle innervation and the adult slow and fast muscle myofibrillar protein composition.

4. Influence of rearing temperature on muscle phenotype

The larval stages of herring experience either increasing (spring spawners) or decreasing (autumn spawners) temperatures during ontogeny. Laboratory studies have shown that most aspects of the ontogenetic changes in body morphology and muscle characteristics in herring larvae vary with rearing temperature. There is strong evidence that temperature produces differential effects on growth and differentiation such that the body length at which particular characters first appear is dependent on temperature. This gives rise to enormous phenotypic variation, some of which impacts on the behaviour and survival of the larval stages.

We have examined ontogenetic changes in morphology and muscle properties in Clyde herring reared at constant temperatures of 5, 8 and 12°C until first feeding, and subsequently at parallel rising temperatures to simulate a seasonal warming (*Figure 3a*). Clyde herring spawn at a depth of 15–25 m between mid-February and early April. The temperature range studied encompasses the extremes of sea temperatures recorded during these months over the last 45 years (*Figure 3b, c*). For simplicity in this chapter we will refer to these different temperature regimes by the starting temperature, that is 5, 8 or 12°C.

Meristic characteristics, including the average number of vertebrae and fin ray counts, are well known to vary with the temperature and salinity experienced during early development (Tåning, 1952). Values for vertebral number in populations of several species of marine fish show good correlations with average sea surface temperatures during the period of egg development (see Hempel and Blaxter, 1961; Brander, 1979). Egg incubation temperature also influences the number and diameter of embryonic muscle fibres in a range of fish species (Stickland *et al.*, 1988; Brooks and Johnston, 1993; Galloway *et al.*, 1998). Vieira and Johnston (1992) reported that the number of embryonic fast muscle fibres in 1-day-old Clyde herring was a function of rearing temperature, such that posterior to the yolk sac there were 311 ± 41 at 15°C, 257 ± 22 at 10°C and 187 ± 22 at 5°C (mean ± SD, six larvae per temperature). However, the average diameter of the inner muscle fibres increased with decreasing temperature so that the total cross-sectional area of muscle was similar at each temperature. Conversely, in subsequent studies variable effects on the number and size distribution of the fast muscle fibres have been observed at hatch, perhaps reflecting egg quality and/or familial differences (Johnston, 1993; Johnston *et al.*, 1998). At hatch, the number of presumptive slow muscle fibres was found to increase on average from around 108 per cross-section at 5°C to 140 per cross-section at 12°C (Johnston, 1993). Studies with the zebrafish (*Rerio danio*) have shown that Hedgehog proteins have an important role in the development of the embryonic slow but not the fast muscle fibres (Devoto *et al.*, 1996; Blagden *et al.*, 1997; Currie and Ingram, 1998). Mutations in the slow-muscle-omitted (*smu*) gene interfere with Hedgehog signalling and *smu* $^{-/-}$ embryos have a 99% reduction in the number of slow muscle fibres and a complete absence of the Engrailed expressing muscle pioneer cells (Barresi *et al.*, 2000). Variations in Hedgehog gene expression with temperature may provide an explanation for differences in the number of slow muscle fibres.

Interspersed between the muscle fibres at hatch are mononuclear cells containing a large central nucleus and relatively little cytoplasm (*Figure 2b*). Immunocytochemical studies with Atlantic salmon (*Salmo salar*) fry showed that around 80% of the mononuclear cells stained for a specific marker of myogenic precursor cells, the c-met tyrosine kinase receptor (Johnston *et al.*, 2000a). It is therefore likely that the majority of the mononuclear cells identified by electron microscopy in herring (cf. Johnston, 1993; see *Figure 4d*) are persistent myoblasts, with the remainder constituting other cell types such as fibroblasts and capillary endothelial stem cells. Rearing temperature

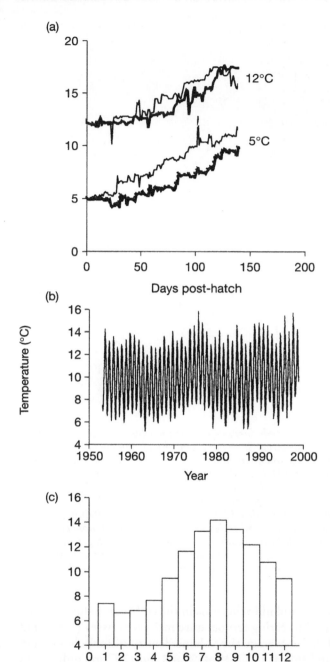

Figure 3. *(a) The thermal regimes used for rearing Clyde herring in 1994 and 1996. Embryos were reared at constant temperatures of 5, 8 and 12°C until hatching, after which the temperature was allowed to rise at a similar rate (only the 5 and 12°C regimes are illustrated). Reproduced from Johnston* et al. *(1998).* J. Exp. Biol. *201: 623–646, with permission of the Company of Biologists Ltd. (b) Average variation in sea surface temperatures in the Clyde between 1953 and 1998 and (c) mean monthly sea surface temperatures for the Clyde over the same time series as in (b). Reproduced from Turrel (1999).* Fisheries Research Services Report no. 9/99. Crown Copyright 1999. *Reproduced with permission of FRS Marine Laboratory, Aberdeen.*

also influenced the density of these mononuclear cells. At hatch, the number of myoblasts per mm² cross-sectional area of muscle in Clyde herring larvae was more than two-fold higher at 8°C than at either 5 or 12°C (*Figure 4b*). This is an interesting result because these myogenic precursor cells are the source of all the nuclei required for growth during the larval and juvenile stages. The ability of the embryonic temperature regime to fix components of the larval phenotype is explored further in Section 6.

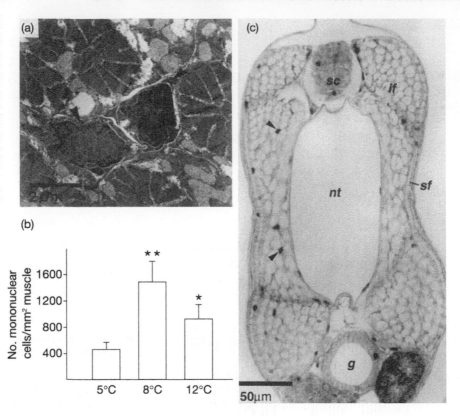

Figure 4. (*a*) *Transmission electron micrograph showing persistent myoblasts (arrowheads) in the fast muscle of a 1-day-old Clyde herring larva. Scale bar 2 μm. (b) The number of myoblasts per trunk cross-section determined from electron micrographs in 1-day-old Clyde herring larvae reared at 5, 8 and 12°C. Data are means ± SD, five fish per temperature. *, ** Significantly different from values at 5°C at the 5 and 1% levels, respectively. Reproduced from Johnston (1993).* Marine Biology ***116****: 363–379, with permission of Springer-Verlag GmbH. (c) Immunoperoxidase staining of a 1-day-old larva after 3 h labelling in seawater containing 5-bromo-2'-deoxyuridine (BrdU). Transverse wax-embedded section stained with a monoclonal antibody to BrdU. Labelled nuclei are black (arrowheads). The scale bar is 50 μm. g, gut; if, fast muscle; sf, slow muscle layer; nt, notochord; sc, spinal cord. Reproduced from Johnston* et al. *(1995).* J. Exp. Biol. ***198****: 1389–1403, with permission of the Company of Biologists Ltd.*

The volume density of myofibrils and mitochondria in the embryonic slow and fast muscle types was found to vary with rearing temperature in 2 year classes (Vieira and Johnston, 1992; Johnston, 1993). On average the volume density of mitochondria in the slow muscle fibres was around 46% at 15°C, 39–41% at 8 and 12°C and only 35–38 at 5°C, perhaps reflecting the activity of the larvae. This contrasts with the normal temperature acclimation response observed in the adult stages of many fish species where several weeks of cold acclimation lead to an increase in mitochondrial density which is thought to represent an attempt to compensate for the adverse effects of low temperature on rates of oxidative phosphorylation and diffusion (Johnston and Maitland, 1980; Egginton and Sidell, 1989).

The body length at which the adult pattern of myofibrillar protein isoforms was established in Clyde herring was found to decrease as rearing temperature increased

(Crockford and Johnston, 1993; Johnston et al., 1997). For example, adult patterns of troponin I and troponin T composition were established in 11 mm TL larvae at 12°C but not until larvae reached 17 mm TL at 5°C. Thus the fast muscle of 11 mm larvae expressed only the adult isoforms at 12°C, but a mixture of embryonic/early larval/and adult isoforms at lower temperatures (Johnston et al., 1997).

The development of muscle innervation was investigated by staining larvae for acetyl cholinesterase (AchE) activity and with an antibody to α-acetylated tubulin to visualize neural processes (Johnston et al., 1997). The development of the multi-terminal pattern of innervation to the slow myotomal muscle was found to take place at longer body lengths as rearing temperature decreased. Thus slow muscle fibres in nine out of 13 larvae of 16–17 mm TL reared at 5°C still only had the focal pattern of innervation 50 days after hatch. The four individuals with some multiply innervated slow fibres had very few endplates (Johnston et al., 1998). In contrast, in larvae at the same length reared at 12°C, the majority of slow fibres had an extensive network of multiterminal and polyneuronal innervation. Under the 8°C regime, 16–17 mm TL larvae had an intermediate number of multiterminally innervated slow fibres. Differences in the extent of multiple innervation to the slow muscle fibres were still marked in 20 mm TL larvae, with the most extensive development of the adult pattern found at the highest temperature (Johnston et al., 1998).

It was also noted that the median fins developed earlier with respect to body length at higher temperatures (Johnston et al., 1998). This was particularly marked for the development of anal fin rays and anal fin ray erector and depressor muscles. For example, under the 5°C regime larvae of 17 mm had on average seven or eight anal fin rays with three or four fin ray muscles showing AchE staining (see Figure 17b in Johnston et al., 1998). In contrast, under the 12°C regime the anal fin was more developed and there were up to 16 fin ray muscles innervated at their distal ends (Figure 17d in Johnston et al., 1998). The fin ray muscles eventually developed a multi-terminal pattern of innervation, which was apparent in all 12°C larvae examined of 20 mm TL and a proportion of the 8°C larvae, but in none of the 5°C larvae of the same length.

5. Post-embryonic muscle growth characteristics

Post-embryonic growth in teleosts involves an increase in the length, diameter and number of muscle fibres. This is different from the situation observed in mammals where the number of muscle fibres is fixed at birth. The post-embryonic increase in muscle fibre number is called fibre recruitment. As the fibres expand in diameter (hypertrophy) they acquire large numbers of nuclei in order to keep the nuclear to cytoplasmic ratio within set limits (see Koumans et al., 1991). The source of nuclei for fibre recruitment and hypertrophy is thought to be a population of persistent myoblasts, a proportion of which correspond to myogenic stem cells. In herring, myogenic progenitors are first observed interspersed between the embryonic slow and fast muscle fibres in the late embryo stage (Figures 2b, 4a). It has been suggested on the basis of ultrastructural observations in several other teleosts species that the myogenic stem cells divide in and originate from the adjacent mesechymal tissue and migrate into the muscle via the myosepta (Stoiber and Sänger, 1996). In juveniles the majority of the myogenic progenitors cells are probably incorporated underneath the basal lamina of the muscle fibres (Veggetti et al., 1990). Muscle stem cells are thought to undergo an asymmetric division to regenerate the stem cell and produce a daughter cell capable of a limited number of further divisions (Schultz, 1996). A working hypothesis for the behaviour of myogenic cells in fish is presented in *Figure 5*. In the absence of direct experimental evidence this model is largely based on what is known about the physiology of growth regulation in mammalian muscle. Hepatocyte growth factor (HGF) has been shown to activate muscle satellite cells in mouse muscle (Cornelison and Wold, 1997; Miller *et al.*, 2000). The receptor for HGF, the c-met tyrosine kinase receptor, is a specific marker for muscle satellite cells in mammals (Cornelison and Wold, 1997) and it also stains mononuclear cells in a range of teleost species (Johnston *et al.*, 1999, 2000a) including Atlantic herring (unpublished results). The majority of satellite cells in fish express one or more myogenic regulatory factors (MRFs) belonging to the MyoD family of basic helix–loop–helix (bHLH) transcription factors (Johnston *et al.*, 1999, 2000a). MyoD and Myf-5 are thought to have a role in myogenic determination whereas myogenin and Myf-6 are expressed later and promote differentiation (Megeney and Rudnicki, 1995). MRFs form heterodimers with other basic helix–loop–helix factors belonging to the E12 and E47 gene families and bind to the promoter regions of many muscle specific genes (reviewed in Watabe, 2001). The growth status of the fish is reflected in the number of times the producer cells divide before terminal differentiation, which is in turn a function of a complex series of

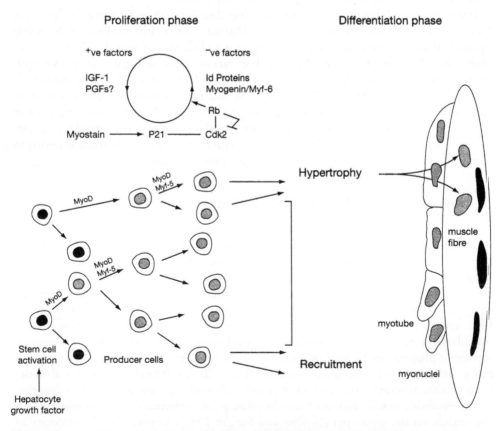

Figure 5. A model for muscle fibre recruitment and hypertrophy and the behaviour of myogenic precursor cells in teleost fish (see text for details). Rb, retinoblastoma protein; Cdk2, cyclin kinase 2; IGF-1, insulin-like growth factor 1; PGFs, peptide growth factors.

antagonistic signals for proliferation and differentiation. Myostatin, a member of the transforming growth factor-β superfamily of regulatory factors is a negative regulator of muscle growth (McPherron et al., 1997). Recent evidence suggests that myostatin functions by controlling the proliferation of muscle precursor cells via the up-regulation of p21, leading to a decrease in the level of Cdk2 protein in myoblasts (Thomas et al., 2000). This in turn leads to the accumulation of hypophosphorylated retinoblastoma protein, which results in the arrest of myoblasts in the G1-phase of the cell cycle (Thomas et al., 2000). MyoD 'knockout' mice show an increase in fibre number and size (McPherron et al., 1997). Evidence from dominant negative myostatin transgenic mice (dnMS) suggests that lower levels of myostatin inhibition may affect fibre hypertrophy whereas higher levels of myostatin inhibition may be required to alter fibre recruitment (Zhu et al., 2000).

The total cross-sectional area of trunk muscle in laboratory reared Clyde herring larvae increased approximately 90-fold (slow fibres) and 160-fold (fast fibres) between hatching and the end of metamorphosis at 37 mm TL (Johnston et al., 1998). The increase in muscle cross-sectional area between 8 and 14–15 mm TL was entirely due to the hypertrophy of the embryonic muscle fibres present at hatching. Recruitment of fast fibres from discrete germinal zones started at the dorsal and ventral apices of the myotome in larvae of 15.6–18.4 mm TL. In larvae longer than 22 mm very small diameter fibres were found scattered throughout the fast muscle, consistent with the activation of satellite cells (Johnston et al., 1998). The recruitment of slow fibres also began when larvae reached 22 mm TL, and the new fibres were added externally to the embryonic superficial slow fibres starting at the level of the major horizontal septum. The relationship between slow and fast fibre number and body length for larvae reared at the 5 and 12°C thermal regimes is shown in *Figure 6a, b*. By the completion of metamorphosis at 37 mm TL there were on average 460 slow muscle fibres per trunk cross-section at 5°C (177 d post-hatch), 523 at 8°C (117 days post-hatch) and 562 at 12°C (101 days post-hatch) (Johnston et al., 1998). At 37 mm TL, the rate of slow fibre recruitment was about 5 fibres day^{-1} at 5°C, and 10 fibres day^{-1} at 8 and 12°C (Johnston et al., 1998). In contrast, the average number of fast fibres per trunk cross-section at 37 mm was estimated to be 4700 at 5°C and 5600 at 12°C, corresponding to recruitment rates of 66 and 103 fibres day^{-1} respectively (Johnston et al., 1998). On average in 50 mm TL juveniles there were about 23% more fast fibres per cross-section at 12°C than at 5°C (Johnston et al., 1998).

The increase in fibre diameter for fast muscle fibres between first feeding and the end of metamorphosis is shown in *Figure 6c*. Shortly after first feeding in 11 mm TL larvae the fast fibres were of relatively uniform size with a peak density at 7 µm (*Figure 6c*). The smoothed curves fitted to the distributions of fibre diameter became progressively broader as the larvae increased in length (*Figure 6c*). The maximum diameter of fast muscle fibres at metamorphosis was around 40µm (*Figure 6c, d*). There was a trend for somewhat greater growth performance by hypertrophy at 12°C than 5°C (Johnston et al., 1998), resulting in a greater maximum fibre diameter for a given length between 20 and 37 mm (*Figure 6d*).

6. Embryonic temperature and larval phenotype

The temperature during development often produces irreversible phenotypic changes in fish and other ectotherms (Johnston et al., 1996). Environment-development interactions can lead either to alternative phenotypes or to a gradient of phenotypes under

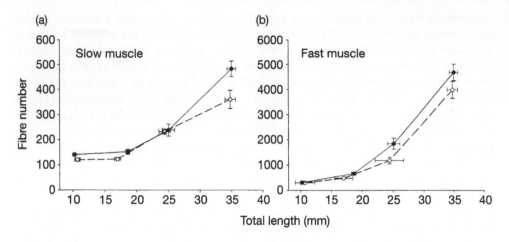

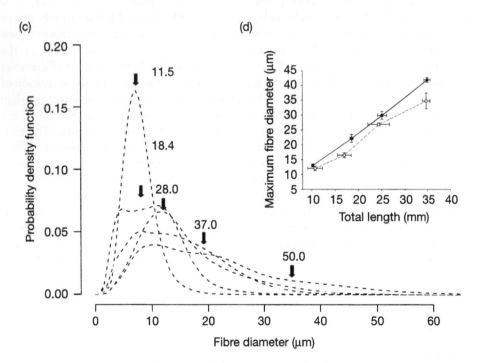

Figure 6. (a, b) *Muscle fibre recruitment in Clyde herring. The relationship between the number of muscle fibres per trunk cross-section at 0.4 L and total length for Clyde herring larvae reared under the 5°C (open circles) and 12°C (closed circles) regimes illustrated in Figure 3a. (a) slow muscle and (b) fast muscle. The values represent means ± SE for 10–12 larvae per sample for the first two sample points and six larvae per sample point for the remaining sample points. Data were replotted from Johnston* et al. *(1998). (c, d) Muscle fibre hypertrophy in Clyde herring. (c) Smooth probability density functions (pdfs) of fast muscle fibre diameter in representative larvae and a 50 mm juvenile reared under the 5°C thermal regime shown in Figure 3a. The curves representing individual fish were fitted using a kernel function as described in Johnston* et al. *(1999). (d) The 95th percentiles calculated from the pdfs of fast muscle fibre diameter for herring larvae reared at 5°C (open circles) and 12°C (closed circles). The values represent means ± SE for 10–12 larvae per sample for the first two sample points and six larvae per sample point for the remaining sample points. Data were replotted from Johnston* et al. *(1998).*

different conditions. Perhaps the best-known example of the former is sex determination. Sex ratio is sensitive to environmental temperature in a number of fish species including the Atlantic silverside (*Menidia menidia*) (Lagomarsino and Conover, 1993) and the pejerry (*Odontesthes bonariensis*) (Strussman *et al.*, 1996). More complex and continuously variable effects of development temperature in fish include effects on pigmentation patterns (Schmidt, 1919), meristic characters (Hempel and Blaxter, 1961) and final body size (Atkinson, 1994).

It seems likely that some of the lack of correspondence between patterns of genetic and morphological variability found in herring populations (Ryman *et al.*, 1984) can be explained on the basis of differences in environmental conditions during early development. Herring eggs experience very different temperatures according to the time and place of spawning. Evidence that muscle growth characteristics varied between spawning stocks in natural populations was obtained by Greer-Walker *et al.* (1972). They found that the number of fast muscle fibres per cross-section ranged from around 34 000 in the Blackwater (estuarine) population to 64 000 in the North Sea Bank stock, and it was suggested that this could be used as a stock identification factor (Greer-Walker *et al.*, 1972).

In order to test the hypothesis that early thermal experience influenced muscle fibre recruitment, Clyde herring from the same families were reared at constant temperatures of 5 and 8°C until first feeding and then transferred to a common temperature regime (Johnston *et al.*, 1998). Fish were sampled on the day of transfer and after 80 days, corresponding to approximately 1400 degree-days of total growth since fertilization in each group. The total cross-sectional area of slow and fast muscle was 80 and 50% greater at the end of the experiment in the 8°C and 5°C groups respectively (Johnston *et al.*, 1998). The superior growth was due to a higher muscle fibre number in the 8°C group than in the 5°C group for both fibre types. Although the maximum fibre diameter was greater in the 8°C group, an analysis of the smoothed distribution of muscle fibre diameter provided little evidence for differences in hypertrophic growth performance (unpublished results). The density of myonuclei was also determined at transfer and after 890 degree-days. The total density of myonuclei was similar in both groups at first feeding but was 36% higher in the 8°C than in the 5°C group at the end of the experiment (Johnson *et al.*, 1998).

Similar experiments were also carried out on the offspring of autumn spawning Manx herring (Johnston *et al.*, 1998). In this case eggs were spawned at higher temperatures than in the spring spawning population and the larvae experienced falling temperatures during ontogeny. The eggs of Manx herring were incubated at either 10 or 13.5°C until hatch. At first feeding the larvae were transferred to ambient temperatures and sampled after 86 days at the common temperature. Manx herring were smaller at first feeding than Clyde herring and grew more slowly. In this case, rates of fast fibre recruitment were significantly faster in the 10°C than in the 13.5°C group (7 and 5 day^{-1} respectively), resulting in 30% more fibres by the end of the experiment (Johnston *et al.*, 1998). We concluded that embryonic temperature does influence muscle fibre recruitment during larval life.

Other results indicate significant population differences in the developmental plasticity of neuromuscular development (Johnston *et al.*, 2001). Myofibril synthesis and the development of acetylcholinesterase activity at the neuromuscular junctions occurred later with respect to embryonic stage at 5°C than at 8 and 12°C in Clyde herring, with much less pronounced differences with temperature in the Blackwater and Manx populations (Johnston *et al.*, 2001). Population differences in the effects of egg incubation temperature and subsequent muscle fibre recruitment patterns have also been reported for Atlantic salmon populations spawning in upland and lowland tributaries of a river system (Johnston *et al.*, 2000a,b).

In laboratory experiments, larval rearing temperature was shown to affect the relative timing of the development of the median fins and their associated nerves and muscles with respect to body length. Recently, we tested the hypothesis that differences in egg incubation temperature alone were sufficient to produce such phenotypic variation in the larval stages (Johnston *et al.*, 2001). Eggs from Clyde herring were incubated at 5 and 12°C and transferred to ambient temperature shortly after hatching (*Figure 7a*). Flexion of the notochord and development of the dorsal and anal fin (*Figure 7b–d*) ray muscles occurred at shorter body lengths in the 12°C than in the 5°C groups, consistent with our original hypothesis. The flexible segmented fin rays and associated muscles in teleosts play an important role in adjusting the stiffness and camber of the fins during swimming and hence the transfer of the propulsive power generated by the muscles to the water (Videler, 1993).

The functional consequences of the variation in the body length at which the median fins formed were investigated by filming the larvae with a high-speed video camera (Johnston *et al.*, 2001). Cruising behaviour differed according to the embryonic temperature regime, with 12°C larvae showing reduced yaw of the head relative to the 5°C larvae, indicative of a more developmentally advanced and effective sub-carangiform style of locomotion. We also studied fast-starts, which are used during escape and prey capture behaviour. Fast-starts are initiated by the Mauthner neurones and involve the bending of the body into an initial 'C' or 'S' shape followed by a contralateral contraction away from the stimulus (*Figure 8a*). In our experiments, over the size range 11.2 – 18.5 mm TL, corresponding to the point when the differences in fin muscle development were most pronounced, the adjusted mean maximum velocity was 24% higher in larvae hatched from 12°C than from 5°C eggs (*Figure 8b*). Herring larvae less than 25 mm TL probably rely on their transparency, rather than locomotor performance to avoid encounters with predators (Blaxter, 1988; Blaxter and Fuiman, 1990). However, differences in maximum velocity during fast-starts may affect feeding success in larvae less than 22 mm TL. Morley and Batty (1996) found temperature altered strike behaviour and velocity during feeding attacks by Clyde and Manx herring larvae on *Artemia* sp. nauplii.

▶ *Figure 7.* Embryonic temperature affects the subsequent innervation to the anal fin ray muscles in Clyde herring. (a) Temperature regimes used in the experiments. The arrows show the time of transfer to the ambient temperature regime at first feeding. (b, c) Whole mount larvae were stained for acetylcholinesterase activity and photographed using differential interference optics. (b) larva reared at 5°C and (c) larva reared 12°C until first feeding and then transferred to ambient seawater temperature which increased from 8° to 11.5°C over the course of the experiment. The total body length of each larva in mm is shown in the bottom right-hand corner of each panel. The arrowhead illustrates a muscle endplate stained for acetylcholinesterase activity. Open arrowheads illustrate actiotrichia strengthening the anal fin. a, anus; frm, fin ray muscle; ms, myosepta. The scale bars are 200 μm. (d) The number of anal fin ray muscles in relation to body length. Open and closed symbols represent fish initially reared at 5° and 12°C, respectively until first feeding and then transferred to ambient seawater temperature. The curves represent first order polynomials fitted to the data. Reproduced from Johnston et al. (2001). Mar. Ecol. Prog. Ser., *213*: 285–300, with permission of Inter-Research, Oldendorf/Luke, Germany.

temperature cycling of about 1.5°C was imposed on a gradual warming as the season advanced (Temple et al., 2000). In order to estimate hatch dates, otolith microincrement analysis was carried out. Successive sampling on 11 May, 18 June and 24 July, 1998 appeared to sample consecutive cohorts, referred to as early-, mid- and late-season larvae respectively. Mean hatch temperatures for the three groups were 7.5 ± 0.9°C (mean ± S.D.), 9.8 ± 1.1°C and 14.9 ± 0.6°C, respectively (Temple et al., 2000). The mid-season larvae were found to be longer for a given estimated age than the early- and late-season larvae (Figure 9b). In addition, at approximately 60 days, the cross-sectional area (CSA) of fast muscle in mid-season larvae was 145% greater than in early-season larvae of the same estimated age due to 60% more muscle fibres and a 22% greater mean fibre diameter. Red muscle showed a similar pattern (Temple et al., 2000). When muscle growth characteristics were examined against length, red muscle CSA, fibre number and mean fibre diameter increased with body length with no significant differences between the groups of larvae. However, both the mid- and late-season larvae had significantly greater white muscle CSAs relative to body length compared to early-season larvae. In the case of the mid-season larvae this was due to greater mean fibre diameter (Figure 9c) and not to any statistically significant differences in white fibre number (Figure 9d).

Laboratory experiments have found that increases in embryonic temperature leads to short, deep bodied larvae (Blaxter, 1992; Vieira and Johnston, 1992). Similarly, for a given age, the late-season Blackwater larvae with the highest hatching temperatures were the shortest group but had a greater CSA for a given length than the early-season larvae (Temple et al., 2000). However, temperature is just one of a number of factors influencing muscle growth in the sea. Survival through the larval stages and recruitment to the adult population is believed to be linked to fast growth rates, which reduce the risk of predatory mortality (Cushing, 1990). Fox et al. (1999) found that the concentration of copepod nauplii in the Blackwater estuary was lowest in early spring. Temple et al. (2000) suggest that this would mean the early-season Blackwater larvae had a lower food supply at first feeding compared to the mid-season group, with potential consequences for muscle fibre growth. Dietary effects on muscle growth characteristics were demonstrated by Galloway et al. (1999), who reported that a diet consisting of a high, compared to a low, ratio of docosahexaenoic acid to eicosapentaenoic acid leads to greater fast muscle CSA in larvae of the cod (Gadus morhua L.). The ratio of cells expressing mycotic regulatory factors (MRFs) to the number of cells immunopositive for c-met was found to vary with nutritional status in herring larvae such that the number of MRF positive cells decreased after 1–3 days of starvation and increased again on re-feeding (Vieira and Johnston, unpublished results). On the other hand measurements of the number and diameter of muscle fibres provide information about feeding opportunity over a time scale of several weeks or months. Measurements of muscle cellularity may prove an additional useful index of larval nutrition in the field, which could be calibrated in the laboratory to provide information on growth history over a wide range of time scales.

8. Conclusions

The Atlantic herring is a temperate species, which has experienced a massive decline in abundance in the last century. Distinct stocks, populations and races can be identified,

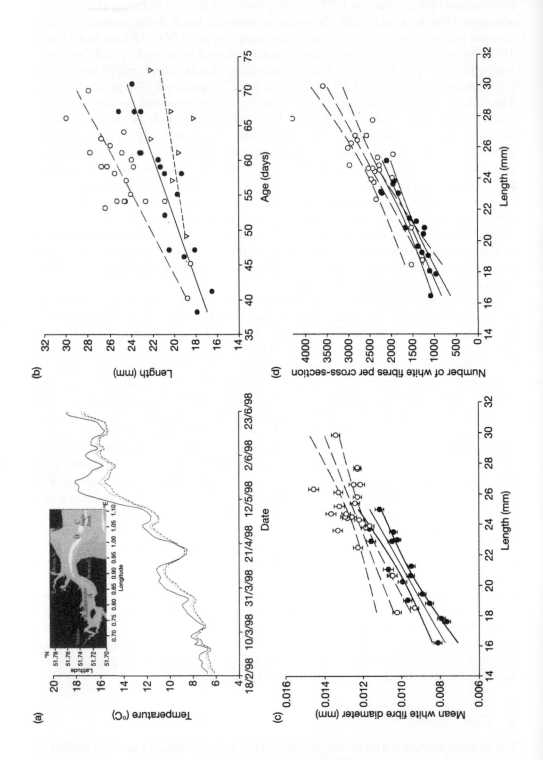

◀*Figure 9.* *(a) Mean daily sub-surface water temperature at three locations within the Blackwater estuary (inset). A, Thirslet (continuous line); B, Bench Head (dotted line); and C, North Eagle (dashed line) during spring 1998. (b) Growth patterns for early- (solid circle, solid line), mid- (open circle, long dashed line) and late- (open triangle, short dashed line) season Blackwater larvae, using age derived from otolith microincrement analysis. (c, d) Relationship between body length (L) and (c) mean (± SE) diameter of white fibres and (d) total number of white muscle fibres, per transverse-section of myotomal muscle in early- (solid circle, solid line) and mid- (open circle, dashed line) season Blackwater larvae; 95% confidence intervals are shown in (c) and (d). Reproduced from Temple et al. (2000). Mar. Ecol. Prog. Ser. 205: 271–281, with permission of Inter-Research, Oldendorf/Luke, Germany.*

---

differing in size and age at maturity and numerous morphological characteristics. Spatial and temporal separation of spawning promotes reproductive isolation between populations. In spite of this, evidence for the genetic structuring of populations is sparse. More than a decade of research has revealed the plasticity of development in this species. Variations in temperature during embryonic development have been shown to affect morphology, muscle growth characteristics and neuromuscular differentiation with significant consequences for behaviour. It seems likely that developmental mechanisms underlie much of the phenotypic variation observed within and between populations in this species.

Acknowledgements

We are grateful to the Natural Environment Research Council for funding this work and to Dr Robert Batty and the Director of the Dunstaffnage Marine Laboratory for cooperation and the provision of facilites. We also thank Marguerite Abercromby for help in preparing the illustrations for this chapter.

References

Altringham, J.D. and Ellerby, D.J. (1999) Fish swimming: Patterns in muscle function. *J. Exp. Biol.* 202: 3397–3403.

Anokhina, L. (1971) Maturation of Baltic and White Sea herring with special reference to variations in fecundity and egg diameter. *Rapp. Cons. Explor. Mer* 160: 12–17.

Atkinson, D. (1994) Temperature and organism size: a biological law for ectotherms? *Adv. Ecol. Res.* 25: 1–58.

Bailey, K.M. and Batty, R.S. (1984) Laboratory study of predations by *Aurelia aurita* on larvae of cod, flounder, plaice and herring: development and vulnerability to capture. *Mar. Biol.* 38: 287–291.

Barresi, M.J.F., Stickney, H.L. and Devoto, S.H. (2000) The zebrafish slow-muscle-omitted gene product is required for Hedgehog signal transduction and the development of slow muscle identity. *Development* 127: 2189–2199.

Batty, R.S. (1984) Development of swimming movements and musculature of larval herring (*Clupea harengus*). *J. Exp. Biol.* 83: 287–291.

Blagden, C.S., Currie, P.D., Ingham, P.W. and Hugues, S.M. (1997) Notochord induction of zebrafish slow muscle mediated by Sonic hedgehog. Genes Devl. 11: 2163–2175.

Blaxter, J.H.S. (1988). Pattern and variety in development. In: *Fish Physiology*, Vol. XIA (eds W.S. Hoar and D.J. Randall). Academic Press, New York, pp. 1–58.

Blaxter, J.H.S. (1992) The effect of temperature on larval fishes. *Neth. J. Zool.* 42: 336–357.

Blaxter, J.H.S. and Fuiman, L.A. (1990) The role of the sensory systems of herring larvae in evading predatory fishes. *J. Mar. Biol. Assoc. UK* 70: 413–427.

Bone, Q. (1964) Patterns of muscular innervation in the lower chordates. *Int. Rev. Neurobiol.* **6:** 99–147.

Bone, Q. (1978) Locomotor muscle. In: *Fish Physiology,* Vol. VII (eds W.S. Hoar and D.J. Randall). Academic Press, New York, pp. 361–424.

Bone, Q., Kicznuik, J. and Jones, D.R. (1978) On the role of different fibre types in fish myotomes at intermediate speeds. *Fishery Bull. Fish Wildlife Serv. US* **76:** 691.

Brander, K. (1979) The relationship between vertebral number and water temperature in cod. *J. Cons. Int. Explor. Mer.* **38:** 286–292.

Brooks, S. and Johnston, I.A. (1993) Influence of development and rearing temperature on the distribution, ultrastructure and myosin sub-unit composition of myotomal muscle fibre types in the plaice *Pleuronectes platessa. Mar. Biol.* **117:** 501–513.

Cornelison, D.D.W. and Wold, B.J. (1997) Single-cell analysis of regulatory gene expression in quiescent and activated mouse skeletal muscle satellite cells. *Devl. Biol.* **191:** 270–283.

Crockford, T. and Johnston, I.A. (1993) Developmental changes in the composition of myofibrillar proteins in the swimming muscles of Atlantic herring *Clupea harengus. Mar. Biol.* **115:** 15–22.

Currie, P.D. and Ingram, P.W. (1998) The generation and interpretation of positional information within the vertebrate myotome. *Mech. Devl.* **73:** 3–21.

Cushing, P.H. (1990) Plankton production and year-class strenght in fish populations: an update of the match/mis-match hypothesis. *Adv. Mar. Biol.* **26:** 249–293.

De Silva, C.D. (1974) Development of the respiratory system in herrring and plaice larvae. In: *The Early Life History of Fish.* (ed J.H.S. Blaxter). Springer, Berlin, pp. 465–485.

Devoto S.H., Melançon, E., Eisen, J.S. and Westerfield, M. (1996) Identification of separate slow and fast muscle precursor cells in vivo, prior to somite formation. *Development* **122:** 3371–3380.

Doyle, M.J. (1977) A morphological staging system for the larval development of the herring, *Clupea harengus* L. *Mar. Biol. Assoc. UK.* **57:** 859–867.

Egginton, S. and Sidell, B.D. (1989) Thermal-acclimation induces adaptive-changes in sub cellular structure of fish skeletal muscle. *Am. J. Physiol.* **256:** R1–9.

El-Fiky, N., Hinterleitner, S. and Wiser, W. (1987) Differentiation of swimming muscles and gills and development of anaerobic power in larvae of cyprinid fish (Pisces Teleostei). *Zoomorphology* **107:** 126–132.

Fox, C.J., Harrop, R. and Winpenny, A. (1999) Feeding ecology of herring (*Clupea harengus*) larvae in the turbid Blackwater Estuary. *Mar. Biol.* **134:** 353–365.

Franklin, C.E., Johnston, I.A., Batty, R.S. and Yin, M.C. (1996) Metabolic recovery in herring larvae following strenuous activity. *J. Fish Biol.* **48:** 207–206.

Galloway, T.F., Kjørsvik, E. and Kryvi, H. (1998) Effect of temperature on viability and axial muscle development in embryos and yolk sac larvae of the Northeast arctic cod (*Gadus morhua*). *Mar. Biol.* **132:** 559–567.

Galloway, T.F., Kjørsvik, E. and Kryvi, H. (1999) Muscle growth and development in Atlantic cod larvae (*Gadus morhua* L.) related to different somatic growth rates. *J. Exp. Biol.* **202:** 2111–2120.

Greer-Walker, M.G., Burd, A.C. and Pull, G.A. (1972) The total number of white skeletal muscle fibres in cross section as a character for stock separation in North Sea herring (*Clupea harengus* L.). *J. Cons. Int. Explor. Mer.* **34:** 238–243.

Hempel, G. and Blaxter, J.H.S. (1961) The experimental modification of meristic characters in herring (*Clupea harengus* L.). *J. Cons. Int. Explor. Mer.* **26:** 336–346.

Henderson, P.A. and Cartwright, G.H. (1980) The dispersal of larval herring (*Clupea harengus*) in the Blackwater estuary, Essex, 1979. Central Electricity Laboratories, Leatherhead, Surrey.

Henderson, P.A.Whitehouse, J.W and Cartwright, G.H. (1984) The growth and mortality of larval herring, *Clupea harengus* L., in the River Blackwater estuary, 1979–1980. *J. Fish Biol.* **24:** 613–622.

Hinterleitner, S., Platzer, U. and Weiser, W. (1987) Development of the activities of oxidative, glycolitic and muscle enzymes during early larval life in three families of freshwater fish. *J. Fish Biol.* **30**: 315–326.

Johnston, I.A. (1981) Structure and function of fish muscles. In: *Vertebrate Locomotion, Symposium of the Zoological Society of London*, Vol. 48 (ed M.H. Day). Zoological Society of London, London, pp. 71–113.

Johnston, I.A. (1993) Temperature influences muscle differentiation and the relative timing of organogenesis in herring (*Clupea harengus*) larvae. *Mar. Biol.* **116**: 363–379.

Johnston, I.A. and Altringham, J.D. (1991) Movement in water: constraints and adaptations. In: *Biochemistry and Molecular Biology of Fishes*, Vol. 1 (eds P.W. Hochachka and T.P. Mommsen.). Elsevier, Oxford, pp. 249–268.

Johnston, I.A. and Horne, Z. (1994) Immunocytochemical investigations of muscle differentiation in the Atlantic herring (*Clupea harengus*: Teleostei). *J. Mar. Biol. Assoc. UK* **74**: 79–91.

Johnston, I.A. and Maitland, B. (1980) Temperature acclimation in crucian carp (*Carassius carassius* L.). Morphometric analysis of muscle fibre ultrastructure. *J. Fish Biol.* **17**: 113–125.

Johnston, I.A., Vieira, V.L.A. and Abercromby, M. (1995) Temperature and myogenesis in embryos of the Atlantic herring *Clupea harengus. J. Exp. Biol.* **198**: 1389–1403.

Johnston, I.A., Vieira, V.L.A. and Hill, J. (1996) Temperature and ontogeny in ectotherms: muscle phenotype in fish. In: *Phenotypic and Evolutionary Adaptation of Organisms to Temperature* (eds I.A. Johnston and A.F. Bennett). Society for Experimental Biology Seminar Series. Cambridge University Press, Cambridge, pp.153–181.

Johnston, I.A., Cole, N.J., Vieira, V.L.A. and Davidson, I. (1997) Temperature and developmental plasticity of muscle phenotype in herring larvae. *J.Exp.Biol.* **200**: 849–868.

Johnston, I.A., Cole, N.J., Abercromby, M. and Vieira, V.L.A. (1998) Embryonic temperature modulates muscle growth characteristics in larval and juvenile herring. *J. Exp. Biol.* **201**: 623–646.

Johnston, I.A., Strugnell, G., McCraken, M.L. and Johnstone, R. (1999). Muscle growth and development in normal-sex-ratio and all-female diploid and triploid Atlantic salmon. *J. Exp. Biol.* **202**: 1991–2016.

Johnston, I.A., McLay, H.A., Abercromby, M. and Robins, D. (2000a) Phenotypic plasticity of early myogenesis and satellite cell numbers in Atlantic salmon spawning in upland and lowland tributaries of a river system. *J. Exp. Biol.* **203**: 2539–2552.

Johnston, I.A., McLay, H.A., Abercromby, M. and Robins, D. (2000b) Early thermal experience has different effects on growth and muscle fibre recruitment in spring- and autumn-running Atlantic salmon populations. *J. Exp. Biol.* **203**: 2553–2564.

Johnston, I.A., Vieira, V.L.A. and Temple, G.K. (2001) Functional consequences and population differences in the developmental plasticity of muscle to temperature in Atlantic herring (*Clupea harengus* L.). *Mar. Ecol. Prog. Sr.* **213**: 285–300.

Jørstad, K.E., Dahle, G. and Paulsen, O.I. (1994). Genetic comparison between Pacific herring (*Clupea pallasi*) and a Norwegian fjord stock of Atlantic herring (*Clupea harengus*). *Can. J. Fish. Aquat. Sci.* **51** (Suppl 1): 233–239.

Koumans, J.T.M., Akster, H.A., Brooms, G.H.R., Lemmens, C.J.J. and Osse, J.W.M. (1991) Numbers of myosatellite cells in white axial muscle of growing fish: *Cyprinus carpio* L. (Teleostei). *Am. J. Anat.* **192**: 418–424.

Lagomarsino, V. and Conover, D.O. (1993) Variation in environmental and genotypic sex-determining mechanisms across a latitudinal gradient in fish, *Menidia menidia. Evolution* **47**: 487–494.

McPherron, A.C., Lawler, A.M. and Lee, S.-J. (1997) Regulation of skeletal muscle mass in mice by a new TGF-β superfamily member. *Nature* **387**: 83–90.

Megeney, L.A. and Rudnicki, M.A. (1995) Determination *versus* differentiation and the MyoD family of transcription factor. *Biochem. Cell Biol.* **73**: 723–732.

Miller, K.J., Thaloor, D., Matteson, S. and Pavlath, G.K. (2000) Hepatocyte growth factor affects satellite cell activation and differentiation in regenerating skeletal muscle. *Am. J. Physiol. Cell Physiol.* **278**: C174–C181.

Morley, S.A. and Batty, R.S. (1996) The effects of temperature on "S-Strike" feeding of larval herring, *Clupea harengus* L. *Mar. Fresh. Behav. Physiol.* **28:** 123–136.

Parrish, B. and Saville, A. (1967) Changes in the fisheries of North Sea and Atlanto-Scadian herring stocks and their causes. *Oceanogr. Mar. Biol. Annu. Rev.* **5:** 409–447.

Rosenberg, R. and Palmén, L.-E. (1982) Composition of herring stocks in the Skagerrak-Kattegat and the relations of these stocks with those of the North Sea and adjacent waters. *Fish. Res.* **1:** 83–104.

Ryman, N.U., Lagercrantz, L., Anderson, R., Chakraborty, R. and Rosemberg, R. (1984) Lack of correspondence between genetic and morphological variability patterns in Atlantic herring (*Clupea harengus*). *J. Hered.* **53:** 687–704.

Schiaffino, S. and Reggiani, C. (1996) Molecular diversity of myofibrillar proteins: gene regulation and functional significance. *Physiol. Rev.* **76:** 371–418.

Schmidt, E. (1919) Racial studies in fish. II. Experimental investigations with *Lebistes reticulatus* (Peters) Regan. *J. Genet.* **8:** 147–153.

Shultz, E. (1996) Satellite cell proliferative compartments in growing skeletal muscles. *Devl. Biol.* **175:** 84–94.

Stickland, N.C., White, R.N., Mescall, P.E., Crook, A.R. and Thorpe, J.E. (1988) The effect of temperature on myogenesis in embryonic development of the Atlantic salmon (*Salmo salar* L.). *Anat. Embryol.* **178:** 253–257.

Stoiber, W. and Sänger, A.M. (1996) An electron microscopic investigation into the possible source of new muscle fibres in teleost fish. *Anat. Embryol.* **194:** 569–579.

Stratoudakis, Y., Gallego, A., Morrison, J.A. (1998) Spatial distribution of developmental egg ages within a herring *Clupea harengus* spawning ground. *Mar. Ecol. Prog. Ser.* **174:** 27–32.

Strussman, C.A., Moriyama, S., Hanke, E.F., Cota, J.C.C. and Takashima, F. (1996) Evidence for thermolabile sex determination in Pejerrey. *J. Fish Biol.* **48:** 643–651.

Tåning, A.V. (1952) Experimental study of meristic characters in fishes. *Biol. Rev.* **27:** 169–193.

Taylor, F.H.C. (1971) Variation in hatching success in pacific herring (*Clupea pallasii*) eggs with water depth, temperature and salinity. *Repp. P.* **160:** 34–41.

Temple, G.K., Fox, C.J., Stewart, R. and Johnston, I.A. (2000) Variability in muscle growth characteristics during the spawning season in a natural population of Atlantic herring (Clupea harengus L.). *Mar. Ecol. Progr. Sr.* **205:** 271–281.

Thomas, M., Langley, B., Berry, C., Sharma, M., Bass, J. and Kambadur, R. (2000) Myostatin, a negative regulator of muscle growth, functions by inhibiting myoblast proliferation. *J. Biol. Chem.* **275:** 40235–40243.

Turan, C.Carvalho, G.R. and Mork, J. (1998) Molecular genetic analysis of Atlanto-Scandian herring (*Clupea harengus*) populations using allozymes and mitochondrial DNA markers. *J. Mar. Biol. Assoc. UK* **78:** 269–283.

Turrel, W.R. (1999) Scottish ocean climate status report 1998. Fisheries Research Services Report no. 9/99.

Veggetti, A., Mascerello, F., Scapolo, P.-A. and Rowlerson, A. (1990) Hyperplastic and hypertrophic growth of lateral muscle in *Dicentrarchus labrax* (L.). *Anat. Embryol.* **182:** 1–10.

Videler, J.J. (1993) *Fish Swimming.* Chapman & Hall, London.

Vieira, V.L.A. and Johnston, I.A. (1992) Influence of temperature on muscle-fibre development in larvae of the herring *Clupea harengus. Mar. Biol.* **112:** 333–341.

Watabe, S. (2000) Myogenic regulatory factors. In *Fish Physiology*, Vol. XVIII (ed I.A. Johnston). Academic Press, San Diego, CA, pp. 19–41.

Wheeler, J.P. and Winters, G.H. (1984) Homing of Atlantic herring in Newfoundland waters as indicated by tagging data. *Can. J. Fish Aquat. Sci.* **41:** 108–117.

Wieser, W. (1995) Energetics of fish larvae, the smallest vertebrates. *Acta Physiol. Scand.* **154:** 279–290.

Zhu, X., Hadhazy, M., Wehling, M., Tidball, J.G. and McNally, E.M. (2000) Dominant negative myostatin produces hypertrophy without hyperplasia in muscle. *FEBS Letters* **474:** 71–75.

6

Lipid dietary effects on environmental stress tolerance

James A. Logue

1. Essential fatty acids in the marine system

The health of an animal and its tolerance to stressful environmental fluctuations is dependant upon dietary input, with under-nourishment and starvation or deficiency of essential elements resulting in reduced fitness, sickness and death. The consequences of a poor diet will be particularly severe during the early life stages where tissue growth, development and differentiation are occurring at a rapid rate and hence the requirement for energy and nutritional building blocks is high. Much research has centred on the nutritional roles of essential fatty acids, those that are necessary for good health but cannot be synthesized in sufficient proportions by the body and are thus dietary requirements. In terrestrial systems these include linoleic acid (18:2n-6) and α-linolenic acid (18:3n-3), which can be elongated and desaturated by animals to form C20 and C22 long chain polyunsaturated fatty acids (LC-PUFA) such as arachidonic acid, (AA, 20:4n-6), eicosapentaenoic acid (EPA, 20:5n-3) and docosahexaenoic acid (DHA, 22:6n-3). Unlike terrestrial systems, however, the marine food chain is rich in n-3 LC-PUFA, particularly DHA and EPA, which are synthesized by the basal phytoplankton. As a result marine fish possess only limited endogenous capacity for their synthesis, having a reduced ability to desaturate and elongate the LC-PUFA precursors (Sargent *et al.*, 1995). The bulk of the requirement must be provided by the diet, thus for marine animals AA, EPA and DHA are essential fatty acids.

2. The biological role of LC-PUFA

2.1 *Eicosanoid synthesis*

n-6 and n-3 LC-PUFA are integral components within biological systems and a deficiency or imbalance in their relative proportions may have severe physiological consequences. AA is the predominant n-6 component of membrane phospholipids and when liberated acts as precursor for a variety of biologically active compounds

---

Environment and Animal Development: Genes, Life Histories and Plasticity, edited by D. Atkinson and M. Thorndyke.

(Whelan, 1996), including the n-6 eicosanoids. Eicosanoids are a group of cellular mediators involved in a whole range of transient cell signalling events (Lands, 1993). n-6 eicosanoids are highly bioactive and promote vigorous responses but their overproduction is implicated in a whole host of pathological conditions and stress-related disorders, including thrombosis, arthritis, asthma, tumour growth, atherosclerosis and immune inflammatory disorders (Kinsella *et al.*, 1990; Lands, 1986, 1993; Sargent *et al.*, 1995). However, the n-3 LC-PUFA, particularly EPA, act to moderate these adverse effects by competing with AA as substrate for the cyclooxygenase enzymes involved in eicosanoid synthesis (Bell *et al.*, 1996; Nordoy *et al.*, 1994). Furthermore, the n-3 eicosanoids produced are not only less bioactive than the n-6 analogues but also compete at cell receptor sites to diminish n-6 eicosanoid signalling (Lands, 1993). As marine animals obtain n-3 and n-6 LC-PUFA from food sources, the dietary ratio of these fatty acids will determine the type of eicosanoids produced (Lands *et al.*, 1992), with greater relative proportions of EPA and consequent production of n-3 eicosanoids acting to alleviate the adverse effects of the AA-derived analogues.

2.2 Nervous system structure and maintenance

DHA is found in high proportions in the membrane phospholipids of the nervous system, particularly in brain synaptosomal and rod outer segment disk membranes. This is especially true for marine species, where DHA can comprise over 60% of total fatty acids in retinal phospholipids (Bell and Dick, 1991; Tocher and Harvie, 1988). High proportions of this fatty acid appear crucial for optimal nervous system function since other LC-PUFA cannot functionally replace DHA in these tissues (Litman and Mitchell, 1996; Salem and Niebylski, 1995) and, once accumulated, levels of DHA are tenaciously conserved during PUFA deficiency (Wiegand *et al.*, 1991) while their proportions may actually increase during starvation as other structural lipids are preferentially metabolised for energy (Salem *et al.*, 1986).

Yet the essential molecular function of DHA within these membranes remains unclear. Evidence suggests high proportions of LC-PUFA may be involved in the temperature adaptation of membrane physical condition. Animals adapted to lower temperatures frequently possess greater proportions of unsaturated fatty acids within their cell membranes (Hazel and Williams, 1990). The incorporation of double bonds within saturated molecules disrupts the optimal packing of the parallel hydrocarbon chains thus promoting disorder within the bilayer and offsetting the rigidifying effects of the cold (Cossins, 1994). LC-PUFA levels are observed to increase in brain synaptic membranes at low temperatures, which in turn is associated with adaptive disordering in the physical properties of this bilayer (Logue *et al.*, 2000b) and consequent reduction in the temperature of synaptic disruption (Cossins *et al.*, 1977). Furthermore, the temperature of thermal block of brain synaptic processes is closely linked to the temperature sensitivity of the whole animal (Friedlander *et al.*, 1976) thus implicating proportions of LC-PUFA within neural membranes in the determination of whole animal thermal tolerance. However, there is doubt as to the effect of highly polyunsaturated fatty acids in this regard. Whereas incorporation of monounsaturated fatty acids in place of saturated fatty acids has a considerable impact in disordering membrane bilayers, there is evidence that further unsaturation has progressively less of an effect such that LC-PUFA may actually promote rather than perturb bilayer packing (Applegate and Glomset, 1991; Hazel, 1995).

The functional role of DHA in nervous tissue may lie in its unique structural properties. It is proposed these highly unsaturated, bulky molecules pack well into bilayers and maintain membrane barrier properties (Baenziger et al., 1992), but possess a high degree of lateral compressibility enabling large lateral area changes of the lipid matrix as would be required during structural conformational changes of membrane-bound proteins (Mitchell et al., 1998). In rod photoreceptor outer segment disk membranes, high proportions of DHA are thought to enable the rapid conformational changes of the rhodopsin photointermediates, conversion of which result in a greater protein volume within the bilayer (Litman and Mitchell, 1996). In addition, the maximally unsaturated structure of DHA imparts the molecule with a low sensitivity to temperature change, thus high proportions of DHA would act to buffer the membrane from the physical effects of temperature fluctuations. DHA is therefore well suited to control bilayer volume and provide optimal conditions for embedded proteins over a wide range of temperatures (Mitchell et al., 1998; Rabinovitch and Ripatti, 1991).

3. Essential fatty acids and nervous system development

Mammalian studies have shown that deficiencies of DHA, particularly during early life stages, have serious consequences on nervous system function. Rats fed an n-3 LC-PUFA-deficient diet possessed reduced proportions of neural tissue DHA, associated with a significantly reduced learning ability when presented with olfactory and spatial tasks (Greiner et al., 1999). Rhesus monkeys deprived of n-3 PUFA during gestation and postnatal development possessed reduced levels of DHA in retinal and cerebral cortex PE lipids and exhibited impaired visual acuity (Neuringer et al., 1986). Similarly, human term and preterm infants fed n-3 LC-PUFA-deficient formulas showed reduced n-3 PUFA in their erythrocyte membrane lipids that was correlated with abnormal rod electroretinograms, poor visual acuity and a significantly reduced neurodevelopmental response (Agostoni et al., 1995; Birch et al., 1992a,b; Hoffman et al., 1993; Uauy et al., 1990). Such functional impairment caused by a dietary insufficiency of essential fatty acids in early life stages may not be rectified by subsequent sufficiency (Crawford, 2000), suggesting some interference with the ontogenetic progression of neuronal cells and tissue into an otherwise functional system. It follows that during the period of rapid larval growth in marine species, a dietary deficiency in DHA may have serious and long-lasting consequences on nervous system development and animal performance.

3.1 Aquaculture of larval marine fish

Many of the studies into dietary PUFA deficiency effects in developing fish have come from the field of aquaculture, where inappropriate diets have caused major problems relating to growth, metamorphosis and survival. Larval marine fish are commonly reared on rotifers or brine shrimp (Artemia) nauplii, which can be grown and maintained easily and at little cost but which do not naturally contain n-3 LC-PUFA. Given the essentiality of these fatty acids to marine fish, successful marine aquaculture requires specific dietary fatty acid supplementation. Typically, fish larvae are fed Artemia that have been previously enriched with high-PUFA lipid emulsions. The lipids are taken up by the Artemia through bioencapsulation and then delivered to the fish larvae when ingested. Effects on fish health are then assessed by comparative short-term stress tests. This established aquacultural model, however, also provides an ideal mechanism to explore the influence of larval lipid deficiencies during early development, and the possible effects of malformed biological systems on an animal's performance and ability to withstand stressful variations in environmental conditions.

3.2 Dietary n-3 LC-PUFA and environmental stress resistance

In recent studies we have reared larval Dover sole, *Solea solea*, on *Artemia* enriched with lipid emulsions containing either high or zero amounts of n-3 LC-PUFA. As with most nutritional studies, the essential fatty acid composition of body tissues is very much influenced by that of the diet. As a percentage of whole body tissue fatty acids, high-PUFA sole possessed 10% DHA and 22% total n-3 LC-PUFA compared to 3% and 12% respectively for the low-PUFA larvae. Stress tests were then performed on the sole to assess the consequences of the different dietary regimes. During these tests fish were placed in experimental tanks at stressful conditions, groups were removed at specific intervals, recovered under ambient conditions and mortality calculated after 24 h. Results for a combined low temperature (3°C) and low salinity (10%) stress are shown in *Figure 1a*. It was found that the PUFA-deficient sole

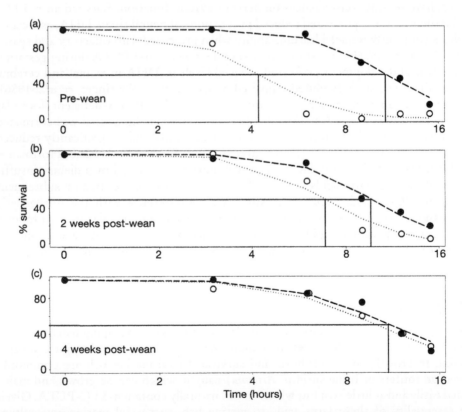

Figure 1. Effect of dietary treatment on stress tolerance of Dover sole subjected to combined low temperature (3°C) and low salinity (10%). Stress tests were performed at pre-weaning (a) and 2 (b) and 4 weeks (c) post-weaning on fish fed high-PUFA (filled circles) or low-PUFA (open circles) enriched Artemia diets. Reproduced from Logue et al. (2000). Lipids 35: 745–755, with permission of AOCS Press.

demonstrated a statistically reduced resistance, 50% mortality evident after 4.3 h compared to 10.6 h for the high-PUFA fish.

When fish had reached a certain size both dietary groups were then weaned onto the same high-PUFA agglomerated feed. Following weaning, tissue levels of LC-PUFA in the previously deficient sole were observed to increase gradually over a period of 2–3 weeks until they matched levels in the high-PUFA fed fish. Stress tests performed following weaning showed this increase in tissue n-3 LC-PUFA was also correlated with a gradually increasing stress resistance (*Figure 1b,c*). Similar tests performed under hypoxic and high-temperature conditions produced identical results. Thus, low dietary n-3 LC-PUFA in larval marine fish was correlated with a reduced tolerance to a range of environmental stresses, but this susceptible phenotype appears not fixed in the early stages of life but is subject to change according to the dietary experiences of the fish, the critical feature being the proportion of n-3 LC-PUFA. Several other observations of improved performance in reared marine larvae following dietary supplementation with n-3 LC-PUFA have been reported. Resistance to and recovery from handling stress was enhanced in larval palmetto bass, *Morone saxatilis chrysops*, following dietary PUFA enrichment (Tuncer *et al.*, 1993) while larval yellowtail, *Seriola quinqueradiata*, deficient in n-3 PUFA showed reduced growth rate and survival (Furuita *et al.*, 1996).

The question therefore is which particular dietary fatty acid or acids are necessary for promoting the stress resistance phenotype? Certainly DHA, 22:6n-3, as discussed above, is a crucial biological component and is generally the dominant LC-PUFA found in the marine food chain. Indeed, in dietary studies on developing marine fish larvae those specifically enriched with DHA possessed increased tolerance to changes in temperature, salinity, hypoxia and exposure to air (Kanazawa, 1997). To establish whether other essential n-3 LC-PUFA could promote resistance in the absence of DHA, we reared batches of sole on *Artemia* enriched with a low-PUFA emulsion to which had been added high proportions of purified EPA, 20:5n-3. Fish fed this diet possessed elevated proportions of EPA in their tissues while levels of DHA were low and comparable to the low-PUFA animals. Yet, these fish displayed similar levels of environmental stress tolerance as DHA fed sole, indicating that high EPA as the solitary dietary n-3 LC-PUFA also confers the resistant phenotype. In contrast, however, whereas larvae of the Japanese flounder, *Paralichthys olivaceus*, fed high levels of EPA or DHA exhibited faster growth and better survival than PUFA-deprived fish, those enriched with DHA were significantly more resistant to high temperature, hypoxia and high salinity than those enriched with EPA, indicating in this case that DHA is superior to EPA in promoting stress tolerance (Furuita *et al.*, 1999; Tago *et al.*, 1999).

4. Mechanisms of dietary-induced stress resistance

It is unlikely that differences in resistance observed with the Dover sole were due to a deficiency in energy provision as both the low- and high-PUFA emulsions contained similar total lipid content and fish growth rates were not significantly different between groups. The mechanisms of this nutritionally-dependant tolerance must relate to the specific biological functions of these LC-PUFA molecules.

4.1 Development of nervous system structure and function

Given evidence for the essential role of DHA in nervous system function and the high proportions normally found within its tissues, the effects of dietary manipulation on neural membrane fatty acid composition have been investigated. The data shown in *Figure 2* are for whole eye phosphatidylethanolamine (PE) lipids in the treated Dover sole. The small size of the eye prevented analysis of retinal fatty acid composition alone but 70% of total eye phospholipids in marine fish reside in the retina (Bell *et al.*, 1995). It can be seen the PUFA-deficient fish possessed much reduced levels of DHA, 12% compared to 35% in the high-PUFA-fed fish. Instead, higher proportions of EPA and other C22, C20 and C18 PUFA are incorporated. Data are also shown in *Figure 2* for sole fed the low-PUFA diet enriched with purified EPA. Interestingly, these fish did not possess greater proportions of EPA than the low-PUFA sole; rather, high dietary EPA appeared to result in an increase in the proportions of DHA. It appears therefore, that EPA cannot directly replace DHA in these neural membranes but, instead, the limited capacity to synthesize DHA from EPA appears to be utilized. The stress resistance phenotype is thus correlated with high proportions of DHA alone in neural tissue, with stress susceptibility associated with low proportions of this particular fatty acid, irrespective of other LC-PUFA that may be present.

Of greater note, however, dietary deficiency of these essential fatty acids has also been correlated with a retarded brain growth rate. Morphological analysis of brain development in the dietary-treated Dover sole revealed that the n-3 LC-PUFA-deficient diet was associated with a 35% reduction in the volume of the thalamus, following allometric correction for body size, from 0.17 to 0.11 mm³. Similar findings have been reported by other studies. The relative volume of the cerebellum in larvae of the Japanese flounder, *Paralichthys olivaceus*, fed a PUFA-deficient diet, was reduced significantly compared to that of EPA and DHA enriched larvae (Furuita *et al.*, 1998). Likewise, whole brain, cerebellum and tectum volumes for DHA and EPA enriched larval yellowtail, *Seriola quinqueradiata*, were significantly larger than for deficient fish, but, in addition, fish supplemented with DHA had significantly greater whole brain volumes than those enriched only with EPA

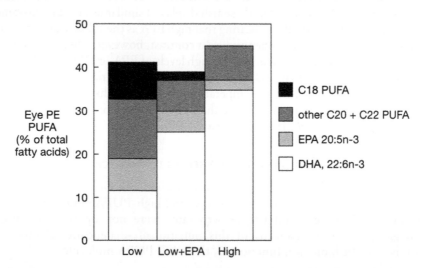

Figure 2. *Effect of dietary treatment on the PUFA content of eye phosphatidylethanolamine lipids in pre-weaned Dover sole. Fish were fed low-PUFA, low-PUFA+EPA or high-PUFA enriched* Artemia *diets. Reproduced from Logue* et al. *(2000). Lipids 35: 745–755, with permission of AOCS Press.*

(Ishizaki *et al.*, 2000), suggesting again an inability of EPA to functionally replace DHA in neural membranes.

Upon further analysis, the reductions in brain volume associated with the PUFA-deprived Dover sole were attributable to a reduction in neuronal cell density, that is, a reduced number of brain cells (*Figure 3*). The neuronal densities for the thalamus, tectum, medulla and cerebellum regions were all significantly reduced in low-PUFA fish, approximating 60–70% of the densities found in the PUFA-enriched sole. The four regions studied comprised approximately 75% of the total brain volume. Reductions in the number and size of hippocampal cells in rat pups resulted in an impaired neurobehavioural development (Ba *et al.*, 1996), while in humans a reduced brain volume or cell number is associated with a variety of psycho-neurological disorders including schizophrenia (Popken *et al.*, 2000), Huntington's disease (Heinsen *et al.*, 1999), autism (Ingram *et al.*, 2000), Alzheimer's disease (Mouton *et al.*, 1994) and development of the Down's Syndrome brain (Busciglio and Yankner, 1995).

As with stress resistance, however, following weaning onto a common PUFA-enriched feed, brain neuronal densities and brain volumes in the low-PUFA sole were seen to recover. These findings have also been reported for reared turbot, *Scophthalmus maximus*, where an elevated DHA content of the weaning diet resulted in a rapid and specific incorporation of DHA into brain phospholipids and a significant increase in brain weight (Castell *et al.*, 1994; Mourente and Tocher, 1992). This is contrary to studies on mammals where recovery of the nervous system after n-3 PUFA deficiency is slow or non-existent (Bourre *et al.*, 1993; Crawford, 2000). However, unlike higher vertebrates, the growth of neural tissue in teleost fish continues throughout life owing to the constant generation of new neuronal cells (Zupanc, 1999).

The debilitating effects of dietary lipid deficiencies are not just related to structural abnormalities of the neural tissue in marine fish, but a functional impairment of the

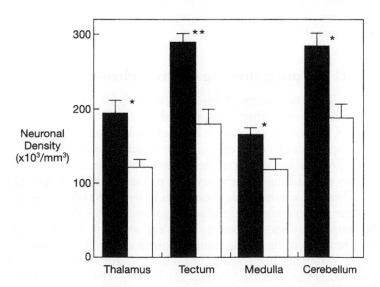

Figure 3. *Neuronal densities of different brain regions of pre-weaned Dover sole fed high-PUFA (filled bars) or low-PUFA (open bars) enriched Artemia diets. Values given are means (n = 5) ± standard errors. *P<0.05 and **P<0.005 (unpublished data from Barlow, Logue, Howell, Howard and Cossins).*

nervous system has been reported by several authors. A good example is the effect of DHA deficiency upon visual performance in juvenile herring (Bell *et al.*, 1995). Deficiency during development resulted in decreased levels of DHA in retinal phospholipids correlated with a reduced feeding ability at low light intensities. This depression in feeding efficiency of the juvenile would have obvious effects upon growth rate, health and subsequent survival. Similarly, dietary DHA deficiency during central nervous system development in the larval yellowtail, *Seriola quinqueradiata*, was associated with an inability to form schools, implicating DHA as a critical factor in the ontogeny of schooling behaviour (Masuda *et al.*, 1999; Masuda and Tsukamoto, 1999). Furthermore, marine flatfish aquaculture is associated with a high incidence of abnormal pigmentation. A proposed cause is an impaired nervous and/or visual system, whereby visual signals in the eye, brain or other regions of the neural network are processed incorrectly. This condition can be improved or resolved by increasing dietary DHA provision (Kanazawa, 1993; Sargent *et al.*, 1999).

4.2 *Levels of n-6 bioactive eicosanoids*

Whole carcass eicosanoid analysis of the dietary treated Dover sole study showed significant differences in the levels of prostaglandin E, with the PUFA-deficient sole (277.3 ± 50.5 pg g^{-1} body tissue) possessing more than double the levels of PGE found in the high-PUFA group (124.3 ± 29.7 pg g^{-1}). This increase in total PGE was due predominantly to elevated levels of the AA-derived n-6 eicosanoid PGE2 (Logue *et al.*, 2000a), that is, the PUFA-deprived fish possessed elevated levels of these bioactive n-6 eicosanoids, over-production of which are implicated in pathophysiological disorders (Lands, 1993). In nutritional studies on Atlantic salmon, feeding smolts a diet enriched with n-6 fatty acids resulted in a high n-6/n-3 acyl chain ratio in body tissues and elevated levels of PGE$_2$ (Bell *et al.*, 1993). Associated with this were severe lesions of the heart tissue and increased mortality when subjected to handling and transportation stress (Bell *et al.*, 1991).

5. Larval diet and programming of stress phenotype

In conclusion, dietary deficiency of essential n-3 LC-PUFA in larval Dover sole promotes the stress-sensitive phenotype, possibly due to sub-optimal nervous system development and/or elevated production of stress-inducing eicosanoids. Stress resistance, however, appears to be recovered in these deficient animals upon feeding of an appropriate PUFA-enriched diet. Yet, there is evidence that more subtle differences in tolerance may persist long into the juvenile phase that are not detected by acute stress tests. Following a similar dietary regime in Dover sole as described above, longterm cold tolerance tests were performed on juveniles several months old. In this instance the temperature was lowered over several weeks to 5°C and mortality was calculated as a function of time. The results showed a clear distinction in cold sensitivity despite no difference being found in the short-term acute tests (Howell *et al.*, 1995). Furthermore, the juvenile fish previously subjected to the PUFA-deficient diet during the larval periods were more susceptible to the effects of handling and transportation, suffering higher mortality (B. Howell, personal communication). Such protracted, more measured trials, rather than the 'short, sharp, shock' treatment, are arguably of greater relevance and have more in common with the stress-inducing environmental fluctuations the fish would encounter in the wild. Thus a canalization or programming effect may exist whereby dietary lipid deficiencies during the larval period can define the environmental stress phenotype of subsequent life stages. An optimal nutritional input during this early phase would be critical for the establishment of a healthy individual.

Another consideration regarding the rearing of marine fish species is the effect of maternal diet on the fitness of the developing embryo and larva. Until first feeding, lipid and all other dietary components are derived solely from the yolk sac, the composition of which is determined by the maternal diet. These first few days of life would see a crucial period of rapid growth and tissue differentiation and therefore optimized feeding of spawning females would result in a greater developmental advantage to the larvae. Indeed, high levels of broodstock dietary n-3 PUFA are positively correlated with egg fertilization and hatching rates, and growth, survival and environmental stress tolerance of the resulting larvae and juveniles (AbiAyad et al., 1997; Czesny and Dabrowski, 1998; Furuita et al., 2000; Rainuzzo et al., 1997).

6. Ecological relevance of dietary lipid-determined stress phenotype

The influence of nutrition on larval and later life stage stress phenotype is not restricted to aquaculture but may exist under natural conditions in the wild. As indicated, marine fish have a requirement for n-3 LC-PUFA in the diet, but there may be instances where such fatty acids are deficient in the food chain. Marine phytoplankton is frequently dominated by a limited number of species which may have very different fatty acid compositions. For instance, whereas dinoflagellates (Dinophyceae) possess large proportions of DHA and EPA, coccolithophores (Haptophyceae) are enriched in DHA but contain little EPA, while diatoms (Bacillariophyceae) are enriched in EPA but deficient in DHA, and green algae (Chlorophyceae) possess negligible amounts of either n-3 LC-PUFA (Ackman et al., 1968; Sargent et al., 1995; Volkman et al., 1981). Indeed, species succession within a bloom can be followed by changes in the phytoplankton fatty acid signature within the water column (Hayakawa et al., 1996; Shamsudin, 1998). Adequate dietary provision of 22:6n-3 in early larval stages of marine fish therefore depends critically upon the species composition of the basal algal community upon which it directly or indirectly feeds. It has not been demonstrated whether plankton deficient in DHA can dominate a bloom at the critical time of larval feeding (Bell and Sargent, 1996), but phytoplankton abundance and composition are subject to change both spatially and temporally. For instance, in the years 1975–1976 the phytoplankton bloom in the English Channel in spring and autumn was dominated by diatoms, in early and late summer by a mixture of diatoms and dinoflagellates and in mid-summer by flagellates and dinoflagellates (Holligan and Harbour, 1977). Commercial fish species such as plaice and herring are known to be winter spawners in the English Channel and southern North Sea such that hatching and larval feeding coincides with the spring bloom (Cushing, 1975). Blooms of DHA-deficient diatoms during the larval period could result in low levels of these fatty acids within body tissues, leading to a reduced general fitness, increased susceptibility to natural environmental stresses and consequent increased mortality rate.

If nutritional lipid deficiencies during early development do indeed determine the stress resistance of later life stages, then the larval diet will have important consequences for recruitment of juveniles into the adult population. This would be particularly relevant for commercially fished species where the pressures of over-fishing may be compounded by a high natural larval and juvenile mortality. Although it is estimated that mortality in high fecundity species such as plaice is 99.995% in the first year of life, fish stocks are often dominated by specific year classes, that is, certain year classes are notable by an under-representation within the population (Barnes and Hughes, 1988). Furthermore, recruitment and stock size often vary quite independently from fishing mortality, indi-cating that changes in the environment are a major cause of variability in commercial fish stock recruitment (Rothschild, 2000). With this in mind larval nutrition could be an important environmental influence on this year-to-year variation (Bell and Sargent, 1996).

Overview

Finally, we should consider the possible effects of global climate change on the plank-tonic dietary lipid intake of larval fish. As a result of global warming, changes in phyto-plankton composition may arise due to proliferation of non-indigenous, warm-tolerant species into higher latitudes (Nehring, 1998). Moreover, it has been proposed that ozone-related increases in UV-B would not only result in a direct reduction in phytoplankton production (Hader, 1996), but that these effects could be species-specific, resulting in both a reduction and alteration in the phytoplankton community composition (Davidson *et al.*, 1996; Ekelund, 1994; Wangberg *et al.*, 1996). There is also evidence that elevated UV radiation may have a direct effect on the lipid synthesis capacity of phytoplankton. Absolute concentrations of phytoplankton fatty acids were found to be lower in the presence of UV-B, due largely to suppression of PUFA synthesis (Goes *et al.*, 1994). In addition, it has been demonstrated that the essential n-3 LC-PUFA of EPA and DHA were particularly susceptible (Hessen *et al.*, 1997). Such changes in phytoplankton community structure and fatty acid compo-sition would alter the quality of the lipid pool available to higher trophic levels, with considerable implications for developing fish larvae, particularly if proportions of the essential PUFA such as DHA and EPA were selectively reduced.

Acknowledgements

This work was supported by grants from the Natural Environmental Research Council (UK), as part of the Developmental Ecology of Marine Animals (DEMA) thematic programme, and the Ministry of Agriculture, Fisheries and Food (UK). The author would like to acknowledge the contributions of Andrew R. Cossins, Vyvian Howard and Liam Barlow (University of Liverpool, UK) and Bari R. Howell (Centre for Environmental, Fisheries and Aquaculture Science, Weymouth, UK) to this work. This work was performed at the CEFAS laboratory, Conway, N. Wales and the labo-ratory for Environmental Gene Regulation, School of Biological Sciences, University of Liverpool, Liverpool, UK.

References

AbiAyad, S., Melard, C. and Kestemont, P. (1997) Effects of n-3 fatty acids in Eurasian perch broodstock diet on egg fatty acid composition and larvae stress resistance. *Aquacult. Int.* 5: 161–168.

Ackman, R.G., Tocher, C.S. and McLachlan, J. (1968) Marine phytoplankter fatty acids. *J. Fish. Res. Board Can.* 25: 1603–1620.

Agostoni, C., Trojan, S., Bellu, R., Riva, E. and Giovannini, M. (1995) Neurodevelopmental quotient of healthy term infants at 4 months and feeding practice: the role of long chain polyunsaturated fatty acids. *Pediatr. Res.* **38**: 262–266.

Applegate, K.R. and Glomset, J.A. (1991) Effect of acyl chain unsaturation on the packing of model diacylglycerols in simulated monolayers. *J. Lipid Res.* **32**: 1645–1655.

Ba, A., Seri, B.V. and Han, S.H. (1996) Thiamine administration during chronic alcohol intake in pregnant and lactating rats: effects on the offspring neurobehavioural development. *Alcohol Alcohol.* **31**: 27–40.

Baenziger, J.E., Jarrell, H.C. and Smith, I.C. (1992) Molecular motions and dynamics of a diunsaturated acyl chain in a lipid bilayer: implications for the role of polyunsaturation in biological membranes. *Biochemistry* **31**: 3377–3385.

Barnes, R.S.K. and Hughes, R.N. (1988) *An Introduction to Marine Ecology*, 2nd Edn. Blackwell Scientific, Oxford.

Bell, J.G., McVicar, A.H., Park, M.T. and Sargent, J.R. (1991) High dietary linoleic acid affects the fatty acid compositions of individual phospholipids from tissues of Atlantic salmon (*Salmo salar*): association with stress susceptibility and cardiac lesion. *J. Nutr.* **121**: 1163–1172.

Bell, J.G., Dick, J.R. and Sargent, J.R. (1993) Effect of diets rich in linoleic or alpha-linolenic acid on phospholipid fatty acid composition and eicosanoid production in Atlantic salmon (*Salmo salar*). *Lipids* **28**: 819–826.

Bell, J.G., Farndale, B.M., Dick, J.R. and Sargent, J.R. (1996) Modification of membrane fatty acid composition, eicosanoid production, and phospholipase A activity in Atlantic salmon (*Salmo salar*) gill and kidney by dietary lipid. *Lipids* **31**: 1163–1171.

Bell, M. and Dick, J. (1991) Molecular species composition of the major diacyl glycerophospholipids from muscle, liver, retina and brain of cod (*Gadus morhua*). *Lipids* **26**: 565–573.

Bell, M. and Sargent, J.R. (1996) Lipid nutrition and fish recruitment. *Mar. Ecol. Prog. Ser.* **134**: 315–316.

Bell, M.V., Batty, R.S., Dick, J.R., Fretwell, K., Navarro, J.C. and Sargent, J.R. (1995) Dietary deficiency of docosahexaenoic acid impairs vision at low light intensities in juvenile herring (*Clupea harengus* L.). *Lipids* **30**: 443–449.

Birch, D.G., Birch, E.E., Hoffman, D.R. and Uauy, R.D. (1992a) Retinal development in very low birth weight infants fed diets differing in omega-3 fatty acids. *Invest. Ophthal. Visual Sci.* **33**: 2365–2376.

Birch, E.E., Birch, D.G., Hoffman, D.R. and Uauy, R. (1992b) Dietary essential fatty acid supply and visual acuity development. *Invest. Ophthal. Visual Sci.* **33**: 3242–3253.

Bourre, J.M., Bonneil, M., Clement, M., Dumont, O., Durand, G., Lafont, H., Nalbone, G. and Piciotti, M. (1993) Function of dietary polyunsaturated fatty acids in the nervous system. *Prostaglandins Leukot. Essent. Fatty Acids* **48**: 5–15.

Busciglio, J. and Yankner, B.A. (1995) Apoptosis and increased generation of reactive oxygen species in Down's syndrome neurons in vitro. *Nature* **378**: 776–779.

Castell, J., Bell, J., Tocher, D. and Sargent, J. (1994) Effects of purified diets containing different combinations of arachidonic and docosahexaenoic acid on survival, growth and fatty acid composition of juvenile turbot (*Scophthalmus maximus*). *Aquaculture* **128**: 315–333.

Cossins, A.R. (1994) Homeoviscous adaptation of biological membranes and its functional significance. In: *Temperature Adaptation of Biological Membranes* (ed A.R. Cossins). Portland Press, London, pp. 63–76.

Cossins, A.R., Friedlander, M.J. and Prosser, C.L. (1977) Correlations between behavioural temperature adaptations of goldfish and the viscosity and fatty acid composition of their synaptic membranes. *J. Comp. Physiol.* **120**: 109–121.

Crawford, M. (2000) Placental delivery of arachidonic and docosahexaenoic acids: implications for the lipid nutrition of preterm infants. *Am. J. Clin. Nutr.* **71**: 275S–284S.

Cushing (1975) *Marine Ecology and Fisheries.* Cambridge University Press, Cambridge.

Czesny, S. and Dabrowski, K. (1998) The effect of egg fatty acid concentrations on embryo viability in wild and domesticated walleye (*Stizostedion vitreum*). *Aquat. Living Res.* 11: 371–378.

Davidson, A., Marchant, H. and Delamare, W. (1996) Natural UV-B exposure changes the species composition of Antarctic phytoplankton in mixed culture. *Aquat. Microb. Ecol.* 10: 299–305.

Ekelund, N. (1994) Influence of UV-B radiation on photosynthetic light response curves, absorption-spectra and motility of four phytoplankton species. *Physiol. Plant.* 91: 696–702.

Friedlander, M.J., Kotchabhakdi, N. and Prosser, C.L. (1976) Effects of cold and heat on behaviour and cerebellar function in goldfish. *J. Comp. Physiol.* 112: 19–45.

Furuita, H., Takeuchi, T., Watanabe, T., Fujimoto, H., Sekiya, S. and Imaizumi, K. (1996) Requirements of larval yellowtail for eicosapentaenoic acid, docosahexaenoic acid, and n-3 highly unsaturated fatty acid. *Fish. Sci.* 62: 372–379.

Furuita, H., Takeuchi, T. and Uematsu, K. (1998) Effects of eicosapentaenoic and docosa-hexaenoic acids on growth, survival and brain development of larval. *Aquaculture* 161: 269–279.

Furuita, H., Konishi, K. and Takeuchi, T. (1999) Effect of different levels of eicosapentaenoic acid and docosahexaenoic acid in *Artemia nauplii* on growth, survival and salinity tolerance of larvae of the Japanese flounder, *Paralichthys olivaceus. Aquaculture* 170: 59–69.

Furuita, H., Tanaka, H., Yamamoto, T., Shiraishi, M. and Takeuchi, T. (2000) Effects of n-3 HUFA levels in broodstock diet on the reproductive performance and egg and larval quality of the Japanese flounder, *Paralichthys olivaceus. Aquaculture* 187: 387–398.

Goes, J., Handa, N., Taguchi, S. and Hama, T. (1994) Effect of UV-B radiation on the fatty acid composition of the marine phytoplankton *Tetraselmis* sp: relationship to cellular pigments. *Mar. Ecol. Prog. Ser.* 114: 259–274.

Greiner, R., Moriguchi, T., Hutton, A., Slotnick, B. and Salem, N. (1999) Rats with low levels of brain docosahexaenoic acid show impaired performance in olfactory-based and spatial learning tasks. *Lipids* 34: S239–S243.

Hader, D. (1996) Effects of enhanced solar UV-B radiation on phytoplankton. *Sci. Mar.* 60: 59–63.

Hayakawa, K., Handa, N., Kawanobe, K. and CS, W. (1996) Factors controlling the temporal variation of fatty acids in piculate matter during a phytoplankton bloom in a marine mesocosm. *Mar. Chem.* 52: 233–244.

Hazel, J.R. (1995) Thermal adaptation in biological membranes: is homeoviscous adaptation the explanation? *Annu. Rev. Physiol.* 57: 19–42.

Hazel, J.R. and Williams, E.E. (1990) The role of alterations in membrane lipid composition in enabling physiological adaptation of organisms to their physical environment. *Prog. Lipid Res.* 29: 167–227.

Heinsen, H., Rub, U., Bauer, M., Ulmar, G., Bethke, B., Schuler, M., Bocker, F., Eisenmenger, W., Gotz, M., Korr, H. and Schmitz, C. (1999) Nerve cell loss in the thalamic mediodorsal nucleus in Huntington's disease. *Acta Neuropathol. (Berl.)* 97: 613–622.

Hessen, D., De Lange, H. and Van Donk, E. (1997) UV-induced changes in phytoplankton cells and its effects on grazers. *Freshwater Biol.* 38: 513–524.

Hoffman, D.R., Birch, E.E., Birch, D.G. and Uauy, R.D. (1993) Effects of supplementation with omega 3 long-chain polyunsaturated fatty acids on retinal and cortical development in premature infants. *Am. J. Clin. Nutr.* 57: 807S–812S.

Holligan, P.M. and Harbour, D.S. (1977) The vertical distribution and succession of phyto-plankton in the Western English Channel. *J. Mar. Biol. Assoc. UK* 57: 1075–1093.

Howell, B.R., Beard, T.W. and Hallam, J.D. (1995) The effects of diet quality on the low teper-ature tolerance of juvenile Dover sole, *Solea solea.* ICES internal report, CM 1995/F13.

Ingram, J.L., Peckham, S.M., Tisdale, B. and Rodier, P.M. (2000) Prenatal exposure of rats to valproic acid reproduces the cerebellar anomalies associated with autism. *Neurotoxicol. Teratol.* 22: 319–324.

Ishizaki, Y., Uematsu, K. and Takeuchi, T. (2000) Preliminary study of the effect of dietary docosahexaenoic acid on the volumetric growth of the brain in larval yellowtail. *Fish. Sci.* **66**: 611–613.

Kanazawa, A. (1993) Nutritional mechanisms involved in the occurrence of abnormal pigmentation in hatchery-reared flatfish. *J. World Aquacult. Soc.* **24**: 162–166.

Kanazawa, A. (1997) Effects of docosahexaenoic acid and phospholipids on stress tolerance of fish. *Aquaculture* **155**: 129–134.

Kinsella, J.E., Lokesh, B., Broughton, S. and Whelan, J. (1990) Dietary polyunsaturated fatty acids and eicosanoids: potential effects on the modulation of inflammatory and immune cells: an overview. *Nutrition* **6**: 24–44 (discussion 59–62).

Lands, W. (1986) *Fish and Human Health*. Academic Press, Orlando, FL.

Lands, W. (1993) Eiscosanoids and health. *Ann. NY Acad. Sci.* **676**: 46–59.

Lands, W.E., Libelt, B., Morris, A., Kramer, N.C., Prewitt, T.E., Bowen, P., Schmeisser, D., Davidson, M.H. and Burns, J.H. (1992) Maintenance of lower proportions of (n–6) eicosanoid precursors in phospholipids of human plasma in response to added dietary (n–3) fatty acids. *Biochim. Biophys. Acta* **1180**: 147–162.

Litman, B.J. and Mitchell, D.C. (1996) A role for phospholipid polyunsaturation in modulating membrane protein function. *Lipids* **31**: S193–S197.

Logue, J.A., Howell, B.R., Bell, J.G. and Cossins, A.R. (2000a) Dietary n-3 long-chain polyunsaturated fatty acid deprivation, tissue lipid composition, ex vivo prostaglandin production, and stress tolerance in juvenile Dover sole (*Solea solea* L.). *Lipids* **35**: 745–755.

Logue, J.A., De Vries, A.L., Fodor, E. and Cossins, A.R. (2000b) Lipid compositional correlates of temperature-adaptive interspecific differences in membrane physical structure. *J. Exp. Biol.* **203** (14): 2105–2115.

Masuda, R. and Tsukamoto, K. (1999) School formation and concurrent developmental changes in carangid fish with reference to dietary conditions. *Environ. Biol. Fish.* **56**: 243–252.

Masuda, R., Takeuchi, T., Tsukamoto, K., Sato, H., Shimizu, K. and Imaizumi, K. (1999) Incorporation of dietary docosahexaenoic acid into the central nervous system of the yellowtail *Seriola quinqueradiata*. *Brain Behav. Evol.* **53**: 173–179.

Mitchell, D.C., Gawrisch, K., Litman, B.J. and Salem, N. Jr (1998) Why is docosahexaenoic acid essential for nervous system function? *Biochem. Soc. Trans.* **26**: 365–370.

Mourente, G. and Tocher, D. (1992) Effects of weaning onto a pelleted diet on docosahexaenoic acid (22:6n-3) levels in brain of developing turbot (*Scophthalmus maximus*). *Aquaculture* **105**: 363–377.

Mouton, P.R., Pakkenberg, B., Gundersen, H.J. and Price, D.L. (1994) Absolute number and size of pigmented locus coeruleus neurons in young and aged individuals. *J. Chem. Neuroanat.* **7**: 185–190.

Nehring, S. (1998) Establishment of thermophilic phytoplankton species in the North Sea: biological indicators of climatic changes? *ICES J. Mar. Sci.* **55**: 818–823.

Neuringer, M., Connor, W.E., Lin, D.S., Barstad, L. and Luck, S. (1986) Biochemical and functional effects of prenatal and postnatal omega 3 fatty acid deficiency on retina and brain in rhesus monkeys. *Proc. Natl Acad. Sci. USA* **83**: 4021–4025.

Nordoy, A., Hatcher, L., Goodnight, S., Fitzgerald, G.A. and Conner, W.E. (1994) Effects of dietary fat content, saturated fatty acids, and fish oil on eicosanoid production and hemostatic parameters in normal men. *J. Lab. Clin. Med.* **123**: 914–920.

Popken, G.J., Bunney, W.E. Jr, Potkin, S.G. and Jones, E.G. (2000) Subnucleus specific loss of neurons in medial thalamus of schizophrenics. *Proc. Natl Acad. Sci. USA* **97**: 9276–9280.

Rabinovitch, A.L. and Ripatti, P.O. (1991) On the conformational, physical properties and functions of polyunsaturated acyl chains. *Biochim. Biophys. Acta* **1085**: 53–62.

Rainuzzo, J., Reitan, K. and Olsen, Y. (1997) The significance of lipids at early stages of marine fish: a review. *Aquaculture* **155**: 103–115.

Rothschild, B. (2000) Fish stocks and recruitment: the past thirty years. *ICES J. Mar. Sci.* **57**: 191–201.

Salem, N. Jr and Niebylski, C.D. (1995) The nervous system has an absolute molecular species requirement for proper function. *Mol. Membr. Biol.* **12**: 131–134.

Salem, N., Kim, H.-Y. and Yergey, J.A. (1986) Docosahexaenoic acid: membrane function and metabolism. In: *Health Effects of Polyunsaturated Fatty Acids in Seafoods* (eds A.P. Simopoulos, R.R. Kifer and R.E. Martin). Academic Press, New York, pp. 263–317.

Sargent, J.R., Bell, M.V., Bell, J.G., Henderson, R.J. and Tocher, D.R. (1995) Origins and functions of n-3 polyunsaturated fatty acids in marine organisms. In: *Phospholipids: Characterisation, Metabolism and Novel Biological Applications* (eds G. Ceve and F. Paltauf). American Oil Chemistry Society Press, Champaign, IL, pp. 248–259.

Sargent, J., McEvoy, L., Estevez, A., Bell, G. and Bell, M. (1999) Lipid nutrition of marine fish during early development: current status and future directions. *Aquaculture* **179**: 217–229.

Shamsudin, L. (1998) Seasonal variation of fatty acid content in natural microplankton from the Tumpat coastal waters of the South China Sea. *Arch. Physiol. Biochem.* **106**: 253–260.

Tago, A., Yamamoto, Y., Teshima, S. and Kanazawa, A. (1999) Effects of 1,2-di-20:5-phosphatidylcholine (PC) and 1,2-di-22:6-PC on growth and stress tolerance of Japanese flounder (*Paralichthys olivaceus*) larvae. *Aquaculture* **179**: 231–239.

Tocher, D.R. and Harvie, D.G. (1988) Fatty acid compositions of the major phosphoglycerides from fish neural tissues – (n-3) and (n-6) polyunsaturated fatty acids in rainbow trout (*Salmo gairdneri*) and cod (*Gadus morhua*) brains and retinas. *Fish Physiol. Biochem.* **5**: 229–239.

Tuncer, H., Harrell, R. and Chai, T. (1993) Beneficial-effects of n-3 hufa enriched *Artemia* as food for larval palmetto bass (*Morone saxatilisxm chrysops*). *Aquaculture* **110**: 341–359.

Uauy, R.D., Birch, D.G., Birch, E.E., Tyson, J.E. and Hoffman, D.R. (1990) Effect of dietary omega-3 fatty acids on retinal function of very low birth weight neonates. *Pediatr. Res.* **28**: 485–492.

Volkman, J.K., Smith, D.J., Eglinton, G., Forsberg, T.E.V. and Corner, E.D.S. (1981) Sterol and fatty-acid composition of 4 marine haptophycean algae. *J. Mar. Biol. Assoc. UK* **61**: 509–527.

Wangberg, S., Selmer, J. and Gustavson, K. (1996) Effects of UV-B radiation on biomass and composition in marine phytoplankton communities. *Sci. Mar.* **60**: 81–88.

Whelan, J. (1996) Antagonistic effects of dietary arachidonic acid and n-3 polyunsaturated fatty acids. *J. Nutr.* **126**: 1086S–1091S.

Wiegand, R.D., Koutz, C.A., Stinson, A.M. and Anderson, R.E. (1991) Conservation of docosahexaenoic acid in rod outer segments of rat retina during n-3 and n-6 fatty acid deficiency. *J. Neurochem.* **57**: 1690–1699.

Zupanc, G.K. (1999) Neurogenesis, cell death and regeneration in the adult gymnotiform brain. *J. Exp. Biol.* **202**: 1435–1446.

7

Influence of environmental factors on the ontogeny of the immune system in turbot

C. Low, I. Taylor, W. Melvin, M.F. Tatner, T.H. Birkbeck and C. Secombes

1. Introduction

Flatfish are of high value in wild fisheries and this has stimulated development of aquaculture for several species (Tilseth, 1990), with production of farmed turbot, *Scophthalmus maximus* now exceeding 2000 tonnes per annum in Europe (Riaza and Hall, 1993). One factor that limits aquaculture production is microbial infection and extensive losses still occur in the early rearing stages of flatfish culture. The immune response in all fish is regulated by temperature and immunological maturity also appears to be related to size, rather than age. Therefore, factors such as food availability and temperature, which affect the growth of larvae, may affect the development of immune organs and mechanisms, as well as the efficiency of the immune response to respond to pathogens. In aquaculture, flatfish are often reared at temperatures greater than those found in natural waters and it is not known whether this is beneficial, neutral or deleterious to immunocompetence in the fish. Similarly, rising temperatures caused by global climate change may affect the immune competence in flatfish in the wild with consequences for these fisheries. Therefore, this review considers available knowledge of the effect of temperature and other factors on the immune development of turbot larvae and larvae of other relevant species.

2. Ontogeny of the lymphoid organs

The ontogeny of the lymphoid tissue in turbot larvae is influenced by genetic factors and environmental factors, such as temperature and diet (Jones, 1972). It is important to determine immune maturity in order to decide the earliest time at which vaccination can take place. In an elegant study by Burrows (1995), the development of lymphoid tissues in turbot was linked to morphological development. These processes were assumed to occur at a similar rate and be similarly influenced by environmental variables. The author used the classification of larval developmental stages of Al-Maghazachi and Gibson (1984) on which to base a similar system for lymphoid ontogeny (*Table 1*). Turbot fry reared at 16°C were sampled from 24 h post-hatch, at regular intervals until day 63. The length, weight and gross morphology of the fry were noted and histological samples were taken to measure corresponding lymphoid ontogeny.

Table 1. The ontogeny of the morphological features (after Burrows, 1995)

Stage	Morphological features	Lymphoid ontogeny	Age/post-hatch (16 °C)
1	Larvae symmetrical; yolk sac present	Thymic, kidney and splenic rudiments present	1–3 days
2	Larvae symmetrical; spine and swim bladder development	First lymphoid cells appear in thymus and kidney	4–10 days
3	Appearance of fin rays; notochord straight	Kidney and spleen increase in size, lymphoid development in kidney	11–12 days
4	Asymmetry and eye migration; notochord slanted dorsally	Thymic cells become lymphoid, lymphoid cells present in splenic white pulp	13–23 days
5	Eye migration complete; spines and swim bladder resorbed	Cortex and medulla regions develop in thymus, melanomacrophage centres appear in spleen	28–63 days

2.1 *Development of the lymphoid organs in turbot*

Stage one (1–3 days). The thymus, kidney and spleen were present by this stage, the thymus consisting of a thickening of the epithelial lining of the dorso-anterior section of the pharynx. This appeared to contain only epithelial-type cells. The kidney occupied the length of the peritoneal cavity; some nephritic tubules were present but there were very few immature lymphoid cells. The presence of haemopoietic stem cells was indicated by a large number of undifferentiated eosinophilic cells. The turbot spleen consisted of a cluster of large, epithelial-like, cells adjacent to the pancreatic rudiment.

Stage two (4–10 days). The thymus was now larger and some lymphoid cells were present. Between the middle of stages 2 and 3, connective tissue formed around the thymus. In the kidney, erythrocytes and lymphoid cells were observed surrounding nephritic tubules, but the anterior kidney had fewer tubules. A thin capsule began to surround the spleen, which had increased in size and shape, and erythrocytes were also visible.

Stage three (11–12 days). The connective tissue surrounding the thymus began to infiltrate the parenchyma. The kidney appeared larger, denser with leucocytes and containing more tubules. The spleen became larger with continued development of the surrounding capsule.

Stage four (13–23 days). The thymus appeared to be fully lymphoid in composition and continued to increase in size. The kidney continued to develop as in stage 3. The spleen was larger and lymphoid cells began to appear with epithelial-like cells, these becoming more obvious throughout stage 4 until red and white pulp regions were established.

Stage five (28–63 days). The thymus was larger and the cells had differentiated into a cortex and a medulla region. As in stage 4, the kidney continued to develop as in stage 3. In the spleen, the first melanomacrophage centre was observed. Development of each organ appeared to be complete by the end of metamorphosis.

In comparison with lymphoid development in other fish, the turbot thymus and kidney were less well developed upon hatching (Burrows, 1995). However, the spleen was present shortly after hatching in turbot, whereas in many other species the spleen develops later (Manning, 1994). The authors suggest that early development of the spleen could indicate a more important role in the immunity of turbot. Comparison between this study and those of other authors is hampered by differences in the environmental temperature ranges of different fish species. In this study, the lymphoid development of the turbot could be assessed using morphological features. However, caution must be exercised since the appearance of lymphoid organs does not mean that they are fully functional.

3. Non-specific defences

Turbot, like other species of bony fish, possess a wide range of non-specific defences involving cells (e.g. granulocytes and monocytes/macrophages) and soluble mediators. Such responses serve as a first line of defence against infection but if they are breached inducible responses are elicited, as seen with inflammation. In contrast to specific defences, mediated by lymphocytes, these responses show no memory component and are similar in magnitude and kinetics on subsequent encounters with the same disease. Thus, whilst they can be temporarily elevated by 'immunostimulants' they cannot be primed for long-term protection.

3.1 *Cell types involved*

In turbot, cells involved in the non-specific defences are present in the blood and a range of lymphoid tissues, with the anterior kidney and spleen being the most studied (Quentel and Obach, 1992). In the blood, where 4.5% of total cells are leucocytes, approximately 30% of these are granulocytes, with the majority (77%) being polymorphonuclear neutrophils (Burrows and Fletcher, 1987). Monocytes are the least frequently found blood cell type, representing only 0.2% of leucocytes (Quentel and Obach, 1992). In the anterior kidney, which has a significant haemopoietic role in turbot (Quentel and Obach, 1992), leucocytes represent 96% of the total cell population whilst in the spleen they represent 46% of total cells. It is difficult to distinguish the different leucocyte types in these tissues but they have served as useful sources of macrophages and neutrophils for *in vitro* studies.

The cells of the non-specific defences have a variety of functions. They can be effector cells that migrate to a site of infection to ingest and kill microbes, using mechanisms involving the production of oxygen and nitrogen free radicals. They can also help to regulate the ensuing immune response by the release of cytokines to augment the non-specific defences and to activate lymphocytes. Many studies in turbot have focused on such responses which are reviewed below.

3.2 *Phagocytosis*

Both splenic and head kidney leucocytes have been shown to be phagocytic in turbot, although the phagocytic rate is very low unless the cells are derived from fish previously injected intraperitoneally with a bacterin (e.g. *Photobacterium (Vibrio) damsela*, *Pasteurella piscicida*) for 2–3 days (Santarem and Figueras, 1994; Figueras *et al.*, 1997). Longer timings or suboptimal bacterin doses (<100 μg antigen) are less effective.

The effect of coating particles with antibody and/or complement (opsonization) on the phagocytic activity of turbot leucocytes has been investigated by Figueras *et al.* (1997). Whilst no change in phagocytic rate occurs using particles opsonized with complement (normal serum), relative to non-opsonized particles, opsonization with specific antibody (heat-inactivated immune serum) or with specific antibody and complement (unheated immune serum) increases the phagocytic rate three-fold. These results suggest that turbot phagocytes have receptors for antibodies which play an important part in facilitating particle ingestion.

As with injection of bacterins, injection of turbot with β-glucans derived from the cell wall of the yeast *Saccharomyces cerevisiae* augments phagocyte activity (Figueras *et al.*, 1998). Injection of β-glucans at the time of bacterin administration, or 2 and 4 days after bacterin administration, elevates phagocyte activity even further than either treatment alone. Phagocyte activity may also be suppressed in turbot, for example incubation of head kidney macrophages with oxytetracycline, an antibiotic used in aquaculture, at concentrations ranging from 0.5 to 50 μg ml^{-1} decreased phagocyte activity (Tafalla *et al.*, 1999a). Since macrophage viability is not influenced by oxytetracycline this was not a toxic effect, but probably resulted from decreased protein synthesis as oxytetracycline is known to permeate mammalian leucocytes and inhibit protein synthesis (Bogert and Kroon, 1982).

3.3 *Respiratory burst*

The respiratory burst is a period of intense oxygen consumption during which phagocytes are able to generate superoxide anion (O_2^-) and a number of reactive derivatives such as hydrogen peroxide and hydroxyl radicals. The production of these reactive oxygen species (ROS) by phagocytes can be stimulated by agents such as phorbol esters, glucans, bacteria and bacterial products. The ROS produced are considered to be toxic for fish bacterial pathogens (Sharp and Secombes, 1993). That turbot head kidney macrophages are capable of undergoing a respiratory burst has been demonstrated by stimulation of the cells with phorbol ester and glucans (Santarem *et al.*, 1997; Castro *et al.*, 1999), the burst being detectable within 1 h using glucan doses ≥50 μg ml^{-1} (Castro *et al.*, 1999).

Glucans have also been shown to prime cells for enhanced respiratory burst activity. Pre-incubation of head kidney macrophages with low concentrations of glucans (1–2.5 μg ml^{-1}) for 3–6 h prior to addition of phorbol ester (as a trigger of the respiratory burst) significantly enhanced the responses, with higher concentrations having an inhibitory effect possibly due to exhaustion of the cells by direct stimulation of the burst (Castro *et al.*, 1999). Similarly, injection of turbot with glucans (50 μg) can increase head kidney macrophage respiratory burst activity, being particularly potent when combined with bacterial O-antigen (Santarem *et al.*, 1997). The increase in the magnitude of the burst after injection was correlated with increased bactericidal activity of head kidney phagocytes for an avirulent strain of *P. damsela* (Santarem *et al.*, 1997). As with phagocytosis, xenobiotics have been shown to suppress respiratory burst activity of turbot phagocytes. Both exposure of fish to polychlorinated biphenyls (PCB)-spiked sediments (Hutchinson *et al.*, 1999) and of cells in culture to oxytetracycline (Tafalla *et al.*, 1999a) resulted in significant inhibition. Interestingly, immunization of PCB-exposed fish overcame the suppressive effect. Host-derived molecules also inhibited respiratory burst activity, (Tafalla *et al.*, 1999b). Supernatants derived from 17 h blood leucocyte cultures inhibit macrophage respiratory burst activity, but if the supernatants were produced in the presence of indomethacin, which prevents prostaglandin release, the effect was neutralized.

3.4 *Nitric oxide production*

Production of nitric oxide (NO), by N-hydroxylation of the guanidino carbon of arginine, is an important defence against some viral, parasitic and bacterial infections. NO is usually produced from macrophages although neutrophil production has been detected in some species (Goff *et al.*, 1996, Wheeler *et al.*, 1997), through the action of an inducible NO synthase (iNOS). iNOS expression is controlled by host cytokines such as interferon gamma (IFN-γ), tumour necrosis factor alpha (TNF-α) and interleukin-1 (IL-1), in addition to pathogen-derived molecules (Nathan, 1992). Other cytokines, including IL-4, IL-10 and transforming growth factor beta (TGF-β), inhibit NO production.

Whilst the iNOS gene in turbot has not been sequenced to date, it has been sequenced in other fish species including rainbow trout, goldfish and catfish (Laing *et al.*, 1996, 1999). Nevertheless, NO production has been assayed in turbot by the detection of generated nitrite in culture supernatants. NO production by turbot macrophages requires induction, with LPS or a combination of LPS and cytokines (Tafalla and Novoa, 2000). Stimulation of cultured cells with LPS at 1–10 μg ml^{-1} for 48 h or 0.01–0.1 μg ml^{-1} for 72 h induced NO production, but earlier timings (24 h) showed no stimulation. However, only macrophage populations derived from 30% of fish responded to LPS alone. Some of the non-responsive macrophage populations can be stimulated by co-culture with LPS and macrophage-activating factor (MAF)-containing supernatants or turbot-interferon-containing supernatants, whilst incubation with human recombinant TNF-α plus LPS resulted in all macrophage populations being stimulated.

NO appears to play an important role in viral haemorrhagic septicaemia virus (VHSV) replication in turbot cells (Tafalla *et al.*, 1999c). This virus, which produces high mortalities of fish in aquaculture, is known to replicate in turbot macrophages. Infection of isolated head kidney macrophages induced NO production within 48 h but inhibitors of iNOS action did not increase VHSV titres, suggesting that endogenously produced NO has no antiviral function. However, exogenously produced NO, from a nitric oxide donor added to the cultures, had a significant antiviral effect when using either turbot macrophages or fibroblasts (Tafalla *et al.*, 1999c).

3.5 *Soluble mediators*

Turbot produce a wide range of soluble mediators involved in non-specific defences. Many of these are key effector molecules that can directly kill pathogens or neutralize their secretions. Other molecules act as regulators, to initiate inflammatory events and control phagocyte action (e.g. cytokines and eicosanoids).

Lysozyme. An example of a directly lytic enzyme present in the blood and lymphoid tissues of turbot is lysozyme. This molecule cleaves the bond between *N*-acetyl glucosamine and *N*-acetyl muramic acid in the bacterial cell wall of Gram-positive bacteria, although purified lysozyme from rainbow trout is also effective against Gram-negative bacteria (Grinde, 1989). Turbot are known to have relatively high concentrations of lysozyme in their kidney (2150 U ml^{-1}; Grinde *et al.*, 1988), and injection of glucans or *V. damsela* O antigen has been shown to elevate blood lysozyme levels (approximately two-fold after 1 week) for up to 3 weeks post-injection (Santarem *et al.*, 1997). Recently a partial sequence for the turbot lysozyme gene has been obtained (*Figure 1*; Tutundjian *et al.*, 1999), and thus study of factors influencing expression are now possible.

Transferrin. Another important effector molecule is transferrin, which plays a role in nutritional immunity by starving microorganisms of iron, which plays a key role in bacterial respiration. Thus, transferrin functions as a bacteriostatic agent by creating an iron-restricted environment for invading microorganisms (Kvingedal and Rotvik, 1993). Transferrin sequences have been obtained for a number of fish species, including a partial sequence (870 bp) from turbot liver cDNA (Low *et al.*, 2000a) which shows high homology to Atlantic cod (89%) and Atlantic salmon (90%) trans-ferrins. The presence of such molecules in blood is reflected in the bactericidal and bacteriostatic activity of turbot serum or plasma, as shown against *V. harveyi* (Tafalla *et al.*, 1999a).

Cytokines. These are simple polypeptides or glycoproteins of less than 30 kDa that act as signalling molecules within the immune system (Thomson, 1998), most of them showing transient production and acting in either an autocrine or paracrine fashion. Biological activity of cytokine-like factors has been demonstrated in many fish species in homologous assay systems (Secombes, 1998). In turbot, a macrophage activating

```
TURBOT      MRCXXXXXXVPVPGAKVFERCEWARLLKRNGMSNYRGISLADWVCLSQWESSYNTRATNR 60

R. TROUT    MRAVVVLLLVAVASAKVYDRCELARALKASGMDGYAGNSLPNWVCLSKWESSYNTQATNR 219

            **         *  *  ***  *** ** **   **  *  *  **  ***** ******* ****

TURBOT      NTDGSTDYGIFQINSRWWCDNGQTP-TSNACGISCSALLTDDVGAAIICAKHVVRDPNGI 119

R. TROUT    NTDGSTDYGIFQINSRYWCDDGRTPGAKNVCGIRCSQLLTADLTVAIRCAKRVVLDPNGI 399

            *************** *** * **   * *** ** *** *    ** *** ** *****

TURBOT      GAWVAWKRHCQGQDLSSYVAGCGV 143

R. TROUT    GAWVAWRLHCQNQDLRSYVAGCGV 471

            ******  *** *** ********
```

Figure 1. *Amino acid sequence of turbot lysozyme. Alignment with the rainbow trout lysozyme sequence shows 69% homology at the amino acid level. Identical residues are indicated by an asterisk.*

factor has been demonstrated in supernatants from ConA-stimulated leucocyte cultures, typical of a type II (gamma) interferon (Tafalla and Novoa, 2000). In addition, type I interferon activity has been shown in supernatants from Poly I:C-stimulated macrophages that could inhibit VHSV replication in fibroblasts (Tafalla and Novoa, 2000). Whilst interferon genes have still to be obtained in fish, interferon-induced genes are known, as with the Mx genes (Leong *et al.*, 1998). In turbot, a 554 bp partial sequence has been obtained for the Mx gene (Robertson *et al.*, 2000), which shows 86% homology to halibut, 81% to salmon and 80% to rainbow trout Mx genes.

Recently, two partial sequences for turbot cytokine genes have been obtained in our laboratory, using degenerate primers to TGF-β and IL-1β that have been cloned in a number of other fish species (Hardie *et al.*, 1998; Zou *et al.*, 1999; Zhan and Kwang, 2000; Fujiki *et al.*, 2000). The function and homologies of these cytokines are given below.

Transforming growth factor β (TGF-β) . The TGF-β superfamily is a group of multi-functional peptides that control the proliferation and differentiation of many cell types (Burt and Law, 1994). Within the immune system TGF-β1 is produced by most leuco-cytes and is a potent modulator of differentiation and an immunosuppressive agent, with the ability to down-regulate the expression of many cytokines and cytokine-induced effects (Letterio and Roberts, 1998). The partial sequence for TGF-β1 obtained from turbot kidney shares high homology with sequences in other bony fish species (*Table 2*).

Table 2. Homology of turbot TGF-β1 with other fish species at the amino acid and nucleotide levels

Species	Amino acid similarity	Nucleotide identity
Rainbow trout	89%	81%
Goldfish	92%	79%
Carp	91%	78%
Plaice	95%	94%

Interleukin 1β. IL-1β is a key mediator in response to microbial invasion and tissue injury which can stimulate the immune response by activating lymphocytes or by inducing the release of other cytokines able to activate macrophages, natural killer cells and lymphocytes. Macrophages are the primary source of IL-1β, although it is produced by a wide variety of other cell types including B lymphocytes and natural killer cells. Using probes to the partial sequence (197 bp) for IL-1β obtained for turbot it was shown that the IL-1β transcript requires induction, as seen *in vitro* after stimulation with LPS or *in vivo* after bacterial challenge (*Figure 2*). Such data strongly suggest that IL-1β is a major player in immune responses of turbot, as in other fish species and mammals.

Eicosanoids. Eicosanoids are released from turbot cells after appropriate stimulation (Tocher *et al.*, 1996), and whilst their role in immune responses of turbot is not well established, prostaglandins released from leucocytes in culture suppress macrophage respiratory burst activity (Tafalla *et al.*, 1999b).

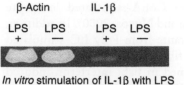

β-Actin   IL-1β

LPS   LPS   LPS   LPS
+   −   +   −

In vitro stimulation of IL-1β with LPS
on turbot head kidney leucocytes for 4 h

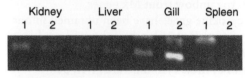

Kidney   Liver   Gill   Spleen
1   2   1   2   1   2   1   2

In vivo stimulation of IL-1β with
Aeromonas salmonicida

Figure 2. PCR *showing induction of IL-1β transcript in* vitro *with LPS, or in vivo after bacterial challenge;* β-*actin was used as a positive control.*

4. Specific defences

4.1 *Lymphocyte populations*

There is substantial evidence that sub-populations of lymphocytes exist in fish analagous to B (bone marrow in mammals) and T (thymus-derived) cells.

T-like cells in fish are defined as being negative for surface immunoglobulin (sIg-) and proliferating in the presence of known T cell mitogens (Partula, 1999). These cells are involved in cell mediated immunity and in assisting B lymphocytes. T-like cells are found in blood and are also associated with mucosal compartments (gut and gills), whereas B cells are more frequently found in blood, spleen and head kidney. T cells can be divided into those that help B cells, by stimulating antibody production (in humans, these carry the CD4 cell marker), and those that attack antigens directly, such as the cytotoxic T cells (identified in humans by the CD8 cell marker; Manning and Nakanishi, 1996). The CD8 cell marker has recently been cloned in rainbow trout (Hansen and Strassburger, 2000) and will provide a valuable means to study T-like cells in teleosts.

T cells possess an antigen receptor, the T cell receptor (TCR), capable of recognizing antigens through variable gene rearrangement. The gene for the TCR has been cloned in some fish species including horned shark, rainbow trout, Atlantic salmon and channel catfish (Rast and Litman, 1994; Partula *et al.*, 1994, 1995; Hordvik *et al.*, 1996; Zhou *et al.*, 1997). In these species, α or β chains (or both) have been described. The TCR α and β chains bind antigens presented by the major histocompatibility complexes (MHC I and II), while δ and γ chains are involved in signal transduction (Partula, 1999). The δ and γ chains have only been described in elasmobranchs, but it is assumed that they are present in teleosts (Charlemagne *et al.,* 1998). All four polypeptide chains consist of Ig-like domains with a variable (V) and constant (C) region, a transmembrane domain and a short intracellular domain (Partula, 1999). In the β chain, diversity (D) and joining (J) segments have also been described; this gene has a translocon arrangement similar to mammalian TCR-β (Partula, 1999).

Variability of the α and β chains is produced by somatic recombination of V and J segments, whereas, in δ and γ chains V, J and D segments are involved (Partula, 1999). In addition, three hypervariable sections exist in the V regions, termed complementarity-determining regions (CDR). The rearrangement of gene segments is mediated by recombination-activating genes (RAG 1 and 2) and a 552 bp sequence for RAG1 has been obtained in turbot in our laboratory (Low *et al.*, 2000b). These excise and ligate DNA around recombination signal sequences (RSS). Variability has been shown in rainbow trout α and β chains, with six different families of Vα identified (<80% homology) and three families of Vβ (Partula *et al.*, 1995, 1996). In addition, the rainbow trout α chain had 32 different J gene segments and 10 Jβ gene segments.

The presence of T lymphocytes in turbot is suggested by the antibody response to the T-dependent antigen, keyhole limpet haemocyanin (Hutchinson *et al.*, 1999). Evidence for T cells in other teleosts includes observations that immunoglobulin (Ig)-lymphocytes are involved in the mixed leucocyte reaction, allograft rejection and production of lymphokines. Furthermore, mRNA encoding the TCR β-chain has been detected in thymocytes and splenocytes of trout (Partula *et al.*, 1995). However, these observations are indirect, and the contribution of T cells is only presumed as yet. However, the production of leucocyte cell lines for channel catfish has allowed studies on leucocyte heterogeneity to be conducted. As well as macrophages, monocytes and B-like cells, T-like cells exist that do not express surface Ig (Chinchar *et al.*, 1994) and demonstrate T cell cytotoxicity (Yoshida *et al.*, 1995).

4.2 *Immunoglobulins, isolation and structure*

In fish, two isotypes of Ig have been shown, IgM and, more recently, IgD (Hordvik *et al.*, 1999). The IgM molecule occurs in fish mainly as a tetramer but can also occur as a monomer or a dimer, with heavy and light chains connected by disulphide bonds. Immunoglobulin is found in serum, mucus secretions, skin and bile, and is also present in the eggs of some fish species. It is possible that monomeric, dimeric and tetrameric forms of IgM in fish may have different functions (Ellis, 2001), for example the tetrameric form can agglutinate antigens, whereas the monomeric form cannot. In addition, tetrameric IgM is more effective in antigen precipitation and complement activation than monomeric IgM (Ellis, 2001), as might be expected by analogy with mammalian immunoglobulins.

The mechanisms for antibody production in fish are reported to be similar to those in mammals, involving B cells, T helper cells and major histocompatibility complexes for the presentation of antigens (Ellis, 2001). In addition, monocytes, macrophages and interleukin-1 (produced by the macrophages) are required to induce an antibody response. For T-dependent antigens, T cells are required to produce antibodies. The T-dependent antigen is processed by macrophages and presented to the T cells, inducing the T cells to produce interleukin which in turn results in antibody-producing plasma cells.

4.3 *Turbot immunoglobulin*

The Ig of turbot, purified using gel filtration and Superose columns (Estevez *et al.*, 1993, 1994) contained two components corresponding to a 78 kDa heavy chain and a 27 kDa light chain. Several other protein bands of approximately 65–70 kDa were present, and probably represented degraded Ig molecules or Ig variants (Estevez *et al.*, 1994). The concentration of Ig in adult turbot serum was found to be 3.35 ± 0.74 mg ml^{-1} for fish weighing 15–25 g and 11.2 ± 1.87 mg ml^{-1} for fish of 1–2 kg. In turbot of

15–25 g, total serum protein concentration was 44.8 ± 4.9 mg ml⁻¹ and for fish 1–2 kg it was 62.6 ± 3.9 mg ml⁻¹. Therefore, Ig represented 7.5 and 18% of the total protein in serum in these fish. The increase in Ig concentration with fish weight may be related to the maturity of the immune system, or be a function of ambient temperature, or nutrition (Ellis, 1988; Zapata *et al.*, 1992; Sánchez *et al.*, 1993). This study also found that the ratio of Ig to total protein is not constant but increases with weight. Monoclonal antibodies to turbot Ig cross-reacted with the sera of a closely related species, brill (*Scophthalmus rhombus*), suggesting similarity between the epitopes of the Ig in these two species, carbohydrate residues on the turbot Ig molecule being important in epitope recognition (Estevez *et al.*, 1994).

4.4 *Vaccination*

Turbot produce specific antibodies in response to a bath challenge with *V. anguillarum* (Estevez *et al.*, 1994, 1995), and adult turbot (5–10 months) vaccinated against vibriosis by the oral route showed less effective protection (70%) from disease than fish vaccinated by the intra-peritoneal route (100%; Dec *et al.*, 1990). Secondary immunization resulted in increased antibody titres, demonstrating immunological memory (Estevez *et al.*, 1994).

   In a recent study, surface positive immunoglobulin cells (sIg+) in turbot were found to be distributed in turbot blood, kidney, spleen and gut, with a small fraction (2%) in the thymus (Fourier-Betz *et al.*, 2000). In the turbot thymus, lymphoid cells which were reactive to the polyvalent rabbit anti-serum for turbot Ig+ tended to cluster around Hassal's corpuscles in the lymphoid outer zone. These B cells may be producing antibodies. However, the great majority of cells in the thymus were small round T-like lymphocytes, although direct evidence would require anti-thymocyte serum (Fourier-Betz *et al.*, 2000). Many Ig+ cells were observed in the kidney scattered through the intertubular renal tissue, sinusoids and nephrons. Ig+ cells were often clustered around melanomacrophage centres, and in the turbot spleen these melanomacrophage centres were observed associated with Ig+ cells, perhaps suggesting an important role of these structures in antigen trapping (Fourier-Betz *et al.*, 2000). However, Ig+ cells were present throughout the parenchyma of the spleen and adjacent to blood vessels. Furthermore, the hindgut was examined for the presence of Ig+ using immunohistochemical staining. A few elongated Ig+ cells were present in the *lamina propria*, but very few were in the epithelial and muscular layer, in agreement with other studies conducted on carp (Rombout *et al.*, 1993). In studies on sea bass, Ig– cells in the midgut outnumbered Ig+ cells, and reacted to a monoclonal antibody against T cells (Abelli *et al.*, 1997).

4.5 *Ontogeny of B and T cells*

The ontogeny of the immune system is a function of fish weight rather than age. This is due to the fact that growth is strongly influenced by temperature (Imsland *et al.*, 1995). In addition, the detection of lymphoid cells in the thymus, kidney and spleen is more relevant to immunocompetence of the fish than the early development of the organs (Tatner, 1996). There is some evidence that immunological maturity and antibody production may be related to a peak in cell number in the thymus (Tatner, 1996) and a critical stage in weight gain in the fish (Bootland *et al.*, 1990). For example, higher antibody titres to *V. anguillarum* were achieved following vaccinations of carp at 128 days than at 99 days. This coincided with a peak in B cell and plasma cell numbers (Koumans van Deipen *et al.*, 1994). Also, immersion vaccination against *V. anguillarum* is ineffective in salmonids weighing below 1 g (Johnson *et al.*, 1982), and the length of protection afforded by vaccination in rainbow trout increased with weight, reaching a maximum after 4 g (Ellis, 1998). This suggests that the lymphocytes involved in immunological memory are not fully functional at early stages. Also, variations in the duration of immunity occur with the concentration of bacterin and the fish species involved.

Antibodies can be used to indicate the maturity of B cells by detecting surface Ig+, but such studies have not yet been conducted on turbot fry. However, studies carried out in carp and Atlantic salmon found that B cells become mature around 8 weeks post-fertilization (Van Loon *et al.*, 1981; Ellis, 1977). B cells appear first in the head kidney, then the spleen, blood, thymus and gut (Scapigliati *et al.*, 1999), and the kidney appears to be the primary lymphoid organ for B cell maturation in fish (Scapigliati *et al.*, 1999).

Immunoglobulins are present in the eggs and fry of some species, including the plaice, *Pleuronectes platessa* (Bly *et al.*, 1986). This occurs through maternal transfer and provides an important barrier against infection of the eggs, and protection to the embryo within. However, levels of maternal Ig decline by the point of first feeding, and thereafter Ig is produced by the fry. Furthermore, maternal transfer of specific immunity has been demonstrated in tilapia by immunizing the mother with protein antigens (Mor and Avtalion, 1989; Brown *et al.*, 1990; Sin *et al.*, 1994). This process has potential use in commercial aquaculture to protect eggs from disease.

The thymus is the first organ to become lymphoid, followed by the kidney and the spleen (Zapata and Cooper, 1990; Tatner, 1996). Allograft rejection is present at an early stage of development in trout, carp and rosy barb (Tatner and Manning, 1983) indicating early maturity of T-like lymphocytes. In addition, T cell ontogeny has been studied in the zebra fish using transcription factors known to be important in higher vertebrates (GATA-1, and 2; Thomson *et al.*, 1998; Trede and Zon, 1998). *In situ* hybridization studies using these probes identified haemopoeisis at 12 h post-fertilization in the intermediate cell mass.

Information on the ontogeny of T-like lymphocytes in turbot is scarce. However, monoclonal antibodies have been produced for Ig- or T-like cells in carp, catfish, sea bass and yellow tail (Scapigliati *et al.*, 1995; Passer *et al.*, 1996; Nishimura *et al.*, 1997; Rombout *et al.*, 1998). In sea bass reared at 16°C, a high number of reactive Ig- cells were located in the gut mucosa at 5 days post-hatch (Scapigliati *et al.*, 1995). Thymocytes reacted to the monoclonal antibody DLT15 at day 30 post-hatch, and 72 h after lymphocytes first appeared in the thymus. T-like lymphocytes increased in the gut and thymus from 44 days post-hatch. However, they were infrequent in the head kidney and spleen. In general, T-like cells become functional rapidly after they are first detected (Tatner, 1996).

RAG 1 and 2 genes are essential for antigen recognition through variability of the TCR; thus, they may be used to indicate the maturity of T-like cells. In zebra fish, RAG 1 and 2 are expressed at 92 h post-fertilization in the thymus (Willet *et al.*, 1997). More recently, partial sequences of RAG 1 have been obtained in turbot (Low *et al.*, 2000b) and will be used to gain a greater understanding of the ontogeny of turbot T and B cells.

4.6 *Influence of environmental conditions on immune ontogeny*

The growth of juvenile turbot is strongly dependent upon water temperature, being greater at higher temperatures, but as the fish grow larger, the optimum growth temperature decreases (Imsland *et al.*, 1995). Furthermore, these authors observed that, upon infection with *Aeromonas* species, higher mortalities occurred in turbot reared at 10°C (17.2%) than in those reared at 16°C (6%). Generally, the higher the temperature the greater the immune response (antibody titres, faster response, allograft rejection rates) that occurs in fish. Conversely, low temperatures result in a slow development of the antibody response and lower antibody titres. Seasonal changes have been observed in the structure and function of the lymphoid organs (Zapata *et al.*, 1996), and the optimum temperature for the immune response is related to the natural temperature range the species experiences (Manning and Nakanishi, 1996).

In channel catfish, the 'permissive range' for immune responses was found to be 27–32°C. Between 17 and 22°C, T cell responses were suppressed and, at around 12°C, B and T cells functions were affected. However, if fish were immunized at permissive temperatures and then subjected to 'non-permissive' temperatures, the immunosuppression was negated, indicating that it is the early part of the antigen response that is temperature sensitive (Avtalion, 1969). Further experiments have led to the theory that T 'virgin cells' are primarily influenced by temperature change, rather than T memory cells or B cells (Vallejo *et al.*, 1992). In trout, MAF production by T cells was suppressed at non-permissive temperatures, but the macrophages could respond when MAF was added (Hardie *et al.*, 1994). In addition, the initial presentation of the antigen is not a temperature-dependent process.

Further understanding of temperature effects has been gained by conducting experiments which involved antigen challenge following sudden drops in temperature (e.g. from 24 to 17°C) or after gradual acclimatization. Rapid change can lead to complete immunosuppression in channel catfish, whereas prior acclimatization to low temperatures lessens the immunosuppression effect (Bly and Clem, 1991). Furthermore, if the same fish were subsequently reared at permissive temperatures, normal secondary responses to T-dependent antigen returned. This suggests that immunization at low temperatures does not result in antigen-specific tolerance.

Physiological studies on the influence of temperature on cell function suggested that changes in the fluidity of T cell plasma membranes were important (Bly and Clem, 1992). This was indicated when the response of T cells to the T cell mitogen, concanavalin A, was improved at low temperatures when oleic acid was added to cell cultures. Oleic acid, being less saturated, allows the T cell membranes to become homeoviscous at low temperatures. Unlike T cells, B cells are able to substitute saturated fatty acids with less saturated fatty acids. Although changes to the plasma membrane occur at low temperatures, they are undetectable by current viscosity probes and are thought to be quantitatively minor (Bly and Clem, 1992).

Cold water fronts in winter can frequently result in 'winter kill' in North American catfish-rearing ponds and it is thought that low temperature immunosuppression can lead to high mortalities in commercial aquaculture. However, other seasonal factors may influence the immune system. For example, photoperiod influences development in turbot (Imsland *et al.*, 1995) and immunizations in trout reared at a constant temperature varied in effectiveness throughout the year (Yamaguchi *et al.*, 1980), indicating endogenous seasonality.

Other environmental factors that affect the immune system include stress, crowding, social aggression and poor water quality (Pickering, 1997). Handling and transport have been found to lead to immunosuppression in channel catfish (Ellsaesser and Clem, 1986). Although there is little information regarding diet and ontogeny of the immune system, low energy levels will result in fewer resources being available for immunity and other physiological processes in the fish (Tatner, 1996). Turbot fed on a restricted diet (0.25 and 0.38% of body weight per day) showed lower growth rates of 60 and 30% when compared to control fish (1% body weight per day). However, when normal feeding was resumed, the starved turbot showed compensatory growth rates (Sæther and Jobling, 1999).

Immune protection has been shown to increase when immunostimulants are added to the diet. These compounds include β1-6 and β1-3 glucans (extracted from yeast and plants), bacterial lipopolysaccharides, chitins and animal extracts. Such immunostimulants boost the non-specific immune system (Anderson, 1996). However, muramyl dipeptide and levamisole stimulate T lymphocytes and vitamin E stimulates both B and T lymphocytes (Jeney and Anderson, 1993; Kodama *et al.*, 1993).

4.7 *Early exposure to pathogens or vaccination*

In many fish species, the immune system becomes functional a few days after hatching (Ellis, 2001) and maturity of the immune system is important if early vaccinations are planned. Information about the response of turbot larvae to early pathogen challenge is not available. However, data is available for other fish species such as carp and trout. When very young carp were immunized with sheep red blood cells, they did not produce antibodies (Ellis, 2001). These carp were tolerant to the red blood cells when a second injection was given at 4 weeks (Ellis, 2001). Young trout (21 days post-hatch), displayed tolerance to the T-dependent antigen, human gamma-globulin, when exposed at 8 weeks. However, trout of the same age did not show tolerance to lipopolysaccharide, a T-independent antigen (Ellis, 2001). This suggests that T helper cells have not yet developed (Ellis, 2001). Tatner (1996) also noted that tolerance has only been observed when the antigen is presented by injection, not in bath challenges, and suggests that such tolerance may be an experimental artefact. Furthermore, the ineffectiveness of immersion vaccines at very early stages in trout was shown by radiolabelling experiments to be due to the exclusion of antigens by the fry (Tatner and Horne, 1984).

In summary, the ontogeny of the lymphoid organs in turbot is similar to that noted in other fish with both the thymus and kidney being present shortly after hatching. However, they appear less well developed than in other species, whilst the ontogeny of spleen is more advanced at early stages. Recently obtained partial sequences for RAG 1 gene in turbot will be used to provide a greater understand the ontogeny of T and B cells in the turbot immune system. Studies on other teleosts suggest that immunological processes will be influenced by environmental variables such as temperature, photoperiod and diet (weight attained). However, experimental data is required to examine this assumption.

References

Abelli, L., Picchietti, S., Romano, N., Mastrolia, L. and Scapigliati, G. (1997) Immunohistochemistry of gut associated lymphoid tissue of the sea bass *Dicentrarchus labrax* (L.). *Fish Shellfish Immunol.* 7: 235–245.

Al-Maghazachi, S.J. and Gibson, R. (1984) The developmental stages of larval turbot *Scophthalmus maximus* L. *J. Exp. Mar. Biol. Ecol.* **82**: 35–51.

Anderson, D.P. (1996) Environmental factors in fish health: Immunological aspects. In: *The Fish Immune System; Organism, Pathogen and Environment* (eds G. Iwama and T. Nakanishi). Academic Press, San Diego, CA.

Avtalion, R.R. (1969) Temperature effect on antibody production and immunological memory in carp (*Cyprinus carpio*) immunised against bovine serum albumin (BSA). *Immunology* **17**: 163–188.

Bogert, C.van den and Kroon, M. (1982) Effect of oxytetracycline on in vivo proliferation of erythroid and lymphoid cells in rat. *Clin. Exp. Immunol.* **50**: 327–335.

Bootland, L.M., Dobos, P. and Stevenson, R.M.W. (1990) Fry age and size effects on immersion immunisation of brook trout *Salvelinus fontinalis* Mitchell, against infectious pancreatic necrosis virus. *J. Fish Dis.* **13**: 125–133.

Bly, J.E. and Clem, L.W. (1991) Temperature-mediated processes in teleost immunity: *in vivo* immunosuppression induced by *in vivo* low temperature in channel catfish. *Vet. Immunol. Immunopathol.* **28**: 365–377.

Bly, J.E. and Clem, L.W. (1992) Temperature and teleost immune functions. *Fish Shellfish Immunol.* **2**: 159–171.

Bly, J.E., Grimm, A.S. and Morris, I.G. (1986) Transmission of passive immunity from mother to young in a teleost fish: haemaglutinating activity in the serum and eggs of plaice, *Pleuronectes platessa* L. *Comp. Biochem. Physiol.* **84A**: 309–314.

Brown, L.L., Albright, L.J. and Evelyn, T.P.T. (1990) Control of vertical transmission of *Renibacterium salmonarium* by injection of antibiotics into maturing female coho salmon, *Oncorhynchus kisutch*. *Dis. Aquat. Organ.* **9**: 127–131.

Burrows, A.S. (1995) Cellular aspects of the immune system of the turbot *Scophthalmus maximus* (L.). Ph.D. Thesis, University of Plymouth.

Burrows A.S. and Fletcher T.C. (1987) Blood leucocytes of the turbot (*Scophthalmus maximus* (L.). *Aquaculture* **67**: 214–215.

Burt, D.W. and Law, A.S. (1994) Evolution of the transforming growth factor-beta super family. *Prog. Growth Factor Res.* **15**: 99–133.

Castro, R., Couso, N., Obach, A. and Lamas, J. (1999) Effect of different β-glucans on the respiratory burst of turbot (*Psetta maxima*) and gilthead seabream (*Sparus aurata*) phagocytes. *Fish Shellfish Immunol.* **9**: 529–541.

Charlemagne, J., Fellah, J.S., De Guerra, A., Kerfourn, F. and Partula, S. (1998) T-cell receptors in ectothermic vertebrates. *Immunol. Rev.* **166**: 87–102.

Chinchar, V.G., Stuge, T., Hogan, R.J., Yoshida, S., Miller, N.W. and Clem, L.W. (1994) Recognition and killing of allogenic and virus-infected lymphoid cells by channel catfish peripheral blood lymphocytes. *Devel. Comp. Immunol.* **18**(Suppl.):1S81.

Dec, C., Angelidis, P. and Baudin Laurencin, F. (1990) Effects of oral vaccination against vibriosis in turbot *Scophthalmus maximus* (L.), and sea bass, *Dicentrarchus labrax* (L.). *J. Fish Dis.* **13**: 369–376.

Ellis, A.E. (1977) Leukocytes of fish: a review. *J. Fish Biol.* **11**: 435–491.

Ellis, A.E. (1988) *Fish Vaccination*. Academic Press, London.

Ellis, A.E. (1998) Immunology of fishes: ontogeny of the immune system. In: *Handbook of Vertebrate Immunology*, (eds P.P. Pastoret, P. Griebel, H. Bazin and A. Govaerts). Academic Press, London.

Ellis, A.E. (2001) The Immunology of Teleosts. In: *Fish Pathology*, 3rd edn. (ed. R.J. Roberts) W.B. Saunders, London, pp.133–150.

Ellasaesser, C.F. and Clem, L.W. (1986) Hematological and immunological changes in channel catfish stressed by handling and transport. *J. Fish Biol.* **28**: 511–521.

Estevez, J., Leiro, J., Sanmartín, M.L. and Ubeira, F.M. (1993) Isolation and characterisation of turbot (*Scophthalmus maximus*) immunoglobulins. *Comp. Biochem. Physiol.* **105A**: 275–281.

Estevez, J., Leiro, J., Santamarina, M.T., Domínguez, J. and Ubeira, F.M. (1994) Monoclonal antibodies to turbot (*Scophthalmus maximus*) immunoglobulins: characterisation and applicability in immunoassays. *Vet. Immunol. Immunopathol.* **41**: 353–366.

Estevez, J., Leiro, J., Santamarina, M.T. and Ubeira, F.M. (1995) A sandwich immunoassay to quantify low levels of turbot (*Scophthalmus maximus*) immunoglobulins. *Vet. Immunol. Immunopathol.* **45**: 165–174.

Figueras, A., Santarem, M.M. and Novoa, B. (1997) Phagocytic activity of turbot (*Scophthalmus maximus*) leucocytes: opsonic effect of antibody and complement. *Fish Shellfish Immunol.* **7**: 621–624.

Figueras, A., Santarem, M.M. and Novoa, B. (1998). Influence of the sequence of administration of β-glucans and a *Vibrio damsela* vaccine on the immune response of turbot (*Scophthalmus maximus* L.). *Vet. Immunol. Immunopathol.* **64**: 59–68.

Fourier-Betz, V., Quentel, C., Lamour, F. and LeVen, A. (2000) Immunocytochemical detection of Ig-positive cells in blood, lymphoid organs and the gut associated lymphoid tissue of the turbot (*Scophthalmus maximus*). *Fish Shellfish Immunol.* **10**: 187–202.

Fujiki, K., Shin, D.-H., Nakao, M. and Yano, T. (2000) Molecular cloning and expression analysis of carp (*Cyprinus carpio*) interleukin-1β, high affinity immunoglobulin E Fc receptor γ subunit and serum amyloid A. *Fish Shellfish Immunol.* **10**: 229–242.

Goff, W.L., Johnson, W.C., Wyatt, C.R. and Cluff, C.W. (1996). Assessment of bovine mononuclear phagocytes and neutrophils for induced L-arginine-dependent nitric oxide production. *Vet. Immunol. Immunopathol.* **55**: 45–62.

Grinde, B. (1989) Lysozyme from rainbow trout, *Salmo gairdneri* Richardson, as an antibacterial agent against fish pathogens. *J. Fish Dis.* **12**: 95–104.

Grinde, B., Lie, O., Poppe, T. and Salte, R. (1988) Species and individual variation in lysozyme activity in fish of interest in aquaculture. *Aquaculture* **68**: 299–304.

Hansen, J.D. and Strassburger, P. (2000) Description of an ectothermic TCR coreceptor, CD8 alpha in rainbow trout. *J. Immunol.* **164**: 3132–3139.

Hardie, L.J., Fletcher, T.C. and Secombes, C.J. (1994) Effect of temperature on macrophage activation and the production of macrophage inhibition factor by rainbow trout (*Oncorhynchus mykiss*) leucocytes. *Devl. Comp. Immunol.* **18**: 57–66.

Hardie, L.J., Laing, K.J., Daniels, G.D., Grabowski, P.S., Cunningham, C. and Secombes, C.J. (1998) Isolation of the first piscine transforming growth factor β gene: analysis reveals tissue specific expression and a potential regulatory sequence in rainbow trout (*Oncorhynchus mykiss*). *Cytokine* **10**: 555–563.

Hordvik, I., Jacob, A.L.J., Charlemagne, J. and Endresen, C. (1996) Cloning of T-cell receptor β chain cDNAs from Atlantic salmon (*Salmo salar*). *Immunogenetics* **45**: 9–14.

Hordvik, I., Thevarajan, J., Samdal, I., Bastani, N. and Krossoy B. (1999) Molecular cloning and phylogenetic analysis of the Atlantic salmon immunoglobulin D gene. *Scand. J. Immunol.* **50**: 202–210.

Hutchinson, T.H., Field, M.D.R. and Manning, M.J. (1999) Evaluation of immune function in juvenile turbot *Scophthalmus maximus* (L.) exposed to sediments contaminated with polychlorinated biphenyls. *Fish Shellfish Immunol.* **9**: 457–472.

Imsland, A.K., Folkvord, A. and Stefansson, S.O. (1995) Growth, oxygen consumption and activity of juvenile turbot (*Scophthalmus maximus* L.) reared under different temperatures and photoperiods. *Nether. J. Sea Res.* **34**(1–3): 149–159.

Jeney, G. and Anderson, D.P. (1993) Enhanced immune response and protection in rainbow trout to *Aeromonas salmonicida* bacterin following prior immersion in immunostimulants. *Fish Shellfish Immunol.* **3**: 51–58.

Johnson, K.A., Flynn, J.K., and Amend, D.F. (1982) Onset of immunity in salmonid fry vaccinated by direct immersion in *Vibrio anguillarum* and *Yerisina ruckeri* bacterins. *J. Fish Dis.* **5**: 197–206.

Jones, A. (1972) Studies of the egg development and larval rearing of turbot, *Scophthalmus maximus* L. and brill, *Scophthalmus rhombus*, in the laboratory. *J. Mar. Biol. Assoc. UK* **52**: 965–986.

Kodama, H., Yoshikatsu, H., Masafumi, M., Tsuyoshi, B. and Azuma, I. (1993) Activation of rainbow trout (*Oncorhynchus mykiss*) phagocytes by muramyl dipeptide. *Dev. Comp. Immunol.* 17: 129–130.

Koumans van Deipen, J.C.E., Taverne-Theile, A.J., van Rrens, B.T.T.M. and Rombout, J.H.W.M. (1994) Immunocytochemical and flow cytometric analysis of B cells and plasma cells in carp *Cyprinus carpio* L.: an ontogenic study. *Fish Shellfish Immunol.* 4: 19–28.

Kvingedal A.M. and Rotvik K. (1993) Cloning and characterization of Atlantic salmon (*Salmo salar*) serum transferrin cDNA. *Mol. Mar. Biol. Biotechnol.* 2: 233–238.

Laing, K.J., Grabowski P.S., Belosevic, M. and Secombes, C.J. (1996) A partial sequence for nitric oxide synthase from a goldfish (*Carassius auratus*) macrophage cell line. *Immunol. Cell Biol.* 74: 374–379.

Laing, K.J., Hardie, L.J., Aartsen, W., Grabowski, P.S. and Secombes, C.J. (1999) Expression of an inducible nitric oxide synthase gene in rainbow trout *Oncorhynchus mykiss. Dev. Comp. Immunol.* 23: 71–85.

Leong, J.-A.C., Trobridge, G.D., Kim, C.H.Y., Johnson, M. and Simon, B. (1998) Interferon-inducible Mx proteins in fish. *Immunol. Rev.* 166: 349–363.

Letterio, J.J. and Roberts, A.B. (1998) Regulation of immune responses by TGF-beta. *Annu. Rev. Immunol.* 16: 137–161.

Low, C.A., Taylor, I., Birkbeck, H., Tatner, M. and Secombes, C.J. (2000a) Partial mRNA sequence for turbot transferrin. Unpublished. EMBL Accession no. AJ277079.

Low C.A., Taylor I., Birkbeck, H., Tatner M., Secombes C.J. (2000b) Partial mRNA sequence of recombinase activating gene 1 (RAG 1) from turbot (*Scophthalmus maximus*). Unpublished. EMBL Accession no. AJ277078.

Manning, M.J. (1994) *Immunology: a Comparative Approach.* (ed R.J. Turner), John Wiley, Chichester.

Manning M.J. and Nakanishi, T. (1996) The specific immune system: cellular defences. In: *The Fish Immune System; Organism, Pathogen and Environment.* (eds G. Iwama and T. Nakanishi). Academic Press, San Diego, CA.

Mor, A. and Avtalion, R.R. (1989) Transfer of antibody activity from immunised mother to embryo in tilapias. *J. Fish Biol.* 37: 249–254.

Nathan, C. (1992) Nitric oxide as a secretory product of mammalian cells. *FASEB J.* 6: 3051–3064.

Nishimura, H., Ikemoto, M., Kawai, K. and Kusuda, R. (1997) Cross-reactivity of anti-yellowtail thymic lymphocyte monoclonal antibody (YeT-2) with lymphocytes from other fish species. *Arch. Histol. Cytol.* 60: 113–119.

Partula, S. (1999) Surface markers of fish T-cells. *Fish Shellfish Immunol.* 9: 241–257.

Partula, S., Fellah, J.S., de Guerra, A. and Charlemagne, J. (1994) Identification of cDNA clones encoding in T-cell receptor β-chain in the rainbow trout (*Oncorhynchus mykiss*). *C.R. Acad. Sci. Paris* 317: 765–770.

Partula, S., de Guerra, A., Fellah, J.S. and Charlemagne, J. (1995) Structure and diversity of T-cell antigen receptor β chain in a teleost fish. *J. Immunol.* 155: 699–706.

Partula, S., de Guerra, A., Fellah, J.S. and Charlemagne, J. (1996) Structure and diversity of T-cell antigen receptor α chain in a teleost fish. *J. Immunol.* 157: 202–212.

Passer, B.J., Chen, C.H., Miller, N. and Cooper, M.D. (1996) Identification of a T lineage antigen in the catfish. *Dev. Comp. Immunol.* 20: 441–450.

Pickering, A.D. (1997) Husbandry and stress. In: *Furunculosis Multidisciplinary Fish Disease Research.* (eds E.M. Bernoth, A.E. Ellis, P.J. Midtlyng, G. Olivier and P. Smith). Academic Press, London.

Quentel, C. and Obach, A. (1992) The cellular composition of the blood and hematopoietic organs of turbot *Scophthalmus maximus* L. *J. Fish Biol.* 41: 709–716.

Rast, J.P. and Litman, G.W. (1994) T-cell receptor gene homologs are present in the most primitive vertebrates. *Proc. Natl Acad. Sci. USA* 91: 9248.

Riaza, A. and Hall, J. (1993) Large scale production of turbot. In *Fish Farming Technology* (eds H. Reinersten, L.A. Dahle, L. Jorgensen and K. Tvinnereim). A.A. Balkema, Rotterdam, pp. 147–148.

Robertsen, B. and Jensen, V. (2000) PCR cloning of a partial turbot (*Scophthalmus maximus*) Mx cDNA. Unpublished. EMBL Accession no. AF245514.

Rombout, J.H.W.M., Taverne-Thiele, A.J. and Villena, M.I. (1993) The gut-associated lymphoid tissue (GALT) of carp (*Cyprinus carpio* L.). *Dev. Comp. Immunol.* 17: 55–66.

Rombout, J.H.W.M., Joosten, P.H.M., Engelsma, M.Y., Vos, N., Taverne, N. and Taverne-Thiele, A.J. (1998) Indication of distinct putative T cell population in mucosal tissue of carp. *Dev. Comp. Immunol.* 22: 63–77.

Sæther, B-S. and Jobling, M. (1999) The effects of ration level on feed intake and growth, and compensatory growth after restricted feeding, in turbot *Scophthlamus maximus* L. *Aquacult. Res.* 30: 647–653.

Sánchez, C.P., Lopez-Fierro, A., Zapata, A.G. and Domínguez, J. (1993) Characterisation of monoclonal antibodies against heavy and light chains of trout immunoglobulins. *Fish Shellfish Immunol.* 3: 237–251

Santarem, M.M. and Figueras, A. (1994) Kinetics of phagocytic activity, plaque-forming cells and specific agglutinins of turbot (*Scophthalmus maximus* L.) immunised with O antigen of *Vibrio damsela* and *Pasteurella piscicida*. *Fish Shellfish Immunol.* 4: 527–537.

Santarem, M., Novoa, B. and Figueras, A. (1997) Effects of β-glucans on the non-specific immune responses of turbot (*Scophthalmus maximus* L.). *Fish Shellfish Immunol.* 7: 429–437.

Scapigliati, G., Mazzini, M., Mastrolia, L., Romano, N. and Abelli, L. (1995) Production and characterisation of a monoclonal antibody against the thymocytes of the sea bass, *Dicentrarchus labrax* (L.) (Teleostea, Percicthydae). *Fish Shellfish Immunol.* 5: 393–405.

Scapigliati, G., Romano, N. and Abelli, L. (1999) Monoclonal antibodies in fish immunology: identification, ontogeny and activity of T- and B-lymphocytes. *Aquaculture* 172: 3–28.

Secombes, C.J. (1998). The phylogeny of cytokines. In: *The Cytokine Handbook* (ed A. Thomson). Academic Press, London.

Sharp, G.J.E. and Secombes, C.J. (1993) The role of reactive oxygen species in the killing of the bacterial fish pathogen *Aeromonas salmonicida* by rainbow trout macrophages. *Fish Shellfish Immunol.* 3: 119–129.

Sin, Y.M., Ling, K.H. and Lam, T.T. (1994) Passive transfer of protective immunity against *Ichthyophthiriasis* from vaccinated mother to fry in tilapias, *Oreochroms aureus*. *Aquaculture* 120: 229–237.

Tafalla, C. and Novoa, B. (2000). Requirements for nitric oxide production by turbot (*Scophthalmus maximus*) head kidney macrophages. *Dev. Comp. Immunol.* 24: 623–631.

Tafalla, C., Novoa, B., Alvarez, J.M. and Figueras, A. (1999a) In vivo and in vitro effect of oxytetracycline treatment on the immune response of turbot, *Scophthalmus maximus* (L.). *J. Fish Dis.* 22: 271–276.

Tafalla, C., Novoa, B. and Figueras, A. (1999b) Suppressive effect of turbot (*Scophthalmus maximus*) leucocyte-derived supernatants on macrophage and lymphocyte function. *Fish Shellfish Immunol.* 9: 157–166.

Tafalla, C., Figueras, A. and Novoa, B. (1999c) Role of nitric oxide on the replication of viral haemorrhagic septicemia virus (VHSV), a fish rhabdovirus. *Vet. Immunol. Immunopathol.* 72: 249–256.

Tatner, M.F. (1996) Natural changes in the immune system of fish. In: *The Fish Immune System; Organism, Pathogen and Environment.* (eds G. Iwama and T. Nakanishi). Academic Press, London.

Tatner, M.F. and Horne, M.T. (1984) The effects of early exposure to the *Vibrio anguillarum* vaccine on the immune response of the fry of the rainbow trout, *Salmo gairdneri* Richardson. *Aquaculture* 41: 193–202.

Tatner, M.F. and Manning M.J. (1983) The ontogeny of cellular immunity in the rainbow trout, *Salmo gairdneri* Richardson, in relation to the stage of development of the lymphoid organs. *Dev. Comp. Immunol.* 7: 69–75.

Thomson, A. (1998) *The Cytokine Handbook*. Academic Press, London.

Thomson, M.A., Ransom, D.G., Pratt, S, J., MacLennan, H., Kieran, M.W., Detrich, H.W.I., Vail, B., Huber, T.L., Paw, B. and Zon, L.I. (1998) The cloche and spadetail genes differentially affect hematopoiesis and vasculogenesis. *Dev. Biol.* **197**: 248–269.

Tilseth, S. (1990). New marine fish species for cold water farming. *Aquaculture* **85**: 235–245.

Tocher, D.R., Bell, J.G. and Sargent, J.R. (1996) Production of eicosanoids derived from 20/4N-6 and 20/5N-3 in primary cultures of turbot (*Scophthalmus maximus*) brain astrocytes in response to platelet activating factor-substance and interleukin-1β. *Comp. Biochem. Physiol.* **115B**: 215–222.

Trede, N.S. and Zon, L.I. (1998) Development of T-cells during fish embryogenesis. *Dev. Comp. Immunol.* **22**: 253–263.

Tutundjian, R., Bullelle, F., Leboulenger, F. and Danger, J.M. (1999) Cloning of the cDNA encoding the turbot (*Scophthalmus maximus*) precursor to lysozyme. Unpublished. EMBL Accession no. AJ250732.

Vallejo, A.N., Miller, N.W. and Clem, L.W. (1992) Cellular pathway(s) of antigen processing in fish APC: effect of varying *in vitro* temperatures on antigen catabolism. *Dev. Comp. Immunol.* **16**: 367–381.

Van Loon, J.J.A., van Oosterom, R. and Van Muiswinkle, W.B. (1981) Development of immune system in carp. In: *Aspects of Developmental Comparative Immunology* (ed J.B. Solomon), Vol. 1. Pergamon, Oxford, pp. 469–470.

Wheeler, M.A., Smith, S.D., Garcia-Cardena, G., Nathan, C.F., Weiss, R.M. and Sessa, W.C. (1997) Bacterial infection induces nitric oxide synthase in human neutrophils. *J. Clin. Invest.* **99**: 110–116.

Willet, C.E., Zapata, A.G., Hopkins, N. and Steiner, L.A. (1997) Expression of zebrafish rag genes during early development identifies the thymus. *Dev. Biol.* **182**: 331–341.

Yamaguchi, N., Teshima, C., Kurashige, S., Saito, R. and Mitsuhashi, S. (1980) Seasonal modulation of antibody formation in rainbow trout, *Salmo gairdneri*. In: *Aspects of Developmental Comparative Immunology* (ed J.B. Solomon), Vol. 1. Pergamon, Oxford.

Yoshida S., Stuge, T.B., Miller, N.W. and Clem, L.W. (1995) Phylogeny of lymphocyte heterogeneity: cytotoxic activity of channel catfish peripheral blood leucocytes directed against allogenic targets. *Dev. Comp. Immunol.* **19**: 71–77.

Zapata, A.G. and Cooper, E.L. (1990) *The Immune System: Comparative Histophysiology*. John Wiley, Chichester.

Zapata, A.G., Torroba, M. and Varas, A. (1992) Seasonal variations in the immune system of lower vertebrates. *Immunol. Today* **13**: 142–147.

Zapata, A.G., Torroba, M., Sacedón, R., Varas, A. and Vincente, A. (1996) Structure of the lymphoid organs of elasmobranchs. *J. Exp. Zool.* **275**: 125–143.

Zhan, Y. and Kwang, J. (2000) Molecular isolation and characterisation of carp transforming growth factor β1 from activated leucocytes. *Fish Shellfish Immunol.* **10**: 309–318.

Zhou, H., Bengtén, E., Miller, N.W., Warr, G.W., Clem, L.W. and Wilson, M.R. (1997) T cell receptor sequences in the channel catfish. *Dev. Comp. Immunol.* **21**: 238.

Zou, J., Cunningham, C. and Secombes, C.J. (1999) The rainbow trout *Oncorhynchus mykiss* interleukin-1β gene has a different organisation to mammals and undergoes incomplete splicing. *Eur. J. Biochem.* **259**: 901–908.

8

Developmental ecology of immunity in crustaceans

Valerie J. Smith, June R.S. Chisholm and Janet H. Brown

1. Introduction

Crustaceans have a hugely important role in marine aquatic communities. They play key ecological roles as bactivorous grazers, epibenthic scavengers or prey. It is as luxury fishery products that they are most familiar, with lobsters, shrimp, crabs and prawns being highly sought after all over the world. While crustacean aquaculture production is somewhat dwarfed by fish in terms of quantity – total world fish culture production being some 14.5 times that of crustacean production – in value fish are only worth 3.6 times more than that of the Crustacea (data from FAO, 1999). Indeed, one shrimp species, *Penaeus monodon,* is the most valuable of all cultured species whilst not even making an appearance on the list in terms of quantity (FAO, 2000). To give a more specific illustration, the UK crustacean fishery is some 59 500 metric tonnes with a value of £106.5 million (MAFF, 1998). The high value of crustaceans is a big incentive for the ever-increasing interest in aquaculture of the group. The tropical shrimp has led in this and has reached a farmed production of some 28% of the world shrimp fishery (including cold water species) at 737 200 metric tonnes in 1998 (Rosenberry, 1998). Most tropical countries with a coastline have some sort of shrimp farming, but the main producers are Thailand (210 000 metric tonnes) and Ecuador (130 000 metric tonnes) (Rosenberry, 1998). Shrimp farming enterprises have spread as far as Europe with some low-intensity farms in Greece, Italy and Albania (Rosenberry, 1998, 1999). While the shrimp is likely to remain the main crustacean farmed on a large scale at all stages of its life cycle, there are many other species that are cultivated for at least part of their life cycle and offer possibilities for further commercial production. Examples include homarid lobsters reared through larval stages for restocking/ranching programmes in UK and USA, while mud crab and spiny lobster are being on-grown from wild-caught juveniles.

As with all farming enterprises, shrimp production has experienced great problems with disease. The production figures from Central and South America for 1998 and 1999 show a significant drop of 17% (Rosenberry, 1999), probably due, at least in part, to the arrival of white spot virus in these regions. Previously, white spot has caused

Environment and Animal Development: Genes, Life Histories and Plasticity, edited by D. Atkinson and M. Thorndyke.
© 2001 BIOS Scientific Publishers Ltd, Oxford.

huge problems in eastern hemisphere shrimp farms. Larvae and juveniles have a greater susceptibility to infection and it is generally accepted that, as the shellfish farming industry is largely dependent on wild-caught broodstock, disease within farms is symptomatic of disease within natural populations. Broodstock are frequently netted from the wild with pre-existing bacterial or viral infections, indicating that disease could be a significant factor in maintaining profit margins in aquaculture. It is highly probable that small changes in local environmental conditions, such as those brought about by global warming or pollution, could affect host susceptibility to microbial infection and thus adversely affect the crustacean shellfish industry. Certainly, the rapid growth rate of larvae means that they are likely to be more vulnerable than adults to physiological disturbance by environmental conditions or changes, perhaps leading to disease (Connor, 1972; De Coursey and Vernberg, 1972; Devresse *et al.*, 1990; Briggs and Brown, 1991; Kondilatos, *et al.*, 2001). Moreover, aquaculture practices may exacerbate disease problems because stock animals are often kept under stressful conditions of overcrowding, high food levels, elevated water temperature and/or with poor water quality. For example, the Taiwanese shrimp farming crash of 1988 was put down to many adverse culture conditions, but one of the main sources was thought to be the culturing of larvae at elevated temperatures to accelerate the culture cycle (Sheeks, 1989). One of the main goals of aquaculture is to minimize larval and juvenile mortality, and whilst, to some degree, disease can be controlled by good husbandry practices, we do not know what long-term effects the early experience of eggs and larvae might have on susceptibility to disease later in life. What we need to know is the way in which early environmental experience affects the subsequent 'health' of the ongrown adult crustaceans and their ability to withstand microbial or other non-self threats to homeostatic integrity.

2. Immunity in crustaceans

For all organisms, susceptibility to disease is a function of pathogen virulence, immunological vigour and environmental conditions. Thus, immune capability is an important determinant of survival, on a par with growth, feeding, success in finding suitable mates, fecundity and behavioural strategies that permit the location of appropriate habitats. It is therefore surprising that immunity is rarely discussed in standard textbooks on comparative physiology and is often omitted from undergraduate courses in animal biology. This is regrettable as it is inappropriate (and unwise) to extrapolate from our mainly medically orientated knowledge of immunity of mammals to that of invertebrates. The most important distinction between invertebrate and higher vertebrate immune processes is that invertebrates do not express specific immunoglobulins (antibodies) or the clonally derived lymphocyte subsets that bring about 'memory-type' immunity. For coelomate invertebrates, host defence against infection rests almost entirely with the innate responses of the circulating blood cells (for recent review of innate immunity, see Boman, 2000), which explains in part why efforts to produce vaccines for invertebrates are misdirected.

The innate responses of invertebrates include several inflammatory-type reactions that are also found in higher vertebrates, but at the biochemical and molecular levels there are many bioactive molecules that are unique to invertebrate phyla. Inflammatory processes shared by invertebrates and higher vertebrates include phagocytosis, induction of blood-based enzyme cascades that generate recognition and activation of defence proteins, cytotoxicity and stimulation of the release of microbicidal agents (Ratcliffe et al., 1985). Others, which appear confined to invertebrates, encompass cellular encapsulation, melanization and blood cell aggregation without plasma gelation (Ratcliffe et al., 1985).

A number of comprehensive reviews detailing the range and character of innate immune responses of crustaceans and other invertebrates have now been published (see, for example, Ratcliffe et al., 1985; Söderhäll and Smith, 1988; Johansson and Söderhäll, 1989; Smith, 1991; Smith and Chisholm, 1992, 2000), so further elaboration is not needed here. The important point for the purpose of this overview is that nearly all the defence strategies shown by crustaceans are mediated, directly or indirectly, through the circulating blood cells (haemocytes). Certainly, the open circulation of many invertebrate species necessitates partitioning of immunologically aggressive molecules into circulating cells, with additional 'self' protection from the effects of these molecules achieved by regulated exocytosis or timely activation of factors by biochemical means.

Considering the important protective role that crustacean haemocytes play in immune defence, it is perhaps surprising that so much confusion exists in the literature with regard to their classification. It has been widely reported that in most crustacean species there are three morphologically, biochemically and functionally distinct haemocyte types, described as hyaline, granular and semi-granular cells (Bauchau, 1981; Söderhäll and Smith, 1983; Hose et al., 1990; Vázquez et al., 1997). In addition it has been observed that the proportion of each type of haemocyte varies considerably between species (Söderhäll and Smith, 1983). However, there is now evidence from ultrastructural studies that crustacean haemocytes may fall into two major categories (hyaline and granular) which arise from two distinct cell lines (Clare and Lumb, 1994), with semi-granular cells representing different stages of maturity in cells from a common granular cell lineage (Martin, 2000). The difficulty still remains that in a number of species, such as Homarus americanus, Panulirus interruptus, Procambarus clarkii and Loxorhyncus grandis, all haemocytes contain cytoplasmic granules. Thus it is proposed that identification of hyaline cells must be based on a suite of cytochemical and functional criteria (Martin, 2000).

Similarly, inconsistencies are found in the described function of the different haemocyte types, which can be separated for analysis by density gradient centrifugation on Percoll gradients (Söderhäll and Smith, 1983). For example, in the shore crab (Carcinus maenas) it is the hyaline cells, lacking obvious intracellular granules, that are actively phagocytic (Söderhäll et al., 1986), whereas the same function is ascribed to the granular or semigranular cells in other crustacean species (Söderhäll et al., 1986). So it appears that with respect to haemocyte classification there is, as yet, no clear pattern that embraces all crustaceans: haemocyte morphology and function is as heterogeneous as the class Crustacea itself and remains a topic for debate.

Regardless of the paradoxes found within the classification system, it is known that both granular and hyaline cells cooperate after immune challenge to form haemocyte clumps that restrict the spread of foreign particles, kill microorganisms and effectively remove infective agents from the general circulation. Haemocyte clumping, caused by the production of adhesive molecules on exposure to non-self material, causes a rapid, but usually temporary, reduction in the number of circulating haemocytes and may result in the appearance of previously cell-bound, immunologically active proteins in the plasma (see review by Smith, 1996).

Both semigranular and granular haemocytes are labile and tend to undergo rapid exocytosis and ultimately lysis upon exposure to bacterial carbohydrates or other non-self materials. The granules released during exocytosis contain a number of bioactive compounds, including opsonins, antimicrobial proteins, antioxidant molecules, clotting factors and components of the prophenoloxidase (proPO) activating system (Söderhäll and Cerenius, 1998).

Of particular interest amongst these molecules are microbicidal proteins. These molecules occur in a wide range of animal groups, from cnidarians to mammals, and they may include enzymes, such as lysozyme (Brown, 1995; Ratcliffe et al., 1985; Fenouil and Roch, 1991; Jollès, 1996; Ito et al., 1999), pigments (Service and Wardlaw, 1985) or peptides (Charlet et al., 1996; Cho et al., 1998; Bulet et al., 1999; Epand and Vogel, 1999; Lehrer and Ganz, 1999; Mitta et al., 2000). Antibacterial peptides, in particular, have become a major focus of research into innate immunity. They have been defined by Boman (1995) as low molecular weight (<10 kDa) proteins that display broad spectrum bactericidal or fungicidal properties and act in a stoichiometric rather than catalytic manner. Their small size makes them readily diffusible to sites of injury or infection and they tend to have low toxicity towards eucaryotic cells (Boman, 1995).

Moreover, they are easily synthesized without complex tissues or cells (Boman, 1995) and can therefore be expressed by larvae or juveniles. To date, over 200 different types have been reported to be present in a variety of species and they may occur in blood cells, fat body, epithelium and/or mucosal membranes (Boman, 1995; Lehrer and Ganz, 1999; Marigomez et al., 1999). Their biological importance is generally perceived to be to protect the host by mediating surface or systemic disinfection (Ganz and Lehrer, 1994; Boman, 1995; Lehrer and Ganz, 1999) and in helping to regulate the normal flora of the skin, internal surfaces, and gut (Bevins, 1994; Boman, 1995), although they may also contribute to the control of surface epibiosis, neuro-endocrine regulation (Ganz and Lehrer, 1994), and tissue organization during meta-morphosis (Ip et al., 1993).

3. Environmental and developmental aspects of immunity in crustaceans

These non-specific responses of the blood cells are widely regarded as the primordial form of host immunity that arose in simple bodied ectothermic invertebrates. There is, therefore, likely to be a strong link between environmental conditions and the innate immune mechanisms. Numerous studies have demonstrated the correlation between immunological vigour in ectotherms with external parameters, such as temperature, season and tidal rhythms (Chisholm and Smith, 1994; Hauton et al., 1995). To date, however, it is unclear how far immunological potential is influenced by phenotypic plasticity (i.e. the extent to which environmental conditioning modulates gene expression or blood cell production), but anecdotal evidence indicates that in crus-taceans, at least, immunity functions 'optimally' at the 'usual' environmental temper-ature of the animal (Chisholm and Smith, 1994). Changes in environmental conditions seem to impose physiological stress on the host which may be manifest as enhanced susceptibility to disease, and thus, by implication, impairment of immunocompetence. This indicates that early environmental conditioning has an impact on subsequent immunological capability in adults, but begs the questions as to how and when this occurs.

As we know little about how environmental conditions affect the general 'health' or well-being of crustaceans, we need a way of evaluating the effect of early, sub-acute environmental disturbance on immunological vigour and homeostatic integrity, coupled with an understanding of the degree of phenotypic plasticity within their physiological systems to environmental conditions. There is no agreement on what constitutes a suitable system to evaluate 'health' in crustaceans, although since these animals have an open-type circulation, the blood represents an ideal tissue for analysis. Without doubt, the haemocytes function in a variety of processes; for example, synthesis of metabolically important substances, transport and storage of materials, moulting, prevention of blood loss through wounding, repair, tanning and deposition of the cuticle, and protection against microbial or parasitic infection (Bauchau, 1981).

With respect to the ontogeny of the host defence systems in crustaceans, there is a great dearth of information. There are no reports of immune capability in larval forms and none addressing when and how the haemocytes are formed during metamorphosis and growth. This paucity of knowledge is due largely to the small size of most larvae which precludes functional *in vitro* or *in vivo* haemocytic analysis. We have a better understanding of the immune systems of insect larvae than of their crustacean relatives because insect larvae (especially lepidopteran) have become widely accepted experimental models (see Boman, 1995). Knowledge of immune ontogeny in insects, however, is not necessarily applicable to other invertebrates because of major differences in anatomy, physiology and ecology.

4. Markers of immune capability in crustaceans

An ideal biomarker of immune development and immunological vigour for any animal needs to be a process or product that: (i) relates directly to immune function; (ii) varies in amount or degree of activity in proportion to immune status or ability; (iii) is easily quantifiable; (iv) is produced or shown by all members of the chosen species of interest, irrespective of age; and (v) is broadly applicable to other, related, species. With crustaceans, workers have explored the potential of a variety of processes or products, such as total or differential haemocyte counts (Smith and Johnston, 1992; Truscott and White, 1990; Smith et al., 1995; Le Moullac et al., 1997), prophenoloxidase activity (Smith and Johnston, 1992; Hauton et al., 1995; Smith et al., 1995; Le Moullac et al., 1997, 1998; Cheng and Chen, 2000) and antibacterial activity (Chisholm and Smith, 1994; Smith et al., 1995). Whilst these have offered glimpses of how environment may affect immune capability, few, unfortunately, have fulfilled all the desired criteria required for application to larval or juvenile forms. Haemocyte counts are complicated by the huge variation between individuals. Measurement of prophenoloxidase or other enzyme activities requires careful standardization of each test sample and quantification of antibacterial activity is time-consuming and lacking in sensitivity.

Instead, we need to pinpoint one or more key defence molecules that can be titred directly and/or identify a haemocytic parameter that more reliably reflects immunological 'fitness'. As the haemocytes play such a central role in host protection against infection in crustaceans, it is reasonable to target the haemocytes or their products for analyses of this type.

Two aspects of haemocyte function offer good possibilities as 'health' markers. One is the expression of antibacterial peptides, the other is the production and maturation of new haemocytes in the haemopoietic tissue. Certainly, it is essential that the animal maintains an adequate population of haemocytes in the circulation so it is critical that the number of fully functional haemocytes is restored as soon as possible after wounding, blood loss or antigenically induced haemocytopaenia (Smith and Ratcliffe, 1980a,b; Smith et al., 1984; Lorenzon et al., 1999). Put simply, immunological vigour in crustaceans is a function of the size of the circulating haemocyte population and the expression of key antimicrobial proteins by these cells.

5. Experimental approaches

Molecular approaches are the most appropriate way to monitor immune development in juveniles, where larval size precludes direct analyses of cellular defence capability in vivo or in vitro. For particular molecules they enable quantification of gene expression rather than measurement of gene products or their biological activities and, for this, techniques such as in situ hybridization, primer extension assays or real time PCR allow us to study expression of specific proteins in vitro.

For cytological studies, quantitative and qualitative evaluation of the circulating blood cell population haemopoiesis and blood cell maturation is best achieved using markers, such as monoclonal antibodies (MABs). Certainly, MABs are a highly specific signpost for potential haemopoietic sites in invertebrates and pioneering work has already been performed by a number of authors. For example, Dyrynda et al. (1997) have raised monoclonal antibodies to mature haemocytes from bivalve molluscs and found that they cross react with cells from disaggregated juveniles (trochophores and 4 or 11 day veligers). In addition, Dikkeboom et al. (1988) have found that MABs recognize haemocyte sub-populations in juvenile and adult snails, Lymnaea stagnalis, while Uyama et al. (1993) have reported that MABs raised to mature blood cells of ascidians cross react with cells in tadpole larvae. Willott et al. (1994) have generated MABs to insect blood cells which appear to show some differences in staining with insects of different ages. Thus, at some point in development, common epitopes between adult and juvenile cells in other invertebrate species can be found, either in the embryonic haemopoietic tissue or in the circulating blood cells. Identifying the point at which these appear provides valuable information about tissue differentiation in invertebrates and identifying the point early in development where there is no cross-reactivity of MABs to cells is indicatative that the haemopoietic (i.e. lymphoid) tissue has yet to differentiate, and the larva is probably without a fully functional cellular immune system. In crustaceans, MABs have been raised against haemocyte populations of Penaeus japonicus by Rodriquez et al., (1995) and P. monodon by Sung et al. (1999), but as yet these have not been employed to study developmental aspects of haemopoietic and immune development.

6. Antibacterial peptides

Although interest in antimicrobial peptides on crustaceans has lagged behind that of insects, chelicerates, ascidians and vertebrates, a number have have been isolated and characterized from decapods (Table 1). The first to be reported was a 6.5 kDa proline-rich protein from haemocytes of the shore crab, Carcinus maenas, purified to homogeneity by Sep Pak C_{18} extraction, gel filtration and reverse-phase HPLC (Schnapp et al., 1996). This co-exists with at least five other constitutive antibiotic proteins of >70 kDa, ca 45 kDa, ca 14 kDa and 11.5 kDa in crab granular haemocytes (Schnapp et al.,

Table 1. Low molecular weight antibacterial proteins in crustaceans

	Penaeidins	Crab 6.5 kDa	Crab 11.5 kDa (Carcinin)	Callinectin
Species	*Penaeus vannamei*	*Carcinus maenas*	*Carcinus maenas*	*Callinectes sapidus*
Molecular mass	5.5–6.5 kDa	6.5 kDa	11.5 kDa	3.7 kDa
Source	Haemocytes	Haemocytes	Granular haemocytes	Haemocytes
Range of activity	Gram +ve bacteria, fungi	Gram +ve and –ve bacteria	Gram +ve marine bacteria	Gram –ve bacteria
Forms	3	1	1	1
Amino acid composition	Proline-rich[a] Cysteine-rich[a]	Proline-rich	Cysteine-rich	Proline-rich
Reference	Destoumieux et al. (1997)	Schnapp et al. (1996)	Relf et al. (1999)	Khoo et al. (1999)

[a] Penaeidins 1, 2 and 3 are proline-rich at the N-terminus and cysteine-rich at the C-terminus.

1996; Relf et al., 1999). The 6.5 kDa peptide is active against both Gram-positive and Gram-negative bacteria *in vitro* and partial amino acid sequencing by Edman degradation has revealed that its active fragment bears a high degree of identity with one of the cathelicidin family of proteins (bactenecin 7) from bovine neutrophils (Schnapp et al., 1996). Measurement of the strength its activity *in vitro* shows that its titre is similar to that of bactenecin 7, but slightly less than a commercially supplied synthetic 3 kDa antibacterial peptide (cecropin A) originally isolated from insects (Schnapp et al., 1996). Curiously, this 6.5 kDa peptide does not appear to be present consistently in C. *maenas* haemocytes, although it is not clear to what extent this relates to season, moult cycle, reproduction, recent natural infection with micro-organisms or other environmental or physiological factors.

By contrast, another antimicrobial protein of molecular mass of 11 534 Da occurs consistently in C. *maenas* granular haemocytes (Relf et al., 1999). This protein (*Table 1*) is heat stable, cationic, hydrophobic and cysteine rich (Relf et al., 1999). Interestingly, it is effective only against Gram-positive bacteria and its activity is salt-dependent, perhaps associated with its mechanism of action, but is fairly potent and is retained at concentrations as low as 10 μg ml^{-1} (Relf et al., 1999).

Partial cDNA sequence has been obtained for this protein (Relf et al., 1999) and used to construct oligonucleotide probes to investigate its expression in different crustacean species. Dot and Northern blots of RNA isolated from a range of crustaceans have revealed the presence of the transcript for the 11.5 kDa crab protein in eight species of crustacean thus far (*Table 2*). Additionally, PCR products have been obtained from a number of these species, using primers designed for the crab sequence. The cloning and sequencing of these products will allow us to determine the degree of homology that exists between species with respect to the gene for this antimicrobial protein.

Table 2. Detection of the transcript for the 11.5 kDa antibacterial protein in crustaceans

Species[a]	RNA/mRNA source[b]	Reaction
Carcinus maenas	Haemocytes	+
	Hepatopancreas	(+)[c]
Macrobrachium rosenbergii	Eggs	
	Larvae	+
	Post-larvae	+
	Adult haemocytes	+
	Adult hepatopancreas	(+)[c]
Liocarcinus depurator	Haemocytes	+
Penaeus indicus	Haemocytes	+
Hyas araneus	Haemocytes	+
Homarus gammarus	Haemocytes	+
	Hepatopancreas	+
Crangon crangon	Whole animal	(+)[c]
Cancer pagurus	Haemocytes	+
	Hepaptopancreas	+
Pacifastacus leniusculus	Haemocytes	+
	Hepatopancreas	+
Nephrops norvegicus	Haemocytes	–

[a] All species are marine apart from *P. leniusculus*, which is a freshwater crayfish, and M. *rosenbergii*, which has the early part of its life cycle spent in brackish water and adult stages in freshwater.
[b] Using dot and Northern blotting, the above reactions were noted when either total RNA (5 g) or mRNA (1 g) from the above species was hybridized with a DIG-labelled synthetic oligonucleotide probe originally designed to detect the transcript for the 11.5 kDa antibacterial protein in the shore crab, *C. maenas*.
[c] (+) indicates that the positive reaction was faint.

The 14 kDa protein with antibacterial activity in a mixed population of haemocytes from *C. maenas* was found to be inhibitory to live *Micrococcus luteus* (Schnapp et al., 1996), but further unpublished investigations by Chisholm and Smith on purified granular haemocytes have confirmed that, although the 14 kDa protein inhibits the growth of *M. luteus*, it does not lyse *M. luteus* cell walls. It appears then that lysozyme does not occur in the granular haemocytes of this crustacean. As regards the hyaline cells, the slow nature of *M. luteus* cell wall lysis caused by hyaline cell extracts (*Figure 1*) suggests some lysozyme-like activity but further confirmatory experiments are required to clarify this.

For another marine decapod, the blue crab, *Callinectes sapidus*, Khoo et al. (1999) have reported the single step purification of a 3.7 kDa basic antibacterial peptide from the haemocytes (*Table 1*). This peptide seems to be the major antimicrobial protein in blue crabs and as partial amino acid sequence data reveal no homology with other known antibacterial peptides, Khoo et al. (1999) have named it callinectin.

The only other crustacean to have been subjected to extensive biochemical and molecular investigation of antibacterial peptides is the shrimp, *Penaeus vannamei*

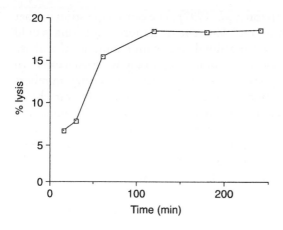

Figure 1. Lysis of Micrococcus luteus *cell wall suspension by* Carcinus maenas *hyaline cell lysate supernatant prepared in 0.06 M phosphate buffer, pH 6.4. Lysis is measured from absorbances read at 570 nm at intervals over a 4 h period and is expressed as a percentage of the control value. These data are related to a typical experiment and repeat experiments showed the same trend.*

(Destoumieux *et al.*, 1997). This animal has been found to contain three members of a new family of antimicrobial proteins (*Table 1*), termed penaeidins, that are active against fungi and Gram-positive bacteria (Destoumieux *et al.*, 1997, 1999). The mature molecules (designated penaeidin 1, 2 or 3) have molecular masses of 5.48, 5.52 or 6.62 kDa, respectively (Destoumieux *et al.*, 1997). They are characterized by a proline-rich N-terminus and a C-terminus containing six cysteine residues engaged in three disulphide bridges and have strong chitin-binding activity (Destoumieux *et al.*, 1997, 2000). They appear to be constitutively synthesized and stored in the circulating haemocytes, and are localized especially in the cytoplasmic granules (Destoumieux *et al.*, 2000). Importantly, levels of penaeidin mRNA in the circulating haemocytes decrease within 3 h of antigenic stimulation of shrimps *in vivo*, coinciding with a rise in the concentration of penaeidin in the plasma (Destoumieux *et al.*, 2000). They also appear to be present in cuticular tissue and must therefore contribute to surface disinfection.

So far, most, if not all, the antimicrobial peptides reported to date appear to be products of single genes, thereby enabling analyses of the expression of the genes encoding these proteins to be a direct measure of immunological vigour. It follows that such molecules represent an ideal way to investigate the developmental ecology of antimicrobial defence in crustaceans and other coelomate invertebrates.

7. Haemopoiesis

With respect to the size and regeneration of circulating blood cells, it is well known that in some invertebrate groups blood cell count reflects the well-being of the whole animal (Smith *et al.*, 1995: Lorenzon *et al.*, 1999). A variety of intrinsic and extrinsic factors can influence the haemocyte titre in crustaceans, either chronically or acutely. For example, exposure to contaminants, changed temperature or handling cause chronic depression of the haemocyte count (Bauchau and Plaquet, 1973; Chisholm and Smith, 1994; Smith *et al.*, 1995), whereas infection with bacteria, moulting or wounding result in more acute changes in blood cell number (Smith and Ratcliffe, 1980a; Hose *et al.*, 1992; Lorenzon *et al.*, 1999). Importantly, there is usually recovery of the haemocyte titre after acute haemocytopenia, bringing the cell count back to the normal or 'resting' level. In C. *maenas* this is usually within 24 h at *ca* 15°C (Smith and

Ratcliffe, 1980a; Smith *et al.*, 1984; Lorenzon *et al.*, 1999). The central question is what sets the normal 'resting' level of haemocytes? Current reductionist thinking would assume that it is genetically determined, but anecdotal information indicates that there is plasticity in the expression of blood cell number in decapod crustaceans. In particular, populations of intermoult decapods may show different 'resting' numbers of blood cells depending on the population sampled. Indeed counts may vary by as much as at least one order of magnitude (Smith, personal observations). Population differences have also been found for haemocyte counts in the shrimp, *Crangon crangon* (Lorenzon *et al.*, 1999; Smith and Johnston, personal observations). These trends hold good even allowing for wide variation in the number of circulating cells between individual animals and for seasonality. Thus it is possible that the normal 'resting' levels in adults is determined by environmental entrainment and that recovery of the haemocyte population following acute stress is a measure of 'health' in terms of immunological fitness.

The mechanisms underlying blood cell recovery are unclear, but they are likely to entail initial mobilization of blood cells from reservoirs in the haemal sinuses and up-regulation of mitosis in the circulating cells and/or the haemopoietic tissue (haemopoiesis). Although there have been some observational studies relating to haemopoietic tissues in crustaceans (Martin *et al.*, 1987, 1993), unfortunately there is a dearth of information about either the trafficking of haemocytes from haemal reservoirs to the circulation or the production of new haemocytes (or stem cells) in the haemopoietic tissue. Some studies provide indirect evidence that crustacean haemocytes may divide in the circulation (Ghiretti-Magaldi *et al.*, 1977; Hose *et al.*, 1992; Sequeira *et al.*, 1996) but direct quantification of up-regulation of mitosis by the haemocytes in the circulation has not been presented. One investigation by Sequeira *et al.* (1996) has reported a change in cell size distribution, as determined by flow cytometry, in the circulating haemocytes of *Penaeus japonicus ca* 5 days after injection of lipopolysaccharide (LPS). This was interpreted as evidence of proliferation within the circulating haemocyte population (Sequira *et al.*, 1996), although the haemocyte type(s) involved in the response were not identified. More recently, preliminary work on swimming crab, *Liocarcinus depurator*, has revealed that the circulating haemocytes show elevated rates of mitosis *in vivo* during recovery from inoculation with LPS (Smith and Hammond, unpublished results).

The phenomenon is important because at present we know little about the cellular and biochemical signals involved in regulating haemocyte number in crustaceans *in vivo*. An understanding of this through knowledge of haemocytic proliferation events is crucial if we are to find ways to promote innate immunity in this ecologically and economically significant group of invertebrates. Moreover, there is a great need for the development of proliferative cell lines from shellfish for *in vitro* work (Smith and Peddie, 1995). Whilst, the techniques for short- and medium-term culture of crustacean haemocytes are improving (Nadala *et al.*, 1993; Walton and Smith, 1999), we do not know yet how these methodologies can be developed for the establishment of immortal cell lines. In other marine invertebrates, principally ascidians, substances that act as mitogens for vertebrate leucocytes have been found to promote proliferation of circulating blood cells *in vitro* (Peddie *et al.*, 1995). Recently, Walton and Smith (1999) have developed a method for culturing the circulating haemocytes from crabs for up to 14 days *in vitro*, thus now enabling studies of crustacean haemocyte proliferation *in vitro*.

8. Developmental ecology of antibacterial defence in crustaceans

We have been investigating the developmental ecology of antibacterial defence in crustaceans by studying expression of the gene encoding the 11.5 kDa antibacterial peptide, originally found in *C. maenas*, in *H. gammarus* and *M. rosenbergii*. The lobster, *H. gammarus*, is a common, commercially caught decapod, which has four larval stages (zoea) that develop over 10–14 days. The tropical prawn, *M. rosenbergii*, by contrast, has 11 distinct zoeal stages, which develop over 17–60 days (depending on environmental conditions). Both species are easy to breed in captivity. Indeed, *M. rosenbergii* is farmed on a commercial scale in many parts of the world, and its biology and development have been extensively studied. The effects of factors, such as diet, on accelerating or retarding the growth cycle have been extensively documented (Devresse *et al.*, 1990) and its various developmental stages have been well characterized in terms of external features and behaviour (Uno and Soo, 1969).

Because of the ease with which *M. rosenbergii* can be cultured, it has been possible to set up experiments designed to examine the effects of diet on the immune capability of the prawns at different stages of development, as measured by the expression of the gene for the antibacterial 11.5 kDa protein described above. In addition we are carrying out bacterial challenge experiments in order to determine the effects of early conditioning, in this case diet, on the immune capability of on-grown adult prawns. Similarly, for *H. gammarus*, we are investigating the effects of temperature on the expression of the same gene, using samples from juveniles and young adults raised on optimal diets at three different temperatures.

So far, for both these species it appears that the transcript for the antibacterial protein is expressed at all stages of the lifecycle that we have tested. Whilst recognizing that environmental factors cannot be viewed in isolation from each other, quantitation of the expression of the transcript for the antibacterial peptide will allow us not only to evaluate the effects of diet or temperature on immune development, but will contribute to the identification of the vulnerable stages of the lifecycle in these crustaceans. This is highly relevant to the commercial culture of crustaceans since identification and control of disease is one of the of the major problems facing the industry and failure in this area results in heavy financial losses.

That antibacterial peptide expression in crustaceans might change during development is supported by the recent work of Mitta *et al.* (2000) on the marine bivalve molluscs, *Mytilus galloprovincalis* and *M. edulis*. Variations in the expression of the genes encoding two types of 4 kDa cysteine-rich, cationic antibacterial peptides were studied in adult mussels subjected to physical stress (shell filing, bacterial challenge *in vivo* or heat shock) and during development. Shell damage and heat shock resulted in up-regulation of gene expression, while bacterial challenge was found to cause down-regulation (Mitta *et al.*, 2000). Furthermore, gene expression increases during development but, interestingly, no gene expression was detected until after larval settlement and metamorphosis (Mitta *et al.*, 2000). In the light of this study, it will be interesting to ascertain if equivalent events occur in decapod crustaceans, which also undergo profound metamorphosis without settling to the substratum for a sessile existence.

9. Methods for assessing larvae and post-larvae

Molecular approaches can be financially prohibitive for studies on wild or farmed crustacean populations. For practical purposes, there is a need for measures of immune fitness/capability in developing crustaceans to be correlated with external features. Rapid, non-invasive assessment of the health and fitness of young crustaceans has become of increasing importance in the shrimp and prawn farming industry: the quality of the animals initially stocked can significantly affect pond yields. Most of the work in developing such systems has been done on post-larvae of farmed penaeid shrimp. Various criteria have been used, either as solo indicators or in concert with a number of factors. These have included simply the stage of development in post-larvae where the greater degree of development shown indicates better chances of survival (Parado-Estepa, 1988) and also correlations of survival with colouration, muscle development and behaviour (Bauman and Jamandre, 1990; Browdy, 1992). Other criteria such as gut fullness, presence of debris on setae, body deformity and muscle condition have also been considered (Villalon, 1991). Stress tests of various types have also been used as a means of assessment (Bauman and Jamandre, 1990; Samocha, et al., 1993; Briggs, 1992; Clifford, 1992; Tackaert et al., 1989)

Because of the limitations of many of these tests on the grounds that they are subjective and difficult to quantify (Fegan, 1992), Tayamen and Brown (1999) have developed a more reliable, versatile and quantifiable scheme based on an Apgar type scale (*Table 3a and b*). This is is based loosely on a scheme devised by Apgar in 1953 and now widely used as a research tool in paediatrics to assess the status of new-born human babies by early observable objective signs (Apgar and James, 1962). The Tayamen and Brown (1999) scheme (*Table 3a and b*) is necessarily a more complex index in order to deal with the numerous larval stages but allows a quantitative assessment of *M. rosenbergii* larvae. This avoids problems of subjectivity and is simple enough to be used both routinely in commercial hatcheries and for research purposes. We are using this scheme in conjunction with measurements of antibacterial peptide gene expression to study whether environmentally related alterations in immunological vigour are reflected in the physical well-being of the developing larvae and post-larvae of *M. rosenbergii.*

10. Future directions

Important areas for future research include firstly the identification of site(s) of synthesis and bioprocessing of marker proteins, such as antibacterial peptides in the adults of crustaceans and, by detecting expression of the gene encoding for these molecules at different larval and post-larval stages, the monitoring of the development of antibacterial immunity during ontogeny; secondly examination of the effect of extrinsic factors (e.g. temperature, diet, salinity) on antibacterial development as assessed by gene expression and immunofluorescence studies with antibodies raised against the active peptide and its precursor. Ideally, linked experiments should measure the 'health' and resistance to experimental infection with pathogens by crustaceans in adulthood. This should enable the determination of how the synthesis, processing and storage of bioactive antibacterial peptides are affected by early environmental influences.

Ontogenesis of haemocyte populations has been studied for a few invertebrates (Dyrynda et al., 1997; Meloni et al., 1999), but for crustaceans there is, at present, little information available. With respect to the circulating cell population, there is a need for clear characterization of changes within the size and composition of the blood cell population in relation to gender, development, moult stage, genetics, diet, temperature and salinity. This will permit tracking of histogenesis of the haemopoietic tissue in larvae and juveniles *in vivo*. Once baseline work has been completed, it should then be possible to compare haemopoietic development in larvae reared under different environmental conditions (e.g. temperature, diet or salinity) to ascertain if and how early environmental experience affects ontogeny of this important tissue. Crucially, this should be linked to determinations of the effects of early (i.e. larval and embryonic) environmental experience on the 'health' of on-grown adults, scored in terms of haemocyte profiles and haemopoiesis to complement parallel studies on immune markers (such as antibacterial peptides).

From the information gained by these studies, it should be possible to establish whether there is a positive correlation between antibacterial peptide synthesis, haemopoiesis and disease resistance in shellfish. This will provide vital information on the ontogeny and genetic regulation of non-specific immune defences of shellfish against pathogens. In terms of epidemiology, the health of populations is directly affected by the health of individuals, so studies such as these should shed light on how early environmental influences alter the success of populations and, in turn, the communities to which they belong. Such knowledge may facilitate breeding or raising of stock animals with enhanced disease resistance, as discussed by Mialhe *et al.* (1995) and Subasinghe *et al.* (1998), while the shellfish industry would benefit from having a measure of disease resistance for assessment of larval quality. In crustaceans, assessment of post-larval quality is considered a vital test test in shrimp farming and has been related already to nutritional status (Tackaert *et al.*, 1989; Briggs and Brown, 1991; Kondilatos *et al.*, 2001). Extending this work to embrace immunological vigour could be invaluable since it would provide a more direct measurement of the relevant factors.

11. Summary and conclusions

In terms of epidemiology, the health of animal populations is directly affected by the health of individuals, so further studies in the area of developmental ecology should shed light on how early environmental influences alter the success of populations and, in turn, the communities to which they belong. Such knowledge may also facilitate breeding or raising of stock animals with enhanced disease resistance. With shellfish, the industry would benefit from having a measure of disease resistance for assessment of larval quality. Assessment of post-larval quality, related already to nutritional status, could be further improved by the development of a measure of immunological vigour.

Acknowledgements

The original work described in the review was supported by a research grant from the Natural Environment Research Council (NERC), UK, under the Developmental Ecology of Marine Animals (DEMA) initative (reference number GST/02/1721). We would like to thank Dr Graham Kemp and Bernie Precious, Biomolecular Sciences, University of St Andrews for their help with the biochemical and molecular aspects of the work. The input of Juliet Relf and Lorian Hartgroves, NERC-supported Ph.D. students, and Susan Ross, NERC-supported research technician, in the Comparative Immunology Group at the Gatty Laboratory is gratefully acknowledged, as is the help of Dr Liz Dyrynda (University of Wales, Swansea) for her comments on the manuscript.

Table 3a. Condition index for evaluating larval quality of Macrobrachium rosenbergii

No.	CRITERIA TO CHECK	SCORE 0	SCORE 1	SCORE 2	1	2	3	4	5	6	7	8	9	10
									s a m p l e					
1.	GUT FULLNESS	gut empty	moderately full (30%–60%)	full gut with faecal strands										
2.	GUT LIPID CONTENT (STATE OF HEPATOPANCREAS)	no lipid vacuoles	very small vacuoles (10%–40%)	relatively full (60%–90%)										
3.	PIGMENTATION (STATE OF CHROMATOPHORES)	no colour/pigments (fully contracted chromatophores)	moderate chromatophores in one area	well dispersed chromatophores (pinkish/amber colouration)										
4.	BODY COLOURATION	gray/dark/bluish on abdominal segment	moderate light orange on abdominal segment	tan/orange/red blend on abdominal segment										
5.	SETATION	disfigured/damaged setae on rostrum, pereopods, telson, uropods	curled/kinked setae on rostrum pereopods, pleopods, telson, uropods	straight/whole setae no deformities on rostrum, pereopods, pleopods, telson, uropods										
6.	MUSCLE TO GUT RATIO	Gut appears wide, muscle narrower on VI abdominal segment	Gut appears narrow and slightly wider muscle	Gut appears narrow and muscle appears thick and wider on VI abdominal segment										
7.	MUSCLE APPEARANCE OF ABDOMEN (APPEARANCE OF ABNORMAL MUSCLES)	opaque/grainy	slightly transparent	clear/transparent smooth										
8.	MELANIZATION (PRESENCE OF BLACK SPOTS)	appendages and parts of body affected	very minor necrosis	no necrosis, absence of black spots										
9.	FOULING ORGANISMS	major parts of body affected	Minor parts of body affected	body clean/absence of protozoans, ciliates,organisms										
10.a	SWIMMING BEHAVIOUR (BETWEEN STAGE VIII TO X)	sluggish/circular motion, erratic movement	moderate movement with head upside down	very active tail first lateral motion/jump-like towards the side										
10.b	PHOTO POSITIVE RESPONSE (BETWEEN STAGE I TO VI)	negative response	slow positive response	fast positive response										
				Total score										
				Larval stage										

Score ratings 0=poor 1=fair 2=excellent

Number of days _____   Batch/Tank No. _____

Date: _____   Scored by: _____

Tables 3a and 3b are reproduced by kind permission of Blackwell Science Ltd, Oxford from Tayaman, M. and Brown, J.H. (1999) A condition index for evaluating larval quality of *Macrobrachium rosenbergii* (de Man). *Aquacult. Res.* 30, 917–922.

Table 3b. Explanatory glossary for condition index

	Score 0	Score 1	Score 2
1	GUT FULLNESS	GUT EMPTY OF FOOD	GUT MODERATELY FULL	FULL GUT WITH FAECAL STRANDS
2	GUT LIPID CONTENT (STATE OF HEPATOPANCREAS)	LARVAE APPEARS THIN, NO LIPID GLOBULES VISIBLE	VERY SMALL GLOBULES VISIBLE IN THE DIGESTIVE GLAND	GLOBULES IN THE DIGESTIVE GLAND VISIBLY FULL
3	PIGMENTATION (STATE OF CHROMATOPHORES)	FULLY CONTRACTED CHROMATOPHORES ASSOCIATED W/DARK BLUISH COLOUR	MODERATE CHROMATOPHORES IN ONE AREA W/LIGHT ORANGE	WELL DISPERSED CHROMATOPHORES W/AMBER/TAN/RED PIGMENTS
4	BODY COLOURATION	GREY/DARK/BLUISH APPEARANCE ON ABDOMINAL SEGMENT	MODERATE LIGHT ORANGE APPEARANCE ON ABDOMINAL SEGMENT	TAN/ORANGE/RED/AMBER LIKE APPEARANCE ON ABDOMINAL SEGMENT
5	SETATION	ROSTRUM DAMAGED/DISFIGURED (Check setae also)	ROSTRUM CURLED BENT/KINKED (Check setae also)	ROSTRUM STRAIGHT/WHOLE (Check setae also)
6	MUSCLE TO GUT RATIO	GUT APPEARS WIDE, MUSCLE THIN IN VI SEGMENT	GUT APPEARS THIN, MUSCLE IN VI SEGMENT WIDE	GUT APPEARS THIN, MUSCLE IN VI SEGMENT WIDER
7	MUSCLE APPEARANCE OF ABDOMEN (APPEARANCE ABNORMAL MUSCLES)	ABDOMINAL MUSCLE OPAQUE/GRAINY	ABDOMINAL MUSCLE SLIGHTLY CLEAR	ABDOMINAL MUSCLE CLEAR/TRANSPARENT, SMOOTH
8	MELANIZATION (PRESENCE OF BLACK SPOTS)	BLACK SPOTS SEEN ON APPENDAGES AND/OR BODY	MINOR BLACK SPOTS ON APPENDAGES/BODY	NO BLACK SPOTS (MELANIZATION)
10a	SWIMMING BEHAVIOUR (BETWEEN STAGE VIII TO X)	SLUGGISH MOTION	MODERATE MOVEMENT	FAST, JUMP LIKE MOTION

DA-BFAR-National Freshwater Fisheries Technology Research Centre, Philippines, Institute of Aquaculture, Stirling, UK

References

Apgar, V. (1953) A proposal for a new method of evaluation of the newborn infant. *Curr. Res. Anesth. Analg.* **32**: 260–267.

Apgar, V. and James, L.S. (1962) Further observations on the newborn scoring system. *Am. J. Dis. Child.* **104**: 419–428.

Bauchau, A.G. (1981) Crustaceans. In: *Invertebrate Blood Cells*, Vol. 2 (eds N.A. Ratcliffe and A.F. Rowley). Academic Press, London, pp. 385–420.

Bauchau, A.G. and Plaquet, J.C. (1973) Variation du nombre des hemocytes chez les crustaces brachyoures. *Crustaceana* **24**: 215–223.

Bauman, R.H. and Jamandre. D. R. (1990) A practical method for determining quality of *Penaeus monodon* (Fabricius) fry for stocking in grow-out ponds. In: *Proceedings of the Aquatech Conference on Technical and Economic Aspects of Shrimp Farming* (eds M.B. New, H. Saram, and T. Singh). Infofish, Kuala Lumpur, pp. 124–137.

Bevins, C.L. (1994) Antimicrobial peptides as agents of mucosal immunity. In: *Antibacterial Peptides*. Ciba Foundation Symposium. John Wiley, Chichester, pp. 186, 250–261.

Boman, H.G. (1995) Peptide antibiotics and their role in innate immunity. *Annu. Rev. Immunol.* **13**: 61–92.

Boman, H.G. (2000) Innate immunity and the normal microflora. *Immunol. Rev.* **173**: 5–16.

Briggs, M.R.P. (1992) A stress test for determining vigour of post larval *Penaeus monodon* (Fabricius). *Aquacult. Fish. Mgmt* **23**: 633–637.

Briggs, M.R.P. and Brown, J.H. (1991) Intensive rearing of postlarval *Penaeus monodon* (Fabricius) in concrete nursery tanks. *Proceedings of 22nd World Aquaculture Conference*, Puerto Rico.

Browdy, C.L. (1992) Review of the reproductive biology of *Penaeus* species: perspectives on controlled shrimp maturation systems for high quality nauplii production. In: *Proceedings of the Special Session on Shrimp Farming* (ed J. Wyban). World Aquaculture Society, Baton Rouge, LA, pp. 22–51.

Brown, P.B. (1995) Physiological adaptations in the gastro-intestinal tract of crayfish. *Am. Zool.* **35**: 20–27.

Bulet, P., Hetru, C., Dimarcq, J.L. and Hoffman, D. (1999) Antimicrobial peptides in insects; structure and function. *Dev. Comp. Immunol.* **4–5**: 329–344.

Charlet, M., Chernysh, S., Philippe, H., Hetru, C., Hoffmann, J.A. and Bulet, P. (1996) Innate immunity-isolation of several cysteine-rich antimicrobial peptides from the blood of a mollusc *Mytilus edulis. J. Biol. Chem.* **271**: 21808–21813.

Cheng, W. and Chen, J.C. (2000) Effects of pH, temperature and salinity on immune parameters of the freshwater prawn, *Macrobrachium rosenbergii. Fish Shellfish Immunol.* **10**: 387–391.

Chisholm, J.R.S. and Smith, V.J. (1994) Variation in antibacterial activity in the haemocytes of the shore crab, *Carcinus maenas*, with temperature. *J. Mar. Biol. Assoc. UK* **74**: 979–982.

Cho, J.H., Park, C.B., Yoon, Y.G. and Sim, S.C. (1998) Lumbricin I, a novel proline-rich antimicrobial peptide from the earthworm: purification, cDNA cloning and molecular characterization. *Biochim. Biophys. Acta Mol. Basis Dis.* **1408**: 67–76.

Clare, A.S. and Lumb, G. (1994) Identification of haemocytes and their role in clotting in the blue crab, *Callinectes sapidus. Mar. Biol.* **118**: 601–610.

Clifford, H.C. (1992) Marine shrimp pond management: a review. In: *Proceedings of the Special Session on Shrimp Farming* (ed J. Wyban), World Aquaculture Society, Baton Rouge, LA, pp. 110–133.

Connor, P.M. (1972) Acute toxicity of heavy metals to some marine larvae. *Mar. Pollut. Bull.* **312**: 190–192.

De Coursey, P.J. and Vernberg, W.B. (1972) Effect of mercury on survival, metabolism and behaviour of larval *Uca pugilator* (Brachyura). *Oikos* **23**: 241–247.

Destoumieux, D., Bulet, P., Loew, D., van Dorsselaer, A. V., Rodriquez, J. and Bachère, E. (1997) Penaeidins, a new family of antimicrobial peptides isolsated from the shrimp, *Penaeus vannamei*, (Decapoda). *J. Biol. Chem.* **272**: 28398–28406.

Destoumieux, D., Bulet, P., Strub, J.M., van Dorsselaer, A. and Bachère, E. (1999) Recombinant expression and range of activity of two penaeidins, antimicrobial peptides from penaeid shrimps. *Eur. J. Biochem.* **266**: 335–346.

Destoumieux, D., Munoz, M., Cosseau, C., Rodriquez, J., Bulet, P., Comps, M. and Bachère, E. (2000) Penaeidins, antimicrobial peptides with chitin-binding activity, are produced and stored in shrimp granulocytes and released after microbial challenge. *J. Cell Sci.* **113**: 461–469.

Devresse, B., Romdhane, M.S., Buzzi, M., Rasowo, J., Leger, P., Brown, J.H., and Sorgeloos, P. (1990) Improved larviculture outputs in the giant freshwater prawn, *Macrobrachium rosenbergii*, fed a diet of *Artemia* enriched with -3 HUFA and phospholipids. *World Aquacult.* **21**: 123–125.

Dikkeboom, R., Tijnagel, J. and Van der Knapp, W. (1988) Monoclonal antibody recognized haemocyte sub-populations in juvenile and adult *Lymnaea stagnalis:* functional characteristics and lectin binding. *Devl. Comp. Immunol.* **12**: 17–32.

Dyrynda, E.A., Pipe, R.K. and Ratcliffe, N.A. (1997) Sub-populations of haemocytes in adult and developing marine mussel, *Mytilus edulis*: identification by the use of monoclonal antibodies. *Cell Tissue Res.* **289**: 527–536.

Epand, R.M. and Vogel, H.J. (1999) Diversity of antimicrobial peptides and their mechanism of action. *Bichim. Biophys. Acta Biomembr.* **1462**: 11–28.

FAO (1999) Aquaculture production statistics 1988–1997. FAO Fisheries Circular No. 815, Revision 11. fidi/c815 (rev 11) Rome, p. 203.

FAO (2000) The state of world fisheries and aquaculture, 1998. Part 1: World review of fisheries and aquaculture. http://www.fao.org/docrep/w9900e/w9900e02.html

Fegan, D.F. (1992) Recent developments and issues in the Penaeid industry. In: *Proceedings of the Special Session on Shrimp Farming* (ed J. Wyban). World Aquaculture Society, Baton Rouge, LA, pp. 55–70.

Fenouil, E. and Roch, P. (1991) Evidence and characterisation of lysozyme in six species of freshwater crayfish from Asacidae and Cambaridae families. *Comp. Biochem. Physiol.* **99B**: 43–49.

Ganz, T. and Lehrer, R.I. (1994) Defensins. *Curr. Opin. Immunol.* **6**: 584–589.

Ghiretti-Magaldi, A., Milanesi, C. and Tognon, G. (1977) Haemopoiesis in Crustacea Decapoda: origin and evolution of haemocytes and cyanocytes of *Carcinus maenas*. *Cell Different.* **6**: 167–186.

Hauton, C., Hawkins, L.E. and Williams, J.A. (1995) Circatidal rhythmicity in the activity of phenoloxidase enzyme in the common shore crab (*Carcinus maenas*). *Comp. Biochem. Physiol.* **111B**: 374–352.

Hose, J.E., Martin, G.G. and Gerard, A.S. (1990) A decapod hemocyte classification scheme integrating morphology, cytochemistry, and function. *Biol. Bull. Mar. Biol. Lab. Woods Hole* **178**: 33–45.

Hose, J.E., Martin, G.G., Tiu, S. and McKrell, N. (1992) Patterns of hemocyte production and release throughout the moult cycle of the penaeid shrimp *Sicyonia ingentis*. *Biol. Bull.* **183**: 185–199.

Ip, Y.T., Reach, M., Engstrom, Y., Kadalayil, L., Cai, H., Crespo, S., Tatei, K. and Levine, M. (1993) *Dif*, a dorsal related gene that mediates immune response in *Drosophila*. *Cell* **75**: 753–763.

Ito, Y., Yoshikaw, A., Hotani, T., Fukuda, S., Sugimura, K. and Imoto, T. (1999) Amino acid sequences of lysozymes newly purified from invertebrates imply wide distribution of a novel class in the lysozyme family. *Eur. J. Biochem.* **259**: 456–461.

Johansson, M.W. and Söderhäll, K. (1989) Cellular immunity in crustaceans and the proPO system. *Parasitol. Today* **5**: 171–176.

Jollès, P. (ed) (1996) *Lysozyme: Model Enzymes in Biochemistry and Biology.* Birkhauser, Basel.

Khoo, L., Robinette, D.W. and Noga, E.J. (1999) Callinectin, an antimicrobial peptide from blue crab, *Callinectes sapidus*, hemocytes. *Mar. Biotechnol.* 1: 44–51.

Kondilatos, G., McEvoy, L. and Brown, J.H. (2001) Rearing of larval *Macrobrachium rosenbergii* (De Man, 1879): investigation of different *Artemia* enrichment and larval condition assessment methods (In press).

Lehrer, R.I. and Ganz, T. (1999) Antimicrobial peptides in mammalian and insect host defence. *Curr. Opini. Immunol.* 11: 23–27.

Le Moullac, G., Le Groumellec, M., Ansquer, D., Froissard, S., Levy, P. and Aquacop, S.T.C. (1997) Haematological and phenoloxidase activity changes in the shrimp *Penaeus stylirostris* in relation to the moult cycle; protection against vibriosis. *Fish Shellfish Immunol.* 7: 227–234.

Le Moullac, G., Soyez, C., Saulnier, D., Ansquer, D., Avarre, J.C. and Levy, P. (1998) Effect of hypoxic strsss on the immune response and the resistance to infection to vibriosis of the shrimp, *Penaeus stylirostris.* Fish Shellfish Immunol. 8: 621–629.

Lorenzon, S., De Guarrini, S., Smith, V.J. and Ferrero, E. A. (1999) Effects of LPS on circulating haemocytes in crustaceans *in vivo. Fish Shellfish Immunol.* 9: 31–50.

MAFF (1998) *United Kingdom Sea Fisheries Statistics.* HMSO, London.

Marigomez, I., Lekube, X. and Cancio, I. (1999) Immunochemical localisation of proliferating cells in mussel digestive gland tissue. *Histochem. J.* 31: 781–788.

Martin, G.G. (2000) A review of haemocyte structure and function based on studies of the penaeid shrimp *Sicyonia ingentis* (Abstract). In: *Responsible Aquaculture in the New Millenium*, International Conference AQUA 2000, Nice, May 2–6. European Aquaculture Society, Oostende, Belgium, p. 444.

Martin, G.G., Hose, J.E. and Kim, J.J. (1987) Structure of haemopoietic nodules in the ridgeback prawn, *Sicyonia ingentis*: light and electron microscopic observations. *J. Morphol.* 192: 193–204.

Martin, G.G., Hose, J.E., Choi, M., Provst, R., Omori, G., McKrell, N. and Lam, G. (1993) Organisation of hematopoietic tissue in the intermolt lobster, *Homarus americanus. J.Morphol.* 216: 65–78.

Meloni, S., Mazzini, M. and Scapigliati, G. (1999) Ontogenesis of hemocytes in the stick insect, *Bacillus rossius* (Rossi) (Phasmatodea, Bacillidae), studied with an anti-hemocyte monoclonal antibody. *Int. J. Insect Morphol. Embryol.* 28: 245–250.

Mialhe, E., Bachère, E., Boulo,V., Cadoret, J.P., Rousseau, C., Cedeno, V., Saraiva, E., Carrera, L. and Colwell, R.R. (1995) Future of biotechnology-based control of disease in marine invertebrates. *Mol. Mar. Biotechnol.* 4: 275–283.

Mitta, G., Hubert, F., Dyrynda, E.A., Boudry, P. and Roch, P. (2000) Mytilin B and MGD2, two antimicrobial peptides of marine mussels: gene structure and expression analysis. *Comp. Devl. Immunol.* 24: 381–393.

Nadala, E.P., Loh, P.C. and Lu, Y. (1993) Primary culture of lymphoid, nerve, and ovary cells from *Penaeus stylirostris* and *Penaeus vannamei. In Vitro Cell Devl. Biol.* 29A: 620–622.

Parado-Estepa, F. D. (1988) Selection, transport and acclimation of prawn fry. In: *Technical Considerations for the Management and Operation of Intensive Prawn Farms* (eds Y.N.Chiu, L.M. Santos and R.O.Juliano). Aquaculture Society, Ilo-Ilo City, Philippines, pp. 81–85.

Peddie, C.M., Riches, A.J. and Smith, V.J. (1995) Proliferation of undifferentiated blood cells from the solitary ascidian, *Ciona intestinalis, in vitro. Devl. Comp. Immunol.* 19: 377–387.

Ratcliffe, N.A., Rowley, A.F., Fitzgerald, S.W. and Rhodes, C.P. (1985) Invertebrate immunity: basic concepts and recent advances. *Int. Rev. Cytol.* 97: 183–350.

Relf, J.M., Chisholm, J.R.S., Kemp, G.D. and Smith, V.J. (1999) Purification and characterisation of a cysteine-rich 11.5 kDa antibacterial protein from the granular haemocytes of the shore crab, *Carcinus maenas. Eur. J. Biochem.* 264: 1–9.

Rodriquez, J., Boulo, V., Mialhe, E. and Bachère, E. (1995) Characterisation of shrimp haemocytes and plasma components by monoclonal antibodies. *J. Cell Sci.* 108: 1043–1050.

Rosenberry, B. (ed) (1998) *World Shrimp Farming, Annual Report.* Shrimp News International, San Diego, CA, p. 328.

Rosenberry, B. (ed) (1999) *World Shrimp Farming, Annual Report.* Shrimp News International, San Diego, CA, p. 320.

Samocha, T.M., Lawrence, A.L. and Bray, W.A. (1993) Design and operation of an intensive nursery raceway system for penaeid shrimp. In: *CRC Handbook of Mariculture, Volume I, Crustacean Aquaculture* (ed J.P. McVey). CRC Press, Boca Raton, FL.

Schnapp, D., Kemp, G. D. and Smith, V.J. (1996) Purification and characterization of a proline-rich antibacterial peptide with sequence similarity to bactenecin 7 from the haemocytes of the shore crab, *Carcinus maenas. Eur. J. Biochem.* **240**: 532–539.

Sequeira, T., Tavares, D. and Arala-Chaves, M. (1996) Evidence for circulating hemocyte proliferation in the shrimp, *Penaeus japonicus. Devl. Comp. Immunol.* **20**: 97–104.

Service, M. and Wardlaw, A.C. (1985) Bactericidal activity of coelomic fluid of the sea urchin, *Echinus esculentus*, on different marine bacteria. *J. Mar. Biol. Assoc. UK* **65**: 133–139.

Sheeks, R.B. (1989) Taiwan's aquaculture – at the crossroads. *Infofish Int.* **6/89**: 38–43.

Smith, V.J. (1991) Invertebrate immunity: phylogenetic, ecotoxicological and biomedical implications. *Comp. Haemat. Int.* **1**: 61–76.

Smith, V.J. (1996) The prophenoloxidase activating system: a common defence pathway for deuterostomes and protostomes? *Adv. Comp. Environ. Physiol.* **23**: 75–114.

Smith, V.J. and Chisholm, J.R.S. (1992) Non-cellular immunity in crustaceans. *Fish Shellfish Immunol.* **2**: 1–31.

Smith, V.J. and Chisholm, J.R.S. (2001) Antimicrobial proteins in crustaceans. In: *Phylogenetic Perspectives on the Vertebrate Immune System* (eds G. Beck, M. Sugumaran and L. Cooper). Kluwer Academic/Plenum Press (in press).

Smith, V.J. and Johnston, P.A. (1992) Differential haemotoxic effect of PCB congeners in the common shrimp, *Crangon crangon. Comp. Biochem. Physiol.* **101C**: 641–649.

Smith, V.J. and Peddie, C.M. (1995) Marine invertebrate blood cell culture. In: *Biology of Protozoa, Invertebrates and Fishes: in Vitro Experimental Models and Applications. Actes de Colloques* Vol. 18, pp. 35–40.

Smith, V.J. and Ratcliffe, N.A. (1980a) Cellular defence reactions of the shore crab, *Carcinus maenas* (L.): clearance and distribution of injected test particles. *J. Mar. Biol. Assoc. UK* **60**: 89–102.

Smith, V.J. and Ratcliffe, N.A. (1980b) Cellular defense reactions of the shore crab, *Carcinus maenas* (L.): *in vivo* haemocytic and histopathological responses to injected bacteria. *J. Invert. Pathol.* **35**: 65–74.

Smith, V.J., Söderhäll, K., and Hamilton, M. (1984) β, 1–3 glucan induced cellular defences in the shore crab, *Carcinus maenas. Comp. Biochem. Physiol.* **77A**: 635–639.

Smith, V.J., Swindlehurst, R.J., Johnston, P.A. and Vethaak, A.D. (1995) Disturbance of host defence capability in the common shrimp, *Crangon crangon* by exposure to harbour dredge spoils. *Aquat. Toxicol.* **32**: 43–58.

Söderhäll, K. and Cerenius, L. (1998) Role of the prophenoloxidase-activating system in invertebrate immunity. *Curr. Opin. Immunol.* **10**: 23–28.

Söderhäll, K. and Smith, V.J. (1983) Separation of the haemocytes of *Carcinus maenas* and other decapod crustaceans and phenoloxidase distribution. *Devl. Comp. Immunol.* **7**: 229–239.

Söderhäll, K. and Smith, V.J. (1988) Internal defence mechanisms in crayfish. In: *The Biology and Culture of Freshwater Crayfish* (eds D. Holdich and R. Lowery). Croom Helm, London, pp. 213–235.

Söderhäll, K., Smith, V.J. and Johansson, M. (1986) Exocytosis and phagocytosis by isolated haemocyte populations of crustaceans: evidence for cell co-operation in the cellular defence reactions. *Cell Tissue Res.* **245**: 43–49.

Subasinghe, R.P., Bartley, D.M., McGladdery, S. and Barg, U. (1998) Sustainable shrimp culture development: biotechnological issues and challenges. In: *Advances in Shrimp*

Biotechnology (ed T.W. Flegel) National Centre for Genetic Engineering and Biotechnology, Bangkok, pp. 13–18.

Sung, H.H., Wu, P.Y. and Song, Y.L. (1999) Characterisation of monoclonal antibodies to haemocyte sub-populations of the tiger shrimp, *Penaeus monodon,* immunochemical differentiation of three major haemocyte types. *Fish Shellfish Immunol.* 9: 167–179.

Tackaert, W., Abelin, P., Dhert, P., Legere, P., Grymonpre, D., Bombeo R. and Sorgeloos, P. (1989) Stress resistance in postlarval penaeid shrimp reared under different feeding procedures. *Proceedings of Aquaculture '89,* Los Angeles, CA, 12–14 February, 1989.

Tayamen, M and Brown, J.H. (1999) A condition index for evaluating larval quality of *M. rosenbergii* (De Mann 1879). *Aquacult. Res.* 30: 917–922.

Truscott, R. and White, K.N. (1990) The influence of metal and temperature stress on the immune system of crabs. *Funct. Ecol.* 4: 455–461.

Uno, Y. and Soo, K.C. (1969) Larval development of *Macrobrachium rosenbergii* reared in the laboratory. *J. Tokyo Univ. Fish.* 55: 179–190.

Uyama, T., Uchiyama, J., Nishikata, T., Satoh, N. and Michibata, H. (1993) The accumulation of vanadium and manifestation of an antigen recognized by a monoclonal-antibody specific to vanadocytes during embryogenesis in a vanadium-rich ascidian. *J. Exp. Zool.* 265: 29–34.

Vázquez, L., Pérez, A., Millán, D., Agundis, C., Martin, G., Cooper, E.L., Lascurain, R. and Zenteno, E. (1997) Morphology of hemocytes from the freshwater prawn, *Macrobrachium rosenbergii. J. Morphol.* 234: 147–153.

Villalon, J. R. (1991) Practical Manual for Semi-intensive Commercial Production of Marine Shrimp. Texas A&M University Sea Grant Program, Galveston, TX, p. 103.

Walton, A. and Smith, V.J. (1999) Primary culture of phagocytic haemocytes from decapod crustaceans. *Fish Shellfish Immunol.* 9: 181–194.

Willott, E., Trenczek, T., Thrower, L.W. and Kanost, M.R. (1994) Immunochemical identification of insect haemocyte populations – monoclonal antibodies distinguish 4 major haemocyte types in *Manduca sexta. Eur. J. Cell Biol.* 65: 417–423.

9

Contaminant-mediated pro-/antioxidant processes and oxidative damage in early life stages of fish

D.R. Livingstone, S.C.M. O'Hara, A. Frettsome and J. Rundle

1. Introduction

A large number of anthropogenic chemicals enter the aquatic environment and are taken up into the tissues of fish and other organisms (Walker and Livingstone, 1992; Walker et al., 1996). The types of contaminants are varied and include polynuclear aromatic hydrocarbons (PAHs), organochlorine pesticides and industrial products, polychlorinated dibenzo-p-dioxins and dibenzofurans, nitroaromatics and other heterocyclic compounds, organophosphate and organometallic compounds, and many metals such as Ag, Cd, Cr, Cu, Fe, Hg, Pb and Zn (Livingstone, 2001). Contaminants may be present in dissolved or particulate form in the sediment, water-column, sea-surface microlayer (SSMC) and food (Cross et al., 1987; Van Veld, 1990; Hardy and Cleary, 1992; Walker and Livingstone, 1992; Livingstone, 1998). Recent measurements on contaminant body-burdens in fish include field studies in waters in or near Egypt (Abouarab et al., 1995), Greece (Albanis et al., 1995), Sweden (Berglund et al., 2000), The Netherlands (Dethlefsen et al., 1996), Morocco (Elhraiki et al., 1994), Chile (Focardi et al., 1996), England (McNeish et al., 1997), France (Roche et al., 2000), Finland (Sinkkonen and Paasivirta, 2000) and Australia (Fabris et al., 1999). The exposure of many early life stages to pollution may be particularly acute due to their interaction with the SSMC and its apparent ability to concentrate contaminants (Word et al., 1987; Liss and Duce, 1997).

Many different mechanisms of toxicity exist for different foreign chemicals (xenobiotics), and a single contaminant may affect its toxicity by more than one such mechanism (Rand, 1995; Walker et al., 1996; Livingstone et al., 2000a). Of recent interest has been evidence to indicate that contaminant-stimulated reactive oxygen species (ROS) production and oxidative stress may be a significant mechanism of toxicity in

Environment and Animal Development: Genes, Life Histories and Plasticity, edited by D. Atkinson and M. Thorndyke.

aquatic organisms, including fish, exposed to pollution (Winston and Di Giulio, 1991; Di Giulio *et al.*, 1995a; Kelly *et al.*, 1998; Livingstone, 2001). Whereas a substantial number of studies on pro-oxidant and antioxidant processes have been carried out on adult fish, much less is known of such processes in their early life stages, particularly in relation to the consequences of exposure to pro-oxidant contaminants for subsequent growth and development. The aims of this chapter are to provide a brief background on the general aspects of pro-oxidant, antioxidant and oxidative damage processes in animals, including adult fish; to review the information known for fish embryos and larvae; and to present results for a study using turbot (*Scophthalmus maximus*) free-feeding larvae to investigate the effects of exposure to selected pro-oxidant chemicals on aspects of ROS production, antioxidant defence, oxidative damage and growth.

2. Pro-oxidant and antioxidant processes in biological systems, including adult fish

2.1 *Nature and sources of ROS*

Molecular oxygen (O_2) is vital for animal life, its tetravalent reduction to water being coupled to the oxidation of food and the generation of energy. However, partial reduction of O_2 results in the formation of potentially toxic reactive chemical species which have been variously collectively termed as ROS, 'oxygen-derived species', oxyradicals, or other descriptors (Kehrer, 1993; Halliwell and Gutteridge, 1999). ROS comprise both radical species, including superoxide anion radical ($O_2^{\cdot-}$; one-electron reduction) and hydroxyl radical (OH·; equivalent to three-electron reduction), and non-radical species, including hydrogen peroxide (H_2O_2; two-electron reduction). ROS are continually produced in biological systems as undesirable bi-products of normal metabolism from various endogenous sources and processes, including certain enzymes (e.g. nitric oxide synthase), auto-oxidation (e.g. of reduced FAD and FMN), and mitochondrial, endoplasmic reticulum and nuclear membrane electron transport.

The reactivity and properties of the different ROS vary considerably, the most potent being OH· that reacts instantly and indiscriminately with virtually all organic molecules (rate constants of 10^8–10^{10} mol^{-1} s^{-1}). However, the production of the different ROS can be inter-related, yielding ultimately the highly damaging OH·. Thus, $O_2^{\cdot-}$ can produce H_2O_2 via dismutation (2 H$^+$ + 2 $O_2^{\cdot-}$ → H_2O_2); and $O_2^{\cdot-}$ and H_2O_2 can react together to yield OH· via a metal-catalysed Haber-Weiss reaction ($O_2^{\cdot-}$ + H_2O_2 → OH· + OH$^-$ + O_2). Such ROS can in turn give rise to other ROS, including the peroxyl radical ($RO_2\cdot$) and alkoxyl radical (RO·), through reaction with other biological molecules. An initial pro-oxidant event can thus give rise to a spreading web of ROS production within a biological system.

2.2 *Pro-oxidant/antioxidant balance, antioxidant defences, oxidative damage and oxidative stress*

In the normal healthy cell, ROS and pro-oxidant products are detoxified and held in check by antioxidant defences, including specific antioxidant enzymes and low molecular weight water-soluble (e.g. vitamin C, reduced glutathione (GSH), carotenoids) and lipid-soluble (e.g. vitamins A and E) free radical scavengers (Halliwell, 1996a; Halliwell and Gutteridge, 1999). The antioxidant enzymes include superoxide dismutase (SOD; EC 1.15.1.1 – converts $O_2\cdot^-$ to H_2O_2), catalase (EC 1.11.1.6 – converts H_2O_2 to water) and glutathione peroxidase (GPX; EC 1.11.1.9 – detoxifies H_2O_2 and organic hydroperoxides utilizing GSH). Thus a balance is thought to exist between ROS production and antioxidant defence, although low levels of oxidative damage, particularly to key biological molecules such as DNA (Wiseman and Halliwell, 1996), protein (Halliwell, 1996b) and lipid (Halliwell and Chirico, 1993) are also always present. However, marked increases in ROS production can tip the balance between pro-oxidant processes and antioxidant defence in favour of the former, resulting in increased oxidative damage to macromolecules and alterations in critical cellular processes such as Ca^{2+} regulation (Darley-Usmar and Halliwell, 1996; Halliwell and Gutteridge, 1999). This oxidative damage may be spread far from its point of cellular origin by the different ROS and the products of oxidation, culminating in a condition of oxidative stress (Halliwell and Chirico, 1993; Halliwell and Gutteridge, 1999). The latter can have dire consequences for health (e.g. in humans and/or other mammals), and it has been implicated in the cause and/or development of a range of diseases, including various cancers, cardiovascular disease and possibly neurodegenerative disease (Wiseman and Halliwell, 1996; Halliwell and Gutteridge, 1999). Exposure to increased ROS production can lead to induction (increased synthesis) of certain protective antioxidant enzymes via interaction with antioxidant responsive elements and resultant transcriptional regulation (Storz *et al.*, 1990; Halliwell and Gutteridge, 1999).

2.3 *ROS production and oxidative stress as a common mechanism of toxicity for a range of natural and pollutant environmental stressors*

Any process which leads to increased ROS production, either directly or indirectly via organic radical formation or other mechanisms, can potentially result in enhanced oxidative stress and biological damage (Halliwell and Chirico, 1993; Halliwell and Gutteridge, 1999). Possible pro-oxidant agents in the environment are many and varied. Natural sources include ROS such as H_2O_2 and OH· produced by photoactivation and other events (Mopper and Zhou, 1990; Price *et al.*, 1992), natural redox cycling transition metals such as Fe (Stohs and Bagchi, 1995) and xenobiotics such as quinones (Mason, 1990), hypoxia and hyperoxia (Jones, 1985). Potential anthropogenic sources include a wide range of contaminants such as redox cycling compounds (quinones, nitroaromatics, nitroamines, bipyridyl herbicides), PAHs (benzene, PAH oxidation products), halogenated hydrocarbons (bromobenzene, dibromomethane, PCBs, lindane), dioxins, pentachorophenol, metal contaminants (Al, As, Cd, Cr, Hg, Ni, Va) and air contaminants (NO_2, O_3, SO_2; Burns, 1993; Lemaire and Livingstone, 1993; Alsharif *et al.*, 1994; Di Giulio *et al.*, 1995a; Halliwell and Gutteridge, 1999). Additional man-related activities or conditions possibly leading to enhanced ROS production include the release of organic matter such as sewage (Fujiwara *et al.*, 1993); the use of ozone (Ritola *et al.*, 2000) and H_2O_2 (ECOTOX, 1993; Arndt and Wagner, 1997; Rach *et al.*, 1998; Gaikowski *et al.*, 1999) and other pro-oxidant chemicals (Culp and Beland, 1996) as disinfectants in aquaculture and other activities; the use of H_2O_2 for re-oxygenation of hypoxic waters (ECOTOX, 1993), and the occurrence of increased UV-radiation (Kagan *et al.*, 1990; Taira *et al.*, 1992; Arfsten *et al.*, 1996; Berghahn, 2000; Choi and Oris, 2000), hypoxia and hyperoxia (Lemaire and Livingstone, 1993).

Organic and metal contaminants can stimulate ROS production by a variety of biochemical mechanisms (Halliwell and Gutteridge, 1999). These include: (i) redox cycling of, for example, quinones (and other compounds, see previous paragraph), catalysed by $NAD(P)H$-dependent flavoprotein reductases, in which the quinone undergoes univalent reduction to the semiquinone which then reacts with O_2 to yield $O_2^{\cdot-}$ and regenerates the parent quinone; (ii) redox reactions of transition metals (e.g. Co, Cr, Ni, Va) with O_2 and other ROS; (iii) autoxidation of particular oxygenases (e.g. cytochrome P450s (CYPs) interacting with PCB substrates which uncouple the enzyme's catalytic cycle; Schlezinger et al., 1999); (iv) enzyme induction (e.g. CYPs, flavoprotein reductases; Strolin-Benedetti et al., 1999); (v) disruption of membrane-bound electron transport (e.g. mitochondrial, microsomal electron transport affected by lipophilic contaminants; Arumugam et al., 1999); and (vi) and depletion of antioxidant defences (e.g. GSH involved in the biotransformation of organic contaminants).

2.4 *Contaminant-stimulated ROS production, oxidative damage, oxidative stress and disease in adult fish*

The same basic scenario of basal and contaminant-stimulated pro-oxidant and antioxidant processes, oxidative damage, oxidative stress and links with disease is indicated for adult fish as for mammals, although much less is known of many of these aspects, in particular *in vivo* events and the relationship of oxidative damage with disease (Livingstone et al., 1994; Livingstone, 2001). The studies in adult fish have largely been carried out on liver (the major organ of biotransformation of xenobiotics), or specialized (hyperoxic) tissues such as the gas gland (Lemaire et al., 1993), and have been the subject of a number of reviews (Di Giulio et al., 1989, 1995a; Winston and Di Giulio, 1991; Lemaire and Livingstone, 1993; Livingstone et al., 1994; Kelly et al., 1998; Livingstone, 2001).

An increased potential for ROS production has been demonstrated in hepatic subcellular fractions both with *in vitro* incubations with organic model compounds and contaminants, for example AH-quinones, lindane (γ-hexachlorocyclohexane), nitroaromatics (Washburn and Di Giulio, 1988, 1989; Lemaire et al., 1994; Lemaire and Livingstone, 1997), anthracene and anthracene/UV interactions (Choi and Oris, 2000), 3, 3', 4, 4'-tetrachlorobiphenyl (Schlezinger et al., 1999, 2000), and with *in vivo* exposures of whole animals to single chemical mixtures (PCBs) in the laboratory, or sampled from contaminated field sites (Livingstone et al., 2000b). The number of species examined include Atlantic flounder (*Platichthys flesus*), bluegill sunfish (*Lepomis macrochirus*), catfish (*Ictalurus punctatus*), large-mouth bass (*Micropterus salmoides*), perch (*Perca fluviatilis*) and rainbow trout (*Oncorhynchus mykiss*), which, combined with the number of compounds examined and mechanistic enzyme studies (Lemaire and Livingstone, 1994, 1997; Lemaire et al., 1996; Kitamura and Tatsumi, 1997) indicate a widespread potential for contaminant-stimulated ROS production in fish (Livingstone, 2001). Antioxidant enzymes and free radical scavengers are found widely in fish (see reviews cited above) and recent examples of new species studied include *Ictalurus nebulosus* (Hai et al., 1997), *Lampanyctus crocodilus* (Capo et al., 1997), gilthead seabream (*Sparus aurata*; Pedrajas et al., 1996), yellow perch (*Perca flavescens*), bream (*Abramis brama*) and tench (*Tinca tinca*; Ciereszko et al., 1999), and various species from marine and fresh waters of Brazil (Filho, 1996; Marcon and Wilhelm, 1999). Other protective enzymes involved in antioxidant function include

DT-diapahorase (quinone: oxidoreductase; EC 1.6.99.2; catalyses the two-electron reduction of quinones to hydroquinones so preventing them from redox cycling) and glutathione S-transferase A (conjugates and detoxifies products (alkenals) of lipid peroxidation) which have been investigated in respectively O. mykiss (Lemaire et al., 1996) and plaice (Pleuronectes platessa; Leaver and George, 1998). Other antioxidants demonstrated, or indicated, in fish are uric acid in seminal plasma (Ciereszko et al., 1999) and metallothionein studied in various fish cell lines (Schlenk and Rice, 1998; Kling and Olsson, 2000). Transient increases in antioxidant enzyme activities have been seen with exposure to contaminants, but overall relatively little is known of the regulation of antioxidant systems in fish in relation to either endogenous or exogenous sources of ROS and other pro-oxidants (Bainy et al., 1996; Pedrajas et al., 1996; Livingstone, 2001; Livingstone and Nasci, 2000; Machala et al., 2000). Recent antioxidant studies have shown changes in antioxidant enzyme activities of O. mykiss with adaptation to seawater or freshwater (Kolayli and Keha, 1999), and in catfish (Heteropneustes fossilis) with exposure to elevated temperature (Parihar et al., 1996, 1997); increased oxidative damage (hepatic lipid peroxidation) in brown bullhead (Ameriurus nebulosus) following depletion of its antioxidant GSH (Ploch et al., 1999); altered GSH and oxidized glutathione levels, indicative of oxidative stress, in blood of Arctic charr (Salvelinus alpinus) with exposure to ozone (Ritola et al., 2000); and have demonstrated a region of the metallothionein-A gene of O. mykiss to be functional in promoting gene transcription following exposure of RTH-149 cells to H_2O_2 (Olsson et al., 1995).

Despite the presence of basal or enhanced antioxidant defence systems, increased levels of oxidative damage occur in fish with laboratory and field exposure to contaminants. Increases in oxidative damage have been seen for all three types of major macromolecules (lipid, protein, DNA) and for single and mixed-contaminants, including both redox cycling (Cu, Fe, nitrofurantoin) and non-redox cycling (Cd, PAHs) contaminants. Examples include: (i) increased lipid peroxidation in various tissues of sea bass (Dicentrarchus labrax) exposed to Cd or Cu (Roméo et al., 2000), African catfish (Clarias gariepinus) exposed to Fe (Baker et al., 1997), dab (Limanda limanda; Livingstone et al., 1993) and I. punctatus (Di Giulio et al., 1993) exposed to sediments containing PAHs and PCBs, S. aurata injected with paraquat, $CuCl_2$, dieldrin or malathion (Pedrajas et al., 1995); and in H. fossilis exposed to bleached kraft paper mill effluent (Fatima et al., 2000); (ii) increased protein oxidation (non-peptide carbonyl group formation) in P. flesus hepatocytes exposed to H_2O_2 (Fessard and Livingstone, 1998) and in livers of animals from contaminated field sites (Livingstone, 2001); and (iii) increased DNA oxidation in sole (Parophrys vetulus) exposed to nitrofurantoin (Nishimoto et al., 1991), O. mykiss exposed to H_2O_2 (Kelly et al., 1992), S. aurata exposed to dieldrin (liver), paraquat (gills), urban or industrial pollution (liver; Rodriguez-Ariza et al., 1999; measured as 8-hydroxy-deoxyguanosine formation in the three studies), P. vetulus from PAH- and PCB-contaminated field sites (Malins et al., 1990; Malins and Gunselman, 1994) and trout (Salvelinus namaycush) from field sites high in iron-ore tailings (Payne et al., 1998; measured as 2,6-diamino-4-hydroxy-formamido-pyrimidine in both studies). Studies have also been carried out where no increase in oxidative damage with exposure to contaminants has been seen, for example no change in lipid peroxidation in liver of I. punctatus exposed to As and methylarsonate (Schlenk et al. 1997), or in DNA oxidation of liver of S. maximus, L. limanda and sole (Solea solea); exposed to nitrofurantoin (Mitchelmore et al., 1996), liver of *S. aurata* exposed to Cu^{2+}, paraquat or malathion (Rodriguez-Ariza *et al.*, 1999), and *L. limanda* from contaminated field sites (Chipman *et al.*, 1992). DNA damage, measured as strand breakage, has been seen in isolated fish cells with exposure to pro-oxidants, viz. hepatocytes of *O. mykiss* (Devaux *et al.*, 1997) and hepatocytes and blood cells of brown trout (*Salmo trutta*; Mitchelmore and Chipman, 1998) exposed to H_2O_2.

Information on the relationship between oxidative stress, disease and fitness in adult fish is extremely limited. Pathological/functional changes have been seen at the cellular and whole-animal level in laboratory studies. At the cellular level, altered intracellular Ca^{2+} levels were seen in head kidney phagocytes and peripheral blood leucocytes of *O. mykiss* with exposure to the well-known oxidative stressor lindane (Betouille *et al.*, 2000a,b), and at the tissue level, gill lesions occurred in *O. mykiss* exposed to H_2O_2 (Speare *et al.*, 1999). At the whole-animal level, respiratory and acid-base pathophysiology (Powell and Perry, 1997) and decreased growth rate (Speare *et al.*, 1999) occurred in *O. mykiss* with exposure to H_2O_2; increased respiratory burst activity of peritoneal and head kidney phagocytes was seen in *H. fossilis* with exposure to bleached kraft paper mill effluent (Fatima *et al.*, 2000); increased lipid peroxidation and decreased growth were seen in *C. gariepinus* fed iron-supplemented diets (Baker *et al.*, 1997); decreased growth rates occurred in killifish or mummichog (*Fundulus heteroclitus*) with exposure to dietary PCBs (Gutjahr-Gobell *et al.*, 1999); and decreased growth and egg production and cessation of spawning were found in zebrafish (*Brachydanio rerio*) exposed to various concentrations of binary mixtures of dichloraniline and the pro-oxidant lindane (Ensenbach and Nagel, 1997). Decreased seminal antioxidant vitamin C concentrations affected sperm quality, resulting in decreased fertilization rate and hatching rate of embryos, in *O. mykiss* (Dabrowski and Ciereszko, 1996). In a comparative study of possible determinants of oxidative stress in two species of fish with known different susceptibilities to cancer, a greater pro-oxidant capacity for microsomal redox cycling of xenobiotics, coupled with a GSH-dependent antioxidant system less able to withstand oxidative challenge, was found in the more cancer-prone *A. nebulosus* compared to the less cancer-prone *I. punctatus* (Di Giulio *et al.*, 1995b). In the field, the oxidized DNA base FapyGua (2,6-diamino-4-OH-5-formamido-pyrimidine) was found in cancerous tissue of liver of sole (*P. vetulus*), indicating a role for oxidative DNA damage in contaminant-caused carcinogenesis (Malins *et al.*, 1990). Additionally, the presence of such oxidized DNA bases in neoplasm-free livers of specimens of *P. vetulus* from a population showing a high incidence of liver cancer was argued to indicate that such bases are part of premalignant changes (Malins and Haimanot, 1991; Malins and Gunselman, 1994).

3. Pro-oxidant and antioxidant processes in early life stages of fish

Fewer studies have been carried out on early life stages than on adult fish, but comparable observations have been made on the uptake of pro-oxidant and other contaminants, the potential for basal and contaminant-stimulated ROS production, and the presence of antioxidant enzyme activities.

3.1 *The uptake and bioaccumulation of contaminants*

Fish early life stages are likely to come in contact with contaminants in particular via the SSMC, but also through other routes including the water-column (Section 1), food sources (Di Pinto and Coull, 1997) and parental transfer (Johnson *et al.*, 1998). The SSMC occurs at the air–water interface of open water; is composed of lipids, proteins, carbohydrates and other molecules; and is able to concentrate contaminants to levels many-fold higher than in the sub-surface waters (Liss and Duce, 1997). Contaminants found in the SSMC include metals (Cd, Cu, Pb, Zn; Barnes *et al.*, 1982), organotins (Cleary, 1991), PCBs (Connolly and Thomann, 1982), phthalates (Davey *et al.*, 1990), PAHs and organochlorines (Hardy, 1982). Lower rates of uptake of contaminants from the water-column are seen for eggs than larvae, but elimination rates are also lower, resulting in similar bioconcentration factors for the two early life stages (Petersen and Kristensen, 1998). The differences in rates of uptake between eggs and larvae appear to be related to the lower rate of transport of xenobiotics across the chorion of the egg compared to the rate of transport across the gill epithelium of the larvae. Time to reach steady-state conditions of uptake of contaminants are shorter in early life stages than in juvenile/adult stages, presumably due to the smaller size of the larvae and therefore the increased total surface of the membranes per unit fish weight. Examples of uptake of contaminants by early life stages include: (i) PCBs from sediments or food (copepods) by juvenile spot (*Leiostomus xanthurus*; Di Pinto and Coull, 1997); (ii) PAHs (naphthalene, phenanthrene, pyrene, benzo[a]pyrene) and PCBs from the water-column by eggs and larvae of *B. rerio* and larvae of cod (*Gadus morhua*), herring (*Clupea harengus*) and *S. maximus* (Petersen and Kristensen, 1998); and (iii) parental transfer of 2,3,7,8-tetrachlorodibenzo-*p*-dioxin (TCDD) to eggs and larvae of brook trout (*Salvelinus fontinalis*; Johnson *et al.*, 1998). As is also observed for adult fish, major factors in determining the degree of bioconcentration of organic contaminants in early life stages are the lipophilicity of the compound and the lipid content of the larvae (Petersen and Kristensen, 1998). A major reason for the high sensitivity of larvae to contaminants compared to juvenile/adult fish has been proposed to be due to differences in bioconcentration kinetics of the two life stages (Petersen and Kristensen, 1998).

3.2 *Contaminant-stimulated ROS production, oxidative damage, oxidative stress and disease in fish early life stages*

A potential for both NAD(P)H-dependent basal and NAD(P)H-dependent contaminant-stimulated ROS-generating potential has been demonstrated for yolk sac larvae of *S. maximus* (Peters *et al.*, 1996). Inhibition studies on 12 000g supernatants of whole-body larvae indicated that the iron/EDTA-mediated formation of OH· occurred via the production of $O_2^{\cdot-}$ and H_2O_2. Basal ROS production of the subcellular fraction was stimulated by a range of redox cycling one- to four-ring aromatic hydrocarbon quinones and was indicated to be stimulated by other xenobiotics, including nitroaroamatics. Antioxidant enzymes have been demonstrated in larvae of several species of fish, including *O. mykiss* (Aceto *et al.*, 1994), *S. maximus* (Peters and Livingstone, 1996), *Dentex dentex* (Mourente *et al.*, 1999), sprat (*Sprattus sprattus*; Peters *et al.*, 1992) and sardine (*Sardinia pilchardus*; Peters *et al.*, 1994). The antioxidant enzyme activities detected comprised catalase, SOD, selenium-dependent glutathione peroxidase (Se-GPX), total GPX (cumene hydroperoxide as substrate), glutathione reductase (EC 1.6.4.2) and DT-diaphorase. The antioxidant vitamin E has been measured in eggs and larvae of *D. dentex* (Mourente *et al.*, 1999).

Endogenous influences on pro-oxidant and antioxidant processes have been indicated from comparison of different early life stages of O. *mykiss* (Aceto *et al.*, 1994), *S. maximus* (Peters and Livingstone, 1996) and *D. dentex* (Mourente *et al.*, 1999). SOD activity in *S. maximus* decreased progressively during development from embryos to 11-day old larvae, indicative of a reduced need to detoxify $O_2 \cdot^-$. This change is generally consistent with the observation that rates of O_2 consumption (and therefore the endogenous potential for $O_2 \cdot^-$ production) increase from fertilization to hatching (Finn *et al.*, 1991) and peak soon after (Rønnestad *et al.*, 1992; Finn *et al.*, 1995). In contrast, catalase, Se-GPX and glutathione reductase activities in *S. maximus* increased progressively from embryos to 11-day-old larvae, indicative of an increased need to metabolize H_2O_2 and organic peroxides, and consistent with observed increased levels of lipid peroxidation. Similar enzyme activity changes, with SOD decreasing and catalase and Se-GPX increasing, from hatching to larvae were also seen for *D. dentex* (Mourente *et al.*, 1999), but, with the exception of catalase, not for O. *mykiss* (Aceto *et al.*, 1994). *D. dentex* also differed from *S. maximus* in showing decreasing levels of lipid peroxidation with development from eggs/embryos to larvae. Environmental impact on pro-oxidant and antioxidant processes in fish larvae have been indicated from field observations on antioxidant enzyme activities of *S. pilchardus* off the North coast of Spain (Peters *et al.*, 1994) and *S. sprattus* in the North Sea (L.D. Peters and D.R. Livingstone, unpublished observations). In the former study, whole-body catalase and SOD activities were highest at sites nearest to the Bilbao estuary.

Fish early life-stages are known to be very sensitive to environmental change (Westernhagen, 1988), including both laboratory (Buhl and Hamilton, 1991) and field (Hall *et al.*, 1993; Cameron *et al.*, 1992) exposure to contaminants. However, very little is known of the direct relationship between contaminant-mediated pro-oxidant processes and biological damage and disease in these organisms. Potential pro-oxidant contaminants exert toxic effects on early life stages, but such pathologies could occur through mechanisms other than, or as well as, enhanced ROS production and oxidative damage. Examples of such contaminants include: (i) TCDD producing DNA damage and apoptotic cell death in vasculature of embryos of medaka (*Orizias latipes*; Cantrell *et al.*, 1996a), oedema in embryos of (parentally exposed) *S. fontinalis* (Johnson *et al.*, 1998), and oedema and other pathologies in embryos of lake trout (*Salvelinus nanamycush*; Guiney *et al.*, 1997); (ii) lindane producing heptocyte pathologies in sac-fry of O. *mykiss* (Sylvie *et al.*, 1996) and lindane/3,4-dichloroaniline mixture reducing growth and egg production in *S. fontinalis* (Ensenbach and Nagel, 1997); and (iii) PCBs producing embryo and larval mortality in zebrafish (*Danio rerio*; Westerlund *et al.*, 2000). Possible pro-oxidant mechanisms linking contaminants such as dioxins and PCBs with biological damage are the induction of CYPs, including CYP1A, and subsequent autoxidation of the enzyme with resultant ROS production (Schlezinger *et al.*, 1999, 2000). Induction of CYP1A in early life stages with exposure to various contaminants, including TCDD, benzo[a]pyrene, SSMC and water-soluble fraction of crude oil, has been observed at the mRNA, protein and/or enzymatic activity levels for a number of fish species, including F. *heteroclitus* (Binder *et al.*, 1985; Nacci *et al.*, 1998), G. *morhua* (Goksøyr *et al.*, 1991), O. *mykiss* (Engwall *et al.*, 1994), fathead minnow (*Pimephales promelas*; Lindström-Seppä *et al.*, 1994), *S. maximus* (Peters and Livingstone, 1995; Peters *et al.*, 1996, 2001), *S. namaycush* (Binder and Lech, 1984; Guiney *et al.*, 1997) and *S. fontinalis* (Johnson *et al.*, 1998). Similar dose-responses of endothelial CYP1A levels and mortality observed for sac-fry of

S. namaycush exposed to TCDD have been argued to indicate that the status of the former may be linked to the TCDD-induced vascular derangements which lead to yolk sac and other oedemas that are associated with sac-fry mortality (Guiney *et al.*, 1997). However, the precise toxic mechanisms remain to be determined (Guiney *et al.*, 1997), although CYP1A-mediated ROS production is a possibility (Schlezinger *et al.*, 1999, 2000). Support for the latter contention is provided by the finding that the antioxidant *N*-acetyl cysteine provided partial protection against TCDD-induced lethality of both a fish cell line (PLHC-1) and embryos of medaka (*Orizias latipes*; Cantrell *et al.*, 1996b).

3.3 Fish embryos, eleutheroembryos and larvae as prime targets for oxidative stress and subsequent effects on growth and development

Alteration of nutritional processes (yolk utilization) are major determinants of early growth, development and survival of fish embryos and eleutheroembryos (Heming and Buddington, 1988). Similarly, deleterious effects on larval stages are likely to have consequences up to and after metamorphosis (Blaxter, 1988). In addition to exposure to pro-oxidant contaminants through the SSMC and other routes (Section 3.1) and the occurrence of contaminant-stimulated ROS production and oxidative stress (Section 3.2), fish early life stages are likely to be prime targets for ROS-mediated structural and functional damage for a number of other reasons. Embryos are largely impervious to solute but will permit uptake of small molecules (Heming and Buddington, 1988) such as possibly H_2O_2. Lipid can constitute up to 36% by dry weight of eggs (Heming and Buddington, 1988), including polyunsaturated fatty acids which are thought to be necessary for embryogenesis and larval development (Tocher *et al.*, 1985; Finn *et al.*, 1996) and which are readily attacked by OH·. Respiration rates increase markedly during early development, usually peaking several days after hatching (Finn *et al.*, 1991, 1995), increasing the potential for ROS production. Finally, early life stages of fish are also extremely sensitive to hypoxia and certain levels of hyperoxia (Rombough, 1988), both of which could lead to enhanced ROS production.

4. Laboratory studies: the effects of short-term exposure of free feeding larvae to pro-oxidant chemicals on pro-oxidant/antioxidant processes, oxidative damage and subsequent growth in turbot (*S. maximus*)

4.1 Rationale and experimental design

Relatively little is known about the relationship between oxidative damage in early life stages of fish and the consequences for subsequent development of the organism. The major aim of this study was to investigate this relationship by exposing free feeding larvae for a limited period (24 h) to known pro-oxidant chemicals and investigating: (i) initial and longer-term effects on pro-oxidant and/or antioxidant processes, and (ii) effects on growth. Turbot (*S. maximus*) was used as the experimental organism because it is hardy and aspects of its pro-oxidant/antioxidant systems have been characterized (Peters and Livingstone, 1996; Peters *et al.*, 1996). Two experiments were carried out, viz. experiment 1 – a 24 h exposure study to range-find (1 – 1000 ppb) non-lethal water-borne concentrations of pro-oxidant chemicals, and to measure pro-oxidant/antioxidant responses; experiment 2 – a 24 h exposure to selected pro-oxidant chemical conditions, followed by time-course measurements of selected pro-oxidant/antioxidant measurements and growth.

The pro-oxidant chemicals chosen were the model redox cycling compound 1,4-naphthoquinone (1,4-NQ), and the organochlorine contaminants PCBs (Arochlor 1254) and the pesticide lindane. Quinones occur as products of PAH combustion (Burns, 1993), and 1,4-NQ is a marked stimulator of NAD(P)H-dependent ROS production by subcellular fractions of larval *S. maximus* (Peters *et al.*, 1996) and adult fish (Lemaire and Livingstone, 1997). Organochlorines such as PCBs and lindane are widespread in the marine environment and both have been shown, or indicated, to be pro-oxidants in larval (Peters *et al.*, 1996) and adult fish (Livingstone *et al.*, 1993; Lemaire *et al.*, 1994; Bainy *et al.*, 1996). Pro-oxidant effects, antioxidant effects and oxidative damage were measured in terms of, respectively, potential for NAD(P)H-dependent ROS production, total oxyradical scavenging capacity (TOSC) for peroxyl radicals, and lipid peroxidation (malondialdehyde (MDA) levels) and protein oxidation (non-peptide carbonyl formation).

4.2 Experimental approach

Fish larvae and experimental design. Post-hatch free feeding *S. maximus* larvae (20 day plus; 7–10 mm length) were obtained from Mannin Seafarms, Isle of Man, UK. Transit time to Plymouth was typically 16 h. On arrival larvae were allowed to equilibrate to local culture facility conditions before transfer into the holding and growing-on aquaculture units (15–16°C; 34–35‰; 1000–1500 lx surface light). Each unit comprised a set of four 65 l tanks within a 750 l recirculating system with an integral, 'under- and over-type' media box filter (containing particulate-, biological- and activated carbon-sections, and a protein-skimmer) as described in Moe (1989). This system supplied long-term high quality seawater to the criteria stipulated in Spotte (1979), namely NH_4^+ and NO_2^- maintained at less than 0.01 mg l^{-1}, NO_3^- at less than 2.8 mg l^{-1}, and pH between 7.7 and 8.3. Salinity was routinely monitored and maintained at 34‰ by addition of distilled water to the sump/filter tank. Specimen tanks and filters were cleaned daily (10% per day replacement of seawater) and fish were fed with algae (*Pavlova lutheri*)-enriched rotifers (*Brachionus plicatilis*) and brine shrimp (*Artemia salina*) nauplii. Larvae were kept 1–2 weeks before being used in the exposure experiments.

Experiment 1. Exposure conditions were the same as the holding conditions, except fish were not fed. Filtered seawater (0.2 μm) was added to each of a series of 3 l volume Pyrex glass exposure vessels (2 l seawater per vessel). Glass Pasteur pipettes were used to provide gentle aeration to the water in all exposure vessels, from a filtered air supply via needle valves. Chemicals (Arochlor 1254, lindane and 1,4-NQ) were dissolved in vehicle solvent (polyoxyethylene-3-methyl ether) at a concentration to give a final exposure concentration of 1, 10, 100 and 1000 ppb when using 0.7 ml of stock solution per exposure vessel. The experiment comprised a control (no additions), vehicle control (solvent added) and 12 contaminant exposure (three chemicals, four concentrations) conditions. Larvae (8–12 mm length), were carefully transferred from the holding system to the exposure vessels (15 larvae per vessel), and their behaviour visually checked and any dead specimens removed at intervals over the next 24 h. After

24 h exposure, larvae were transferred to damp (seawater) absorbent tissue and excess water allowed to drain, frozen in liquid nitrogen and stored at $-80°C$ prior to biochemical analysis.

Experiment 2. Exposure procedures were essentially as for experiment 1, except that ethanol was used as the vehicle solvent at an application volume of 20 μl l^{-1} (v/v). Exposure concentrations of the chemicals were 100 and 1000 ppb Arochlor 1254, 10 and 20 ppb lindane, and 10, 25 and 50 ppb 1,4-NQ. Control and vehicle control exposures were also carried out. Sampling and storage procedures for biochemical analysis were as for experiment 1. After the initial 24 h exposure period, larvae were placed briefly in clean seawater before careful transfer back to the aquaculture system where they were grown on under the feeding conditions described above.

Tissue preparation, biochemical analyses and growth measurement. Frozen whole larvae (pools of three (experiment 1) or individuals (experiment 2)) were weighed, pulverized to powder at liquid nitrogen temperature in disposable polypropylene tubes with a Teflon tipped plunger, and homogenized on ice in 4 (experiment 1) or 1.5 (experiment 2) ml of 10 mM Tris–HCl containing 170 mM KCl, pH 7.6 using an electrically-driven Potter Elvehjem homogeniser. Homogenates were centrifuged at 1200g for 30 min at 4°C and the resulting supernatants decanted into a number of aliquots for biochemical analyses. Aliquots were either used immediately (lipid peroxidation), or plunged into liquid nitrogen before storage at $-80°C$ prior to analysis. Typical sample volumes were 600 μl for lipid peroxidation assay (mixed with 60 μl of 0.2% butylated hydroxytoluene in ethanol), 500–1000 μl for protein oxidation assay (mixed with 50 μl of protease inhibitor preparation (Sigma, UK), 100 μl of 1 M Tris-HCl pH 7.5 and 12 μl of 200 mM EDTA pH 7.0), 250 μl or more for TOSC assay, and 50 μl for total protein determination. Total protein concentration was determined by the colorimetric assay of Lowry *et al.* (1951).

NAD(P)H-dependent ROS production. The potential for NAD(P)H-dependent ROS production was determined by the iron/EDTA-mediated oxidation of the scavenging agent 2-keto-4-methiolbutyric acid (KMBA) to ethylene by OH· (produced both directly and from O_2·$^-$ and H_2O_2) essentially as described by Peters *et al.* (1996). Formation of ethylene was measured by gas chromatography. Assay incubations were performed in glass vials ($\sim$24 ml volume) closed with butyl rubber-sealed aluminium screw-caps. Assays were carried out at 25°C, and incubations contained in a final volume of 1 ml were 100 mM KH_2PO_4/K_2HPO_4 pH 7.5, 10 mM $MgCl_2$, 1 mM NaN_3, 75 μM $FeCl_3$/150 μM EDTA, 10 mM KMBA, 0.3 mM of NADH or NADPH, and 100 μg sample protein. At timed intervals after initiation of the reaction by addition of NAD(P)H, 1 ml headspace air samples were removed by syringe and injected into 2 m $\times$ 3 mm o.d. packed column of Porapak Q (80–100 mesh) maintained at 60°C in a Varian Star 3400 C_X gas chromatograph and equipped with flame ionization detector. Nitrogen and hydrogen flow rates were 30 ml min^{-1}, air flow rate was 300 ml min^{-1}, and injector and detector oven temperatures were respectively 120 and 200°C. Ethylene was quantified by reference to a standard ethylene gas.

Total oxyradical scavenging capacity. TOSC for peroxyl radicals was measured as the degree of inhibition of chemically-generated peroxyl radical formation, the latter being quantified by the oxidation of KMBA to ethylene, essentially as described by Winston *et al.* (1998). Incubations were performed at 37°C in glass 24 ml volume vials closed with butyl rubber-sealed aluminium screw-caps. Incubations mixtures contained in a final volume of 1 ml, 100 mM KH_2PO_4/K_2HPO_4 pH 7.5, 200 μM KMBA and 50 μg of sample protein (or equivalent volume of sample buffer). Incubations were started by the addition of the peroxyl radical-generating reagent, 2,2′-azobis-amidinopropane dihydrochloride (Fluka Chemical GmbH). One millilitre samples of headspace air were taken at intervals after this and ethylene measured by gas chromatography as described for the NAD(P)H-dependent ROS production assay. The area beneath the peak height versus time curve was determined for incubations with sample and without sample (control), and TOSC quantified as: TOSC = 100 − ((area under sample curve/area under control curve) × 100).

Lipid peroxidation. The assay was based on the measurement of MDA by reaction with thiobarbituric acid (TBA) and separation of the $(TBA)_2$-MDA adduct by high performance liquid chromatography (HPLC) as described in Chirico (1994), with modifications, as follows. Duplicate 250 μl aliquots of sample were mixed with 625 μl of 0.44 M H_3PO_4, left at room temperature for 10 min, followed by addition of 250 μl of 0.6% TBA and heating at 90°C for 30 min. Incubated mixtures were cooled to room temperature, mixed with 606 μl methanol, and centrifuged at 10 000g at 20°C for 20 min. The sample supernatants, reagent blank and a series dilution of standards (1,1,3,3-tetramethoxy-propane in 40% ethanol; similarly processed as for samples) were analysed by HPLC. Ten microlitres aliquots of reagent blank, standard or sample preparation were injected onto a 150 × 4.6 mm Spherisorb 5ODS2 column (HPLC Technology, Macclesfield, UK) fitted with a Jour-Guard C18 guard column (Fisher Scientific, Loughborough, UK) maintained at 32.5°C. The elution buffer was 65:35 of 50 mM K_3PO_4 pH 7.0: methanol at a flow rate of 1 ml min^{-1}, and the $(TBA)_2$-MDA adduct was detected with a Jasco 821-FP HPLC fluorescence spectrofluorometer (excitation 529 nm, emission 545 nm).

Protein oxidation. Non-peptide carbonyl groups were determined essentially as described by Reznick and Packer (1994) and Evans *et al.* (1999). Samples were mixed with 10% streptomycin sulphate (to give a final concentration of 1%), stood at room temperature for 30 min and centrifuged at 1200g for 10 min. Resulting supernatants were sub-divided into two equal aliquots and each were treated as follows. To one aliquot was added 2 M HCl and to the other 10 mM dinitrophenylhydrazine in 2 M HCl (both reagents added in ratio of 1:4 sample:reagent, v/v) and both mixtures were left at room temperature for 1 h. The duplicate samples were then treated with 20% trichloroacetic (1:1 v/v), left at room temperature for 15 min, centrifuged at 1200g for 10 min, and the resulting protein pellets washed with 1.5 ml of 10% trichloroacetic acid. The washing/centrifugation step was repeated and the resulting pellet treated with 1.0 ml of 1:1 v/v ethylacetate acetate:ethanol, mixed and centrifuged at 1200g for 10 min. The washing/centrifugation step was repeated twice and the final pellet dissolved in 1.5 ml 6 M guanidine hydrochloride/20 mM KH_2PO_4 pH 2.3. The protein concentration of the sample was determined by absorbance at 280 nm of the HCl-only treated aliquot against a reference containing 6 M guanidine hydrochloride/20 mM KH_2PO_4 and quantified using a standard curve of bovine serum albumen (0.05–2.0 mg ml^{-1}) in 6 M guanidine hydrochloride/20 mM KH_2PO_4. The carbonyl content of the sample pairs were assayed by determining the difference in absorbance for the range 410–320 nm between the HCl-only and the dinitrophenylhydrazine-treated aliquots, using a molar absorption coefficient for the latter of 22×10^3 M^{-1} cm^{-1}.

Measurement of growth. Growth measurements were taken from 7 days post-exposure to the chemicals. Growth of larvae was determined by direct measurement with a 1 mm square grid graph paper sealed between $150 \times 150 \times 1.5$ mm sheets of glass. Ten or fewer larvae were carefully transferred to a 140 mm Petri dish containing about 50 ml seawater and the dish placed on the measuring grid. The length in millimetres between the tip of the mouth and the caudal peduncle was recorded for each sampled fish. The fish were then carefully returned to the aquaculture system.

Sample numbers and statistics. The number of samples are given in the text. Values are expressed as means ± SEM. Groups of values were compared by one-way analysis of variance or Student's paired *t*-test (Sokal and Rohlf, 1981). Levels of confidence less than or equal to 0.05 were considered statistically significant.

4.3 *Experimental analysis*

No or negligible deaths of *S. maximus* larvae were observed in the control or vehicle control conditions during the 24 h exposure periods. After exposure of larvae to a water-borne chemical concentration range of 1, 10, 100 and 1000 ppb, mixed PCBs were not lethal after 24 h exposure at any concentration, whereas 100 and 1000 ppb of both lindane and 1,4-NQ were highly toxic, causing death after approximately only 30 min (1000 ppb lindane) or 8 h (other conditions; experiment 1, see *Figures 1–3*). Lindane and 1,4-NQ were not lethal at 1 or 10 ppb. After a similar period of 24 h exposure, a reduced concentration of lindane of 20 ppb produced 20% mortality after 24 h, whereas 25 and 50 ppb 1,4-NQ were still highly lethal, resulting in total mortality after about 24 h (experiment 2, see *Figure 4*).

No statistically significant differences in any biochemical or growth measurement between the control and vehicle control conditions were seen for experiment 1 or 2 and the pooled data for the two conditions are presented as control. No increases in NADH-dependent (*Figure 1a*) or NADPH-dependent (*Figure 1b*) ROS production by whole-body 1100g supernatants were seen following exposure of larvae to any chemical. However, significant decreases in both NADH- and NADPH-dependent ROS production were seen, or indicated, under conditions causing lethality in larvae, viz. 100 and 1000 ppb lindane and 1,4-NQ (*Figure 1*). A depression of NADH-dependent ROS production at 10 ppb PCBs followed by a linear increase from 10 to 100 ppb PCBs was indicated, but was not statistically significant (*Figure 1*). Slight increases in TOSC values were indicated at certain concentrations of PCBs (10 and 1000 ppb) and 1,4-NQ (1 and 10 ppb), but the results were not statistically significant and no relationship with increasing chemical concentration was evident (*Figure 2*). No significant increases in lipid peroxidation were seen for any chemical compared to control (*Figure 3*). However, trends of increasing lipid peroxidation with increasing chemical concentration were clearly indicated for lindane and 1,4-NQ, whereas PCBs showed the unusual pattern of lower levels of malondialdehyde compared to the control at low concentrations of PCBs followed by increasing lipid peroxidation with increasing PCB concentration. No differences in protein oxidation were seen with

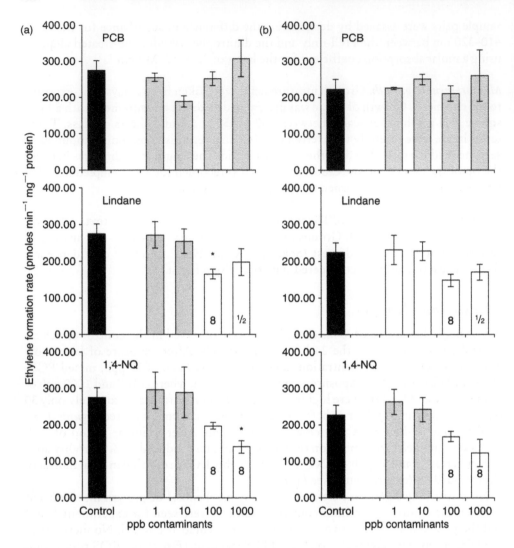

Figure 1. Experiment 1 – effect of 24 h exposure of turbot (S. maximus) larvae to water-borne Arochlor 1254 (PCB), lindane or 1,4-naphthoquinone (1,4-NQ) on NADH-dependent (a) and NADPH-dependent (b) ROS production (iron/EDTA-mediated KMBA oxidation to ethylene) of whole-body 1100 g supernatants; white boxes, animals died (hour of death given in box); means ± SEM (n=3, or 6 for control); *p < 0.05.

PCB exposure and in most of the other exposure conditions (data not shown), with the exception of 100 ppb 1,4-NQ (3.98 ± 0.57) and 1000 ppb lindane (4.11 ± 0.32) which were higher than the control condition (2.53 ± 0.35; values in nmol carbonyl moiety mg^{-1} protein).

Changes in lipid peroxidation for experiment 2 are presented in *Figure 4*. About a 1-fold increase in lipid peroxidation occurred in larvae following 24 h exposure to 100 ppb PCBs, whereas no change was seen at 1000 ppb PCBs. After 24 h exposure, no increase in lipid peroxidation was seen at either 10 or 20 ppb lindane, but a trend of increasing lipid peroxidation was indicated with exposure to 10–50 ppb 1,4-NQ, with a maximal

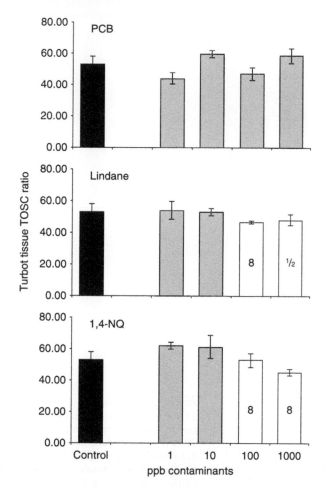

Figure 2. Experiment 1 – effect of 24 h exposure of turbot (S. maximus) larvae to water-borne Arochlor 1254 (PCB), lindane or 1,4-naphthoquinone (1,4-NQ) on total oxyradical scavenging capacity (TOSC) of whole-body 1100 g supernatants; other details as for Figure 1.

increase of about 1-fold at 50 ppb, although these animals were moribund at collection. Following transfer to clean water, no differences in lipid peroxidation were seen between previously chemically exposed and control animals at either 7 or 49 days after exposure. Levels of lipid peroxidation in all animals decreased about 4-fold between days 7 and 49 post-exposure.

The results for growth for experiment 2 are presented in *Figures 5* (absolute length) and 6 (net increase in mean length). From day 0 to day 42 post-exposure, there was about 50% mortality of fish, but no consistent differences were evident between any of the conditions (data not shown). The mean length of control larvae at day 7 post-exposure were less than animals from any exposed condition, but the differences were not statistically significant. The trend line of growth rate, expressed in terms of absolute length, appeared to be reduced between 15 and about 40 days compared to controls following exposure to 100 ppb PCBs, 1000 ppb PCBs or 10 ppb 1,4-NQ, such that smaller fish were produced after 40 days post-exposure, but the differences were not statistically significant (*Figure 5*). Such a trend was not indicated for lindane (*Figure 5*). However, when calculated in terms of net increase in mean fish length, all the growth curves for exposed conditions were below that for the control (*Figure 6*).

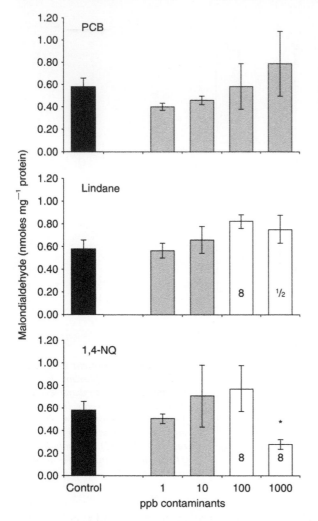

Figure 3. Experiment 1 – effect of 24 h exposure of turbot (S. maximus) larvae to water-borne Arochlor 1254 (PCB), lindane or 1,4-naphthoquinone (1,4-NQ) on lipid peroxidation (malondialdehyde levels) of whole-body 1100 g supernatants; other details as for Figure 1.

The fish in the control and 1000 ppb PCBs exposed conditions were grown on in clean water for a further 42 days (i.e. up to 84 days post-exposure) and a reduced growth rate was still evident for the PCBs condition, viz. net increases in mean length of 6.7 (control) and 3.9 (PCB-exposed) mm.

5. Summary and overview

The fact that contaminants can affect the higher-order characteristics of fish early life stage development, such as growth, pathology and survival, is well established (Section 3.2). Demonstrating that oxidative stress can lead directly to such effects is problematical because increased ROS formation is thought, at least in mammalian systems, to often accompany tissue injury through a range of processes, such as metal ion release and altered intracellular Ca^{2+} regulation (Halliwell and Gutteridge, 1999). Thus, increased oxidative damage and a condition of oxidative stress may either be the cause or the consequence of a particular impacted condition. The criteria for implicating

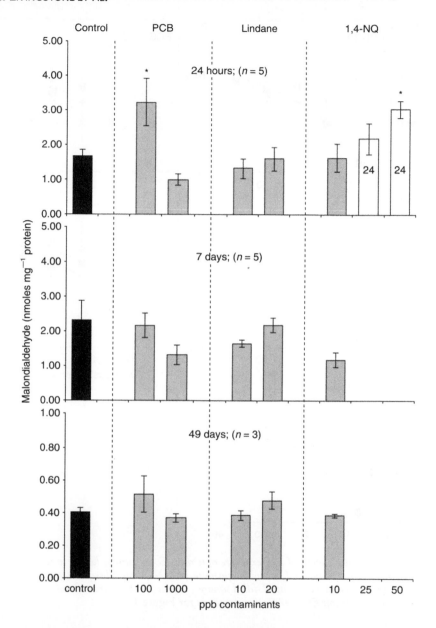

Figure 4. Experiment 2 – effect of 24 h exposure of turbot (S. maximus) larvae to water-borne Arochlor 1254 (PCB), lindane or 1,4-naphthoquinone (1,4-NQ) on lipid peroxidation (malondialdehyde levels) of whole-body 1100 g supernatants at various times after exposure; other details as for Figure 1, except n = 5 or 3, or 10 or 6 for control.

ROS formation as a causative agent are several, but include 'that the time-course of formation should be consistent with the time-course of tissue injury' (Halliwell and Gutteridge, 1999). A principle aim of this study therefore was to see whether contaminant-mediated alterations in pro-oxidant and antioxidant processes might precede changes in growth characteristics.

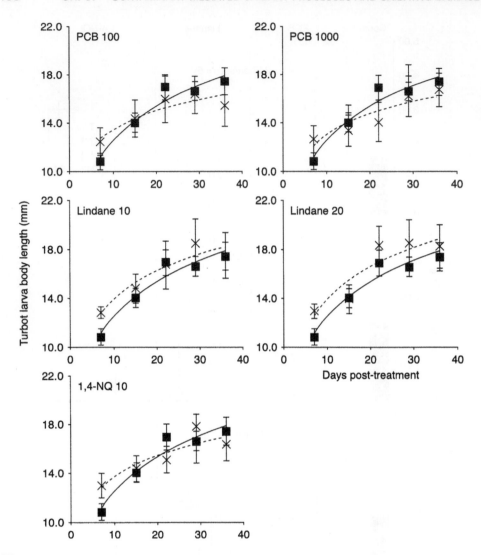

Figure 5. Experiment 2 – effect of 24 h exposure of turbot (S. maximus) *larvae to water-borne Arochlor 1254 (PCB), lindane or 1,4-naphthoquinone (1,4-NQ) on subsequent growth expressed as absolute length measurements; control, filled squares, exposed, cross; curves fitted logarithmetically (Microsoft Excel 97); other details as for* Figure 4.

Organic contaminants, including PCBs and other organochlorines, are known to be taken up by fish larvae from the water-column (Section 3.1) and to produce a number of deleterious effects on early life stages of fish (Section 3.2). The observed lethality responses of S. *maximus* larvae to lindane are roughly similar to those observed for a range of other larval and adult fish species, albeit at longer exposure periods of 96 h compared to 24 h, viz. (i) no affect on hatchability or fry survival up to about 20 ppb (IPCS, 1991), and (ii) an LC_{50} of about 50 ppb (range ~ 20–90 ppb) for adult fish species (IPCS, 1991), compared to the observed 20% lethality at 20 ppb, and 100% lethality at 100 and 1000 ppb lindane, for S. *maximus* larvae. Lethality was also seen for 1,4-NQ down to concentrations of 25 ppb, but no comparative data are available for

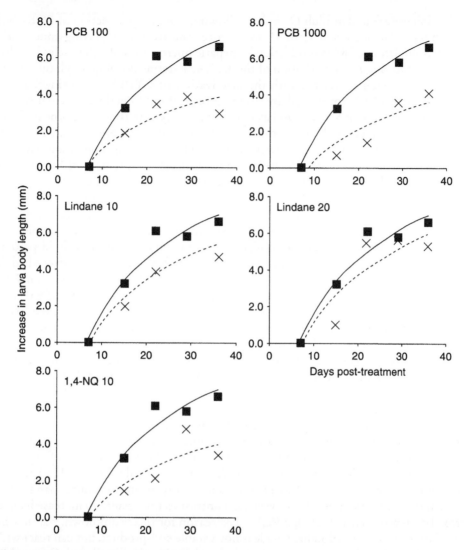

Figure 6. Experiment 2 – effect of 24 h exposure of turbot (S. maximus) larvae to water-borne Arochlor 1254 (PCB), lindane or 1,4-naphthoquinone (1,4-NQ) on subsequent growth expressed as increase in mean length; control, filled squares, exposed, cross; curves fitted logarithmetically (Microsoft Excel 97).

this chemical. Overall, pro-oxidant and antioxidant responses in experiment 1 were not obvious, with the exception of changes in lipid peroxidation and protein oxidation. Although larvae were collected as soon as possible after becoming moribund, the reduced potential for NAD(P)H-dependent ROS production at higher concentrations of lindane and 1,4-NQ (100 and 1000 ppb) could be due to tissue breakdown and enzyme loss following death. Increases in NAD(P)H-dependent ROS production have been seen *in vitro* in subcellular fractions of *S. maximus* larvae following addition of 1,4-NQ and other contaminants (Peters *et al.*, 1996), but no information is available on the effects of *in vivo* exposure to contaminants on this parameter. Increases in hepatic microsomal and/or cytosolic potential for NAD(P)H-dependent ROS production were found in adult *O. mykiss* following a single i.p. injection of PCBs, but the increases did not occur until 15 weeks after the treatment (Livingstone et al., 2000b). Stimulation of ROS production by hepatic microsomes of fish species has also been seen following *in vitro* addition of the 3, 3′, 4, 4′-tetrachlorobiphenyl, the generation of ROS being correlated with the inactivation of CYP1A (Schlezinger et al., 1999, 2000). CYP1A (measured by 7-ethoxyresorufin O-deethyase activity) was induced about 5-fold in *S. maximus* larvae following exposure to water-borne lindane, but after 48 h (Peters and Livingstone, 1995). No changes were seen in the total antioxidant capacity to scavenge peroxyl radicals, which is at least consistent with the observed lack of change in NAD(P)H-dependent ROS potential, although it is important to realize that neither of these measurements represent a complete picture of antioxidant defence or ROS production (Livingstone, 2001). Increases in lipid peroxidation in *S. maximus* larvae with 24 h exposure to the three chemicals were indicated, but the results were not statistically significant due to the small sample size ($n = 3$). No other information is available on contaminant effects on lipid peroxidation in fish larvae, but increases in lipid peroxidation have been seen in liver of adult fish with exposure over various time periods to a range of chemicals (Section 2.4), including PCBs, PAHs (Di Giulio et al., 1993; Livingstone et al., 1993) and organochlorine pesticides (Pedrajas et al., 1995). The increases in protein oxidation (non-peptide bond free carbonyl formation) at 100 ppb 1,4-NQ and 1000 ppb lindane could be related to death of the larvae, but it may be significant that such changes were not observed for other lethal exposure conditions, viz. 1000 ppb 1,4-NQ and 100 ppb lindane. Increases in protein oxidation have been seen in liver of adult *P. flesus* with contaminant exposure, but this was either in *in vitro* subcellular and hepatocyte studies using a free radical-generating system (Fessard and Livingstone, 1998), or in field studies of unknown exposure periods (Livingstone, 2001).

Based on the above results, emphasis in experiment 2 was focused on lipid peroxidation following initial exposure and over time, and on the consequences of exposure for subsequent growth of *S. maximus* larvae. An approximate doubling of lipid peroxidation was seen at 100 ppb PCBs, which was a markedly greater response than for the same exposure in experiment 1. However, in contrast to this result, no increase in lipid peroxidation was seen at 1000 ppb PCBs. The reason for this is unknown, but may be related to the fact that malondialdehyde is not a stable end-product, but can react with other biological molecules such as proteins and DNA (Halliwell and Gutteridge, 1999). Increased lipid peroxidation was again indicated or seen with exposure to 1,4-NQ, but caution is required with this result because the animals were dead at the higher 1,4-NQ concentrations. The decreases in the level of lipid peroxidation in all conditions with growth of the larvae probably reflects the increase in the contribution of non-lipid materials, such as skeletal muscle, to total body mass. Although, growth rates expressed as change in absolute length were not statistically different between exposed and control conditions, the net increase in mean length over time was clearly lower for exposed compared to control *S. maximus*, the reduction in growth being more marked for PCBs and 1,4-NQ than lindane. Decreased growth has been seen in adult *F. heteroclitus* with exposure to dietary PCBs (Gutjahr-Gobell et al., 1999), but no or little effect on growth was found for adult *P. promelas* or *S. fontinalis* with chronic exposure to, respectively, 9 and 16 ppb lindane (IPCS, 1991). Decreased growth was found for *B. rerio*, but this was for a binary mixtures of lindane with dichloraniline (Ensenbach and Nagel, 1997).

In conclusion, many chemicals can have multiple molecular mechanisms of toxicity that may contribute to different degrees to higher-order biological effects (Rand, 1995; Livingstone, 2001). Although the results of the S. maximus study are far from conclusive, the occurrence of increased lipid peroxidation at 100 ppb PCBs was correlated with a subsequent reduced growth rate of the larvae, possibly indicating a role for oxidative stress in higher-order impact. However, the absence of lipid peroxidation in all the other conditions in which reduced growth was also seen, including 1000 ppb PCBs, suggests that oxidative stress may be just one of multiple toxic effects. Such problems can only be resolved in future by focussing both on the criteria stipulated to identify oxidative stress as a causative agent (see Halliwell and Gutteridge, 1999) and on chemicals which can be shown to act predominantly mainly via pro-oxidant mechanisms.

Acknowledgements

The authors gratefully acknowledge helpful advice from Dr S. Baynes (CEFAS, Conway, UK) in the initial setting-up of the S. maximus aquaculture facility. The experimental work on turbot (S. maximus) larvae was funded by NERC DEMA (Developmental Ecology of Marine Animals) thematic grant GST/02/1713.

References

Abouarab, A.A.K., Gomaa, M.N.E., Badawy, A. and Naguib, K. (1995) Distribution of organochlorine pesticides in the Egyptian aquatic ecosystem. Food Chem. 54: 141–146.

Aceto, A., Amicarelli, F., Saccetta, P., Dragani, B., Bucciarelli, T., Masciocco, L., Miranda, M. and Di Ilio, C. (1994) Developmental aspects of detoxifying enzymes in fish. Free Rad. Res. 21: 285–294.

Albanis, T.A., Hela, D.G. and Hatzilakos, D. (1995) Organochlorine residues of eggs of Pelecanus crispus and its prey in wetlands of Amvrakikos Gulf, North-western Greece. Chemosphere 31: 4341–4349.

Alsharif, N.Z., Schleuter, W.J. and Stohs, S.J. (1994) Stimulation of NADPH-dependent reactive oxygen species formation and DNA damage by 2,3,7,8-tetrachlorodibenzo-p-dioxin in rat peritoneal lavage cells. Arch. Environ. Contam. Toxicol. 26: 392–397.

Arfsten, D.P., Schaeffer, D.J. and Mulveny, D.C. (1996) The effects of near ultraviolet radiation on the toxic effects of polycyclic aromatic hydrocarbons in animals and plants. Ecotox. Environ. Safety 33: 1–24.

Arndt, R.E. and Wagner, E.J. (1997) The toxicity of hydrogen peroxide to rainbow trout Oncorhynchus mykiss and cutthroat trout Oncorhynchus clarkii fry and fingerlings. J. World Aquacult. Soc. 28: 150–157.

Arumugam, N., Thanislass, J., Ragunath, K., Niranjali Devaraj, S. and Devaraj, H. (1999) Acrolein-induced toxicity – defective mitochondrial function as a possible mechanism. Arch. Environ. Contam. Toxicol. 36: 373–376.

Bainy, A.C.D., Saito, E., Carvalho, P.S.M. and Junqueira, V.B.C. (1996) Oxidative stress in gill, erythrocytes, liver and kidney of Nile tilapia (Oreochromis niloticus) from a polluted site. Aquat. Toxicol. 34: 151–154.

Baker, R.T.M., Martin, P. and Davies, S.J. (1997) Ingestion of sub-lethal levels of iron sulphate by African catfish affects growth and tissue lipid perioxidation. Aquat. Toxicol. 40: 51–61.

Barnes, R.K., Batley, G.E. and Sharp, J.H. (1982) Heavy-metal enrichment on the surface microlayer of the nepean Hawkesbury river system. Aust. J. Mar. Freshw. Res. 33: 417–430.

Berghahn, R. (2000) Response to extreme conditions in coastal areas: biological tags in flatfish otoliths. Mar. Ecol. Prog. Ser. 192: 277–285.

Berglund, O., Larsson, P., Ewald, G. and Okla, L. (2000) Bioaccumulation and differential partitioning of polychlorinated biphenyls in freshwater, planktonic food webs. *Can. J. Fish Aquat. Sci.* **57**: 1160–1168.

Betouille, S., Duchiron, C. and Deschaux, P. (2000a) Lindane increases *in vitro* respiratory burst activity and intracellular calcium levels in rainbow trout (*Oncorhynchus mykiss*) head kidney phagocytes. *Aquat. Toxicol.* **48**: 211–221.

Betouille, S., Duchiron, C. and Deschaux, P. (2000b) Lindane differently modulates intracellular calcium levels in two populations of rainbow trout (*Oncorhynchus mykiss*) immune cells: head kidney phagocytes and peripheral blood leucocytes. *Toxicology* **145**: 203–215.

Binder, R.L. and Lech, J.J. (1984) Xenobiotics in gametes of Lake Michigan lake trout (*Salvelinus namaycush*) induce hepatic monooxygenase activity in their offspring. *Fund. Appl. Toxicol.* **4**: 1042–1054.

Binder, R.L., Stegeman, J.J. and Lech, J.J. (1985) Induction of cytochrome P450-dependent monooxygenase systems in embryos and eleutheroembryos of the killifish *Fundulus heteroclitus*. *Chem. Biol. Interact.* **55**: 185–202.

Blaxter, J.H.S. (1988) Pattern and variability in development. In: *Fish Physiology* (eds W.S. Hoar and D.J. Randall), Vol. XI, *The Physiology of Developing Fish. Part A Eggs and Larvae*. Academic Press, New York, pp. 1–58.

Buhl, K.J. and Hamilton, S.J. (1991) Relative sensitivity of early life stages of Arctic grayling, coho salmon, and rainbow trout to nine congeners. *Ecotox. Environ. Safety* **22**: 184–197.

Burns, K.A. (1993) Evidence for the importance of including hydrocarbon oxidation products in environmental assessment studies. *Mar. Pollut. Bull.* **26**: 77–85.

Cameron, P., Berg, J., Dethlefsen, V. and Westernhagen, H. von (1992) Developmental defects in pelagic embryos of several flatfish species in the southern North Sea. *Neth. J. Sea Res.* **29**: 239–256.

Cantrell, S.M., Lutz, L.H., Tillitt, D.E. and Hannink, M. (1996a) Embryotoxicity of 2,3,7,8-tetrachlorodibenzo-*p*-dioxin (TCCD): the embryonic vasculature is a physiological target for TCCD-induced DNA damage and apoptotic cell death in medaka (*Orizias latipes*). *Toxiol. Appl. Pharmac.* **141**: 23–34.

Cantrell, S.M., Hannink, M. and Tillitt, D.E. (1996b) *N*-acetyl cysteine provides partial protection against TCCD-induced lethality in fish embryos. *Mar. Environ. Res.* **42**: 113–118.

Capo, C., Stroppolo M.E., Galtieri, A., Lania, A., Costanzo, S., Petruzzelli R., Calabrese, L., Polticelli, F. and Desideri A. (1997) Characterization of Cu, Zn superoxide dismutase from the bathophile fish, *Lampanyctus crodolilus*. *Comp. Biochem. Physiol.* **117B**: 404–407.

Chipman, J.K., Marsh, J.W., Livingstone, D.R. and Evans, B. (1992) Genetic toxicity in the dab (*Limanda limanda*) from the North Sea. *Mar. Ecol. Prog. Ser.* **91**: 121–126.

Chirico, S. (1994) High performance liquid chromatography-based thiobarbituric acid tests. In: *Methods in Enzymology*, Vol. 233. Academic Press, New York, pp. 314 – 318.

Choi, J. and Oris, J.T. (2000) Evidence for oxidative stress in bluegill sunfish (*Lepomis macrochirus*) liver microsomes simultaneously exposed to solar ultraviolet radiation and anthracene. *Environ. Toxicol. Chem.* **19**: 1795–1799.

Ciereszko, A., Dabrowski, K., Kucharczyk, D., Dobosz, S., Goryczko, K. and Glogowski, J. (1999) The presence of uric acid, an antioxidative substance, in fish seminal plasma. *Fish Physiol. Biochem.* **21**: 313–315.

Cleary, J.J. (1991) Organotin in the marine surface microlayer and sub-surface waters of Southwest England: relation toxicity thresholds and the UK environmental quality standard. *Mar. Environ. Res.* **32**: 213–222.

Connolly, J.P. and Thomann, R.V. (1982) Calculated contribution of surface microlayer PCB contamination of Lake Michigan lake trout. *J. Great Lakes Res.* **8**: 367–375.

Cross, J.N., Hardy, J.T., Rose, J.E., Hershelman, G.P., Antrim, L.D., Gossett, R.W. and Crecelius, E.A. (1987) Contaminant concentrations and toxicity of sea-surface microlayer near Los Angeles, California. *Mar. Environ. Res.* **23**: 307–323.

Culp, S.J. and Beland, F.A. (1996) Malachite green: a toxicological review. *J. Am. Coll. Toxicol.* 15: 219–238.

Dabrowski, K. and Ciereszko, A. (1996) Ascorbic acid protects against male infertility in a teleost fish. *Experientia* 52: 97–100.

Darley-Usmar, V. and Halliwell, B. (1996). Blood radicals. Reactive nitrogen species, reactive oxygen species, transition metal ions, and the vascular system. *Pharm. Res.* 13: 649–662.

Davey, E.W., Perez, K.T., Soper, A.E., Lackie, N.F., Morrison, G.E., Johnson, R.L. and Heltshe, J.F. (1990) Significance of the surface microlayer to the environmental fate of di-(2-ethylhexyl)phthalate predicted from marine mesocosms. *Mar. Chem.* 31: 231–270.

Dethlefsen, V., Soffker, K., Buther, H. and Damm, U. (1996) Organochlorine compounds in marine organisms from the international North Sea incineration area. *Arch. Fish. Mar. Res.* 44: 215–242.

Devaux, A., Pesonen, M. and Monod, G. (1997) Alkaline comet assay in rainbow trout hepatocytes. *Toxicol. in Vitro* 11: 71–72.

Di Giulio, R.T., Washburn, P.C., Wenning, R.J., Winston, G.W. and Jewell, C.S. (1989) Biochemical responses in aquatic animals: a review of oxidative stress. *Environ. Toxicol. Chem.* 8: 1103–1123.

Di Giulio, R.T., Habig, C. and Gallagher, E.P. (1993) Effects of Black Rock Harbor sediments on indices of biotransformation, oxidative stress, and DNA integrity in channel catfish. *Aquat. Toxicol.* 26: 1–22.

Di Giulio, R.T., Benson, W.H., Sanders, B.M. and Van Veld, P.A. (1995a) Biochemical mechanisms: metabolism, adaptation, and toxicity. In: *Fundamentals of Aquatic Toxicology. Effects, Environmental Fate, and Risk Assessment* (ed G. Rand). Taylor and Francis, London.

Di Giulio, R.T., Behar, J.V., Carlson, D.B., Hasspieler, B.M. and Watson, D.E. (1995b) Determinants of species susceptibility to oxidative stress: a comparison of channel catfish and brown bullhead. *Mar. Environ. Res.* 39: 175–179.

Di Pinto, L.M. and Coull, B.C. (1997) Trophic transfer of sediment-associated polychlorinated biphenyls from meiobenthos to bottom-feeding fish. *Environ. Toxicol. Chem.* 16: 2568–2575.

ECOTOX (1993) *Joint Assessment of Commodity Chemicals no. 22. Hydrogen Peroxide*, CAS no. 7722–84–1. European centre for Ecotoxicology and Toxicology of Chemicals Publications, Brussels, 141pp.

Elhraiki, A., Elalami, M., Kessabi, M., Buhler, D.R. and Benard, P. (1994) Pesticide residues in seafood products from the Mediterranean coastal waters of Morocco. *Toxicol. Environ. Chem.* 41: 21–30.

Engwall, M., Brunström, B., Brewer, A. and Norrgren, L. (1994) Cytochrome P-4501A induction by a co-planar PCB, a PAH mixture, and PCB-contaminated sediment extracts following microinjection of rainbow trout sac-fry. *Aquat. Toxicol.* 30: 311–324.

Ensenbach, U. and Nagel, R. (1997) Toxicity of binary chemical mixtures: effects on reproduction of zebrafish (*Brachydanio rerio*). *Arch. Environ. Contam. Toxicol.* 32: 204–210.

Evans, P., Lyras, L. and Halliwell, B. (1999) Measurement of protein carbonyls in human brain tissue. In: *Methods in Enzymology*, Vol. 300. Academic Press, New York, pp. 145–156.

Fabris, G., Theodoropoulos, T., Sheehan, A. and Abbott, B. (1999) Mercury and organochlorines in black bream, *Acanthopagrus butcheri*, from the Gippsland Lakes, Victoria, Australia: evidence for temporal increases in mercury levels. *Mar. Pollut. Bull.* 38: 970–976.

Fatima, M., Ahmad, I., Sayeed, I., Athar, M. and Raisuddin, S. (2000) Pollutant-induced over-activation of phagocytes is concomitantly associated with peroxidative damage in fish tissues. *Aquat. Toxicol.* 49: 243–250.

Fessard, V. and Livingstone, D.R. (1998) Development of Western analysis of oxidised proteins as a biomarker of oxidative damage in liver of fish. *Mar. Environ. Res.* 46: 407–410.

Filho, D.W. (1996) Fish antioxidant defences – a comparative approach. *Braz. J. Med. Biol. Res.* 29: 1735–1742.

Finn, R.N., Fyhn, H.J. and Evjen, M.S. (1991) Respiration and nitrogen metabolism of Atlantic halibut eggs (*Hippoglossus hippoglossus*). *Mar. Ecol.* 108: 11–19.

Finn, R.N., Widdows, J. and Fyhn, H.J. (1995) Calorespirometry of developing embryos and yolk-sac larvae of turbot (Scophthalmus maximus). Mar. Biol. 122: 157–163.

Finn, R.N., Fyhn, H.J., Henderson, R.J. and Evjen, M.S. (1996) The sequence of catabolic substrate oxidation and enthalpy balance of developing embryos and yolk-sac larvae of turbot (Scophthalmus maximus L.). Comp. Biochem. Physiol. 115A: 134–151.

Focardi, S., Fossi, C., Leonzio, C., Corsolini, S. and Parra, O. (1996) Persistent organochlorine residues in fish and water birds from the Biobio river, Chile. Environ. Monit. Assess. 43: 73–92.

Fujiwara, K., Ushiroda T., Takeda K., Kumamoto Y. and Tsubota, H. (1993) Diurnal and seasonal distribution of hydrogen peroxide in seawater of the Seto inland sea. Geochem. J. 27: 103–115.

Gaikowski, M.P., Rach, J.J. and Ramsey, R.T. (1999) Acute toxicity of hydrogen peroxide treatments to selected lifestages of cold-, cool- and warm-water fish. Aquaculture 178: 191–207.

Goksøyr, A., Solberg, T.S. and Serigstad, B. (1991) Immunochemical detection of cytochrome P4501A1 induction in cod larvae and juveniles exposed to a water-soluble faction of North Sea crude oil. Mar. Pollut. Bull. 22: 122–127.

Guiney, P.D., Smolowitz, R.M., Peterson, R.E. and J.J. Stegeman (1997) Correlation of 2,3,7,8-tetrachlorodibenzo-p-dioxin induction of cytochrome P4501A in vascular endothelium with toxicity in early life stages of lake trout. Toxicol. Appl. Pharmac. 143: 256–273.

Gutjahr-Gobell, R.E., Black, D.E., Mills, L.J., Pruell, R.J., Taplin, B.K. and Jayaraman, S. (1999) Feeding the mummichog (Fundulus heteroclitus) a diet spiked with non-ortho- and mono-ortho-substituted polychlorinated biphenyls: accumulation and effects. Environ. Toxicol. Chem. 18: 699–707.

Hai, D.Q., Varga S.I. and Matkovics, B. (1997) Organophosphate effects on antioxidant system of carp (Cyprinus carpio) and catfish (Ictalurus nebulosus). Comp. Biochem. Physiol. 117C: 83–88.

Hall, L.W. Jr, Ziegenfuss, M.C., Fischer, S.A., Sullvan, J.A. and Palmer, D.M. (1993) The influence of contaminant and water quality conditions on larval striped bass in the Potomac River and upper Chesapeake Bay in 1990: an in situ study. Arch. Environ. Contam. Toxicol. 24: 1–10.

Halliwell, B. (1996a) Antioxidant characterization. Methodology and mechanism. Biochem. Pharmac. 49: 1341–1348.

Halliwell, B. (1996b) Commentary. Oxidative stress, nutrition and health. Experimental strategies for optimization of nutritional antioxidant intake for humans. Free Rad. Res. 25: 57–74.

Halliwell, B. and Chirico, S. (1993) Lipid peroxidation: its mechanism, measurement, and significance. Am. J. Clin. Nutr. 57 (Suppl.): 715S–725S.

Halliwell, B. and Gutteridge, J.M.C. (1999) Free Radicals in Biology and Medicine. Oxford University Press, Oxford.

Hardy, J.T. (1982) The sea surface microlayer biology, chemistry and anthropogenic enrichment. Prog. Oceanogr. 11: 307–328.

Hardy, J.T. and Cleary, J.J. (1992) Surface microlayer contamination and toxicity in the German Bight. Mar. Ecol. Prog. Ser. 91: 203–210.

Heming, T.A. and Buddington, R.K. (1988) Yolk absorption in embryonic and larval fishes. In: Fish Physiology (eds W.S. Hoar and D.J. Randall), Vol. XI, The Physiology of Developing Fish. Part A Eggs and Larvae. Academic Press, New York, pp. 1–58.

IPCS (1991) Environmental Health Criteria 124, Lindane. World Health Organization, Genea, 208pp.

Johnson, R.D., Tietge, J.E., Jensen, K.M., Fernandez, J.D., Linnum, A.L., Lothenbach, D.B., Holcombe, G.W., Cook, P.M., Christ, S.A., Lattier, D.L. and Gordon, D.A. (1998) Toxicity of 2,3,7,8-tetrachlorodibenzo-p-dioxin to early life stage brook trout (Salvelinus fontinalis) following parental dietary exposure. Environ. Toxicol. Chem. 17: 2408–2421.

Jones, D.P. (1985) The role of oxygen concentration in oxidative stress. In: *Oxidative Stress* (ed H. Sies). Academic Press, New York.

Kagan, J., Wang, T.P., Benight, A.S., Tuveson, R.W., Wang, G.-R. and Fu, P.P. (1990) The phototoxicity of nitropolycyclic aromatic hydrocarbons of environmental importance. *Chemosphere* **20:** 453–466.

Kehrer, J. (1993) Free radicals as mediators of tissue injury and disease. *Crit. Rev. Toxicol.* **23:** 21–48.

Kelly, J.D., Orner, G.A., Hendricks, J.D. and Williams, D.E. (1992) Dietary hydrogen peroxide enhances hepatocarcinogenesis in trout: correlation with 8-hydroxy-2'-deoxyguanosine levels in liver DNA. *Carcinogenesis* **13:** 1639–1642.

Kelly, S.A., Havrilla, C.M., Brady, T.C., Abramo, K.H. and Levin, E.D. (1998) Oxidative stress: established mammalian and emerging piscine model systems. *Environ. Health Perspect.* **106:** 375–384.

Kitamura, S. and Tatsumi, K. (1997) Purification of NADPH-linked and NADH-linked quinone reductases from liver of sea bream, *Pagrus major. Comp. Biochem. Physiol.* **118B:** 675–680.

Kling, P.G. and Olsson, P.E. (2000) The involvement of differential metallothionein expression in free radical sensitivity of RTG-2 and CHSE-214 cells. *Free Radic. Biol. Med.* **28:** 1628–1637.

Kolayli, S. and Keha, E. (1999) A comparative study of antioxidant enzyme activities in freshwater- and seawater-adapted rainbow trout. *J. Biochem. Mol. Toxicol.* **13:** 334–337.

Leaver, M.J. and George, S.G. (1998) A piscine glutathione *S*-transferase which efficiently conjugates the end-products of lipid peroxidation. *Mar. Environ. Res.* **46:** 71–74.

Lemaire, P. and Livingstone, D.R. (1993) Pro-oxidant/antioxidant processes and organic xenobiotic interactions in marine organisms, in particular the flounder *Platichthys flesus* and mussel *Mytilus edulis. Trends Comp. Biochem. Physiol.* **1:** 1119–1150.

Lemaire, P. and Livingstone, D.R. (1994) Inhibition studies on the involvement of flavoprotein reductases in menadione-and nitrofurantoin-stimulated oxyradical production by hepatic microsomes of flounder (*Platichthys flesus*). *J. Biochem. Toxicol.* **9:** 87–95.

Lemaire, P. and Livingstone, D.R. (1997) Aromatic hydrocarbon quinone-mediated reactive oxygen species production in hepatic microsomes of the flounder (*Platichthys flesus* L.). *Comp. Biochem. Physiol.* **117C:** 131–139.

Lemaire, P., Viarengo, A., Canesi, L. and Livingstone, D.R. (1993) Pro-oxidant and antioxidant processes in gas gland and other tissues of cod (*Gadus morhua*). *J. Comp. Physiol.* **163B:** 477–486.

Lemaire, P., Matthews, A., Förlin, L. and Livingstone, D.R. (1994) Stimulation of oxyradical production of hepatic microsomes of flounder (*Platichthys flesus*) and perch (*Perca fluviatilis*) by model and pollutant xenobiotics. *Arch. Environ. Contam. Toxicol.* **26:** 191–201.

Lemaire, P., Sturve, J., Förlin, L. and Livingstone, D.R. (1996) Studies on aromatic hydrocarbon quinone metabolism and DT-diaphorase function in liver of fish species. *Mar. Environ. Res.* **42:** 317–321.

Lindström-Seppä, P., Korytko, P.J., Hahn, M.E. and Stegeman, J.J. (1994) Uptake of waterborne 3,3',4,4'-tetrachlorobiphenyl and organ and cell-specific induction of cytochrome P4501A in adult and larval fathead minnow *Pimephales promelas. Aquat. Toxicol.* **28:** 147–167.

Liss, P.S. and Duce, R.A. (eds) (1997) *The Sea Surface and Global Change.* Cambridge University Press, Cambridge.

Livingstone, D.R. (1998) The fate of organic xenobiotics in aquatic ecosystems: quantitative and qualitative differences in biotransformation by invertebrates and fish. *Comp. Biochem. Physiol.* **120A:** 43–49.

Livingstone, D.R. (2001) Contaminant-stimulated reactive oxygen species production and oxidative damage in aquatic organisms. *Mar. Pollut. Bull.* (in press).

Livingstone, D.R. and Nasci, C. (2000) Biostransformation and antioxidant enzymes as potential biomarkers of contaminant exposure in goby (*Zosterisessor ophiocephalus*) and mussel (*Mytilus gallorpvincialis*). In: *Man and the Biosphere Series* (Series ed J.N.R. Jeffers), Vol. 25, *The Venice Lagoon Ecosystem. Inputs and Interactions Between Land and Sea* (eds P. Lasserre and A. Marzollo). Parthenon, Carnforth, pp. 357–373.

Livingstone, D.R., Lemaire, P., Matthews, A., Peters, L., Bucke, D. and Law, R.J. (1993) Pro-oxidant, antioxidant and 7-ethoxyresorufin O-deethylase (EROD) activity responses in liver of dab (*Limanda limanda*) exposed to sediment contaminated with hydrocarbons and other chemicals. *Mar. Pollut. Bull.* **26**: 602–606.

Livingstone, D.R., Förlin, L. and George, S. (1994) Molecular biomarkers and toxic consequences of impact by organic pollution in aquatic organisms. In: *Water Quality and Stress Indicators in Marine and Freshwater Systems: Linking Levels of Organisation* (ed D.W. Sutcliffe). Freshwater Biological Association, Ambleside, UK, pp. 154–171.

Livingstone, D.R., Chipman, J.K., Lowe, D.M., Minier, C., Mitchelmore, C.L., Moore, M.N., Peters, L.D. and Pipe, R.K. (2000a) Development of biomarkers to detect the effects of organic pollution on aquatic invertebrates: recent molecular, genotoxic, cellular and immunological studies on the common mussel (*Mytilus edulis* L.) and other mytilids. *Int. J. Pollut.* **13**: 56–91.

Livingstone, D.R., Mitchelmore, C.L., O'Hara, S.C.M., Lemaire, P., Sturve, J. and Förlin, L. (2000b) Increased potential for NAD(P)H-dependent reactive oxygen species production of hepatic subcellular fractions of fish species with *in vivo* exposure to contaminants. *Mar. Environ. Res.* **50**: 57–60.

Lowry, O.H., Rosebrough, N.J., Farr, A.L. and Randall, R.J. (1951) Protein measurement with the Folin reagent. *J. Biol. Chem.* **193**: 265–275.

Machala, M., Ulrich, R., Neca, J., Vykusova, B., Kolatrova, J., Machova, J. and Svobodova, Z. (2000) Biochemical monitoring of aquatic pollution: indicators of dioxin-like toxicity and oxidative stress in the roach (*Rutilus rutilus*) and chub (*Leuciscus cephalus*) in the Skalice river. *Vet. Med.* **45**: 55–60.

Malins, D.C. and Gunselman, S.J. (1994) Fourier-transform infrared spectroscopy and gas chromatography-mass spectrometry reveals a remarkable degree of structural damage in DNA of wild fish exposed to toxic chemicals. *Proc. Natl Acad. Scien. USA* **91**: 13038–13041.

Malins, D.C. and Haimanot, R. (1991) The etiology of cancer: hydroxyl radical-induced DNA lesions in histologically normal livers of fish from a population with liver tumors. *Aquat. Toxicol.* **20**: 123–130.

Malins, D.C., Ostrander, G.K., Haimanot, R. and Williams, P. (1990) A novel DNA lesion in neoplastic livers of feral fish: 2,6-diamino-4-hydroxy-5-formamidopyrimidine. *Carcinogenesis* **11**: 1045–1047.

Marcon, J.L. and Wilhelm, D. (1999) Antioxidant processes of the wild tambaqui, *Colossoma macropomum* (Osteichthyes, Serrasalmidae) from the Amazon. *Comp. Biochem. Physiol.* **123C**: 257–263.

Mason, R.P. (1990) Redox cycling of radical anion metabolites of toxic chemicals and drugs, and the Marcus theory of electron transfer. *Environ. Health Perspect.* **87**: 237–243.

McNeish, A.S., Johnson, M.S. and Leah, R.T. (1997) Methyl lindane and other analogues of hexachlorocyclohexane in dab and plaice from the Mersey Estuary. *Aquat. Toxicol.* **40**: 11–22.

Mitchelmore, C.L. and Chipman, J.K. (1998) Detection of DNA strand breaks in brown trout (*Salmo trutta*) hepatocytes and blood cells using the single cell gel electrophoresis (comet) assay. *Aquat. Toxicol.* **41**: 161–182.

Mitchelmore, C.L., Chipman, J.K., Garcia Martinez, P., Lemaire, P., Peters, L.D. and Livingstone, D.R. (1996) Normal status of hepatic 7-ethoxyresorufin O-deethylase (EROD) activity, antioxidant enzymes and DNA oxidation in turbot (*Scophthalmus maximus*) and other flatfish species following exposure to nitroaromatic compounds. *Mar. Environ. Res.* **42**: 329–333.

Moe, M.A. Jr (1989) *The Marine Aquarium Reference. Systems and Invertebrates.* Green Turtle Publications, Plantation, FL.

Mopper, K. and Zhou, X. (1990) Hydroxyl radical photoproduction in the sea and its potential impact on marine processes. *Science* 250: 661–664.

Mourente, G., Tocher, D.R., Diaz, E., Grau, A. and Pastor, E. (1999) Relationships between antioxidants, antioxidant enzyme activities and lipid peroxidation products during early development in *Dentex dentex* eggs and larvae. *Aquaculture* 179: 309–324.

Nacci, D., Coiro, L., Kuhn, A., Champlin, D., Munns, W. Jr., Specker, J. and Cooper, K. (1998) Nondestructive indicator of ethoxyresorufin-O-deethylase activity in embryonic fish. *Environ. Toxicol. Chem.* 17: 2481–2486.

Nishimoto, M., Roubal, W.T., Stein, J.E. and Varanasi, U. (1991) Oxidative DNA damage in tissues of English Sole (*Parophrys vetulus*) exposed to nitrofurantoin. *Chem. Biol. Interact.* 80: 317–326.

Olsson, P.E., Kling, P., Erkell, L.J. and Kille, P. (1995) Structural and functional-analysis of the rainbow trout (*Oncorhynchus mykiss*) metallothionein-a gene. *Eur. J. Biochem.* 230: 344–349.

Parihar, M.S., Duby, A.K., Javeri, T. and Prakash, P. (1996) Changes in lipid peroxidation, superoxide dismutase activity, ascorbic acid and phospholipid content in liver of freshwater catfish *Heteropneustes fossilis* exposed to elevated temperature. *J. Them. Biol.* 21: 323–330.

Parihar, M.S., Javeri, T., Hemnani, T., Dubey, A.K. and Prakash, P. (1997) Responses of superoxide dismutase, glutathione peroxidase and reduced glutathione antioxidant defencesin gills of freshwater catfish (*Heteropneustes fossilis*) to short-term elevated temperature. *J. Them. Biol.* 22: 151–156.

Payne, J.F., Malins, D.C., Gunselman, S., Rahimtula, A. and Yeats, P.A. (1998) DNA oxidative damage and vitamin a reduction in fish from a large lake in Labrador, Newfoundland, contaminated with iron-ore mine tailings. *Mar. Environ. Res.* 46: 289–294.

Pedrajas, J.R., Peinado J. and Lopez-Barea, J. (1995) Oxidative stress in fish exposed to model xenobiotics. Oxidatively modified forms of Cu, Zn-superoxide dismutase as potential biomarkers. *Chem. Biol. Interact.* 98: 267–282.

Pedrajas, J.R., Lopez-Barea, J. and Peinado J. (1996) Dieldrin induces peroxisomal enzymes in fish (*Sparus aurata*) liver. *Comp. Biochem. Physiol.* 115C: 125–131.

Petersen, G.I. and Kristensen, P. (1998) Bioaccumulation of lipophilic substances in fish early life stages. *Environ. Toxicol. Chem.* 17: 1385–1395.

Peters, L.D. and Livingstone, D.R. (1995) Studies on cytochrome P4501A1 in early and adult life stages of turbot (*Scophthalmus maximus* L.). *Mar. Environ. Res.* 39: 5–9.

Peters, L.D. and Livingstone, D.R. (1996) Development of antioxidant enzyme activities in embryologic and early larval stages of turbot (*Scophthalmus maximus* L.). *J. Fish Biol.* 49: 986–997.

Peters, L.D., Coombs, S.H., McFadzen, I., Solé, M., Albaigés, J. and Livingstone, D.R. (1992) Toxicity studies, 7-ethoxyresorufin O-deethylase (EROD) and antioxidant enzymes in early life stages of turbot (*Scophthalamus maximus* L.) and sprat (*Sprattus sprattus* L.). Intergovernmental Commission for the Exploration of the Sea C.M. 1992/E: 12.

Peters, L.D., Porte, C., Albaigés, J. and Livingstone, D.R. (1994) 7-Ethoxyresorufin O-deethylase (EROD) and antioxidant enzyme activities in larvae of sardine (*Sardinia pilchardus*) from the north coast of Spain. *Mar. Pollut. Bull.* 28: 299–304.

Peters, L.D., O'Hara, S.C.M. and Livingstone, D.R. (1996) Benzo[a]pyrene metabolism and xenobiotic-stimulated reactive oxygen species generation by subcellular fractions of larvae of turbot (*Scophthalmus maximus* L.). *Comp. Biochem. Physiol.* 114C: 221–227.

Peters, L.D., O'Hara, S.C.M., Cleary, J. and Livingstone, D.R. (2001) Toxicity and elevation of 7-ethoxyresorufin O-deethylase activity in turbot (*Scophthalmus maximus* L.) larvae exposed to contaminated sea surface microlayer. *Bull. Environ. Contam. Toxicol.* (in press).

Ploch, S.A., Lee, Y.P., MacLean, E. and Di Giulio, R.T. (1999) Oxidative stress in liver of brown bullhead and channel catfish following exposure to *tert*-buytl hydroperoxide. *Aquat. Toxicol.* 46: 231–240.

Price, D., Worsfold, P.J. and Mantoura, R.F.C. (1992) Hydrogen peroxide in the marine environment: cycling and methods of analysis. *Trends Analyt. Chem.* 11: 379–384.

Powell, M.D. and Perry, S.F. (1997) Respiratory and acid-base pathophysiology of hydrogen peroxide in rainbow trout (*Oncorhynchus mykiss* Walbaum). *Aquat. Toxicol.* 37: 99–112.

Rach, J.J., Gaikowski, M.P., Howe, G.E. and Schreier, T.M. (1998) Evaluation of the toxicity and efficacy of hydrogen peroxide treatments on eggs of warm and cool water fishes. *Aquaculture* 165: 11–25.

Rand, G. (ed) (1995) *Fundamentals of Aquatic Toxicology. Effects, Environmental Fate, and Risk Assessment.* Taylor and Francis, London.

Reznick, A.Z. and Packer, L. (1994) Oxidative damage to proteins: spectrophotometric method for carbonyl assay. In: *Methods in Enzymology*, Vol. 233. Academic Press, New York, pp. 357–363.

Ritola, O., Lyytikainen, T., Pylkko, P., Molsa, H. and Lindström-Seppä, P. (2000) Glutathione-dependent defence system and monooxygenase enzyme activities in Arctic charr *Salvelinus alpinus* (L.) exposed to ozone. *Aquaculture* 185: 219–233.

Roche, H., Buet, A., Jonot, O. and Ramade, F. (2000) Organochlorine residues in European eel (*Anguilla anguilla*), Crucian carp (*Carassius carassius*) and catfish (*Ictalurus nebulosus*) from Vaccares Lagoon (French National Nature Reserve of Camargue) – effects on some physiological parameters. *Aquat. Toxicol.* 48: 443–459.

Rodriguez-Ariza, A., Alhama, J., Diaz-Mendez, F.M. and Lopez-Barea, J. (1999) Content of 8-oxodG in chromosomal DNA of *Sparus aurata* fish as biomarker of oxidative stress and environmental pollution. *Mutat. Res. Genet. Toxicol. Environ. Mutagen.* 438: 97–107.

Rombough, P.J. (1988) Respiratory gas exchange, aerobic metabolism, and effects of hypoxia during early life. In: *Fish Physiology* (eds W.S. Hoar and D.J. Randall) Vol. XI, *The Physiology of Developing Fish. Part A Eggs and Larvae.* Academic Press, New York, pp. 59–161

Roméo, M., Bennani, N., Gnassia-Barelli, M., LaFaurie, M. and Girard, J.P. (2000) Cadmium and copper display different responses towards oxidative stress in the kidney of the sea bass *Dicentrarchus labrax*. *Aquat. Toxicol.* 48: 185–194.

Rønnestad, I., Fyhn, H.J. and Gravningen, K. (1992) The importance of free amino acids to the energy metabolism of eggs and larvae of turbot (*Scophthalmus maximus*). *Mar. Ecol.* 114: 517–525.

Schlenk, D. and Rice, C.D. (1998) Effect of zinc and cadmium treatment on hydrogen peroxide-induced mortality and expression of glutathione and metallothionein in a teleost hepatoma cell line. *Aquat. Toxicol.* 43: 121–129.

Schlenk, D., Wolford, L., Chelius, M., Steevens, J. and Chan, K.M. (1997) Effect of arsenite, arsenate, and the herbicide monosodium methyl arsonate (MSMA) on hepatic metallothionein expression and lipid peroxidation in channel catfish. *Comp. Biochem. Physiol.* 118C: 177–183.

Schlezinger, J.J., White, R.D. and Stegeman, J.J. (1999) Oxidative inactivation of cytochrome P-450 1A (CYP1A) stimulated by 3, 3', 4, 4'-tetrachlorobiphenyl: production of reactive oxygen by vertebrate CYP1As. *Mol. Pharmac.* 56: 588–597.

Schlezinger, J.J., Keller, J., Verbrugge, L.A. and Stegeman, J.J. (2000) 3, 3', 4, 4'-tetrachlorobiphenyl oxidation in fish, bird and reptile species: relationship to cytcohrome P450 1A inactivation and reactive oxygen production. *Comp. Biochem. Physiol.* 125C: 273–286.

Sinkkonen, S. and Paasivirta, J. (2000) Polychlorinated organic compounds in the arctic cod liver: trends and profiles. *Chemosphere* 40: 619–626.

Sokal, R.R. and Rohlf, F.J. (1981) *Biometry. The Principles and Practice of Statistics in Biological Research.* W.H. Freeman, San Francisco, CA.

Speare, D.J., Carvajal, V. and Horney, B.S. (1999) Growth suppression and branchitis in trout exposed to hydrogen peroxide. *J. Comp. Pathol.* 120: 391–402.

Spotte, S. (1979) *Fish and Invertebrate Culture. Water Management in Closed Systems*, 2nd edn. Wiley Interscience, New York.

Stohs, S.J. and Bagchi, D. (1995) Oxidative mechanisms in the toxicity of metal ions. *Free Rad. Biol. Med.* 18: 321–336.

Storz, G., Tartaglia, L.A. and Ames, B.N. (1990) Transcriptional regulator of oxidative stress-inducible genes: direct activation by oxidation. *Science* 248: 189–194.

Strolin-Benedetti, M., Brogin, G., Bani, M., Oesch, F. and Hengstler, J.G. (1999) Association of cytochrome P450 induction with oxidative stress *in vivo* as evidenced by 3-hydroxylation of salicyclate. *Xenobiotica* 29: 1171–1180.

Sylvie, B.R., Pairault, C., Vernet, G. and Boulekbache, H. (1996) Effect of lindane on the ultrastructure of the liver of the rainbow trout, *Oncorhynchus mykiss*, sac-fry. *Chemosphere* 33: 2065–2079.

Taira, J., Mimura, K., Yoneya, T., Hagi, A., Murukami, A. and Makimo, K. (1992) Hydroxyl radical formation by UV-irradiated epidermal cells. *J. Biochem.* 111: 693–695.

Tocher, D.R., Fraser, A.J., Sargent, J.R. and Gamble, J.C. (1985). Fatty acid composition of phospholipids and neutral lipids during embryonic and early larval development in Atlantic herring (*Clupea harengus* L.). *Lipids* 20: 69–74.

Van Veld, P. A. (1990) Absorption and metabolism of dietary xenobiotics by the intestine of fish. *Rev. Aquat. Sci.* 2: 185–203.

Walker, C.H. and Livingstone, D.R. (eds) (1992) *Persistent Pollutants in Marine Ecosystems.* SETAC Special Publication. Pergamon Press, Oxford.

Walker, C.H., Hopkin, S.P., Sibley, R.M. and Peakall, D.B. (1996) *Principles of Ecotoxicology.* Taylor and Francis, London.

Washburn, P.C. and Di Giulio, R.T. (1988) Nitrofurantoin-stimulated superoxide production by channel catfish (*Ictalurus punctatus*) hepatic microsomal and soluble fractions. *Toxicol. Appl. Pharmac.* 95: 363–377.

Washburn, P.C. and Di Giulio, R.T. (1989) Stimulation of superoxide production by nitrofurantoin, *p*-nitrobenzoic acid and *m*-dinitrobenzoic acid in hepatic microsomes of three species of freshwater fish. *Environ. Toxicol. Chem.* 8: 171–180.

Westerhagen, H. von (1988) Sublethal effects of pollutants on fish eggs. In: *Fish Physiology* (eds W.S. Hoar and D.J. Randall), Vol. XI, *The Physiology of Developing Fish. Part A Eggs and Larvae.* Academic Press, New York, pp. 253–346.

Westerlund, L., Billsson, K., Andersson, P.L., Tysklind, M. and Olsson, P.E. (2000) Early life-stage mortality in zebrafish (*Danio rerio*) following maternal exposure to polychlorinated biphenyls and estrogen. *Environ. Toxicol. Chem.* 19: 1582–1588.

Winston, G.W. and Di Giulio, R.T. (1991) Prooxidant and antioxidant mechanisms in aquatic organisms. *Aquat. Toxicol.* 19: 137–161.

Winston, G.W., Regoli, F., Dugas, A.J. Jr, Fong, J.H. and Blanchard, K.A. (1998) A rapid gas chromatographic assay for determining oxyradical scavenging capacity of antioxidants and biological fluids. *Free Rad. Biol. Med.* 24: 480–493.

Wiseman, H. and Halliwell, B. (1996) Damage to DNA by reactive oxygen and nitrogen species: role in inflammatory disease and progression to cancer. *Biochem. J.* 313: 17–29.

Word, J.Q., Hardy, J.T., Crecelius, E.A. and Kiesser, S.L. (1987) A laboratory study of the accumulation and toxicity of contaminants at the sea surface from sediments proposed for dredging. *Mar. Environ. Res.* 23: 325–338.

10

Endocrine disrupters, critical windows and developmental success in oyster larvae

Helen E. Nice, Mark Crane, David Morritt and Michael Thorndyke

1. Introduction

Recent years have witnessed an increasing focus on endocrine disruption in vertebrates while, in contrast, comparatively few studies have assessed the effect of endocrine disrupters on invertebrates (Pinder et al., 1998). 4-Nonylphenol (4-NP) is a known endocrine disrupter that is widespread in the aquatic environment (Nimrod and Benson, 1996), reaching levels as high as 180 μg l^{-1} in British rivers and 5.2 μg l^{-1} in British estuaries (Blackburn and Waldock, 1995). It belongs to the group of chemicals known as the alkylphenols, which are produced during the breakdown of alkyphenol polyethoxylates (APEs; Ahel et al., 1991). APEs are non-ionic surfactants produced from a range of industrial processes worldwide including the production of emulsifiers, wetting agents, dispersing agents and detergents in household products, and agricultural and industrial applications (Marcomini et al., 1990).

Toxicological effects of the endocrine disrupter 4-NP have been recorded for adults of the common mussel, *Mytilus edulis* and include decreased byssus strength and reduced scope for growth (Granmo et al., 1989). In addition 4-NP has been shown to have an oestrogenic mode of action on various aquatic vertebrates such as eelpout *Zoarces viviparus* (Christiansen et al., 1998), Atlantic salmon *Salmo salar* (Madsen et al., 1997), rainbow trout *Oncorhynchus mykiss* (Jobling and Sumpter, 1993) and roach *Rutilus rutilus* (Routledge et al., 1998).

Potentially endocrine-disrupting effects reported for aquatic invertebrates include oocyte malformations in *Gammarus pulex* (Gross et al., 2001); decreased offspring fecundity and metabolic elimination of testosterone in *Daphnia magna* (Baldwin et al., 1997); and a negative effect on the production of resting eggs, female offspring and a developmental deformity, also in *D. magna* (Shurin and Dodson, 1997). The specific mode of action in many of these studies is unknown, mainly because comparatively

---

Environment and Animal Development: Genes, Life Histories and Plasticity, edited by D. Atkinson and M. Thorndyke.

little is known about the endocrine systems of aquatic invertebrates compared to those of aquatic vertebrates.

Studies on the effects of endocrine disrupting contaminants on bivalve larval development are scarce. This is unfortunate because these animals are of vital importance to the aquaculture industry and are clearly susceptible to such contaminant damage. For example, a single exposure to a range of concentrations of 4-NP applied immediately after fertilization resulted in delayed development to D-shape and caused a significant decrease in survival of *Crassostrea gigas* (Nice *et al.*, 2000).

It was also shown that a single dose of 4-NP (100 μg l⁻¹) in a static system caused up to 73% of D-shaped larvae to develop with the condition known as 'convex hinge' within the first 48 h post-fertilization (p.f.). Lower concentrations of 4-NP also caused this deformity but in lower numbers. For example, concentrations between 0.1 and 10 μg l⁻¹ resulted in 5–20% of D-shaped larvae developing with a convex hinge (Nice *et al.*, 2000; *Figure1*).

1.1 *Experimental approach*

In this chapter we explore some life history effects of 4-NP on developmental success in the Pacific Oyster *Crassostrea gigas*. While not a native species it is now cultivated widely in the UK and represents an important component of the aquaculture industry. Here the effects of 4-NP are described following exposure of larvae during either one of two critical periods or 'windows' in their developmental programme. The period of exposure in each case was 48 h, following which the animals were returned to clean seawater. The exposure periods selected were: (a) the point of change between veliger and veliconcha stages at days 7–8 p.f.; and (b) eyespot stage at days 23–24 p.f., which is the stage immediately preceding settlement and metamorphosis.

These stages are important because they are periods when substantial and significant changes in larval morphology take place and therefore involve intense cellular reprogramming, reorganization and organogenesis. The eyespot stage, for example, is just prior to metamorphosis (Galtsoff, 1964), a stage when considerable tissue reorganization takes place under the control of both endogenous neuro-endocrine factors and external chemical cues (Kennedy *et al.*, 1996).

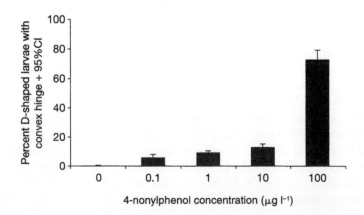

Figure 1. Percentage larvae developing with a convex hinge 48 h post-fertilization. Reproduced from Nice et al. (2000). Mar. Pollut. Bull. *40: 491–496, with permission of Elsevier Science.*

All original data presented result from trials carried out at Seasalter Hatchery, Whistable, Kent. Developing larvae of *C. gigas* were prepared and exposed to 4-NP following standard procedures (Nice *et al.*, 2000).

1.2 *Critical windows*

The concept of 'critical windows' of development was put forward by Lucas (1991) as the critical or sensitive period when an early stimulus or environmental challenge in an organisms life results in long-term change in the structure and function of an organism. Such 'critical windows' or 'critical periods' of development can be identified across the animal kingdom for a range of specific developmental events.

Many critical windows have been documented in fish development, particularly in species used for aquaculture. For example, halibut larvae, *Hippoglossus hippoglossus*, metamorphose when the larvae are between 16 and 21 mm in length with thyroxine as the primary endogenous regulator (Solbakken *et al.*, 1999). Clearly any interference with thyroxine biosynthesis and activity during this period may disrupt metamorphosis.

Temperature during the critical early developmental period of herring, *Clupea harengus*, determines the relative timing and degree of myogenesis resulting in significant phenotypic variation in the swimming muscles of later larval stages (see Johnston *et al.*, Chapter 5). Similarly, the critical period of morphogenetic activity in four species of leafhopper: *Nephotettix cincticeps*, *N. nigropictus*, *N. virescens* and *Recilia dorsalis* was found to be limited to a 24 h period before and after the fourth larval moult. During this time, the larvae exposed to the juvenile-hormone analogue NC-170 did not develop into normal adults but into secondary supernumerary larvae, which were unable to complete their subsequent moult and, as a result, died (Miyake *et al.*, 1991).

2. Larval responses to 4-nonylphenol

2.1 *Survival*

When larvae were exposed at days 7–8 p.f. a significant difference between treatments was noted post-exposure ($F_{3, 384} = 49.408, p < 0.001$). Larval density was significantly lower in the 100 μg l^{-1} treatment than in the 1 μg l^{-1} treatment and in either control ($p < 0.001$ for all cases). Larval density in the 1 μg l^{-1} treatment was also significantly lower than in either control ($p < 0.001$ for seawater control; $p < 0.01$ for solvent control). There was no significant difference in larval density between seawater and solvent controls ($p > 0.05$; *Figure 2*).

This effect is similar to that seen with embryonic exposure in the same species (Nice *et al.*, 2000). Similar findings have been recorded in *Crassostrea* larvae exposed to a range of toxicants including the metals mercury, silver, copper, nickel (Calabrese *et al.*, 1977) and the organotin tributyltin (Thain and Waldock, 1986; Rexroad, 1987; Roberts *et al.*, 1987).

It has been suggested that larval stages undergoing rapid cellular differentiation are highly susceptible to inappropriate metal-binding (Kennedy *et al.*, 1996) and reports that metals are capable of inhibiting RNA-polymerase (Novello and Stirpe, 1969; Hidalgo *et al.*, 1976), an enzyme that is central to gene expression, are consistent with the enhanced sensitivity of early life stages.

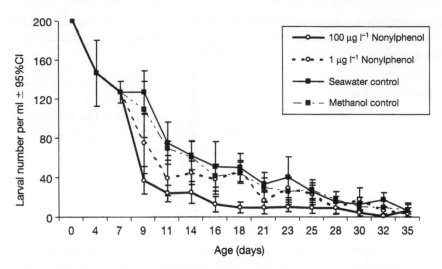

Figure 2. Larval survival after a 48 h exposure to 4-nonylphenol at days 7–8 post-fertilization.

In the case of tributyltin poisoning, the mortalities and retarded growth recorded in larvae and spat of both the native oyster, *Ostrea edulis* (Alzieu, 1986) and *Crassostrea* (Thain and Waldock, 1986; Rexroad, 1987; Roberts *et al.*, 1987) are thought to be brought about by a quite different mechanism to that for inorganic metals. Tributyltin exhibits a behaviour based on lipophilic characteristics rather than on affinities for charged ligands and freely enters the soft lipid-rich bodies of animals (Kennedy *et al.*, 1996).

2.2 *Growth*

Growth, represented by increase in hinge length, peaked at approximately 23 days p.f. in normal control larvae. This is because larvae reach a maximum length and then either settle or eventually die (Kennedy *et al.*, 1996). In this investigation many were starting to settle at 23 days p.f. and for the purposes of the investigations reported here, growth was only measured in free-swimming larvae in the water column.

Larvae exposed to both low and high concentrations of 4-NP at days 7–8 p.f. peaked in length between 28 and 30 days p.f., indicating a developmental delay compared to controls. Growth differences in this treatment group were already detectable at days 9 and 11 p.f. ($F_{3,32} = 25.05$, for day 9; $F_{3,32} = 23.19$ for day 11; $p < 0.001$ for all cases), that is, effects were seen as early as one or two days following exposure to 4-NP and continued to be evident until settlement (*Figure 3*).

A developmental delay has also been seen in earlier life history stages (Nice *et al.*, 2000) where larvae showed delays of up to 16 h in reaching D-stage. The delays illustrated in *Figure 3* are longer, that is up to 7 days and it should be noted that even at their peak size, exposed larvae are still smaller than control larvae at their peak size just before settlement.

Larvae exposed at eyespot stage showed no significant differences in growth because at 23 days p.f. they are almost competent for metamorphosis and typically show little increase in larval size after this stage (*Figure 4*).

The longer term consequences of such developmental delays are important since one of the main factors contributing to the mortality of early juvenile invertebrates is a

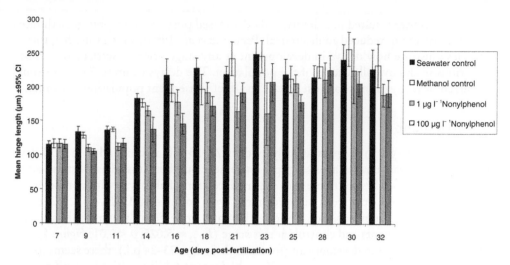

Figure 3. *Larval growth over time for oysters exposed during the period days 7–8 post-fertilization where growth is represented by mean hinge length (μm).*

delay prior to metamorphosis (Hunt and Scheibling, 1997). Delays due to heavy metal toxicity have been investigated in association with commercial aquaculture (Kennedy *et al.*, 1996) and delays in larval growth have been recorded for *C. gigas* exposed to cadmium, copper, lead and zinc (Fichet *et al.*, 1998). The same has also been observed for *C. virginica* exposed to cadmium (Roesijadi *et al.*, 1996).

A longer pelagic phase in an organism's life cycle makes it more vulnerable to predation by planktivorous predators and disease. A 5-day delay to settlement and metamorphosis caused a 63% decline in the overall number of *C. virginica* spat (Drinnan and Stallworthy, 1979). Delays to metamorphosis of 3–5 days in the barnacle,

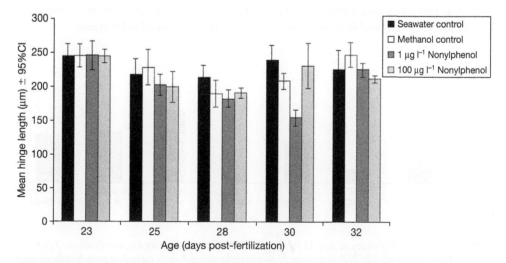

Figure 4. *Larval growth over time for oysters exposed during the period days 23–24 post-fertilization where growth is represented by mean hinge length (μm).*

Balanus amphitrite resulted in a dramatically depressed post-metamorphic growth rate (which translated to reduced adult survival; Pechenik *et al.*, 1993). Taken together, these examples illustrate how delays in development at any stage prior to settlement can be deleterious to the overall success of the population. The delays of up to 7 days seen in the current investigation are almost certain to affect subsequent population dynamics.

3. Settlement

Settlement can begin in *C. gigas* as early as days 23–25 p.f. and extend over a 10 day period. For this reason when assessing spat settlement numbers it is advisable to allow a 10 day window from first settlement to ensure detection of all larvae with competence to settle. In our tests, spat from each treatment were counted at 35 days p.f. A significant difference in the number of larvae settling and undergoing metamorphosis was seen between treatments and controls at this stage ($F_{6, 12}$ = 5.025, p <0.01; *Figure 5*).

When exposed later in development (for example at days 23–24 p.f.), there seems to be an 'all or nothing' effect with both 1 and 100 μg l^{-1} doses of 4-NP resulting in significantly lower spat densities than the controls (p <0.05 for both concentrations). From this, one can predict that the presence of 4-NP at eyespot stage (regardless of concentration between 1 and 100 μg l^{-1}) will significantly impair settlement and metamorphosis. An interesting point here is that, when exposure is early in life history (days 7–8 p.f.), only the higher 4-NP concentration affected the number that reached settlement (p <0.05). However, the low spat density in this situation is probably attributable, in part at least, to high larval mortality rather than an effect on settlement *per se*. From such data it can be concluded that a later life history exposure, for example at eyespot stage, is far more critical in terms of settlement and metamorphosis than an earlier exposure such as during days 7–8 p.f.

3.1 *Settlement factors*

The chemical structures of the compounds that naturally induce metamorphosis in most invertebrate species remain to be identified. Oyster larvae, however, have been the subject of considerable interest because of their commercial importance.

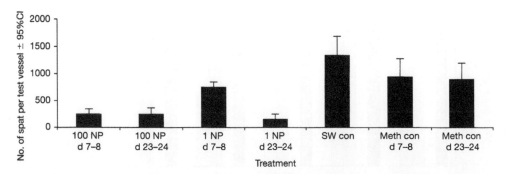

Figure 5. *Mean spat density at day 35 for 4-nonylphenol treatments and controls where 1 NP = 1 μg l^{-1} 4-nonylphenol; 100 NP = 100 μg l^{-1} 4-nonylphenol; d 7–8 = days 7–8 post-fertilization; d 23–24 = days 23–24 post-fertilization; SW con = seawater control and Meth con = methanol solvent control.*

It has been shown that chemical removal of the organic outer layers of the shell of C. virginica reduced the settlement of conspecific larvae on that substrate while aqueous extracts of adult animals enhanced settlement (Crisp et al., 1967). Particularly high levels of settlement occurred on tiles treated with lyophilized aqueous extracts of whole oysters or material from extracts of whole oysters that had been partitioned in diethyl ether (Keck et al., 1971). Settlement has also been found to be stimulated by the water held between the valves of living adult oysters and a protein-containing fraction purified from this water promoted larval settlement (Hidu, 1969). The protein had a molecular mass greater than 10 kD and contained iodinated amino acids (Veitch and Hidu, 1971). Acetazolamide, which is an inhibitor of carbonic anhydrase, promoted settlement of the New Zealand oyster Ostrea lutaria, although its mechanism of action remains unexplained (Nielson, 1973). Muller (1973) observed that the settlement of planula larva of the hydrozoan, Hydractinia echinata seemed to be induced by 'leakage-products' from marine gram-negative bacteria. The active factor was identified as a polar lipid that could be partitioned from a cell-free 'leakage solution' into chloroform and was unstable.

The idea of a settlement factor being present in, or emitted from, bacterial films has been developed through various studies on a range of invertebrates including larvae of the scyphozan, Cassiopea andromeda (Neumann, 1979) and the spirorbid polychaete Janua brasiliensis (Kirchman et al., 1982). More recent experiments with the Eastern and Pacific oysters, C. virginica and C. gigas, have shown that bacterial films found in association with juveniles enhance larval settlement and metamorphosis (Weiner et al., 1985; Bonar et al., 1986). Unknown chemical inducers in the supernatants of cultures from the bacteria Alteromonas colwelliana and Vibrio cholerae are responsible for the onset of settlement behaviour in C. gigas (Fitt et al., 1990). A gram-negative, pigment-forming bacterium, designated LST, was isolated from oysters and oyster-holding tanks and enhanced the settlement of C. virginica (Weiner et al., 1985). The active compounds isolated from films of LST were L-3,4-dihydroxyphenylalanine (L-DOPA), melanin precursors and melanin itself (Bonar et al., 1985, 1986).

Later work has shown that L-DOPA is converted into the neurotransmitter dopamine by C. gigas larvae (Coon and Bonar, 1987). This is important because dopamine is present in the neural tissue of many molluscs, including oysters, and has been further established as a neurotransmitter in a number of species including the bivalves Mercenaria mercenaria and Mytilus edulis; and the gastropods, Aplysia californica and Planorbis corneus (Gospe, 1983). L-DOPA is also a precursor to other neuroactive compounds such as the catecholamines, norepinephrine, epinephrine and octopamine; and of melanin pigments (Coon and Bonar, 1985). It is a water-soluble amino acid and the capacity of invertebrate larvae to transport amino acids into their bodies has been clearly established (Jaeckle and Manahin, 1989). It seems likely that these neuroactive compounds stimulate responses by influencing the larval nervous system either directly or indirectly by diffusing internally rather than by acting on epithelial chemoreceptors (Coon et al., 1990; Hirata and Hadfield, 1986).

A model has been proposed by Coon and Bonar (1985) for C. gigas settlement and metamorphosis that incorporates induction of settlement (a behavioural response where the larval foot is extended beyond the shell margin to allow it to crawl along the substratum); and induction of metamorphosis (a morphogenetic response which incorporates the loss of the velum and growth of the shell and gill). Separate cues have been hypothesized for the onset of each (Bonar et al., 1990; Coon et al., 1990).

In controlled investigations with *C. gigas*, however, L-DOPA has been shown to induce both settlement and metamorphosis (Coon and Bonar, 1985; Coon *et al.*, 1990; Beiras and Widdows, 1995) as has acetylcholine (Beiras and Widdows, 1995). Epinephrine has been shown to induce metamorphosis in *C. gigas* even overriding the usual preceding settlement and attachment behaviour (Coon and Bonar, 1985, 1986, 1987; McAnally-Salas *et al.*, 1989; Beiras and Widdows, 1995), a technique used in the aquaculture industry to produce cultchless (non-attached) spat. It is known that epinephrine induces such an effect by interacting with endogenous α_1-adrenergic receptors (Coon and Bonar, 1987). Norepinephrine also induces metamorphosis of *C. gigas* without settlement and attachment but is less potent and acts more slowly (Coon *et al.*, 1986). This work reinforces the hypothesis that metamorphosis and settlement are not inextricably linked, that is, they can be independently induced and experimentally separated. It should be noted that even though the two processes have been isolated experimentally, with the intention of producing cultchless spat for aquaculture, the *natural* inducer would probably never operate in isolation and induce settlement behaviour prior to competence to undergo metamorphosis (Coon *et al.*, 1990).

The ability to settle and metamorphose in response to appropriate environmental stimuli is acquired between 24 and 48 h after eyespot development in *Crassostrea* larvae (Galtsoff, 1964; Waller, 1981). It is thought that the physiological mechanisms by which larvae reach this stage (i.e. become 'competent', as defined by Burke, 1983) are controlled by the nervous system (Burke, 1983). Acquisition of competence in oysters is a multi-step process (Coon *et al.*, 1990) rather than the single step process assumed by most invertebrate models. These steps include:

(i)   responsiveness to the environmental settlement inducer;
(ii) secondary cues associated with the substrate;
(iii) endogenous metamorphic inducers.

It would be logical to assume, therefore, that 4-NP will interfere with one or more of these steps, rendering oyster larvae unable to metamorphose (thus incompetent).

3.2 *Settlement inducers*

It should be noted that many organisms are able to delay metamorphosis beyond the time they first become physiologically competent to metamorphose. This may happen where the environmental settlement inducer is absent. Pechenik (1990) considers some of the adaptive benefits associated with being able to delay metamorphosis. The ability to postpone metamorphosis in the absence of suitable benthic conditions should promote the placement of individuals into those habitats most likely to support future survival and successful reproduction (Meadows and Campbell, 1972; Crisp, 1974). *Mytilus edulis* larvae can delay metamorphosis for a few days if the appropriate substrate is not available (Bayne, 1965), while barnacle cyprids, *Balanus balanoides* can delay their metamorphosis for several weeks without compromising juvenile survival (Knight-Jones, 1953). Similarly, larvae of the nudibranch, *Phestilla sibogae* can delay metamorphosis for up to 21 days without adverse effects on subsequent development or adult fecundity (Miller, 1988) and delayed metamorphosis in the gastropod, *Crepidula fornicata*, did not appear to disadvantage on-growing juveniles (Pechenik and Eyster, 1989).

Rejection of a substrate, and subsequent delay to metamorphosis, is demonstrated in a study by Billinghurst *et al.* (1998) on cyprid larvae of the barnacle *Balanus amphitrite*, which also showed settlement inhibition by acute exposure to 4-NP. One of the explanations given by the authors for this inhibition of settlement was that 4-NP modified the physical nature of the surface of the test vessels given its affinity for surfaces. In their investigation (Billinghurst *et al.*, 1998), the larvae recovered following transfer to clean water. The possibility that 4-NP inhibited settlement due to a surface modification of the test vessels in the study of *C. gigas* described here can be ruled out because after the 48 h exposure period all vessels were cleaned. If it were simply a surface modification that was preventing settlement (or causing the larvae to delay settlement) then recovery would have been noted after transfer to clean water as in the barnacle experiment (Billinghurst *et al.*, 1998). Whilst settlement continued successfully in all control vessels in the current study, it was clearly still inhibited in treated vessels since a 10 day recovery period (post-exposure) was allowed to pass before spat densities were recorded. Moreover, since metamorphosis occurred in control vessels from day 25 p.f., the containers used must have provided appropriate substrates. Similar containers are widely used at Seasalter and other commercial oyster hatcheries and produce no problems as a substrate for settlement and metamorphosis.

The strategy of delaying metamorphosis in the absence of an appropriate substrate has its costs. Whilst having the ability to delay metamorphosis once competent can provide advantages to the organism, the post-metamorphic fitness of an individual will decline as metamorphosis is delayed beyond a certain point (Pechenik, 1985). Abnormal metamorphosis or poor juvenile survival has been reported for a number of species including the polychaete *Capitella* sp., whose metamorphosis can normally be delayed with no deleterious effect for up to 5 days at 15°C once the animal has become competent. However, at 20°C, a delay of 3 days or more substantially reduces juvenile survival (Butman *et al.*, 1988; Pechenik and Cerulli, 1989). Poor juvenile survival has also been reported following delayed metamorphosis in the mitraria larvae of *Owena fusiformis* (Wilson, 1932), the sand dollar, *Dendraster excentricus* (Highsmith and Emlet, 1986), abalone larvae *Haliotis rufescens* (Morse *et al.*, 1979) and an estuarine crab, *Chasmagnathus granulata* (Gebauer *et al.*, 1999).

3.3 *Secondary cues associated with substrate*

Bacterial films have been shown to enhance settlement of many invertebrate larvae including oysters (Zobell and Allen, 1935; Crisp, 1967; Bonar *et al.*, 1985, 1986; Fitt *et al.*, 1989). 4-NP might, therefore, destroy or damage the bacteria in the culture vessels and, in this respect, it has been reported to have bactericidal effects on *Bacillus subtilus*, *Micrococcus luteus*, *Pseudomonas aeruginosa* and *Staphylococcus aureus* (Okai *et al.*, 2000). If this were so then 4-NP would be acting indirectly to influence the settlement behaviour of *C. gigas*. This seems unlikely because cleaning the vessels at the end of the exposure period would have also removed bacteria from the control vessels. Settlement, however, was not inhibited in the control vessels. In addition, although bacterial populations are likely to have re-established in the 10 day recovery period, settlement was still not observed in the treatment vessels after this time. In a previous study, Crisp (1967) showed that 45 h is ample time for bacterial films to reform after cleaning and a surface is likely to develop a film within minutes of being placed in seawater (Crisp, 1974).

3.4 Endogenous metamorphic inducers

It is possible that 4-NP affects the endogenous metamorphic inducers described by Coon et al. (1986). In this way, competent larvae landing on a suitable surface are unable to interpret (or transduce) subsequent cues to metamorphose and thus start swimming again. In this scenario, 4-NP may be affecting metamorphosis alone and not settlement. One mechanism might be that 4-NP inhibits activation of the α_1-adrenergic receptors known to be involved in the natural metamorphic process (Coon et al., 1986; Bonar et al., 1990).

In Coon et al.'s (1990) model it is suggested that dopamine release at the nerve terminal is triggered by environmental cues and it is this dopamine release that then triggers an 'Integration Centre' (a hypothetical construct) that assimilates sensory input from the environment, then acts to initiate and sustain settlement behaviour. Assuming this model to be viable, 4-NP may also be acting in one or both of the following ways: by blocking environmental cues (see above); or by inhibiting dopamine release at the nerve terminal which in turn inhibits settlement and subsequent meta-morphosis. An important point to consider here is that if the mechanism involves inhi-bition of dopamine release, then it appears to be a permanent inhibition because 4-NP exposure was for only 48 h yet the larvae were prevented from settling for the remaining period (a further 10 days). Another possibility is that the window for dopamine release is restricted to the first 48 h following the onset of eyespot stage and competence.

A further possible explanation for the inhibition of barnacle settlement by 4-NP (Billinghurst et al., 1998) is that the larvae were suffering from narcosis. Organic cont-aminants have been found to be a common cause of narcosis in aquatic organisms (Christie and Crisp, 1966; Crisp et al., 1967; Verhaar et al., 1996). In a typical response the affected organisms sink to the bottom in a narcotized condition until conditions recover. At this point, narcosis-suffering animals tend to recover normal swimming behaviour. The oyster larvae in the trials described earlier in this chapter showed no signs of narcosis; and as noted previously, there was no recovery period with respect to settlement and metamorphosis when returned to clean water.

Metamorphosis was completely inhibited in C. gigas raised in a copper-enriched sediment (Phelps and Warner, 1990) probably due to the inhibition of RNA-polymerase (Novello and Stirpe, 1969; Hidalgo et al., 1976). The metamorphic rate was reduced when competent pediveligers were exposed to mercury for 48 h prior to the addition of the metamorphosis inducer epinephrine (Beiras and His, 1994). However, the results of this study may be regarded as somewhat inconclusive, as the authors did not rule out the possibility of starvation affecting the metamorphic rate. C. virginica settlement was significantly inhibited by the presence of biofilms containing copper and zinc when compared to settlement on non-contaminated control biofilms (Chang et al., 1996).

In contrast to these responses, a number of earlier studies have shown an abnor-mally rapid metamorphosis in response to sub-lethal doses of toxic substances. This was seen in C. virginica in the presence of copper salts (Prytherch, 1934), but was refuted by Korringa (1940). Huxley (1928) showed accelerated metamorphosis in Echinus plutei exposed to dilute mercuric chloride. C. gigas larvae have been reported to begin settlement behaviour in the presence of ammonia gas (NH_3) dissolved in seawater (Bonar et al., 1990; Coon et al., 1990; Fitt et al., 1990). This response is believed to be the result of increased intracellular pH, rather than a specific response to NH_3 (Coon et al., 1990). NH_3 was originally identified as the active agent isolated from bacterial supernatants (Bonar et al., 1990) although this claim has not been confirmed (Fitt et al., 1990). In addition, although Coon et al. (1990) suggest it may play a role in oyster settlement in the natural environment, NH_3 occurs naturally in the water column at concentrations at least an order of magnitude lower than those required for the induction of settlement behaviour.

Other factors reported to affect settlement and metamorphosis in marine invertebrates include above and below normal levels of oxygen (hypoxia and microxia; Baker and Mann, 1994a,b) and salinity fluctuations (of less than 10 ppt; Richmond and Woodin, 1996). Dissolved oxygen and salinity were recorded on every sampling day throughout the trials on C. gigas reported here and no differences between the treated and control vials were detectable.

Whatever mechanisms underlie the ability of 4-NP to inhibit settlement and metamorphosis, a critical window for the cues involved seems to fall within the period 23–24 days p.f. (eyespot stage and the following 48 h). When metamorphosis is inhibited completely, even if the animals were able to survive in their pelagic, larval form, it is unlikely that they would ever reach reproductive maturity. Thus, the ability of 4-NP to inhibit settlement completely will have serious consequences for the population dynamics of oysters growing in areas prone to 4-NP contamination.

It should be noted that the exposures described here (100 and 1 µg l⁻¹ 4-NP) were those present at the beginning of the exposure periods. Measurement of actual concentrations by GCMS show a substantial reduction of real concentrations with time (*Table 1*). Thus, after 6 h, 4-NP levels were 9 and <1 µg l⁻¹ respectively (detection limit 1 µg l⁻¹). This indicates that the effects seen were in fact induced by declining levels of 4-NP.

Possible reasons for the loss of 4-NP from the water column include degradation and adsorption to test vessel surfaces.

In summary, with respect to 'critical exposure periods', C. gigas larval growth is mainly affected early in development prior to eyespot stage. This is a proven period of vulnerability when the transformation from veliger to veliconcha is taking place. Another critical period for 4-NP exposure is at the later eyespot stage when settlement and metamorphosis are affected. As yet, the mechanisms for the responses seen remain unknown but indications are that they involve, at least in part, interference with native endocrine and/or neuroendocrine regulatory pathways, perhaps in some cases by mimicking endogenous or exogenous chemical cues, or both.

4. Data analysis

Analysis of larval survival data was performed by means of a General Linear Model followed by a Tukey multiple comparison test. Analysis of larval growth and spat data was performed using Analysis of Variance followed by a Tukey test.

Table 1. Measured 4-nonylphenol concentrations (µg l⁻¹) in the water column over time during each exposure period

Treatment	Time (h)		
	0	6	48
Control	0	0	0
1 µg l⁻¹ 4-nonylphenol	1	<1	<1
100 µg l⁻¹ 4-nonylphenol	100	9	2

Acknowledgements

Much of the original work reported here was supported by NERC DEMA Thematic Grant GST/02/1738 and a NERC studentship to HN. We are also very grateful to John Bayes at Seasalter Shellfish, Whitstable, Kent for advice on larval husbandry and for putting the hatchery facilities at our disposal and to Rupert Marshall for his help in the production of this chapter.

References

Ahel, M., Giger, W. and Schaffner, C. (1991) Behaviour of detergent derived organic chemicals in the Glatt river 1. Alkylphenol polyethoxylates and their metabolites. In: *Organic Micropollutants in the Aquatic Environment*. Commission of the EC, Brussels.

Alzieu, C. (1986) The detrimental effects on oyster culture in France – evolution since antifouling paint regulation. *Oceans* 86: 1130–1134.

Baker, S.M. and Mann, R. (1994a) Description of metamorphic phases in the oyster *Crassostrea virginica* and effects of hypoxia on metamorphosis. *Mar. Ecol. Prog. Ser.* 104: 91–99.

Baker, S.M. and Mann, R. (1994b) Feeding ability during settlement and metamorphosis in the oyster *Crassostrea virginica* (Gmelin, 1791) and the effects of hypoxia on postsettlement ingestion rates. *J. Exp. Mar. Biol. Ecol.* 181: 239–253.

Baldwin, W.S., Graham, S.E., Shea, D. and LeBlanc, G.A. (1997) Metabolic androgenization of female *Daphnia magna* by the xenoestrogen 4-nonylphenol. *Environ. Toxicol. Chem.* 16: 1905–1911.

Bayne, B.L. (1965) Growth and the delay of metamorphosis of the larvae of *Mytilus edulis*. *Ophelia* 2: 1–47.

Beiras, R. and His, E. (1994) Effects of dissolved mercury on embryogenesis, survival, growth and metamorphosis of *Crassostrea gigas* oyster larvae. *Mar. Ecol. Prog. Ser.* 113: 95–103.

Beiras, R. and Widdows, J. (1995) Induction of metamorphosis in larvae of the oyster *Crassostrea gigas* using neuroactive compounds. *Mar. Biol.* 123: 327–334.

Billinghurst, Z., Clare, A.S., Fileman, T., McEvoy, J., Readman, J. and Depledge, M.H. (1998) Inhibition of barnacle settlement by the environmental oestrogen 4-nonylphenol and the natural oestrogen 17 B oestradiol. *Mar. Pollut. Bull.* 36: 833–839.

Blackburn, M.A. and Waldock, M.J. (1995) Concentrations of alkylphenols in rivers and estuaries in England and Wales. *Water Res.* 29: 1623–1629.

Bonar, D.B., Coon, S.L., Weiner, R.M. and Colwell, R.R. (1985) Induction of oyster metamorphosis by bacterial products and biogenic amines. *Bull. Mar. Sci.* 37: 763.

Bonar, D.B., Weiner, R.M. and Colwell, R.R. (1986) Microbial invertebrate interactions and potential for biotechnology. *Microb. Ecol.* 12: 101–110.

Bonar, D.B., Coon, S.L., Walch, M., Weiner, R.M. and Fitt, W. (1990) Control of oyster settlement and metamorphosis by endogenous and exogenous chemical cues. *Bull. Mar. Sci.* 46: 484–498.

Burke, R.D. (1983) The induction of metamorphosis of marine invertebrate larvae: stimulus and response. *Can. J. Zool.* 61: 1701–1719.

Butman, C.A., Grassle, J.P. and Buskey, E.J. (1988) Horizontal swimming and gravitational sinking of *Capitella sp.* larvae: implications for settlement. *Ophelia* 29: 43–57.

Calabrese, A., MacInnes, J.R., Nelson, D.A. and Miller, J.E. (1977) Survival and growth of bivalve larvae under heavy metal stress. *Mar. Biol.* 41: 179–184.

Chang, E.Y., Coon, S.L., Walch, M. and Weiner, R. (1996) Effects of *Hyphomonas* PM-1 biofilms on the toxicity of copper and zinc to *Crassostrea gigas* and *Crassostrea virginica* larval set. *J. Shellfish Res.* 15: 589–595.

Christiansen, T., Korsgaard, B. and Jespersen, A. (1998) Effects of nonylphenol and 17 beta-oestradiol on vitellogenin synthesis, testicular structure and cytology in male eelpout *Zoarces viviparus*. *J. Exp. Biol.* 201: 179–192.

Christie, A.O. and Crisp, D.J. (1966) Toxicity of aliphatic amines to barnacle larvae. *Comp. Biochem. Physiol.* **18**: 59–69.

Coon, S.L. and Bonar, D.B. (1985) Induction of settlement and metamorphosis of the Pacific oyster, *Crassostrea gigas*, by L-DOPA and catecholamines. *J. Exp. Mar. Biol. Ecol.* **94**: 211–221.

Coon, S.L. and Bonar, D.B. (1986) Norepinephrine and dopamine content of larvae and spat of the Pacific oyster, *Crassostrea gigas*. *Biol. Bull.* **171**: 632–639.

Coon, S.L. and Bonar, D.B. (1987) Pharmacological evidence that alpha-adrenoreceptors mediate metamorphosis of the Pacific oyster, *Crassostrea gigas*. *Neuroscience* **23**: 1169–1174.

Coon, S.L., Bonar, D.B. and R.M.W. (1986) Chemical production of cultchless oyster spat using epinephrine and norepinephrine. *Aquaculture* **58**: 255–262.

Coon, S.L., Fitt, W.K. and Bonar, D.B. (1990) Competence and delay of metamorphosis in the Pacific oyster *Crassostrea gigas*. *Mar. Biol.* **106**: 379–387.

Crisp, D.J. (1967) Chemical factors inducing settlement in *Crassostrea virginica*. *J. Animal Ecol.* **36**: 329–335.

Crisp, D.J. (1974) *Chemoreception in Marine Organisms* (eds P.T. Grant and A.M. Mackie). Academic Press, London, pp. 177–265.

Crisp, D.J., Christie, A.O. and Ghobashy, A.F.A. (1967) Narcotic and toxic action of organic compounds on barnacle larvae. *Comp. Biochem. Physiol.* **22**: 629–649.

Drinnan, R.E. and Stallworthy, W.B. (1979) Oyster larval populations and assessment of spatfall, Bideford River, P.E.I. *Can. Fish. Mar. Serv. Tech. Rep.* 797.

Fichet, D., Radenac, G. and Miramand, P. (1998) Experimental studies of impacts of harbour sediment resuspension to marine invertebrate larvae: bioavailability of Cd, Cu, Pb and Zn and toxicity. *Mar. Poll. Bull.* **36**: 509–518.

Fitt, W.K., Labare, M.P., Fuqua, W.C., Walch, M., Coon, S.L., Bonar, D.B., Colwell, R.R. and Weiner, R.M. (1989) Factors influencing bacterial production of inducers of settlement behaviour of larvae of the oyster *Crassostrea gigas*. *Microb. Ecol.* **17**: 287–298.

Fitt, W.K., Coon, S.L., Walch, M., Weiner, R.M., Colwell, R.R. and Bonar, D.B. (1990) Settlement behaviour and metamorphosis of oyster larvae (*Crassostrea gigas*) in response to bacterial supernatants. *Mar. Biol.* **106**: 389–394.

Galtsoff, P.S. (1964) *The American Oyster* (Crassostrea virginica), Vol. 64. United States Department of the Interior, Washington, DC.

Gebauer, P., Paschke, K. and Anger, K. (1999) Costs of delayed metamorphosis:reduced growth and survival in early juveniles of an estuarine grapsid crab, *Chasmagnathus granulata*. *J. Exp. Mar. Biol. Ecol.* **238**: 271–281.

Gospe, S.M., Jr. (1983) Studies of dopamine pharmacology in molluscs. *Life Sci.* **33**: 1945–1957.

Granmo, A., Ekelund, R., Magnusson, K. and Berggren, M. (1989) Lethal and sublethal toxicity of 4-nonylphenol to the common mussel (*Mytilus edulis* L.) *Environ. Pollut.* **59**: 115–127.

Gross, M.Y., Maycock, D.S., Thorndyke, M.C., Morritt, D. and Crane, M. (2001) Abnormalities in sexual development of the amphipod *Gammarus pulex* found below sewage treatment works. *Environ. Toxicol. Chem.* (in press).

Hidalgo, H.A., Koppa, V. and Bryan, S.E. (1976) Effect of cadmium on RNA-polymerase synthesis in rat liver. *FEBS Lett.* **64**: 159–162.

Hidu, J. (1969) Gregarious setting in the American oyster *Crassostrea virginica*. *Chesapeake Sci.* **10**: 85–92.

Highsmith, R.C. and Emlet, R.B. (1986) Delayed metamorphosis:effect on growth and survival of juvenile sand dollars. *Bull. Mar. Sci.* **39**: 347–361.

Hirata, K.Y. and Hadfield, M.G. (1986) The role of choline in metamorphic induction of *Phestila*. *Comp. Biochem. Physiol. C* **84**: 15–21.

Hunt, H.L. and Scheibling, R.E. (1997) Role of early post-settlement mortality in recruitment of benthic marine invertebrates. *Mar. Ecol. Prog. Ser.* **155**: 269–301.

Huxley, J.S. (1928) Experimentally induced metamorphosis in *Echinus*. *Am. Nat.* **62**: 363–377.

Jaeckle, W.B. and Manahin, D.T. (1989) Feeding by a "nonfeeding" larva: uptake of dissolved amino acids from seawater by lecithodtrophic larvae of the gastropod *Haliotis rufescens*. *Mar. Biol.* **103**: 87–94.

Jobling, S. and Sumpter, J.P. (1993) Detergent components in sewage efflutent are weakly oestrogen to fish: An in vitro study using rainbow trout (*Oncorhynchus mykiss*) hepatocytes. *Aquat. Toxicol.* **27**: 361–372.

Keck, R., Maurer, D., Kauer, J.C. and Sheppard, W.A. (1971) Chemical stimulants affecting larval settlement in the American oyster. *Proc. Nat. Shellfish. Assoc.* **61**: 24–28.

Kennedy, V.S., Newell, R.I.E. and Eble, A. (1996) *The Eastern Oyster* Crassostrea virginica. Maryland Sea Grant, Maryland.

Kirchman, D., Graham, S., Reish, D. and Mitchell, R. (1982) Bacteria induce settlement and metamorphosis of *Janua brasiliensis*. *J. Exp. Mar. Biol. Ecol.* **56**: 153–163.

Knight-Jones, E.W. (1953) Decreased discrimination during setting after prolonged planktonic life in larvae of *Spirorbis borealis*. *J. Mar. Biol. Assoc. UK* **32**: 237–345.

Korringa, P. (1940) Experiments and observations on swarming, pelagic life and setting in the European Flat Oyster, *Ostrea edulis*. *Arch. Neerl. Zool.* **5**: 1–249.

Lucas, A. (1991) Programming by early nutrition in man. In: *The Childhood Environment and Adult Disease* (eds G. R. Bock and J. Whelan). John Wiley, Chichester, pp. 38–55.

Madsen, S.S., Mathiesen, A.P. and Korsgaard, B. (1997) Effects of 17 beta-estradiol and 4-nonylphenol on smoltification and vitellogenesis in Atlantic Salmon (*Salmo salar*) *Fish Physiol. Biochem.* **17**: 303–312.

Marcomini, A., Pavoni, B., Sfriso, A. and Orio, A.A. (1990) Persistent metabolites of alkylphenol polyethoxylates in the marine environment. *Mar. Chem.* **29**: 307–323.

McAnally Salas, L., Cavazos Lliteras, H. and Salas, G., A. (1989) Effect of epinephrine, norepinephrine and L-DOPA on the settlement and metamorphosis of larvae of *Crassostrea gigas*. *Ciencs Mar* **15**: 85–103.

Meadows, P.S. and Campbell, J.I. (1972) Habitat selection by aquatic invertebrates. *Adv. Mar. Biol.* **10**: 271–382.

Miller, S.E. (1988) Effect of larval duration on post-larval lifespan and fecundity of a nudibranch mollusc. *Am. Zool.* **28**: 171.

Miyake, T., Haruyama, H., Ogura, T., Mitsui, T. and Sakurai, A. (1991) Effects of insect juvenile-hormone active NC-170 on metamorphosis, oviposition and embryogenesis in leafhoppers. *J. Pest. Sci.* **16**: 441–448.

Morse, D.E., Hooker, N., Duncan, H. and Jensen, L. (1979) Gamma-aminobutyric acid, a neurotransmitter, induces planktonic abalone to settle and begin metamorphosis. *Science* **204**: 407–410.

Muller, W.A. (1973) Metamorphose-Induktion bei Planulalarven.1.Der bakterielle Induktor. *Wilhelm Roux Arch Entwicklungsmech. Org.* **173**: 107–121.

Neumann, R. (1979) Bacterial induction of settlement and metamorphosis in the planula larvae of *Cassiopea andromeda*. *Mar. Ecol. Prog. Ser.* **1**: 21–28.

Nice, H.E., Thorndyke, M.C., Morritt, D., Steele, S. and Crane, M. (2000) Development of *Crassostrea gigas* larvae is affected by 4-nonylphenol. *Mar. Pollut. Bull.* **40**: 491–496.

Nielson, S.A. (1973) Effect of acetazolamide on larval settlement of *Ostrea lutaria*. *Veliger* **16**: 667.

Nimrod, A.C. and Benson, W.H. (1996) Environmental estrogenic effects of alkylphenol ethoxylates. *Crit. Rev. Toxicol.* **26**: 335–364.

Novello, F. and Stirpe, F. (1969) The effects of copper and other ions on the ribonucleic acid polymerase activity of isolated rat liver nuclei. *Biochemistry* **111**: 115–119.

Okai, Y., Higashi-Okai, K., Machida, K., Nakamura, H., Nakayama, K., Fijita, K., Tanaka, T. and Taniguchi, M. (2000) Protective effects of alpha-tocopherol and beta-carotene on para-nonylphenol-induced inhibition of cell growth, cellular respiration and glucose-induced proton extrusion of bacteria. *FEMS Microbiol. Lett.* **187**: 161–165.

Pechenik, J.A. (1985) Delayed metamorphosis of marine molluscan larvae:current status and directions for future research. *Am. Malacolog. Bull. Spec. Edn.* **1**: 85–91.

Pechenik, J.A. (1990) Delayed metamorphosis by larvae of benthic marine invertebrates:does it occur? Is there a price to pay? *Ophelia* **32**: 63–94.

Pechenik, J.A. and Cerulli, T.R. (1989) Does delayed metamorphosis influence growth and reproduction of the marine benthic polychaete *Capitella* sp.? *Am. Zool.* **29**: 173.

Pechenik, J.A. and Eyster, L.S. (1989) Influence of delayed metamorphosis on the growth and metabolism of young *Crepidula fornicata* (Gastropoda) juveniles. *Biol. Bull.* **176**: 14–24.

Pechenik, J.A., Rittschof, D. and Schmidt, A.R. (1993) Influence of delayed metamorphosis on survival and growth of juvenile barnacles *Balanus amphitrite. Mar. Biol.* **115**: 287–294.

Phelps, H.L. and Warner, K.A. (1990) Estuarine sediment bioassay with pediveliger larvae (*Crassostrea gigas*) *Bull. Environ. Contam. Toxicol.* **44**: 197–204.

Pinder, L.C.V., Pottinger, T.G., Billinghurst, Z. and Depledge, M.H. (1998) *Endocrine function in aquatic invertebrates and evidence for disruption by environmental pollutants.* Technical Report E67, Environment Agency Endocrine Modulators Steering Group, Windermere, UK.

Prytherch, H.F. (1934) The role of copper in the settling, metamorphosis and distribution of the American oyster, *Ostrea virginica. Ecol. Monogr.* **4**: 49–107.

Rexroad, M. (1987) Ecotoxicity of tributyltin. *Oceans* 87: 1443–1455.

Richmond, C.E. and Woodin, S.A. (1996) Short-term fluctuations in salinity: effects on planktonic invertebrate larvae. *Mar. Ecol. Prog. Ser.* **133**: 167–177.

Roberts, M.H., Bender, M.E., De Lisle, P.F., Sutton, H.C. and Williams, R.L. (1987) Sex ratio and gamete production in American oysters exposed to tributyltin in the laboratory. In: *Proceedings of the International Organotin Symposium,* Vol. 4. Institute of Electrical Engineers, New York, pp. 1471–1476.

Roesijadi, G., Hansen, K.M. and Unger, M.E. (1996) Cadmium-induced metallothionein expression during embryonic and early larval development of the mollusc *Crassostrea virginica. Toxicol. Appl. Pharmac.* **140**: 356–363.

Routledge, E.J., Sheahan, D., Desbrow, C., Brighty, G.C., Waldock, M. and Sumpter, J.P. (1998) Identification of estrogenic chemicals in STW effluent. 2. In vivo responses in trout and roach. *Environ. Sci. Technol.* **32**: 1559–1565.

Shurin, J.B. and Dodson, S.I. (1997) Sublethal toxic effects of cyanobacteria and nonylphenol on environmental sex determination and development in *Daphnia. Environ. Toxicol. Chem.* **16**: 1269–1276.

Solbakken, J.S., Norberg, B., Watanabe, K. and Pittman, K. (1999) Thyroxine as a mediator of metamorphosis of Atlantic halibut, *Hippoglossus hippoglossus. Environ. Biol. Fishes* **56**: 53–65.

Thain, J.E. and Waldock, M.J. (1986) The impact of tributyltin (TBT) antifouling paints on molluscan fisheries. *Water Sci. Technol.* **18**: 193–202.

Veitch, F.P. and Hidu, H. (1971) Gregarious setting in the American oyster *Crassostrea virginica* Gmelin:1. Properties of a partially purified "setting factor". *Chesapeake Sci.* **12**: 173–178.

Verhaar, H.J. M., Ramos, E.U. and Hermens, J.L. (1996) Classifying environmental pollutants. 2: Separation of class 1 (baseline toxicity) and class 2 (polar narcosis) type compounds based on chemical descriptions. *J. Chemomet.* **10**: 149–162.

Waller, T.R. (1981) Functional morphology and development of veliger larvae of the European oyster, *Ostrea edulis. Smithsonian Contrib. Zool.* **328**: 1–70.

Weiner, R.M., Segall, A.M. and Colwell, R.R. (1985) Characterization of a marine bacterium associated with *Crassostrea virginica. Appl. Environ. Microbiol.* **49**: 83–90.

Wilson, D.P. (1932) On the mitraria larva of *Owena fusiformis. Phil. Trans. R. Soc.* **221**: 231–334.

Zobell, C.E. and Allen, E.C. (1935) The significance of marine bacteria in the fouling of submerged surfaces. *J. Bacteriol.* **29**: 239–251.

11

Interrupted development in aquatic organisms: ecological context and physiological mechanisms

Steven C. Hand, Jason E. Podrabsky, Brian D. Eads and Frank van Breukelen

1. Introduction

Living in aquatic habitats, particularly those of an ephemeral nature, can require that animals tolerate prolonged bouts of physical extremes like severe dehydration and oxygen deprivation. Across diverse groups of animals, survival under these conditions is facilitated by interrupting their developmental programmes and entering states of dormancy. The biological advantages of interrupted development have been appreciated by numerous authors. Particularly in the case of resting eggs, the abilities to recolonize an environment and disseminate the species to new locations are well known (King, 1980; Elgmork, 1980; Pourriot and Snell, 1983). These developmentally arrested forms can serve to synchronize the actively feeding or growing stage of the organism with favourable environmental conditions (Crowe and Clegg, 1973). In parasitic species, a propagule may represent an obligate stage for the transmission to a specific host. For other organisms, production of encysted forms may allow for a change between asexual and sexual reproduction or for the avoidance of predators (Hairston, 1987).

A feature that often, but not always, accompanies developmental arrest is energy conservation through metabolic down-regulation. A coordinated depression of catabolic and anabolic processes is a common trait of organisms that enter and recover from states of metabolic depression and may be essential in order to maintain energy balance for prolonged periods (Hochachka and Guppy, 1987; Storey and Storey, 1990; Guppy et al., 1994; Hand and Hardewig, 1996; Hand, 1999). Biosynthesis of macromolecules is one of the most energetically expensive processes in the cell. For example, estimates indicate that protein synthesis may account for 12–40% of the basal metabolism of mammalian cells, tissues, or whole animals (Buttgereit et al., 1992; Buttgereit

Environment and Animal Development: Genes, Life Histories and Plasticity, edited by D. Atkinson and M. Thorndyke. © 2001 BIOS Scientific Publishers Ltd, Oxford.

and Brand, 1995; Rolfe and Brown, 1997) and for as much as 60–90% of the total
oxygen consumption of various cell types from fish (Pannevis and Houlihan, 1992;
Smith and Houlihan, 1995). Fuery *et al.* (1998) report that over half (52%) of the
metabolic depression observed in liver tissue isolated from the estivating frog,
Neobatrachus centralis, is due to a reduction in protein synthesis. The rate of protein
synthesis is depressed by 92% in isolated turtle hepatocytes when exposed to anoxia,
which can account for nearly 36% of the metabolic depression observed (Land *et al.,*
1993).

In this perspectives chapter, our goal is to illustrate the developmental and metabolic principles above by drawing on current data from our laboratory that address two
forms of interrupted development – diapause in annual killifish embryos and anoxia-
induced quiescence in encysted embryos of the brine shrimp.

2. Diapause in an annual killifish

2.1 *Ecology and natural history*

Populations of the annual killifish, *Austrofundulus limnaeus,* persist in ephemeral pond
habitats by producing drought-tolerant, diapausing embryos (Wourms, 1972a,b).
Diapause is a state of developmental arrest that is promoted by endogenous cues and
typically precedes the onset of unfavourable environmental conditions (Hand, 1991;
Hand and Podrabsky, 1999). Thus embryos may enter diapause even under conditions
considered optimal for development. Diapause differs from quiescence (Section 3,
below), which is a state of developmental and metabolic arrest promoted directly by
exposure to a stressful environmental condition (anoxia, dehydration, etc.). There are
intermediate states displayed by some organisms that fall in between these two defini-
tions. Diapausing embryos of *A. limnaeus* may remain dormant for months to years
(Wourms, 1972b; Podrabsky, 1999). While there may be three distinct stages of
diapause in some annual killifish (Wourms, 1972b), in *A. limnaeus* diapause interrupts
development at only two distinct developmental stages (diapause II and III;
Podrabsky, 1999; Podrabsky and Hand, 1999). Diapause II occurs in an embryo
possessing 38–40 pairs of somites, the foundations of the central nervous system, optic
and otic cups, and a functional tubular heart (Wourms, 1972a; Podrabsky, 1999).
Diapause III occurs in the fully developed embryos just prior to hatching and has been
compared to estivation in African lungfish (Wourms, 1972b). In *A. limnaeus,* diapause
II embryos exhibit a profound depression of metabolism (Podrabsky and Hand, 1999)
and appear to be very resistant to environmental stresses such as anoxia and dehydration. A 90% depression of metabolism is associated with entry into diapause II in
embryos of *A. limnaeus,* and a simultaneous down-regulation of ATP-producing and
ATP-consuming processes is indicated during this transition in metabolic rate by the
maintenance of stable levels of ATP (Podrabsky and Hand, 1999).

2.2 *Arrest of protein synthesis*

As mentioned above, protein synthesis is a major energy sink, which if depressed could
save valuable energy stores. Rates of protein synthesis are substantially depressed by
over 93% in embryos of *A. limnaeus,* as this annual killifish enters diapause (*Table 1*;
Podrabsky and Hand, 2000). Cycloheximide inhibits 36% of the oxygen consumption

Table 1. Depression of protein synthesis in embryos of the annual killifish, Austrofundulus limnaeus *(after Podrabsky and Hand, 2000)*

Developmental stage (days post-fertilization)	Protein synthesis (cpm μg^{-1} DNA)[a]
2 (developing)	2033 ± 374[*]
8 (developing)	3117 ± 295[†]
24 (diapausing)	340 ± 53[‡]
42 (diapausing)	203 ± 53[‡]

[a] Incorporation of radiolabelled amino acids into acid-precipitable protein, normalized to the level of DNA content of embryos. For methods used to label the amino acid pool, see Podrabsky and Hand (2000).
Values are means ± SEM ($n = 4$). Values with different superscript symbols are statistically different from each other.

and heat dissipation in developing embryos 8 days post-fertilization (8 days p.f.). The equivalent inhibition of oxygen consumption and heat dissipation by cycloheximide suggests a largely aerobic metabolism for embryos at 8 days p.f. In contrast, statistically significant inhibition by cycloheximide is not detectable for either oxygen consumption or heat dissipation in embryos during diapause II (24 days p.f.), which indicates a negligible contribution of protein synthesis to ATP turnover (*Table 2*; Podrabsky and Hand, 2000). Taken together, these data indicate that arrest of protein synthesis results in a 36% savings in cellular energy expenditure during diapause.

Depression of protein synthesis during diapause is presumably induced by endogenous cues and appears to be independent of the energy status of the cells, because high ATP:ADP ratios and adenylate energy charge exist during development and diapause II in these embryos (Podrabsky and Hand, 1999). A similar situation apparently exists in diapausing embryos of the brine shrimp, *Artemia franciscana*, which maintain high levels of adenylates and have reduced rates of protein synthesis (Drinkwater and Crowe, 1987; Drinkwater and Clegg, 1991; Clegg *et al.*, 1996). Despite high ATP:ADP ratios, AMP levels are elevated in diapausing embryos of *A. limnaeus*, and a strong correlation has been shown between increases in the AMP:ATP ratios and a decrease in both oxygen consumption and heat dissipation (Podrabsky and

Table 2. Cycloheximide inhibition of respiration and heat dissipation in embryos of Austrofundulus limnaeus *(after Podrabsky and Hand, 2000)*

Developmental stage	Percentage inhibition of respiration	Percentage inhibition of heat dissipation
8	35.8 ± 1.7	36.3 ± 2.7
16	18.3 ± 0.98	23.6 ± 2.2
24	−8.8 ± 5.0	−4.8 ± 3.5

Values are means ± SEM ($n = 3$).

Hand, 1999). There is also a strong negative correlation between the rate of protein synthesis and AMP:ATP ratios during early development and diapause in *A. limnaeus* (Podrabsky and Hand, 2000). Lefebvre *et al.* (1993) report a similar relationship between increases in adenosine 5′-monophosphate (AMP) and decreases in the rate of protein synthesis in hepatocytes exposed to hypoxia. Elevated levels of AMP have also been shown to inhibit protein synthesis despite the presence of high levels of ATP and guanosine 5′-triphosphate (GTP) in lysates prepared from rabbit reticulocytes (Mosca *et al.*, 1983). The effect of AMP could be mediated by the AMP-activated protein kinase (AMPK, for a review see Hardie and Carling, 1997; Hardie *et al.*, 1998), perhaps through phosphorylation of key initiation and elongation factors of translation (Podrabsky and Hand, 1999). Mechanistic data that link AMPK to the regulation of protein synthesis during metabolic depression are currently not available, but this possibility deserves future attention.

Restricting energy expenditure undoubtedly increases the time that embryos can survive in a state of developmental arrest. Data for these embryos and previous contributions from Clegg *et al.* (1996) for brine shrimp embryos and Joplin *et al.* (1990) for diapausing insects suggest that depression of protein synthesis is a common mechanism for reduction of ATP turnover during diapause. This conclusion agrees with numerous studies that illustrate a depression in the rate of protein synthesis during metabolic depression associated with quiescence (Hand and Hardewig, 1996). A number of similarities between metabolic depression associated with diapause and quiescence are beginning to emerge. These similarities may help elucidate mechanisms which govern large-scale changes in cellular metabolism associated with cellular proliferation and morphogenesis.

2.3 *Desiccation tolerance*

Survival under desiccating conditions is accomplished not by the adult or juvenile forms, but by diapausing embryos of *A. limnaeus*. Although there are no data describing the distribution of *A. limnaeus* embryos in pond sediments, they are probably buried too shallowly (a few centimetres) to access moisture resources often found deeper in the soil. Diapausing embryos of *A. limnaeus* are probably exposed to a highly variable and unpredictable environment that includes intense dehydration pressures for extended periods while encased in the dry mud. The life history of *A. limnaeus* and the nature of their ephemeral pond habitat suggest that embryos must either enter a state of anhydrobiosis (Crowe and Clegg, 1973, 1978), during which removal of virtually all cellular water is tolerated, or else evaporative water loss is dramatically reduced. Mechanisms for the prevention of water loss are typically effective at mild or moderate dehydration pressures (above 98% relative humidity (R.H.), water potentials above –2.8 MPa). Thus, it was surprising to discover that embryos of *A. limnaeus* survive desiccating conditions by substantially reducing water loss and retaining a large portion of their water (Podrabsky *et al.*, 2001). Surface soils (<0.5 m) in desert regions can become severely dried. For example, an ephemeral stream bed in the Chihuahuan desert has been reported to be extremely dry, with water potentials reaching –13 to –16 MPa after long periods (months) without rain (Scanlon, 1994). Many amphibians and fish that inhabit xeric or ephemeral environments depend on mechanisms that reduce water loss for survival (Fishman *et al.*, 1992; Mayhew, 1965; Podrabsky, 1999). However, these mechanisms are only effective under relatively mild dehydration stresses. In constrast, diapause II embryos of *A. limnaeus* retain water under extremely desiccating conditions for over 113 days at 25°C (Podrabsky *et al.*, 2001). Such high resistance to desiccation is unprecedented among aquatic vertebrates.

Of the developmental stages studied, diapause II is the most resistant to water loss; approximately 50% of the initial water is retained even after 32 days of exposure to 50% R.H. The rate of water loss is high in diapause II embryos over the first few days of exposure to desiccating conditions, but after the initial 4 days, water loss is reduced to low values. Water loss is indistinguishable from zero after 32 days. Embryonic stages that are less dehydration-resistant (dispersion–reaggregation stage and diapause III) continue to lose water at a significant rate until dehydrated. The early and rapid loss of water in diapause II embryos appears to come from the perivitelline fluid. The volume of the embryonic tissues and yolk does not change appreciably during the initial phase of water loss, while the perivitelline space is greatly reduced. Using a measured egg diameter of 1.73 ± 0.026 mm and an average yolk diameter of 1.46 ± 0.022 mm (Podrabsky, 1999) the volume of the perivitelline space is estimated to be 1.082 µl. This value corresponds to approximately 0.76 mg of water, when one considers the composition of perivitelline fluid (Hamor and Garside, 1977). At 75.5% R.H., diapause II embryos lose on average 0.76 mg of water embryo^{-1} after 16 days of exposure to dehydrating conditions. Therefore, all of the water lost during the initial phases of dehydration appears to come from the perivitelline space and not from the embryo or yolk. It is appropriate to note that the water retained by diapause II embryos exposed for 8 days to 75.5% R.H. is freezable and therefore likely to be bulk water. Differential scanning calorimetry indicates a large heat flow near 0°C during warming of samples previously cooled to –40°C, which indicates the melting of water that froze during cooling (Podrabsky *et al.*, 2001).

For all developmental stages, survival of dehydrating conditions is highest in 85% R.H., intermediate in 75.5% R.H., and lowest in 50% R.H (Podrabsky *et al.*, 2001). Diapause II embryos have the highest survivorship, with 40% surviving for over 100 days at 75.5% R.H. Diapause III embryos show the least resistance (100% mortality after 8 days at 75.5% R.H.), and embryos at the dispersion-reaggregation stage are all dead after 16 days of exposure to 75.5% R.H. Contact of embryos with a wetted surface early in the dehydration process is required to obtain the survivorship data observed.

2.4 *Mechanisms for water retention*

Damage to the egg envelope during water stress, either mechanically or by fungal infection, causes rapid water loss and death of embryos. Thus, the egg envelope is required for survival under desiccating conditions. Dehydration greatly increases the intermolecular β-sheet content of the egg envelope proteins, a process that is reversible upon rehydration. These β-sheet structures are indicative of increased inter-molecular contacts and interactions (Chiti *et al.*, 1999; Dong *et al.*, 1990; Fink, 1998; Lansbury, 1999) that could serve to decrease the water permeability of the egg envelope. However, based on biochemical measurements, infra-red spectroscopy, and far UV circular dichroism (CD) spectroscopy, the structure of the proteins that comprise the egg envelope does not change across early development, diapause II or diapause III (Podrabsky *et al.*, 2001). Thus, other properties are required to explain the ontogenetic reduction in water loss observed specifically for diapause II embryos.

We speculate that secretion of substances into the perivitelline space during development could play a role in the dehydration resistance of diapause II embryos of *A. limnaeus*. Constituents like protein, lipid, polysaccharides, mucopolysaccharides and polysialoglycoproteins – all present in the perivitelline space (Hamor and Garside, 1977; Kitajima *et al.*, 1986; Laale, 1980) – could prevent water loss by creating a hydrophobic barrier upon drying. It is also possible that such constituents could vitrify (form a glass; Crowe *et al.*, 1998) when severely desiccated, thus substantially slowing water loss from the embryonic and yolk compartments. The latter possibility is supported by observations that the outer layers of embryos become brittle and glass-like after the initial dehydration period. Whatever the precise mechanism for reducing water loss might be, our results indicate that water retention is the key to high survivorship of diapausing embryos under desiccating conditions (Podrabsky *et al.*, 2001).

Finally, an unexpected result from our studies was that the egg envelope of *A. limnaeus* is composed of proteins with characteristics of amyloid fibrils often associated with human diseases (Lansbury, 1999). The conclusion is substantiated by the positive staining and green birefringence observed with Congo red, the fibrillar structure of the aggregates demonstrated with TEM, and the characteristic intermolecular β-sheet observed in the secondary structure (Podrabsky *et al.*, 2001). This occurrence of a nonpathological amyloid fibril is one of only a few described in nature.

3. Anoxia-induced quiescence in *Artemia* embryos

Another remarkable example of developmental and metabolic arrest is that observed in embryos of the brine shrimp *Artemia franciscana* under anoxia. These encysted, gastrula-stage embryos enter a reversible state of quiescence that can last for 4 years or more (Clegg, 1997; for reviews of quiescence see Hochachka and Guppy, 1987; Storey and Storey, 1990; Hand, 1991, 1998; Guppy and Withers, 1999; Hand and Hardewig, 1996; Hand and Podrabsky, 1999). Many catabolic and anabolic processes are depressed, which results in energy flows that are very low compared to aerobic values (Hand and Gnaiger, 1988; Hontoria *et al.*, 1993; Hand, 1995). The embryos may even approach an ametabolic state (Clegg, 1997). Considerable experimental evidence shows that the rapid and acute acidification of intracellular pH (pH 7.7–7.9 to 6.7) that occurs as these embryos enter anoxia (Busa *et al.*, 1982; Kwast *et al.*, 1995; Clegg *et al.*, 1995) plays a key role in the metabolic arrest. A depression of protein synthesis occurs during anoxia in *A. franciscana* embryos (Clegg and Jackson, 1989; Hofmann and Hand, 1990, 1994) that is mediated through translational control. However, the fate of transcription during anoxia-induced quiescence has only been addressed recently.

3.1 *Arrest of transcription and mRNA turnover*

Nuclear compartment. Hardewig *et al.* (1996) demonstrated that the ontogenetic increase in actin mRNA is prevented by exposure of *A. franciscana* embryos to anoxia and that the levels of this message are stable for several hours, as is also the case for total polyadenylated mRNA (Hofmann and Hand, 1992). One explanation for these patterns could be simultaneous arrest of nucleic acid synthesis and degradation, similar to that documented for protein synthesis and degradation in these embryos under anoxia (cf. Clegg and Jackson, 1989; Anchordoguy *et al.*, 1993; Anchordoguy and

Hand, 1994, 1995). Transcription only accounts for a modest portion of the cellular energy budget (typically 1–10% of basal metabolism in mammalian cells; Rolfe and Brown, 1997), but considering the large metabolic depression required to survive prolonged anoxia, even a 1% drain on cellular energy stores might not be sustainable.

We measured transcriptional activity in nuclei isolated from both normoxic and anoxic embryos, as well as those exposed to aerobic acidosis and to carbon monoxide under anoxia (van Breukelen *et al.*, 2000). Organellar run-on assays offer the advantage of providing a snapshot of the levels of *in vivo*-initiated transcripts at the moment of nuclear isolation (Lohr and Ide, 1983). Transcriptional activity in *A. franciscana* embryos during pre-emergence development was unchanged over 10 h at room temperature under normoxia. However, exposure of embryos to only 4 h of anoxia resulted in a 79.3 ± 1% decrease in levels of *in vivo*-initiated transcripts, and after 24 h of anoxia, transcription was depressed 88.2 ± 0.7 % compared to normoxic controls (mean ± SE, $n = 3$; *Figure 1*; van Breukelen *et al.*, 2000). Initiation was fully restored after 1 h of normoxic recovery.

Incubation of embryos under the artificial condition of aerobic acidosis (60% CO_2:40% O_2) is known to depress pH_i of embryos to 6.8 (Busa and Crowe, 1983) under fully aerobic conditions, a value comparable to that measured in embryos after 1 h of anoxia. Intracellular ATP concentration remains high and unchanged for several hours of aerobic acidosis, which serves to segregate this factor from pH_i (Carpenter and Hand, 1986; Anchordoguy and Hand, 1995). When nuclei were isolated from embryos given 4 h of normoxic incubation followed by an additional 4 h exposure to aerobic acidosis, initiation was depressed 55.2 ± 3.6% compared to the normoxic value (*Figure 2a*). Thus, over half of the transcriptional arrest under anoxia can be ascribed to acidification of pH_i. In addition to a direct influence on the transcriptional machinery, the proximal effect of pH acidification on initiation could involve the recently reported pH-dependent translocation of p26, a small heat shock protein, to the nuclear compartment of *A. franciscana* embryos (Clegg *et al.*, 1994; Liang *et al.*, 1997; Liang and MacRae, 1999). Under anoxia, approximately half of the total embryonic p26 is transported from the cytoplasm to the nucleus, and this process is reversed during aerobic recovery. Acidic pH favours the accumulation of p26 in the nucleus and re-alkalinization of embryos favours removal of p26 from the nucleus. It may well be that

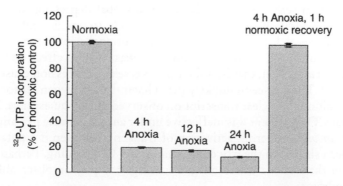

Figure 1. Arrest of transcriptional initiation due to exposure of A. franciscana embryos to anoxia. Embryos received a 4 h incubation under normoxia prior to anoxic bouts, and recovery was accomplished by returning embryos to normoxia. Transcriptional activity was measured in isolated nuclei. Values represent means ± SE, n=3 (after van Breukelen et al., 2000).

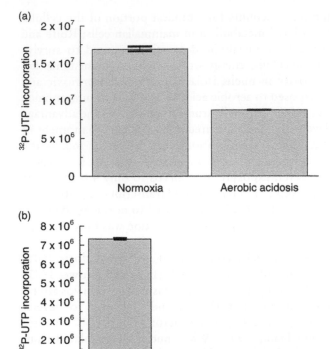

Figure 2. (a) *Depression of transcription by incubating* A. franciscana *embryos under the artificial condition of aerobic acidosis (60% CO_2:40% O_2) for 4 h, compared to embryos given 4 h of normoxia alone. Transcriptional activity was measured in isolated nuclei.* (b) *Exposure of embryos to anoxia (medium equilibrated with 100% N_2) compared to carbon monoxide under anoxia (medium equilibrated with 80% CO:20% N_2). All values represent means $\pm$ SE, n=3 (after van Breukelen et al., 2000).*

this protein is involved in some manner in protein stabilization or perhaps the transcriptional arrest under anoxia (Clegg *et al.*, 1995). Nevertheless, other factors are required to fully explain the depression of initiation seen *in vivo* during anoxia, and we also considered activity of haem-based oxygen sensors.

Ligation of the porphyrin haem with CO is sufficient to convert haemoproteins to the oxy conformation in the absence of oxygen (Goldberg *et al.*, 1988). Thus, if a haem-based oxygen sensor were involved in the global depression of transcription under anoxia, the addition of 80% carbon monoxide to anoxic embryos would be expected to reverse at least some of the transcriptional depression. However, the addition of carbon monoxide to anoxic embryos was without effect on transcription (van Breukelen *et al.*, 2000), that is, it did not reverse the depression caused by anoxia alone (*Figure 2b*). Thus, it seems unlikely that a haem-based oxygen sensor is pivotal in the global inhibition of nuclear transcription observed during quiescence. However, it is possible that CO treatment was ineffective under anoxia because pH_i and adenylates were too low *in vivo* to permit activation of transcription even if a putative oxygen sensor had been shifted to the oxy conformation by CO binding. Perhaps it is under hypoxia rather than anoxia where oxygen sensing may play an observable role in *A. franciscana* embryos.

When these transcriptional data are combined with the finding that mRNA levels are unchanged for at least 6 h of anoxia (Hofmann and Hand, 1992; Hardewig *et al.*, 1996), it is clear that mRNA half-life is extended at least 8.5-fold compared to the aerobic embryos embryo (where $t_{1/2}$ = [RNA]·ln 2·synthesis rate^{-1}). mRNA pools in the embryo would markedly decline if message stability were not increased simultaneously in the face of the transcriptional arrest during anoxia. One biological advantage of stable mRNA pools under anoxia would be the quick resumption of protein synthesis as soon as oxygen is returned to the embryo.

Mitochondrial compartment. The other subcellular location that deserves consideration in the context of transcriptional arrest is the mitochondrial matrix. Nuclear regulation of mitochondrial transcription primarily involves mitochondrial transcription factor A (mtTFA; Fischer and Clayton, 1988), a nuclear-encoded protein that stimulates polycistronic transcription of the mitochondrial genome *in vitro* (Shadel and Clayton, 1993). The promoter of the mtTFA gene is activated in turn by nuclear respiratory factor 1 (Virbasius and Scarpulla, 1994). Thus, regulating the availability of mtTFA to the mitochondrion would be sufficient in principle to control mitochondrial transcription, at least for slow cellular responses. However, we suggest such a system is not sufficiently responsive to explain the rapid and acute modulation of gene expression in *A. franciscana* embryos under anoxia (minutes to hours). Faithful transcription by isolated rat liver mitochondria for up to 6 h (Enríquez *et al.*, 1996) suggests that mtTFA is provisioned in large excess, which is inconsistent with speedy regulation via changes in mtTFA levels. Rapid cessation of gene expression when confronted with anoxia may be critical for preserving cellular energy stores in these embryos. Thus we suggest that local control (vs nuclear-mediated control) over mitochondrial transcription during anoxia may be quite important for rapid shifts in organellar gene expression.

For example, very large changes in ATP levels and intracellular pH occur in several minutes during oxygen limitation (Busa *et al.*, 1982; Kwast *et al.*, 1995; Clegg *et al.*, 1995; Anchordoguy and Hand, 1994), and these signals have a marked effect on mitochondrial transcription. It should be noted that this extramitochondrial acidification promotes a change in matrix pH of equal or greater magnitude (Kwast and Hand, 1996a,b). The influence of nucleotide concentrations on transcriptional rate is complex, in that interdependence is observed among some nucleotide species. Particularly noteworthy is the observation that the strong ATP inhibition observed at subsaturating concentrations of uridine 5'-triphosphate (UTP) is diminished when UTP concentration is increased to support maximal transcription rate. The pH-dependency of *in organello* transcription indicates pronounced proton sensitivity across the physiologically relevant range for these embryos. The extramitochondrial pH optimum is 7.9, which matches the *in vivo* pH$_i$ of embryos under normoxic conditions (7.7–7.9; Busa *et al.*, 1982; Kwast *et al.*, 1995). Transcription is inhibited over 80% by lowering extramitochondrial pH to 6.3, the lowest value measured for anoxic embryos *in vivo* (Busa *et al.*, 1982). Thus, it appears that pH is one factor that may serve to depress mitochondrial transcription during anoxia-induced quiescence. Unpublished observations (B. Eads and S. Hand) suggest that limitation of molecular oxygen may also serve as an agent for local control.

The depression of transcription and extension of mRNA half-lives in both the nucleo-cytoplasmic and mitochondrial compartments are concordant with limited energy availability. Transcriptional down-regulation is facilitated by decreases in pH$_i$ and ATP concentration that occur under anoxia, and their resulting influence on initiation and perhaps elongation. At this juncture available evidence does not indicate the involvement of a haemoprotein oxygen sensor that down-regulates nuclear transcription. In vertebrates the up-regulation of critical suites of genes serves a homeostatic role during anoxic bouts of limited duration (Semenza et al., 1994). In contrast, the global arrest seen in A. franciscana precludes normal metabolic functioning, but it extends the tolerance to anoxia dramatically, which may be more critical when the variable physical environment of these animals is considered.

3.2 Ubiquitin-dependent pathway for protein degradation

Because protein synthesis is severely depressed under anoxia in A. franciscana embryos (Clegg and Jackson, 1989; Hofmann and Hand, 1990, 1994; Kwast and Hand, 1993, 1996a,b), a continued degradation of proteins could conceivably deplete cellular stores of key regulatory enzymes. This depletion could jeopardize the embryo's ability to recover from its quiescent state upon return to oxygenated conditions. Thus, one might predict that protein degradation would be inhibited in the absence of oxygen.

While there are multiple proteolytic mechanisms available to most cells, it has been suggested that ubiquitin-mediated proteolysis (for review see Ciechanover and Schwartz, 1994; Goldberg, 1995; Mykles, 1998) is responsible for degrading virtually all regulatory proteins (Ciechanover et al., 1984; Rock et al., 1994). A basic feature of the ubiquitin-dependent proteolytic system is that proteins are tagged for degradation by covalently coupling them to ubiquitin. Measurement of ubiquitin-conjugated proteins demonstrated a sharp decline in level during entrance into anoxia-induced quiescence (Anchordoguy and Hand, 1994, 1995). Conjugate concentrations were only 7% of normoxic values following 24 h of anoxia. This result suggests that the conditions imposed by anoxia inhibit ubiquitin conjugation and probably proteolysis. However, until recently, depression of the rate of ubiquitin-dependent proteolysis per se had not been demonstrated for A. franciscana embryos under anoxia.

We developed an assay for monitoring ATP/ubiquitin-dependent proteolysis in order to establish the presence of this degradation mechanism in A. franciscana embryos and to describe characteristics that may regulate its function during anoxia-induced quiescence (van Breukelen and Hand, 2000). For lysates experimentally depleted of adenylates, supplementation with ATP and ubiquitin stimulated protein degradation rates by $92 \pm 17\%$ (mean $\pm$ SE) compared to control rates (Figure 3). The stimulation by ATP was maximal at a concentration of 11 μM. In the presence of ATP and ubiquitin, ubiquitin-conjugated proteins were produced by lysates during the course of the 4 h assays, as detected by Western blotting. Acute acidification of lysates to values approximating the intracellular pH observed under anoxia completely inhibited ATP/ubiquitin-dependent proteolysis (Figure 4a,b). Depressed degradation was also observed under conditions where net ATP hydrolysis occurred. These results suggest that ATP/ubiquitin-dependent proteolysis is markedly inhibited under cellular conditions promoted by anoxia. The process itself is energetically expensive, so down-regulating it in this manner would be advantageous when energy is restricted.

Inhibition of proteolysis during quiescence may be one critical factor that increases macromolecular stability, which may ultimately govern the duration of embryo survival under anoxia. For A. franciscana embryos, it seems that gene expression and macromolecular turnover are cellular costs incompatible with survival, and the need for continued development and metabolic function is secondary.

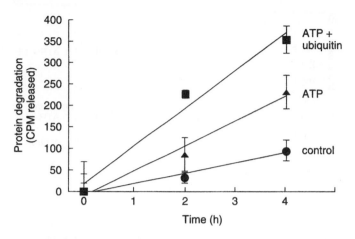

Figure 3. Stimulation of protein degradation in lysates of A. franciscana *embryos by ATP and ubiquitin. Circles represent reaction mixtures in the absence of ATP and without exogenous ubiquitin. Triangles represent reaction mixtures containing 1 mM ATP, which was held constant by an ATP regenerating system. Squares represent reactions with 1 mM ATP and 24 µM exogenous ubiquitin. All reactions contained approximately 5.8 mg ml*$^{-1}$ protein and were conducted at pH 7.9. Values represent means ± SE, n=3 (after van Breukelen and Hand, 2000).*

4. Summary

The ecological importance of resting stages has been emphasized by a number of authors (Clutter, 1978; Crowe and Clegg, 1973; Danks, 1987; Elgmork, 1980; Ellner *et al.*, 1999; Hairston, 1987; Hand, 1991; Hand and Hardewig, 1996; King, 1980; Lees, 1955; Pourriot and Snell, 1983; Tauber *et al.*, 1986). The advantages can be profound in the context of 'egg banks' in marine and freshwater habitats (Hairston *et al.*, 1996, 2000). Sediment cores from marine environments have uncovered viable copepod and rotifer embryos that have apparently withstood severe hypoxia/anoxia for 10–40 years; these forms are now thought to serve as egg banks for future generations (Marcus *et al.*, 1994). Reports on diapausing eggs from freshwater copepods indicate that even longer survival times are possible (Hairston *et al.*, 1995). Eventually we may gain a better appreciation of how certain stages in the life cycle of aquatic animals survive environmental conditions that destroy the vast majority of living organisms.

Hopefully it is clear from this brief chapter that only a small fraction of the cases of interrupted development in aquatic animals has been investigated at the physiological and biochemical levels. Consequently, current notions about the existence of pleiotropic mechanisms governing metabolic and developmental arrest during dormancy in these organisms are limited. Our understanding of processes that *regulate* the entry and breakage of diapause is still rudimentary. Only through a comparative approach can these unifying principles be discovered. Nevertheless, information already gained from studies of fish diapause and anoxia-induced quiescence has begun to provide a picture of the types of mechanisms operative. Without metabolic energy conservation, survival under such conditions would be greatly reduced. Results should have implications for a variety of biological systems that experience stress resulting from temperature extremes, desiccation, and oxygen deprivation.

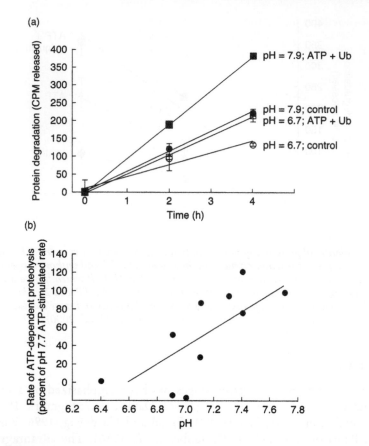

Figure 4. *Effects of pH on proteolysis. (a) Circles represent reaction mixtures in the absence of ATP and without exogenous ubiquitin. Squares represent reaction containing 1 mM ATP and 100 µg ml⁻¹ ubiquitin. Closed symbols represent reactions conducted at pH 7.9, and open symbols were at pH 6.7. Values represent means ± SE, n=3. (b) Effects of pH on ATP/ubiquitin-stimulated protein degradation. Data were pooled from three preparations and are expressed as a percentage of the rate observed at pH 7.7 (after van Breukelen and Hand, 2000).*

Acknowledgements

This work was supported by NSF grant IBN-9723746 to SCH.

References

Anchordoguy, T.J. and Hand, S.C. (1994) Acute blockage of the ubiquitin-mediated proteolytic pathway during invertebrate quiescence. *Am. J. Physiol.* **267**: R895–R900.

Anchordoguy, T.J. and Hand, S.C. (1995). Reactivation of ubiquitination in *Artemia franciscana* embryos during recovery from anoxia-induced quiescence. *J. Exp. Biol.* **198**: 1299–1305.

Anchordoguy, T.J., Hofmann, G.E. and Hand, S.C. (1993) Extension of enzyme half-life during quiescence in *Artemia* embryos. *Am. J. Physiol.* **264**: R85–R89.

Busa, W.B. and Crowe, J.H. (1983) Intracellular pH regulates transitions between dormancy and development of brine shrimp (*Artemia salina*) embryos. *Science* **221**: 366–368.

Busa, W.B., Crowe, J.H. and Matson, G.B. (1982) Intracellular pH and the metabolic status of dormant and developing Artemia embryos. Arch. Biochem. Biophys. 216: 711–718.

Buttgereit, F. and Brand, M.D. (1995) A hierarchy of ATP-consuming processes in mammalian cells. Biochem. J. 312: 163–167.

Buttgereit, F., Brand, M.D. and Muller, M. (1992) ConA induced changes in energy metabolism of rat thymocytes. Biosci. Rep. 12: 381–385.

Carpenter, J.F. and Hand, S.C. (1986) Arrestment of carbohydrate metabolism during anaerobic dormancy and aerobic acidosis in Artemia embryos: determination of pH-sensitive control points. J. Comp. Physiol. 156B: 451–459.

Chiti, F., Webster, P., Taddei, N., Clark, A., Stefani, M., Ramponi, G. and Dobson, C.M. (1999) Designing conditions for in vitro formation of amyloid protofilaments and fibrils. Proc. Natl Acad. Sci. USA 96: 3590–3594.

Ciechanover, A. and Schwartz, A.L. (1994) The ubiquitin-mediated proteolytic pathway: mechanisms of recognition of the proteolytic substrate and involvement in the degradation of native cellular proteins. FASEB J. 8: 182–191.

Ciechanover, A., Finley, D. and Varshavsky, A. (1984) Ubiquitin dependence of selective protein degradation demonstrated in the mammalian cell cycle mutant ts85. Cell 37: 57–66.

Clegg, J.S. (1997) Embryos of Artemia franciscana survive four years of continuous anoxia: the case for complete metabolic rate depression. J. Exp. Biol. 200: 467–475.

Clegg, J.S. and Jackson, S.A. (1989) Aspects of anaerobic metabolism of Artemia cysts. In: The Cellular and Molecular Biology of Artemia Development (eds A.H. Warner, J. Bagshaw and T. MacRae). Plenum Press, New York, pp. 1–15.

Clegg, J.S., Jackson, S.A. and Warner, A.H. (1994). Extensive intracellular translocations of a major protein accompany anoxia in embryos of Artemia franciscana. Exp. Cell Res. 212: 77–83.

Clegg, J.S., Jackson, S.A., Liang, P. and MacRae, T.H. (1995) Nuclear-cytoplasmic translocation of protein p26 during aerobic–anoxic transitions in embryos of Artemia franciscana. Exp. Cell Res. 219: 1–7.

Clegg, J.S., Drinkwater, L.E. and Sorgeloos, P. (1996) The metabolic status of diapause embryos of Artemia franciscana (SFB). Physiol. Zool. 69: 49–66.

Clutter, M.E. (ed) (1978) Dormancy and Developmental Arrest. Academic Press, New York.

Crowe, J.H. and Clegg, J.S. (1973) Anhydrobiosis. Dowden, Hutchinson and Ross, Stroudsburg, PA.

Crowe, J.H. and Clegg, J.S. (1978) Dry Biological Systems. Academic Press, New York.

Crowe, J.H., Carpenter, J.F. and Crowe, L.M. (1998) The role of vitrification in anhydrobiosis. Annu. Rev. Physiol. 60: 73–103.

Danks, H.V. (1987) Insect Dormancy: an Ecological Perspective. Biological Survey of Canada Monograph Series 1. Tyrell Press, Gloucester, Ontario.

Dong, A., Huang, P. and Caughey, W.S. (1990) Protein secondary structures in water from second-derivative amide I infrared spectra. Biochemistry 29: 3303–3308.

Drinkwater, L.E. and Clegg, J.S. (1991) Experimental biology of cyst diapause. In: Artemia Biology (eds R.A. Browne, P. Sorgeloos and C.N.A. Trotman). CRC Press, Boca Raton, FL, pp. 93–117.

Drinkwater, L.E. and Crowe, J.H. (1987) Regulation of embryonic diapause in Artemia: environmental and physiological signals. J. Exp. Zool. 241: 297–307.

Elgmork, K. (1980) Evolutionary aspects of diapause in fresh-water copepods. In: Evolution and Ecology of Zooplankton Communities (ed W.C. Kerfoot). University Press of New England, Hanover, pp. 411–417.

Ellner, S.P., Hairston, N.G., Kearns, C.M. and Babai, D. (1999) The roles of fluctuating selection and long-term diapause in microevolution of diapause timing in a freshwater copepod. Evolution 53: 111–122.

Enríquez, J., Fernández-Silva, P., Pérez-Martos, A., López-Pérez, M. and Montoya, J. (1996) The synthesis of mRNA in isolated mitochondria can be maintained for several hours and is inhibited by high levels of ATP. Eur. J. Biochem. 237: 601–610.

Fink, A.L. (1998) Protein aggregation: folding aggregates, inclusion bodies and amyloid. *Folding Des.* **3**: R9–R23.

Fischer, R.P. and Clayton, D.A. (1988) Purification and characterization of human mitochondrial transcription factor 1. *Mol. Cell. Biol.* **8**: 3496–3509.

Fishman, A.P., Galante, R.J., Winokur, A. and Pack, A.I. (1992) Estivation in the African Lungfish. *Proc. Am. Phil. Soc.* **136**: 61–72.

Fuery, C.J., Withers, P.C., Hobbs, A.A. and Guppy, M. (1998) The role of protein synthesis during metabolic depression in the Australian desert frog *Neobatrachus centralis*. *Comp. Biochem. Physiol.* **119**A: 469–476.

Goldberg, A.L. (1995) Functions of the proteasome: the lysis at the end of the tunnel. *Science* **268**: 522–524.

Goldberg, M.A., Dunning, S.P. and Bunn, H.F. (1988) Regulation of the erythropoietin gene: evidence that the oxygen sensor is a heme protein. *Science* **242**: 1412–1415.

Guppy, M. and Withers, P. (1999) Metabolic depression in animals: physiological perspectives and biochemical generalizations. *Biol. Rev.* **74**: 1–40.

Guppy, M., Fuery, C.J. and Flanigan, J.E. (1994) Biochemical principles of metabolic depression. *Comp. Biochem. Physiol.* **109**B: 175–189.

Hairston, N.G. Jr. (1987) Diapause as a predator-avoidance adaptation. In: *Predation. Direct and Indirect Impacts on Aquatic Communities* (eds W.C. Kerfoot and A. Sih). University Press of New England, Hanover, pp. 281–290.

Hairston, N.G. Jr., Van Brunt, R.A. and Kearns, C.M. (1995) Age and survivorship of diapausing eggs in a sediment egg bank. *Ecology* **76**: 1706–1711.

Hairston N.G., Kearns C.M. and Ellner, S.P. (1996) Zooplankton egg banks as biotic reservoirs in changing environments. *Limnol. Oceanogr.* **41**: 1087–1092.

Hairston N.G., Hansen A.M. and Schaffner, W.R. (2000) The effect of diapause emergence on the seasonal dynamics of a zooplankton assemblage. *Freshwater Biol.* **45**: 133–145.

Hamor, T. and Garside, E.T. (1977) Quantitative composition of the fertilized ovum and constituent parts in the Atlantic salmon *Salmo salar* L. *Can. J. Zool.* **55**: 1650–1655.

Hand, S.C. (1991) Metabolic dormancy in aquatic invertebrates. In: *Advances in Comparative and Environmental Physiology* (ed R. Gilles). Springer, Berlin, pp. 1–50.

Hand, S.C. (1995) Heat flow is measurable from *Artemia franciscana* embryos under anoxia. *J. Exp. Zool.* **273**: 445–449.

Hand, S.C. (1998) Quiescence in *Artemia franciscana* embryos: reversible arrest of metabolism and gene expression at low oxygen levels. *J. Exp. Biol.* **201**: 1233–1242.

Hand, S.C. (1999) Calorimetric approaches to animal physiology and bioenergetics. In: *Handbook of Thermal Analysis and Calorimetry, Vol. 4, Life Sciences* (ed R.B. Kemp). Elsevier Science, Amsterdam, pp. 469–510.

Hand, S.C. and Gnaiger, E. (1988) Anaerobic dormancy quantified in *Artemia* embryos: a calorimetric test of the control mechanism. *Science* **239**: 1425–1427.

Hand, S.C. and Hardewig, I. (1996) Downregulation of cellular metabolism during environmental stress: mechanisms and implications. *Annu. Rev. Physiol.* **58**: 539–563.

Hand, S.C. and Podrabsky, J.E. (1999) Bioenergetics of diapause and quiescence in aquatic animals. *Thermochim. Acta* **349**: 31–42.

Hardewig, I., Anchordoguy, T.J., Crawford, D.L. and Hand, S.C. (1996) Profiles of nuclear and mitochondrial encoded mRNAs in developing and quiescent embryos of *Artemia franciscana*. *Mol. Cell. Biochem.* **158**: 139–147.

Hardie D.G. and Carling, D. (1997) The AMP-activated protein kinase. Fuel gauge of the mammalian cell? *Eur. J. Biochem.* **246**: 259–271.

Hardie, D.G., Carling, D. and Carlson, M. (1998) The AMP-activated/SNF1 protein kinase subfamily: metabolic sensors of the eukaryotic cell? *Annu. Rev. Biochem.* **67**: 821–855.

Hochachka, P.W. and Guppy, M. (1987) *Metabolic Arrest and the Control of Biological Time*. Harvard University Press, Cambridge, MA.

Hofmann, G.E. and Hand, S.C. (1990) Arrest of cytochrome-c oxidase synthesis coordinated with catabolic arrest in dormant Artemia embryos. Am. J. Physiol. 25: R1184–R1191.

Hofmann, G.E. and Hand, S.C. (1992) Comparison of messenger RNA pools in active and dormant Artemia franciscana embryos: evidence for translational control. J. Exp. Biol. 164: 103–116.

Hofmann, G.E. and Hand, S.C. (1994) Global arrest of translation during invertebrate quiescence. Proc. Natl Acad. Sci. USA 91: 8492–8496.

Hontoria, F., Crowe, J.H., Crowe, L.M. and Amat, F. (1993) Metabolic heat production by Artemia embryos under anoxic conditions. J. Exp. Biol. 178: 149–159.

Joplin, K.H., Yocum, G.D. and Denlinger, D.L. (1990) Diapause specific proteins expressed by the brain during pupal diapause of the flesh fly, Sarcophaga crassipalpis. J. Insect Physiol. 36: 775–783.

King, C.E. (1980) The genetic structure of zooplankton populations. In: Evolution and Ecology of Zooplankton Communities (ed W.C. Kerfoot). University Press of New England, Hanover, pp. 315–328.

Kitajima, K., Inoue, Y. and Inoue, S. (1986) Polysialoglycoproteins of Salmonidae fish eggs. Complete structure of 200-kDa polysialoglycoprotein from the unfertilized eggs of rainbow trout (Salmo gairdneri). J. Biol. Chem. 261: 5262–5269.

Kwast, K.E. and Hand, S.C. (1993) Regulatory features of protein synthesis in isolated mitochondria from Artemia embryos. Am. J. Physiol. 265: R1238–R1246.

Kwast, K.E. and Hand, S.C. (1996a) Oxygen and pH regulation of protein synthesis in mitochondria from Artemia franciscana. Biochem. J. 313: 207–213.

Kwast, K.E. and Hand, S.C. (1996b). Acute depression of mitochondrial protein synthesis during anoxia. J. Biol. Chem. 271: 7313–7319.

Kwast, K.E., Shapiro, J.I., Rees, B.B. and Hand, S.C. (1995) Oxidative phosphorylation and the realkalinization of intracellular pH during recovery from anoxia in Artemia franciscana embryos. Biochim. Biophys. Acta 1232: 5–12.

Laale, H.W. (1980) The perivitelline space and egg envelopes of bony fishes: a review. Copeia 1980: 210–226.

Land, S.C., Buck, L.T. and Hochachka, P.W. (1993) Response of protein synthesis to anoxia and recovery in anoxia tolerant hepatocytes. Am. J. Physiol. 265: R41–R48.

Lansbury, P.T. Jr. (1999) Evolution of amyloid: what normal protein folding may tell us about fibrillogenesis and disease. Proc. Natl Acad. Sci. USA 96: 3342–3344.

Lees, A.D. (1955) Physiology of Diapause in Arthropods. Cambridge University Press, Cambridge.

Lefebvre, V.H.L., Van Steenbrugge, M., Beckers, V., Roberfroid, M, and Buc-Calderon, P. (1993) Adenine nucleotides and inhibition of protein synthesis in isolated hepatocytes incubated under different pO_2 levels. Arch. Biochem. Biophys. 304: 322–331.

Liang, P. and MacRae, T.H. (1999) The synthesis of a small heat shock/alpha-crystallin protein in Artemia and its relationship to stress tolerance during development. Devl. Biol. 207: 445–456.

Liang, P., Amons, R., Clegg, J.S. and MacRae, T.H. (1997) Molecular characterzation of a small heat shock/α-crystallin protein in encysted Artemia embryos. J. Biol. Chem. 272: 19051–19058.

Lohr, D. and Ide, G.I. (1983) In vitro initiation and termination of ribosomal RNA transcription in isolated yeast nuclei. J. Biol. Chem. 258: 4668–4671.

Marcus, N.H., Lutz, R., Burnett, W. and Cable, P. (1994) Age, viability, and vertical distribution of zooplankton resting eggs from an anoxic basin: evidence of an egg bank. Limnol. Oceanogr. 39: 154–158.

Mayhew, W.W. (1965) Adaptations of the amphibian, Scaphiopus couchi, to desert conditions. Am. Midland Nat. 74: 95–109.

Mosca J.D., Wu, J.M. and Suhadolnik, R.J. (1983) Restoration of protein synthesis in lysed rabbit reticulocytes by the enzymatic removal of adenosine 5′-monophosphate with either AMP deaminase or AMP nucleosidase. Biochemistry 22: 346–354.

Mykles, D. (1998) Intracellular proteinases of invertebrates: calcium-dependent and proteasome/ubiquitin-dependent systems. Int. Rev. Cytol. 184: 157–289.

Pannevis, M.C. and Houlihan, D.F. (1992) The energetic cost of protein synthesis in isolated hepatocytes of rainbow trout (*Oncorhynchus mykiss*). *J. Comp. Physiol.* **162B**: 393–400.

Podrabsky, J.E. (1999) Husbandry of the annual killifish *Austrofundulus limnaeus* with special emphasis on collection and rearing of embryos. *Environ. Biol. Fish.* **54**: 421–431.

Podrabsky, J.E. and Hand, S.C. (1999) The bioenergetics of embryonic diapause in an annual killifish, *Austrofundulus limnaeus*. *J. Exp. Biol.* **202**: 2567–2580.

Podrabsky, J.E. and Hand, S.C.(2000) Depression of protein synthesis during diapause in embryos of the annual killifish, *Austrofundulus limnaeus*. *Physiol. Biochem. Zool.* **73**: 799–808.

Podrabsky, J., Carpenter, J.F. and Hand, S.C. (2001) Survival of water stress in annual fish embryos: dehydration avoidance and egg envelope amyloid fibers. *Am. J. Physiol.* **280**: R123–R131.

Pourriot, R. and Snell, T.W. (1983) Resting eggs in rotifers. *Hydrobiologia* **104**: 213–224.

Rock, K.L., Gramm, C., Rothstein, L., Clark, K., Stein, R., Dick, L., Hwang, D. and Goldberg, A.L. (1994) Inhibitors of the proteasome block the degradation of most cell proteins and the generation of peptides presented on MHC Class I molecules. *Cell* **78**: 761–771.

Rolfe, D.F.S. and Brown, G.C. (1997) Cellular energy utilization and molecular origin of standard metabolic rate in mammals. *Physiol. Rev.* **77**: 731–758.

Scanlon, B.R. (1994) Water and heat fluxes in desert soils 1. Field studies. *Water Resour. Res.* **30**: 709–719.

Semenza, G.L., Roth, P.H., Fang, H.M. and Wang, G.L.(1994) Transcriptional regulation of genes encoding glycolytic enzymes by hypoxia-inducible factor 1. *J. Biol. Chem.* **269**: 23757–23763.

Shadel, G.S. and Clayton, D.A. (1993) Mitochondrial transcription initiation: variation and conservation. *J. Biol. Chem.* **268**: 16083–16086.

Smith, R.W. and Houlihan, D.F. (1995) Protein synthesis and oxygen consumption in fish cells. *J. Comp. Physiol.* **165B**: 93–101.

Storey, K.B. and Storey, J.M. (1990) Metabolic rate depression and biochemical adaptation in anaerobiosis, hibernation, and estivation. *Q. Rev. Biol.* **65**: 145–174.

Tauber, M.J., Tauber, C.A. and Masaki, S. (1986) *Seasonal Adaptations of Insects*. Oxford University Press, New York.

van Breukelen, F. and Hand, S.C. (2000) Characterization of ATP-dependent proteolysis in embryos of the brine shrimp, *Artemia franciscana*. *J. Comp. Physiol.* **170B**: 125–133.

van Breukelen, F., Maier, R. and Hand, S.C. (2000) Depression of nuclear transcription and extension of mRNA half-life under anoxia in *Artemia franciscana* embryos. *J. Exp. Biol.* **203**: 1123–1130.

Virbasius, J.V., and Scarpulla, R.C. (1994) Activation of the human mitochondrial transcription factor A gene by nuclear regulatory factors: a potential regulatory link between nuclear and mitochondrial gene expression in organelle biogenesis. *Proc. Natl Acad. Sci. USA* **91**: 1309–1313.

Wourms, J.P. (1972a) The developmental biology of annual fishes I. Stages in the normal development of *Austrofundulus myersi* Dahl. *J. Exp. Zool.* **182**: 143–168.

Wourms, J.P. (1972b) The developmental biology of annual fishes III. Pre-embryonic and embryonic diapause of variable duration in the eggs of annual fishes. *J. Exp. Zool.* **182**: 389–414.

12

Interrupted development: the impact of temperature on insect diapause

David L. Denlinger

1. Introduction

The low temperature of winter is probably the ultimate factor driving the evolution of insect diapause. Although the temperatures themselves may be survivable, the seasonal disappearance of host plants and other food resources necessitate a mechanism for interrupting development. Diapause in some form or other is the norm for insects. Relatively few species in the temperate zones lack the ability to enter an overwintering diapause. This interruption of development can occur in any developmental stage, but the stage is specific for each species. Thus, the silkworm always diapauses as an embryo, the European corn borer as a larva, the flesh fly as a pupa, and the Colorado potato beetle as an adult. A few species with long life cycles, more common at high latitudes, may diapause in two or more successive stages of the life cycle. Regardless of the developmental stage of diapause, this interruption is characterized by a halt in development and a suppression of metabolic activity. Some species will enter diapause at a certain developmental stage regardless of the environmental conditions (obligate diapause), but for most species the decision to enter diapause is dependent on cues received from the environment (facultative diapause). The cues used to programme diapause are usually received far in advance of the actual diapause stage, thus offering a period for physiological preparation. Like mammalian hibernators, insects destined for diapause commonly sequester an abundance of high-energy reserves prior to the onset of diapause.

Several good reviews present an ecological perspective on diapause (Tauber et al., 1986; Danks, 1987). The involvement of photoperiodic clocks in regulation of diapause is reviewed by Saunders (1982) and Takeda and Skopik (1997), and the hormonal aspects of diapause regulation are summarized in Denlinger (1985). In this review I highlight the role of temperature in insect diapause and briefly discuss its role in regulating the timing of diapause and in influencing the physiological adaptations that are essential for successfully interrupting development. Many of the examples will be drawn from experiments with flesh flies (*Sarcophaga* sp.), the model system used

Environment and Animal Development: Genes, Life Histories and Plasticity, edited by D. Atkinson and M. Thorndyke.
© 2001 BIOS Scientific Publishers Ltd, Oxford.

most extensively in my laboratory. Flies in this genus rely exclusively on a pupal diapause for overwintering, and even some of the tropical species enter a pupal diapause during the cooler seasons of the year.

2. Contribution of temperature to diapause induction, maintenance and termination

Temperature itself is a rather poor seasonal indicator. Temperature variations within the day, as well as day-to-day variations, present an unreliable seasonal cue. It is thus not surprising that the primary environmental cue used for diapause induction is photoperiod, not temperature. However, this does not mean that temperature is not important. Although low temperature alone will not induce diapause in the temperate species (Denlinger, 1972a), it clearly augments the effect of photoperiod. For populations of *S. bullata* and *S. crassipalpis* from 40°N day lengths shorter than approximately 13.5 h of light/day promote the induction of diapause when these signals are received during the photosensitive period, the last two days of embryonic life and the first two days of larval life (Denlinger, 1971, 1972a). However, under conditions of short day length at 25°C only 85% of the pupae enter diapause. Lowering the temperature to 18°C increases the incidence of diapause to nearly 100%. Quite commonly, low temperature will increase the diapause-promoting effect of short days (*Figure 1*). In the *Sarcophaga* example the critical day length is not altered by temperature (*Figure 1a*), but in some species, for example *Pieris brassicae*, lowering the temperature also lengthens the critical day length, thus implying that the insect will enter diapause at a longer day length if the temperature is low (*Figure 1b*). This low temperature enhancement of diapause expression appears to be the consequence of slowing the rate of development (*Figure 2*). The fly appears to count the number of short days it has

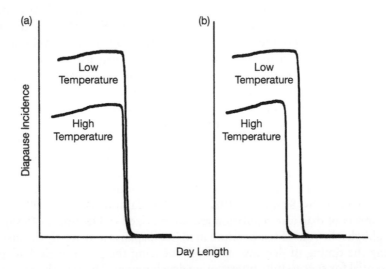

Figure 1. *Low temperature can impact the photoperiodic response curve in two ways. In some cases (a), the critical photoperiod remains the same but the incidence of diapause expressed at short day lengths is higher. In other instances (b) the critical photoperiod is also shifted toward longer day lengths at low temperature.*

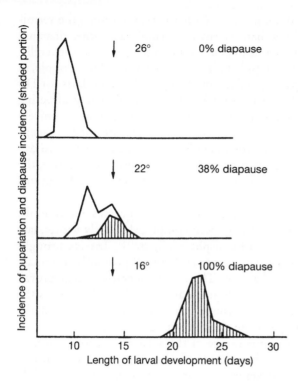

Incidence of pupariation and diapause incidence (shaded portion)

26°   0% diapause

22°   38% diapause

16°   100% diapause

10   15   20   25   30

Length of larval development (days)

Figure 2. *The effect of temperature on the time of puparium formation (pupariation) and incidence of pupal diapause induction in the flesh fly* Sarcophaga argyrostoma. *Arrow indicates 14 days, the number of short days required for diapause induction. At lower temperatures development is slower, thus more individuals are exposed to the 14 short days required to enter pupal diapause. Reproduced from Kerkut and Gilbert, eds (1985)* Comprehensive Insect Physiology Biochemistry and Pharmacology, *Vol 8. With permission of Elsevier Science.*

received, and the higher the number of short days, the greater the incidence of diapause (Saunders, 1971). The slower development rate at low temperature thus increases the number of short days the fly will receive during its photosensitive period and thereby boosts the incidence of diapause.

Constant temperatures, the conditions commonly used in laboratory experiments, are, of course, not relevant ecologically. Night temperatures are usually considerably cooler than temperatures during the day. Coupling a low temperature with the scotophase and a high temperature with the photophase usually results in a higher diapause incidence than observed when the insect is exposed to the mean of the two extremes (Beck, 1991). Thermoperiod can sometimes even substitute for photoperiod as a meaningful signal for diapause induction if the insect is held under conditions of constant darkness (Saunders, 1973).

In some geographic regions, temperature cues may completely supplant photoperiodic cues as the environmental signal for diapause induction. Such is the case for tropical flesh flies living near the equator. In Nairobi, Kenya, 1° south of the equator, only a 7 min difference distinguishes the longest day from the shortest day of the year. Yet, flies from this region still enter a pupal diapause, in this case during July and August, the coolest portion of the year (Denlinger, 1974). In this environment the flies are not influenced by photoperiod, but low temperature exposure during larval development can shunt the pupae toward diapause. Diapause appears to be widespread in tropical species, but within 5°N or S of the equator other cues, such as temperature, commonly replace day length as the environmental token programming the insect for diapause (Denlinger, 1986).

Temperature also plays an important role in determining the duration of diapause. The physiological and biochemical events that culminate in the termination of diapause are referred to as diapause development. With a few exceptions, the termination of diapause is not triggered by a distinct environmental cue, rather diapause development proceeds at a certain rate that is dependent upon temperature. Generally, the rate of diapause development is more rapid at higher temperatures. For example, in the flesh fly, *S. crassipalpis*, pupal diapause persists for 70 days at 25°C but lasts for 118 days at 17°C (Denlinger, 1972a). This sort of temperature dependency affecting diapause duration is fairly common among insects from a wide species range, as nicely documented by Tauber *et al.* (1986). There are, of course, a number of notable exceptions. The giant silkmoth, *Antheraea pernyi*, for example, does rely on a change in day length to terminate its pupal diapause (Williams and Adkisson, 1964). In this case, the long day lengths of spring are detected by the pupal brain and trigger the hormonal events culminating in the onset of development.

Temperature plays another critical role in the post-diapause processes. Commonly, the events of diapause development have actually been completed in early winter, late December or early January in the northern hemisphere (*Figure 3*). Insects brought from the field into warm conditions in the laboratory at that time promptly terminate diapause, but in the field this does not happen. The low temperatures that prevail outside at that time of year are below the developmental threshold and suppress the progression of development. The oxygen consumption rate remains low, and the insects are morphologically indistinguishable from their diapausing counterparts. Then, in response to the elevation of temperature in the spring the progression of development is initiated, but progress is slow due to the low temperatures that still prevail. Thus, the latency period that characterizes overwintering diapauses is mainly needed to bridge the interval from late summer or early autumn until the cold of winter sets in. At that point, the insect is actually ready to develop but the low temperature prevents development from ensuing. In the flesh fly example, *S. bullata* from 40°N enters pupal diapause in late August to early October. The latency period inherent to diapause prevents diapause termination until January. At that time,

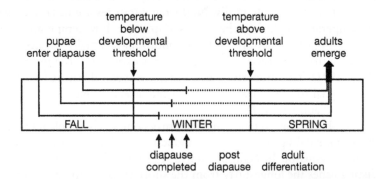

Figure 3. *The role of low winter temperatures in synchronizing post diapause development in the spring. In this schematic diagram, based on the flesh fly* Sarcophaga bullata *(Denlinger, 1972a, b), pupae enter diapause over a period of several months in the fall. The latency inherent to diapause prevents further development during the remainder of the fall. Pupae will complete the processes of diapause development at different times during the winter, but they then remain locked into a post-diapause period by the low temeratures that prevail during winter and then simultaneously initiate adult differentiation when higher temperatures return in the spring. Thus, pupae that enter diapause at quite different times in the fall can complete development and emerge as adults at the same time in the spring.*

diapausing pupae held outside will promptly terminate diapause when brought inside, but under outside conditions no signs of adult differentiation can be noted until mid-April. This interval is commonly referred to as the post-diapause period. When the springtime temperatures rise above the developmental threshold, adult differentiation is initiated. Accumulation of sufficient heat needed for the completion of this process is not received until mid-May. Remarkably, flies that enter pupal diapause over a period of a month or more in the autumn will emerge within a day or so in mid-May (Denlinger, 1972b). Diapause thus serves another very significant function as a synchronizer of the life cycle. Flies that have entered diapause at very different times also complete diapause development at different times, but the low temperature of winter serves to synchronize the population and thus allows them all to progress at the same rate in response to warm temperatures in spring.

In summary, temperature commonly serves to augment photoperiodic cues to regulate the induction of diapause, although in some cases low temperature alone can provide the environmental trigger for diapause. Once diapause has been entered, temperature is the main factor regulating the rate of diapause development, the events that culminate in diapause termination. Even after diapause development has been completed, the low temperatures of winter suppress the onset of post-diapause development until more permissive temperatures prevail in the spring.

3. Relationship between diapause and cold-hardiness

Insects that overwinter in diapause are invariably cold hardy, but what is the relationship between these two events? Simply being developmentally arrested does not imply cold hardiness. Some species are cold hardy without being in diapause, and others, especially those that enter a summer diapause or tropical diapause, may be in diapause without being cold hardy. I envision a relationship between diapause and cold hardiness as shown in *Figure 4* (Denlinger, 1991).

Cold hardiness without diapause can be seen in some species such as the mealworm, *Tenebrio molitor* (Patterson and Duman, 1978), and to a limited extent in non-diapausing stages of species that do have a diapause, for example in non-diapausing flesh flies that have been reared at low temperatures (Chen *et al.*, 1987b). In addition,

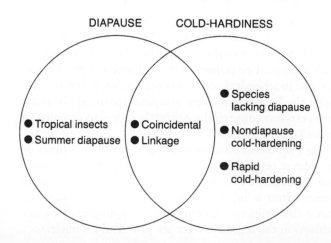

Figure 4. Relationship between diapause and cold hardiness. The two events may be expressed independently or in association with each other. When associated, the relationship may be coincidental or linked. Adapted from Denlinger (1991).

we have demonstrated that insects in diverse developmental stages can undergo rapid cold hardening, a form of cold hardening that can be attained within minutes of exposure to moderately low temperatures (Chen *et al.*, 1987a; Lee *et al.*, 1987). For example, the flesh fly, *S. crassipalpis*, cannot survive direct exposure to –10°C, but if it first is exposed to 0°C for 10 min or more it readily survives the –10°C exposure. A very rapid generation of the cryoprotectant glycerol is at least one of the physiological mechanisms associated with rapid cold hardening. The speed of this response suggests that insects can very quickly track changes in temperature and respond accordingly, even if they are not in diapause.

The most impressive adaptations to low temperature, however, are noted in diapausing individuals. Usually, it is only in this state that insects can survive for a long time at temperatures approaching their supercooling point (see Bale and Walters, Chapter 18). Supercooling points in the range of –20 to –30°C are not at all uncommon for diapausing individuals (Lee, 1991). The majority of species, those that are freeze-intolerant, can survive down to the supercooling point during diapause but not below. A few species, the freeze-tolerant ones, can actually survive at temperatures below the supercooling point. Such species usually have much higher supercooling points, and thus freeze solid at temperatures just a few degrees below 0°C.

As depicted in *Figure 4*, the association between diapause and cold hardiness can be either a direct linkage or a coincidental relationship. Direct linkage is exemplified by the pupal diapause of flesh flies (Adedokun and Denlinger, 1984). In this example, the entry into diapause is consistently associated with an increase in cold hardiness. The two cannot be separated. If the fly is in diapause, it will be cold hardy. In contrast, diapause and the expression of cold hardiness can be separated in the European corn borer, *Ostrinia nubilalis* (Hanec and Beck, 1960) and a number of other species. In the example of the European corn borer, the fifth instar larva enters diapause in early autumn, but at this time of the year it is not yet cold hardy. The environmental stimulation elicited by exposure to low temperatures is required for the larva to finally achieve cold hardiness. Thus, in this case two distinct environmental cues operate to regulate these two events. First, the larva responds to short day lengths by entering diapause, and a bit later, it becomes cold hardy in response to the onset of low temperatures.

4. Physiological responses to low temperature during diapause

A suite of responses contribute to protection of insects from injury at low temperatures during diapause (reviewed by Leather *et al.*, 1993; Lee, 1991; Denlinger and Lee, 1998). In brief, these responses include behavioural activities that direct the insect to a protected microenvironment, physiological responses that minimize the presence of ice nucleators in the insect body, and biochemical responses that generate cyroprotectants or other agents that protect the body against low temperature injury. In many cases it is not known whether the responses observed are directly linked to diapause or to the low temperatures that prevail during winter.

The first line of defence, seeking a protected hibernaculum, presumably is a function of entering diapause rather than a direct response to low temperature. This behavioural response is a major component of the diapause syndrome and is often the first indication of impending diapause. Movement into a hibernaculum is normally preceded by cessation of feeding, an event that is also frequently accompanied by a purging of the gut. Food and associated microorganisms in the digestive tract are powerful ice nucleators, and voiding these materials enhances supercooling (Cannon and Block, 1988; Lee *et al.*, 1998). A reduction in body water is also frequently associated with overwintering (Zachariassen, 1991; Danks, 2000). Less body water implies less risk of ice formation and also results in higher concentrations of cryoprotective agents as the solute concentration declines.

One of the most common adaptations to low temperature is the accumulation of low molecular weight polyols, sugars and amino acids. Glycerol and sorbitol are the polyols most widely used as cryoprotectants, trehalose is the most common sugar, and alanine and proline are the amino acids most commonly elevated. The accumulation of these cryoprotectants enhances cold tolerance in several ways. Species that cannot tolerate freezing use the cryoprotectants to lower the supercooling point and thus avoid freezing (Duman *et al.*, 1995). In freeze-tolerant species the cryoprotectants cause a colligative depression of the melting point. This, in turn, reduces the amount of ice that can form at a certain sub-zero temperature and consequently reduces cellular dehydration (Karow, 1991). Although diapause itself may prompt the synthesis of these cryoprotective agents, the response is frequently temperature dependent: progressively higher concentrations of the cryprotectants are generated as temperatures drop.

Several haemolymph proteins also play key roles in cold hardening (Duman *et al.*, 1991). Antifreeze proteins, also known as thermal hysteresis proteins, were first discovered in cold water fish but now have been isolated in several beetles and a few other insect species (Walker *et al.*, 2001). These proteins, molecular mass of 14–20 kDa, depress the freezing point of water by a non-colligative mechanism but leave the melting point unchanged. This difference between the freezing point and melting point is referred to as thermal hysteresis. In the presence of thermal hysteresis proteins the freezing point may be lowered 5–6°C below the melting point, thus enabling the insect to increase its supercooling capacity. It appears to achieve this effect by masking ice nucleators present in the haemolymph. Synthesis of the thermal hysteresis proteins is seasonal. Both short day length and low temperature promote the synthesis of these proteins by the fat body in the autumn. The proteins persist in the haemolymph throughout the winter months and disappear in response to long day lengths in the spring.

Ice nucleator proteins represent another class of proteins that contribute to cold hardening in some species. Unlike the thermal hysteresis proteins, the ice nucleator proteins function to facilitate the organization of water molecules into embryonic crystals which, in turn, seed the supercooled solution and promote freezing. Such proteins, large lipoproteins described from crane flies (Neven *et al.*, 1989) and a few other freeze-tolerant species (Duman *et al.*, 1995), promote freezing at relatively high sub-zero temperatures. Initiating ice nucleation at these rather high temperatures presumably restricts ice formation to extracellular spaces, thus reducing injury to the cells. The rate of water movement out of the cells is also decreased, thus reducing the potential of osmotic shock.

Proteins are not the only ice nucleators utilized by freeze tolerant insects. A recent study of the gall fly, *Eurosta solidaginis*, suggests that spherules of calcium phosphate located in the Malpighian tubules serve as ice nucleation sites to promote whole body freezing in overwintering larvae of this species (Mugnano *et al.*, 1996).

Thus, a number of adaptations contribute to low-temperature survival during diapause. While some of these responses are clearly promoted by low temperature regardless of the insect's developmental status, others are likely to be developmentally regulated events directly linked to the diapause programme.

5. Diapause-regulated genes

The distinct attributes of diapause already suggest that the expression of different sets of genes may distinguish diapause and non-diapause development. In addition to different patterns of gene expression associated with different mechanisms of cold hardiness, we might predict differences in the expression of genes associated with energy metabolism and structural modifications distinct to diapause (e.g. extra cuticular wax deposition to enhance waterproofing, etc.). However, of greater potential interest are regulatory proteins, those that are actually involved in evoking, maintaining and terminating the diapause state. In seeking such proteins we have focused on the brain. Brain transplantations and other experimental approaches clearly point to the brain as the center regulating the diapause response (Williams, 1946; Bowen *et al.*, 1984; Denlinger, 1985; Giebultowicz and Denlinger, 1986). Our first approach was to compare the proteins being synthesized in brains of diapausing and non-diapausing flesh fly pupae (Joplin *et al.*, 1990). With pulse labelling and two-dimensional electrophoresis we were able to resolve over 300 proteins from brains of non-diapausing pupae, while only 180 proteins were observed in gels of diapausing brains. This result suggested that, indeed, fewer genes are being expressed in the brain of a diapausing pupa, but of special interest was a group of approximately 14 well-labelled proteins present in the diapausing brain but not in their non-diapausing counterparts. The discovery that unique proteins are synthesized in the brain during diapause implies that diapause is characterized not merely by a shut-down in gene expression but also by the expression of a novel set of genes.

To search for the identity of diapause-specific genes we used subtractive hybridization to isolate several cDNAs from brains of fly pupae that were either up-regulated or down-regulated during diapause (Flannagan *et al.*, 1998). This search yielded four diapause up-regulated clones and seven clones that were down-regulated during diapause. Several of these clones have now been examined in some detail.

5.1 *Diapause up-regulated genes in the brain*

One of the most prominent diapause up-regulated clones encodes for a 23 kDa small heat shock protein (Yocum *et al.*, 1998). The transcript is not evident prior to diapause but is highly up-regulated at the onset of diapause and transcription persists throughout the duration of diapause (*Figure 5*). Then, at diapause termination,

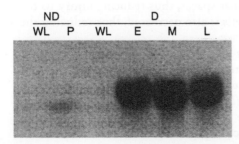

Figure 5. Up-regulation of the transcript for a small heat shock protein (hsp23) during pupal diapause (D) in the flesh fly Sarcophaga crassipalpis. During the wandering larval (WL) and pupal (P) stage of flies not programmed for diapause (ND), expression at 20°C is low, but the gene is highly up-regulated throughout early (E), mid (M), and late (L) diapause. Reproduced from Yocum et al. (1998) Insect. Biochem. Mol. Biol. 28: 667–682, with permission of Elsevier Science.

expression declines quickly. Although this transcript is highly expressed in non-diapausing individuals in response to heat shock or cold shock, no such environmental stresses are required for expression in diapausing pupae. The gene is expressed simply in association with the entry into diapause. In the diapausing pupae, high or low temperature exposures do not cause a higher level of expression. A very similar response is evident for the dominant heat shock protein, hsp70. As noted for hsp23, the transcript encoding the 70 kDa heat shock protein is highly up-regulated at the onset of diapause, is expressed throughout diapause, and then promptly declines in expression within hours after diapause is terminated (Rinehart *et al.*, 2000). To see expression of stress protein genes for such a prolonged period is quite unusual. In most instances the stress proteins are expressed quickly in response to a stress and then, just as quickly, disappear when the stress has been removed. Prolonged expression is known to reduce cell growth and impair the progression of development (Feder *et al.*, 1992; Krebs and Feder, 1997). Yet, in this case, the genes encoding these stress proteins are apparently up-regulated for months! Why this is so remains unclear. A protective role against low temperature injury is certainly a possibility, but the stress proteins could also be involved in invoking and maintaining the cell cycle arrest associated with diapause. Several experiments have correlated the expression of heat shock proteins and cell cycle arrest (Ireland and Berger, 1982; Cheney and Shearn, 1983). The expression of these genes during diapause may represent one of the controlling elements assuring that development will remain arrested.

5.2 *Diapause down-regulated genes in the brain*

Diapause down-regulated genes are potentially just as important as those that are diapause up-regulated. The first down-regulated gene that we isolated from flesh flies was proliferating cell nuclear antigen (*PCNA*), a cell cycle regulator (Tammariello and Denlinger, 1998). During diapause, the cells of the brain are locked into a G_0/G_1 cell cycle arrest (*Figure 6*) that correlates with the down-regulation of *PCNA*. *PCNA* is highly expressed after diapause is terminated but not during diapause. By contrast, other cell cycle regulators that we have examined, *cyclin E, p21, p53*, are expressed equally in diapausing and non-diapausing flies. These results suggest the possibility that down-regulation of *PCNA* may be critical for the cell cycle arrest associated with diapause. The search is currently underway for other genes that may be expressed differently in the brains of diapausing and non-diapausing flesh fly pupae.

Another approach we have pursued in our search for diapause-related gene expression is to investigate the expression of genes known to be associated with the hormones that regulate diapause. In flesh flies, the hormonal basis for diapause is an ecdysteroid deficiency. The brain fails to release prothoracicotropic hormone, the neuropeptide that stimulates the prothoracic gland to produce the ecdysteroids needed to promote development. In the absence of ecdysteroids the pupa is locked into diapause. Ecdysteroids exert their effect on target tissues by binding to two receptors, ecdysone receptor (EcR) and ultraspiracle (USP; Henrich and Brown, 1995). Since these receptors are directly linked to the action of ecdysteroid, the critical hormone regulating diapause, we used probes for these genes to examine their expression patterns in the brain in association with diapause (Rinehart *et al.*, 2001). As the fly entered pupal diapause EcR and USP were both strongly expressed in the brain,

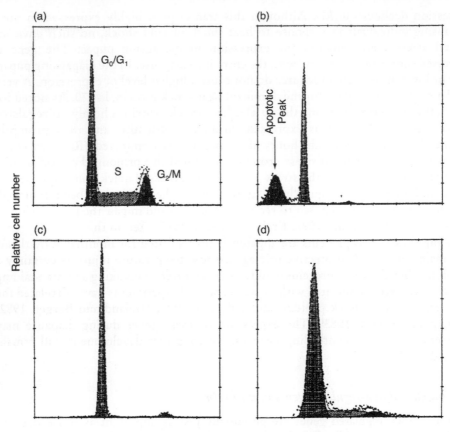

Figure 6. *A G_0/G_1 cell cycle arrest in the brain of the flesh fly* Sarcophaga crassipalpis *during pupal diapause. Representative flow cytometric diagrams showing percentage of cells in G_0/G_1, S and G_2/M from (a) prediapausing third instar larvae, (b) 10 day diapausing pupae, (c) 20-day diapausing pupae, and (d) post-diapause pharate adults three days after diapause termination. Reproduced from Tammariello and Denlinger (1998)* Insect. Biochem. Mol. Biol. **28**: 83–89, *with permission of Elsevier Science.*

but thereafter the pattern differed. Expression of EcR persisted throughout diapause with little change and increased further when diapause ended. Expression of USP, by contrast, declined at the beginning of diapause, remained low for several months and then increased as the time of diapause termination approached. We suspect that this return of USP near the end of diapause may be a critical step permitting the pupa to become responsive to ecdysteroids.

Diapause-regulated proteins are also evident in the brain of the gypsy moth, *Lymantria dispar* (Lee *et al.*, 1998). This species completes its embryonic development (pharate first instar larva) and then enters an overwintering diapause before hatching out of the egg. One conspicuous 45 kDa brain protein declines at the onset of diapause, remains low throughout diapause, and then is again highly expressed as the end of diapause is approached. Partial amino acid sequence determination suggested that the protein is actin. This pattern of expression appears to be unique to the brain, and its reappearance is most likely needed for the progression of development.

5.3 *Diapause-specific expression in non-neural tissues*

Tissues other than the brain may also yield gene products unique to diapause. Although proteins expressed in tissues other than the brain are less likely to exert a regulatory action, such proteins may also play roles critical to the success of diapause. Several such proteins have been described. In the gypsy moth, several gut proteins have distinct expression patterns in relation to diapause (Lee and Denlinger, 1996). Most notable is a 55 kDa protein that is specific to the mid and hind gut during diapause. This 55 kDa gut protein is highly expressed in early diapause and then gradually declines in expression as diapause progresses. The identity of the 55 kDa gut protein is still unknown, therefore it is premature to speculate on function.

In the embryonic diapause of the silkworm, *Bombyx mori*, a gene encoding for a member of the ETS family of proteins is expressed beginning approximately 20 days after oviposition and continuing for an additional 20 days (Suzuki *et al.*, 1999). When diapause is artifically terminated, the gene is quickly down-regulated. Earlier experiments with the silkworm reported that a single recessive mutation of a gene on chromosome 11 (*pnd*) is involved in diapause determination (Yamamoto *et al.*, 1978). Mutant homozygous embryos never enter diapause. Suzuki *et al.* (1999) suggest that the ETS-related gene they have isolated probably operates downstream from *pnd*, but the sequence of the *pnd* gene is unknown, thus the two genes may prove to be the same.

One of the most conspicuous distinctions is evident in the pattern of protein storage and utilization. Storage proteins, known as diapause-associated proteins, are particularly abundant in the haemolymph and fat body of many diapausing Lepidoptera (e.g. Brown and Chippendale, 1978; Brown, 1980; Salama and Miller, 1992; Palli *et al.*, 1998) and Coleoptera (Koopmanschap *et al.*, 1992). These proteins, produced by the fat body, usually become abundant in the haemolymph shortly before the onset of diapause. They persist in the haemolymph throughout diapause and then are utilized in the differentiation of adult tissues when diapause is terminated. In non-diapausing individuals these same proteins are also evident The difference is simply that during diapause these storage proteins persist in the haemolymph for a long time before being used, while they are utilized immediately for development in the non-diapausing individuals. Thus, there is little evidence to suggest that the diapause-associated proteins are actually made or utilized during diapause, and the gene expression associated with synthesis of such proteins is actually a pre-diapause event, as demonstrated in the spruce budworm, *Choristoneura fumiferana* (Palli *et al.*, 1998).

6. Role of low temperature in elevating the expression of select genes during diapause

Several genes are not up-regulated by diapause but are up-regulated during diapause in response to low temperature. The transcript encoding the cognate form of the 70 kDa heat shock protein, hsc70, is of this type (Rinehart *et al.*, 2000). It is not up-regulated when the flesh fly pupa enters diapause but is quickly up-regulated when the diapausing pupa is exposed to low temperature. Interestingly, high temperature does not elicit the response, only low temperature. The response of yet another stress protein, hsp90, appears to be similar to the response of hsc70. The transcript encoding hsp90 is not up-regulated by diapause, but low temperature exposure will elicit expression of this transcript in diapausing pupae (Rinehart and Denlinger, 2000). Thus, the stress proteins in flesh flies are involved with diapause in two distinct manners. Expression of both hsp23 and 70 are developmentally up-regulated by entry into diapause, while expression of hsc70 and hsp90 is up-regulated in response to low temperature during diapause.

In the gypsy moth, a period of chilling is essential for termination of the pharate first instar larval diapause. Eggs that are deposited and held at 25°C fail to break diapause, but after 100 days of chilling at 5°C diapause is readily broken. Prior to chilling the eggs are not cold hardy. Thus, this is a species in which diapause and cold hardiness are regulated separately (coincidental relationship as shown in *Figure 4*). In this case, low temperature exposure enables the diapausing pharate larvae to generate stress proteins in response to low (or high) temperature (Denlinger *et al.*, 1992). The major stress protein, a 75 kDa protein, was expressed much more highly in pharate larvae that had first been chilled for 65 days than in unchilled individuals.

Quite possibly a number of genes associated with cold hardening will fall into this category of showing higher expression during diapause because there are clearly many cases in which agents associated with cold hardening (polyols, sugars, antifreeze proteins, etc.) are produced in greater abundance in response to low temperature. However, in many instances it is not clear whether it is a direct response to low temperature. For example, expression of antifreeze proteins in field-collected samples of the beetle *Dendroides canadensis* increases during the fall, reaches a peak in December and then gradually declines toward spring (Andorfer and Duman, 2000). Although this response is possibly due to changes in temperature, the change could certainly also be in response to some other feature of the overwintering environment or it may simply be a developmental programme inherent to the beetle.

7. Conclusions

It is evident that temperature has a major impact on insect diapause. For temperate region species it probably is the ultimate factor driving the evolution of diapause. Although there are some instances in which low temperature alone serves as an environmental cue promoting the induction of diapause, its role on diapause induction is usually secondary to the role of photoperiod. Yet, once diapause has been entered temperature plays a critical role in regulating the rate of diapause development and in determining the precise timing to diapause termination and the initiation of post-diapause development.

Arresting development, however, does not by itself assure survival, and a successful period of overwintering also requires adaptations to ensure cold hardiness. In some cases the expression of cold hardiness is a component of the diapause syndrome but in other cases the two appear to be regulated separately. Low temperature is likely to be a common environmental cue boosting the expression of genes associated with cold hardening, but very little is actually known about the molecular mechanisms that may link diapause and cold hardiness.

The insect brain coordinates diapause, and a unique pattern of gene expression can be detected within the brain. Far fewer genes are expressed in the brain during diapause, but diapause is not simply a shut-down in gene expression. Several genes are up-regulated or uniquely expressed in the brain during diapause. Examples of diapause up-regulated genes include those that encode for certain stress proteins and a few other genes involved in protective roles that may be essential for overwinter survival. Among the diapause down-regulated genes are the cell cycle regulator proliferating cell nuclear antigen and a gene that encodes one of the ecdysone receptor proteins. Diapause-specific expression of several additional genes has also been noted in tissues other than the brain. At this point, the expression patterns of only a few genes have been examined in association with diapause. Many more genes remain to be discovered. The exciting challenge that remains is to identify more of the key players and to construct a gene hierarchy that will satisfactorily explain how insect development can be so effectively interrupted by diapause.

Understanding the molecular underpinnings of diapause also offers new tools for probing insect seasonality. The expression patterns of specific genes have potential as biomarkers that can be used to monitor the seasonal progression of insect development. Genes that are turned on or off late in diapause may prove to be especially useful for forecasting seasonal distributions and the timing of outbreaks.

References

Adedokun, T.A. and Denlinger, D.L. (1984) Cold-hardiness: a component of the diapause syndrome in pupae of the flesh flies, *Sarcophaga crassipalpis* and *S. bullata*. *Physiol. Entomol.* 9: 361–364.

Andorfer, C.A. and Duman, J.G. (2000) Isolation and characterization of cDNA clones encoding antifreeze proteins of the pyrochroid beetle *Dendroides canadensis*. *J. Insect Physiol.* 46: 365–372.

Beck, S.D. (1991) Thermoperiodism. In: *Insects at Low Temperature* (eds R.E. Lee, Jr and D.L. Denlinger). Chapman and Hall, New York, pp. 199–228.

Bowen, M.F., Bollenbacher, W.E. and Gilbert, L.I. (1984) *In vitro* studies on the role of the brain and prothoracic glands in the pupal diapause of *Manduca sexta*. *J. Exp. Biol.* 108: 9–24.

Brown, J.J. (1980) Haemolymph protein reserves of diapausing and non-diapausing codling moth larvae, *Cydia pomonella* (L) (Lepidoptera: Tortricidae). *J. Insect Physiol.* 26: 487–491.

Brown, J.J. and Chippendale, G.M. (1978) Juvenile hormone and a protein associated with the larval diapause of the southwestern corn borer, *Diatraea grandiosella*. *Insect Biochem.* 8: 359–367.

Cannon, R.J.C. and Block, W. (1988) Cold tolerance of microarthropods. *Biol. Rev.* 63: 23–77.

Chen, C.-P., Denlinger, D.L. and Lee, R.E. Jr (1987a) Cold shock injury and rapid cold-hardening in the flesh fly, *Sarcophaga crassipalpis*. *Physiol. Zool.* 60: 297–304.

Chen, C.-P., Denlinger, D.L. and Lee, R.E. Jr (1987b) Responses of non-diapausing flesh flies (Diptera: Sarcophagidae) to low rearing temperatures: development rate, cold tolerance and glycerol concentrations. *Ann. Entomol. Soc. Am.* 80: 790–796.

Cheney, C.M. and Shearn, A. (1983) Developmental regulation of *Drosophila* imaginal disc proteins: Synthesis of a heat shock protein under non-heat-shock conditions. *Devl. Biol.* 95: 325–330.

Danks, H.V. (1987) *Insect Dormancy: an Ecological Perspective*. Biological Survey of Canada (Terrestrial Arthropods), Ottawa.

Danks, H.V. (2000) Dehydration in dormant insects. *J. Insect Physiol.* 46: 837–852.

Denlinger, D.L. (1971) Embryonic determination of pupal diapause induction in the flesh fly *Sarcophaga crassipalpis* Macquart. *J. Insect Physiol.* 17: 1815–1822.

Denlinger, D.L. (1972a) Induction and termination of pupal diapause in *Sarcophaga* flesh flies. *Biol. Bull.* 142: 11–24.

Denlinger, D.L. (1972b) Seasonal phenology of diapause in the flesh fly *Sarcophaga bullata*. *Ann. Entomol. Soc. Am.* 65: 410–414.

Denlinger, D.L. (1974) Diapause potential in tropical flesh flies. *Nature* 252: 223–224.

Denlinger, D.L. (1985) Hormonal control of diapause. In: *Comprehensive Insect Physiology Biochemistry and Pharmacology* (eds G.A. Kerkut and L.I. Gilbert). Pergamon Press, Oxford, pp. 353–412.

Denlinger, D.L. (1986) Dormancy in tropical insects. *Ann. Rev. Entomol.* 31: 239–264.

Denlinger, D.L. (1991) Relationship between cold-hardiness and diapause. In: *Insects at Low Temperature* (eds R.E. Lee Jr and D.L. Denlinger). Chapman and Hall, New York, pp. 174–198.

Denlinger, D.L. and Lee, R.E. Jr (1998) Physiology of cold sensitivity. In: *Temperature Sensitivity in Insects and Application in Integrated Pest Management* (eds G.J. Hallman and D.L. Denlinger). Westview Press, Boulder, CO, pp. 55–95.

Denlinger, D.L., Lee, R.E. Jr, Yocum, G.D. and Kukal, O. (1992) Role of chilling in the acquistion of cold tolerance and the capacitation to express stress proteins in diapausing pharate larvae of the gypsy moth, *Lymantria dispar*. *Arch. Insect Biochem. Physiol.* 21: 271–280.

Duman, J.G., Xu, L., Neven, L., Tursman, D. and Wu, D.W. (1991) Hemolymph proteins involved in insect subzero-temperature tolerance: ice nucleators and anti-freeze proteins. In: *Insects at Low Temperature* (eds R.E. Lee and D.L. Denlinger). Chapman and Hall, New York, pp. 94–127.

Duman, J.G., Olsen, T.M., Yeung, K.L. and Jerva, F. (1995) The roles of ice nucleators in cold tolerant invertebrates. In: *Biological Ice Nucleation and its Applications* (eds R.E Lee, G.J. Warren and L.V. Gusta). American Phytopathological Society, St Paul, MN, pp. 201–219.

Feder, J.H., Rossi, J.M., Solomon, J., Solomon, N. and Linquist, S. (1992) The consequences of expressing hsp70 in *Drosophila* cells at normal temperatures. *Genes Devl.* 6: 1402–1413.

Flannagan, R.D., Tammariello, S.P., Joplin, K.H., Cikra-Ireland, R.A., Yocum, G.D. and Denlinger, D.L. (1998) Diapause-specific gene expression in pupae of the flesh fly *Sarcophaga crassipalpis*. *Proc. Natl Acad. Sci. USA* 95: 5616–5620.

Giebultowicz, J.M. and Denlinger, D.L. (1986) Role of the brain and ring gland in regulation of pupal diapause in the flesh fly, *Sarcophaga crassipalpis*. *J. Insect Physiol.* 32: 161–166.

Hanec, W. and Beck, S.D. (1960) Cold hardiness in the European corn borer, *Pyrausta nubilalis* (Hubn.). *J. Insect Physiol.* 5: 169–180.

Henrich, V.C. and Brown, N.E. (1995) Insect nuclear receptors: a developmental and comparative perspective. *Insect Biochem. Mol. Biol.* 25: 881–897.

Ireland, R.C. and Berger, E.M. (1982) Synthesis of the low molecular weight heat shock proteins stimulated by 20-hydroxyecdysone in a cultured *Drosophila* cell line. *Proc. Natl Acad. Sci. USA* 79: 855–859.

Joplin, K.H., Yocum, G.D. and Denlinger, D.L. (1990) Diapause specific proteins expressed in the brain during the pupal diapause of the flesh fly, *Sarcophaga crassipalpis*. *J. Insect Physiol.* 36: 775–783.

Karow, A.M. (1991) Chemical cryoprotection of metazoan cells. *BioScience* 41: 155–160.

Koopmanschap, A.B., Lammers, H. and de Kort, C.A.D. (1992) Storage proteins are present in the haemolymph from larvae and adults of the Colorado potato beetle. *Arch. Insect Biochem. Physiol.* 20: 119–133.

Krebs, R.A. and Feder, M.E. (1997) Deleterious consequences of Hsp70 overexpression in *Drosophila melanogaster* larvae. *Cell Stress Chaperones* 2: 60–71.

Leather, S.R., Walters, K.F.A. and Bale, J.S. (1993) *The Ecology of Insect Overwintering*. Cambridge University Press, Cambridge.

Lee, K.-Y. and Denlinger, D.L. (1996) Diapause-regulated proteins in the gut of pharate first instar larvae of the gypsy moth, *Lymantria dispar*, and the effect of KK-42 and neck ligation on expression. *J. Insect Physiol.* 42: 423–431.

Lee, K.-Y., Hiremath, S. and Denlinger, D.L. (1998) Expression of actin in the central nervous system is switched off during diapause in the gypsy moth, *Lymantria dispar*. *J. Insect Physiol.* 44: 221–226.

Lee, R.E. Jr (1991) Principles of low temperature tolerance. In: *Insects at Low Temperature* (eds R.E. Lee Jr and D.L. Denlinger). Chapman and Hall, New York, pp. 17–46.

Lee, R.E. Jr, Chen, C.-P. and Denlinger, D.L. (1987) A rapid cold-hardening process in insects. *Science* 238: 1415–1417.

Mugnano, J.A., Lee, R.E. and Taylor, R.T. (1996) Fat body cells and calcium phosphate spherules induce ice nucleation in the freeze-tolerant larvae of the gall fly *Eurosta solidaginis* (Diptera, Tephritidae). *J. Exp. Biol.* 199: 465–471.

Neven, L.G., Duman, J.G., Low, M.G., Sehl, L.C. and Castellino, F.J. (1989) Purification and characterization of an insect hemolymph lipoprotein ice nucleator: evidence for the importance of phosphatidylinositol and apolipoprotein in the ice nucleator activity. *J. Comp. Physiol.* 159: 71–82.

Palli, S.R., Ladd, T.R., Ricci, A.R., Primavera, M., Mungrue, I.N., Pang, A.S.D. and Retnakaran, A. (1998) Synthesis of the same two proteins prior to larval diapause and pupation in the spruce budworm, *Choristoneura fumiferana*. *J. Insect Physiol.* 44: 509–524.

Patterson, J.L. and Duman, J.G. (1978) The role of the thermal hysteresis factor in *Tenebrio molitor* larvae. *J. Exp.Biol.* 74: 37–45.

Rinehart, J.P. and Denlinger, D.L. (2000) Heat-shock protein 90 is down-regulated during pupal diapause in the flesh fly, *Sarcophaga crassipalpis*, but remains responsive to thermal stress. *Insect Mol. Biol.* 9: 641–645.

Rinehart, J.P., Yocum, G.D. and Denlinger, D.L. (2000) Developmental upregulation of inducible hsp70 transcripts, but not the cognate form, during pupal diapause in the flesh fly, *Sarcophaga crassipalpis*. *Insect Biochem. Mol. Biol.* 30: 515–521.

Rinehart, J.P., Cikra-Ireland, R.A., Flannagan, R.D. and Denlinger, D.L. (2001) Expression of ecdysone receptor is unaffected by pupal diapause in the flesh fly, *Sarcophaga crassipalpis*, while its dimerization partner, USP, is down-regulated. *J. Insect Physiol.* (in press).

Salama, M.S. and Miller, T.A. (1992) A diapause associated protein of the pink bollworm *Pectinophora gossypiella* Saunders. *Arch. Insect Biochem. Physiol.* 21: 1–11.

Saunders, D.S. (1971) The temperature-compensated photoperiodic clock 'programming' development and pupal diapause in the flesh fly, *Sarcophaga argyrostoma*. *J. Insect Physiol.* 17: 801–812.

Saunders, D.S. (1973) Thermoperiodic control of diapause in an insect: theory of internal coincidence. *Science* 181: 358–360.

Saunders, D.S. (1982) *Insect Clocks,* 2nd edn. Pergamon Press, Oxford.

Suzuki, M., Terada, T, Kobayashi, M. and Shimada, T. (1999) Diapause-associated transcription of *BmEts*, a gene encoding an ETS transcription factor homolog in *Bombyx mori*. *Insect Biochem. Mol. Biol.* 29: 339–347.

Takeda, M. and Skopik, S.D. (1997) Photoperiodic time measurement and related physiological mechanisms in insects and mites. *Annu. Rev. Entomol.* 42: 323–349.

Tammariello, S.P. and Denlinger, D.L. (1998) G_0/G_1 cell cycle arrest in the brain of *Sarcophaga crassipalpis* during pupal diapause and the expression pattern of the cell cycle regulator, proliferating cell nuclear antigen. *Insect Biochem. Mol. Biol.* 28: 83–89.

Tauber, M.J., Tauber, C.A. and Masaki, S. (1986) *Seasonal Adaptations on Insects*. Oxford University Press, Oxford.

Walker, U.K., Kuiper, M.J., Tyshenko, M.G., Doucet, D., Graether, S.P., Liou, Y.-C., Sykes, B.D., Jia, Z., Davies, P.L. and Graham, L.A. (2001) Surviving winter with anti-freeze proteins: studies on budworms and beetles. In: *Insect Timing: Circadian Rhythmicity to Seasonality* (eds D.L. Denlinger, J.M. Giebultowicz and D.S. Saunders) Elsevier, Amsterdam, pp. 199–211.

Williams, C.M. (1946) Physiology of insect diapause: the role of the brain in the production and termination of pupal dormancy in the giant silkworm *Platysamia cecropia*. *Biol. Bull.* 90: 234–243.

Williams, C.M. and Adkisson, P.L. (1964) Physiology of insect diapause. XVI. An endocrine mechanism for the photoperiodic control of pupal diapause in the oak silkworm *Antheraea pernyi*. *Biol. Bull.* 127: 511–525.

Yamamoto, T., Goto, T. and Hirobe, T. (1978) Genetic studies of the pigmented and non-diapause egg mutant in *Bombyx mori*. *J. Sericulture Sci. Jpn.* **47**: 181–185.

Yocum, G.D., Joplin, K.H. and Denlinger, D.L. (1998) Upregulation of a 23 kDa small heat shock protein transcript during pupal diapause in the flesh fly, *Sarcophaga crassipalpis. Insect Biochem. Mol. Biol.* **28**: 667–682.

Zachariassen, K.E. (1991) The water relations of overwintering insects. In: *Insects at Low Temperature* (eds R.E. Lee Jr and D.L. Denlinger). Chapman and Hall, New York, pp. 47–63.

13

Development and hatching in cephalopod eggs: a model system for partitioning environmental and genetic effects on development

P.R. Boyle, L. Noble, A.M. Emery, S. Craig, K.D. Black and J. Overnell

1. Introduction

The success of the egg, developmental and larval stages of marine animals is critical to completion of the life cycle and maintenance of their populations. Much of the variability in the density and distribution of marine invertebrates and fish, particularly those with strongly seasonal reproductive activity, can be understood in terms of the viability and fitness of the egg and larval stages in determining the strength of recruitment to growth and adult stages. Marine organisms commonly develop through several delicate larval phases, vulnerable to physical and biological conditions in the environment. Variability in the duration and success of these phases contributes in large measure to the 'match or mis-match' (Cushing, 1982) coincidence of the demands of the hatchling cohort with optimum conditions for growth and survival.

Embryonic development may be perturbed by sublethal environmental insults which leave a morphological record in the hatchling, ranging from barely noticeable asymmetries of minor characters to gross malformations of major organs. The study of such teratologies can prove important in determining the ecological threshold and effects of environmental pollutants, from toxins (Fent, 1992; Fent and Meier, 1994) and their degradation products (Polifka et al., 1996) to heat (Fisher et al., 1995), proximately on embryo development and ultimately on population recruitment. A first step towards the formulation of sensitive techniques for analysis of developmental abnormalities in commercial species is to recognize the contribution of the genotype to morphological stability.

Environment and Animal Development: Genes, Life Histories and Plasticity, edited by D. Atkinson and M. Thorndyke.
© 2001 BIOS Scientific Publishers Ltd, Oxford.

Here we examine the basis of variation in incubation time and hatchling size in squid (*Loligo forbesi*) and its potential as a model system for quantifying the relative contributions of local environment and genotype on the timing and success of embryonic development stages. We review this approach based on a programme using squid populations from the Scottish west coast.

2.   Cephalopod reproduction

2.1 *Spawning*

The eggs of the modern cephalopods (octopuses, cuttlefish and squid) are spawned in one of two basic modes. For those species which are mainly restricted to coastal and shelf habitat, eggs are attached to the seabed or other hard surfaces. Octopus eggs are bound individually by a short stalk into strings, which are then attached to rock surfaces in groups. Typically the female octopus will deposit her spawn under an overhang or within an enclosed space that will enable her to gain some protection from predators during the time she will spend guarding and brooding the eggs. Cuttlefish enclose each egg in a tough protective external coating, often pigmented black from the ink-sac secretions. These eggs form distinctive clusters of lozenge-shaped capsules attached to rock surfaces, often disguised among the many other encrusting organisms.

The loliginid squid (order Teuthoidea, family Loliginidae) ensheathe their eggs with material from the nidamental glands, embedding them into finger-like capsules each containing some 90–120 eggs. Large numbers of these conspicuous white capsules are attached by the female to hard surfaces. Individual females progressively deposit their capsules at the same site in concentrated bouts of spawning. They may also be joined by other females to produce large 'mops' of egg strings, and in those species which aggregate densely at breeding, e.g. *Loligo opalescens*, the spawn deposits may cover large areas of seabed (McGowan, 1954). Some loliginid species such as *Loligo reynaudii* (Sauer and Smale, 1993) and *Loligo plei* (Vecchione, 1988) are less dependant on solid surfaces for egg-laying and may simply insert the egg capsule into sandy or gravel sediment. Boletzky (1987) compares the various forms of encapsulation of cephalopod embryos and the mechanisms of its formation. Less is known about spawning by squid species which are mainly pelagic in habit. Species such as *Illex illecebrosus* and *Todarodes pacificus* (family Ommastrephidae) are described in captivity to extrude their eggs in a continuous fragile gelatinous mass, expanding on production by absorption of water to form masses up to 1 m in diameter (O'Dor *et al.*, 1982a,b; Bower and Sakurai, 1996). Apparently these masses are almost neutrally buoyant in seawater and probably 'float' at density discontinuities between water masses (O'Dor and Balch, 1985).

2.2 *Development*

The large size and robust nature of the encapsulated eggs of coastal cephalopods has prompted studies of the embryological development in many species. Naef (1923, 1928) described their embryonic stages and formalized the main steps in the process into a series of numbered stages from the earliest cell division through to hatching. In a modified form his scheme provides the framework for comparative studies (Arnold,

1965; Baeg *et al.*, 1992; Boletzky, 1999; Mangold *et al.*, 1971; Segawa *et al.*, 1988; Villaneuva, 1995) and reviews (Arnold, 1971; Boletzky, 1987).

Modern cephalopods share embryological characteristics with other molluscs and protostomes (Willmer, 1990). From the first cell division of the zygote, a fundamentally determinate cleavage pattern means that the outcomes of the cell lines are fixed very early in development. There is the possibility of mapping the fates of regions of yet undivided cells leading to the descriptive term 'mosaic' development. Mistakes or damage during these early cleavage stages result in developmental errors and abnormalities. Cephalopods differ from most other molluscs by virtue of the very large yolk content of the ovum. The yolk mass remains uncleaved, forcing the initial development of the embryo as a plate or cap of cells at the animal pole of the zygote. They also differ from other molluscs in that development is direct, there are no specialized larval forms, and the hatchlings emerge as miniature adults often with the residual yolk sac still to be absorbed (*Figure 1*).

2.3 *Life cycle*

In these short-lived species the time for embryonic development may be a significant fraction of the total lifespan of the individual. During this phase the nutritional requirements of the embryo are provided by the yolk supplied by the female. Larger-egged species, therefore, take longer to develop between spawning and hatching (Mangold, 1963) and emerge as a larger, more competent hatchling. At an early stage in development, pulsation of the yolk sac is seen, followed by activity of the heart and blood vessels. At later stages, mechanical stimulation will cause whole body movements of the developing embryo and spasmodic chromatophore expansion. It appears that movements of the *Loligo* embryo are chemically inhibited by a substance within the embryonic fluid (Weischer and Marthy, 1983), and that the egg string jelly of loliginids also has a generalized effect inhibiting ciliary action (Atkinson, 1973).

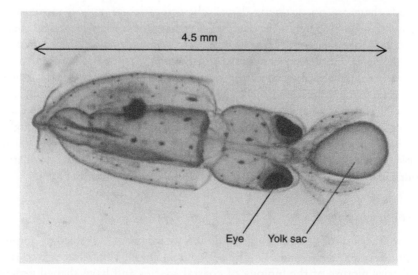

Figure 1. *Photomicrograph of a stage 30 embryo/hatchling of* Loligo forbesi.

Embryonic development times vary considerably between the different types of cephalopod egg ranging from 5–10 days for small unprotected eggs, to 50–100 days for large yolky-egged species. Whatever the optimum development time for a species, the temperature at which development takes place is a very significant variable, with higher temperature shortening development time and producing smaller hatchlings. Many studies have established specific time–temperature relationships for different species (Boyle, 1983; McMahon and Summers, 1971; O'Dor et al., 1982b; Sakurai et al., 1995, 1996).

Although few data are available, the influence of temperature in accelerating or retarding embryonic development in the field is expected to be a key determinant in the life cycle. The interaction between the timing of egg-laying and the development temperature clearly will fix the time of hatching which, in turn, determines the conditions for survival and growth met by the hatchling. Especially at low (Boletzky, 1994) or fluctuating seasonal temperatures (Boyle et al., 1995), this interval may be a critical component in completion of the life cycle. The influence of temperature continues as a main determinant of the growth rate of the hatchling. Forsythe (1993) models how this effect may so accelerate the growth rate between hatching and juvenile stages that later-hatched broods may overtake those hatched earlier, thus dominating the early recruitment to the adult population.

2.4 Hatchlings

Few studies have been made of the stimuli inducing hatching. In laboratory conditions, in Loligo sp. (Paulij, 1990) and Sepia (Paulij et al., 1991), photoperiod is important and the majority of hatching occurs during the hours of darkness. At hatching the larval cephalopod ingests the remainder of the yolk into the stomach. This provides the energy source for the first few days of active life until the transition to external feeding.

Within a day or so of hatching, the young cephalopod is a competent predator on first presentation of live food and will attack and consume prey of appropriate sizes. In most cephalopods there are no distinctly 'larval' characteristics or radical metamorphoses during post-larval development. Although the term 'larva' is still in wide usage, Young and Harman (1988) introduced the term 'paralarva' to accommodate the ecological rather than morphological differences between the early young stages and adult forms. Many species have been hatched in captivity and reared through to adult stage (for reviews see Hanlon and Hixon, 1983; Hanlon, 1990). The main problem for laboratory rearing is the provision of live food of an acceptable kind in the rapidly increasing quantities required by the growing brood and this has been the main obstacle to routine culture, although some advances have been made in the managed provision of crab zoeae as food (Villanueva, 1994). Nevertheless, laboratory studies provide the main source of information on behaviour and development of the paralarvae (Mangold and Boletzky, 1985).

2.5 Population biology

Studies on the reproduction and population biology of a selected number of cephalopod species (Boyle and Ngoile, 1993; Boyle, 1983, 1987) have given rise to a fairly cohesive picture of the biology of coastal species and have shaped generalizations about their life cycle. In squid of the family Loliginidae there is seasonality of breeding, more-or-less defined geographical spawning areas, patterns of migration within coastal waters, an annual life cycle and evidence for semelparity or terminal spawning followed by massive post-spawning mortality. These life cycle character-istics contribute to large inter-annual variability in loliginid populations. The population in any year depends almost entirely on the success of recruitment, there being almost no overlap of generations (Boyle, 1990; Boyle and Pierce, 1994a,b; Boyle and Boletzky, 1996).

The common squid species of British coastal waters, *Loligo forbesi*, has typically complex population structures (Pierce *et al.*, 1994) and recruitment patterns (Collins *et al.*, 1997). Some of this variability may be due to diversity in patterns of repro-duction, specifically the time course of spawning, size at breeding, and whether there are several spawning episodes (Boyle *et al.*, 1995). Two distinct size modes in the male population at maturity, which are not age-related, suggest there may be components of the population with different growth performance arising from mixed populations at breeding or variable growth cohorts within the same gene pool. In other loliginid species large- and small-bodied male individuals are seen competing to mate with a single female over the spawning beds (Hanlon and Messenger, 1996). This association between large dominant and small 'sneaker' males offers a possible explanation of the population size modes discovered in male *L. forbesi* and a partial mechanism for main-taining a complex population structure in an essentially seasonal semelparous species. Microsatellite typing of embryos of this species (Shaw and Boyle, 1997) has proven multiple paternity of a sample of the 90–120 eggs within a single female's egg string, the female being fertilized by at least two male squid. Many females contribute their eggs to a single egg cluster (estimated fecundity ~20 000 per female).

Another feature of this species is the considerable sexual dimorphism in adult size. Males typically reach a mean final size some 50% greater than their female contempo-raries, and show correspondingly higher growth rates in the fished population (Boyle and Ngoile, 1993). Since the sexes are not obviously segregated by habitat the presumption is of genetically determined growth differences which may be detectable in the embryonic stages.

3. Rationale

The rationale of these investigations was to utilize the large size, well-defined develop-mental stages, and accessibility of the cephalopod embryo to examine the basis of vari-ation in incubation time and hatchling size; and to use molecular markers to quantify where possible the relative contributions of genotype, resource utilization, and local environment in developing squid, specifically to test whether the maternal and/or paternal genetic contribution to development rates could be detected and to assess the significance of plasticity of development rate to the highly variable recruitment patterns of cephalopod fisheries.

4. Material sources

Egg masses of the squid, *Loligo forbesi*, collected in the field within 24 h of spawning, were incubated at constant temperatures at Dunstaffnage Marine Laboratory and the emergent hatchlings collected daily. Contacts were made with creel fishermen working around Seil Island, Argyll and Loch Ewe and the Summer Islands. Creel fishermen set and recover their creels (traps set on the bottom for lobster and crab) on a daily basis. Liaison with the fishermen provided details of egg laying as well as the timing of the spawning event to within 24 h and allowed the start of development to be closely estimated.

5. The developmental environment

Collection and incubation procedures are summarised in *Figure 2*.

5.1 *Incubation methods*

All egg strings were held and maintained in the aquarium facilities of Dunstaffnage Marine Laboratory, Oban, Argyll. Egg strings were collected in masses, attached together at one end – these masses are termed 'mops'. These 'mops' can be contributed by several females laying at the same site. After acclimation to tank temperatures, egg strings were attached to circles of plastic garden mesh with twine. Suspending them in this way mimicked natural conditions. Air stones placed in each of the tanks provided adequate aeration and water movement.

5.2 *Temperature effects on incubation period*

The mean incubation interval of single egg strings ranges from 32 days (16°C) to 130 days (8°C). More surprisingly, the range of mean incubation interval of egg strings held at a single temperature (12°C) ranged between 45 and 64 days (*Figure 3*).

As well as this wide inter-string variation there was normally intra-string variation of about 20 days between the incubation times of the first and last individual hatched (*Figure 4*). This degree of variation in incubation interval, coupled with the normal range of environmental conditions, will clearly make a major contribution to spreading

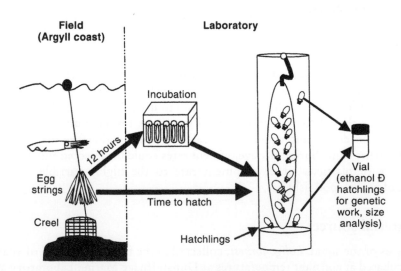

Figure 2. Schematic diagram illustrating the field and laboratory methods for data collection from eggs and hatchlings of Loligo forbesi.

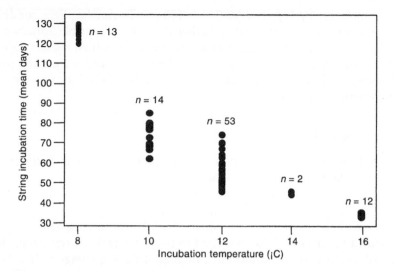

Figure 3. *Summary of data on incubation time (from egg deposition to hatching) for* Loligo forbesi *reared at a range of incubation temperatures. Mean days to hatching for all eggs in a single string (single data point) show considerable inter-string variation at a constant temperature.*

the risk that emergent hatchlings from a spawning bed, and even those from a single female, will meet unfavourable conditions.

5.3 *Light intensity*

Diurnal light regimes imposed a strong rhythm of hatching pattern with the majority of hatchlings emerging during the hours of darkness. There is considerable variation in

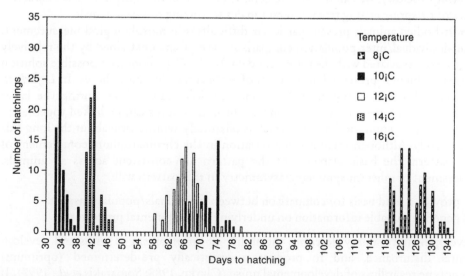

Figure 4. Numbers of hatchlings emerging from selected strings held at a range of incubation temperatures showing inter- and intra-string variation in numbers of days to hatching (incubation time) at a constant temperature.

the depth of spawning by *L. forbesi* (*ca* 5 to >100 m). To simulate light conditions at this range of depths and to investigate whether there was a possible influence on the total duration of development, egg strings were maintained in high and low light levels. No significant differences in incubation interval were detected, suggesting that over the depth range used by spawning squid the average light intensity was not a factor influencing emergence patterns.

5.4 *Mechanical agitation*

Expansion and loosening of the chorion precedes normal hatching. Artificial stimuli such as lowered salinity may also lead to the same effect. In comparison to the field environment, the aquarium conditions were relatively undisturbed. To test whether mechanical agitation is also a necessary stimulus to hatching, egg batches were incubated in two tanks which differed only by the constant mild mechanical agitation supplied to one. A modified mechanical shaker was used to move the racks of egg strings forward and backwards in the tank and resulted in swinging of the egg strings within the tubes. The control tank was set up at the same temperature but without the shaker. There were no significant differences in the time to hatching or the proportion successfully hatched, but at the point of hatching mechanical stimulation was an important factor in assisting eclosion and allowing the hatchling to free itself from the egg capsule.

6. Hatchling quality

6.1 *Morphometry and meristics*

The length and weight of hatchlings were measured for all individuals emerging from selected strings. The range of these measurements can be used to compare the hatchling cohort of different strings and different females. Combined with analysis of paternity (below), we can investigate genetic effects on the morphological characteristics of offspring arising from different males fertilizing the same female.

Soft-bodied animals present particular difficulties in morphological measurement. Inter-individual/inter-population comparisons are often constrained by the relatively large error associated with measurement data. In the Cephalopoda, a possible solution to this problem is presented in the form of embryonic chromatophores. In *L. forbesi*, chromatophores first appear, and become physiologically active, during the latter stages of embryogenesis. These 'founder' chromatophores can be linked together to form a simple geometrical pattern which is bilaterally symmetrical about the long axis of the body. Although there is some variation in the chromatophore complement of this pattern, the basic elements of the pattern are consistent across all animals. Assessment of inherent symmetry/asymmetry in this pattern will:

(i) provide a solid basis for comparison between individuals/populations; and
(ii) generate valuable information on underlying developmental processes.

Developmental stability describes the ability of an organism to withstand developmental disturbance, and to produce a genetically pre-determined (optimum) phenotype regardless of developmental noise (Clarke, 1988; Somarakis *et al.*, 1997). If stress occurs during development, it is argued that energy is diverted from the least essential control processes and that such disturbances often manifest in phenotypic characters as small random departures from some form of symmetry. Fluctuating asymmetry is the most common measure used to assess developmental stability in bilaterally symmetrical characters. Much of the interest stems from its potential as an indicator of organism fitness, the basic premise being that both sides of a bilaterally symmetrical trait are controlled by the same genome: any non-directional differences between sides must therefore reflect developmental disturbance. In this manner, high-quality individuals are identified by low levels of asymmetry, and low quality individuals by relatively elevated levels of asymmetry.

A digital camera linked to image analysis software is used to capture a dorsal and ventral image for each animal. Images are then sub-divided into arm, head and mantle regions and the total chromatophore number is recorded for each body region. Software then determines the colour (red/yellow) of each chromatophore. Finally, image area (mm²) is measured to provide a basis for size corrections between individuals. Images are processed (e.g. colour channels extracted, contrast enhanced) to facilitate analysis. Individual chromatophores are coded (singularly or in pairs) from anterior to posterior on the head and mantle. This coding ensures that relative chromatophore asymmetry can be assessed both between chromatophores and between animals.

Cephalopod chromatophore patterns and the ability to compare them quantitatively between individuals potentially offers a powerful tool for investigation of errors occurring during development.

6.2 *Yolk composition*

There are no published data yet describing the proportions and status of protein, lipid or carbohydrate in squid eggs, embryos or hatchlings. Yolk sacs are virtually depleted at hatching, thus squid are required to hunt actively and capture prey immediately after emergence. Being sophisticated visual feeders, squid hatchlings must have fully functional eyes from the moment of hatching. Both of the fatty acids, docosahexaenoic acid and eicosapentaenoic acid, are deemed essential in the phospholipids of nervous tissue and eyes of fish. For these reasons, there is likely to be a high provision of such fatty acids supplied to squid eggs.

Completely new data has been obtained describing the fatty acid composition of the cephalopod yolk. Both docosahexaenoic and eicosapentaenoic acids have been shown to make up approximately 25 and 10% of the total fatty acids, respectively. Significant differences have been found in the lipid quality of eggs from different sites (Oban and Loch Ewe). Differences in total lipid and fatty acid profiles of eggs, yolks, embryos, hatchlings and some adult tissues of the squid *L. forbesi* are in progress.

As part of the programme to isolate and characterize the zinc- and copper-binding proteins in the yolk, gel filtration chromatography has been applied to the eggs and adult tissues. Eggs are richer in zinc than copper and both elements appear bound to high molecular weight proteins.

7. Hatchling genotype

7.1 *Genetic relationships considered*

In order to identify the genetic basis of traits, it is essential to identify the relationships between the individuals in the investigation. In practice this means identifying the full-/half-sibling relationships between embryos or hatchlings from squid egg strings. Egg strings attached to fishing creel lines offer a convenient source of freshly laid strings but it is not possible to associate the parents with any string collected in this manner. However, it is certain that all individuals within an egg string share the same mother because egg strings are encapsulated as they are laid. Multiple mating has been shown to lead to multiple paternity of individuals within an egg string, so that the genetic relationship between each individual is either full- or half-sibling (Shaw and Boyle, 1997). An egg string can therefore provide a full-/half-sibling 'progeny array' ideal for genetic investigations, but further analysis using genetic markers is necessary to identify individual relationships.

7.2 Microsatellite markers

We are using previously published and novel microsatellite DNA markers to identify the sibling relationships of individuals within an egg string (Shaw, 1997; Emery et al., 2000). Microsatellites are widely interspersed repetitive DNA sequences. Each microsatellite is composed of a 2–6 nucleotide DNA sequence repeated several times (see Moxon and Wills, 1999). 'Slippage' mutations cause variation in the number of repeats (microsatellites can also be described as VNTR markers, variable number of tandem repeats). The mutation events lead to a number of alleles at each microsatellite locus, due to variation in the number of repeat units found in the microsatellite array. These alleles are inherited according to simple Mendelian rules, except when further mutations occur. The alleles for each individual can be identified by using PCR with oligonucleotide primers specific for the non-repetitive DNA flanking the variable microsatellite array. Each individual has two alleles, one of which is inherited from the mother, the other from the father. Highly variable microsatellite markers are required for investigation of full-/half-sibling relationships.

7.3 Identification of sibling groups within egg strings

When one male fertilizes all of the eggs in an egg string, genotyping every embryo or hatchling within it will reveal a maximum total of four alleles per locus for the whole egg string, two of which may be expressed in any particular individual. As each individual inherits one allele from each parent, the genotypes of the parents at a particular locus may be reconstructed based on the segregation patterns found in the offspring (*Figure 5a*). It is not possible to identify which genotype corresponds to the father or the mother.

When several males fertilize an egg string, all individuals within the string share the same mother but not necessarily the same father. This means that every individual has one of the two maternal alleles, but the paternal alleles are found at a lower frequency, depending on the number of offspring corresponding to each father (*Figure 5b*). This pattern of allele segregation allows the maternal alleles to be identified. Ambiguities arise only when both maternal alleles are shared by one or more of the fathers. The remaining paternal alleles cannot be associated with particular fathers unambiguously except when the frequencies of offspring associated with each father differ significantly. To establish paternity groups, it is necessary to examine paternal alleles from several loci to identify groups of associated paternal alleles. This can done empirically (see below), or more quantitatively using Baysian statistical approaches currently under development.

- environmental variables and stressors cause rate changes and result in recognizable abnormalities and asymmetries;
- prospects for experimental interventions throughout the developmental period;
- microsatellite markers available for 15 variable loci;
- variable mating strategies resulting in single and multiple paternity strings of eggs;
- methodology developed for the assignment of paternity to individual embryons and hatchlings;
- many quantifiable phenotypic traits available based on size, morphometry, meristics and chromatophore patterns.

In summary, a model system for the partitioning of environmental and genetic effects on development.

Acknowledgements

We are grateful to Graham McPheat (Seil Island) and Phil McLachlan (Loch Ewe) for collection and transport of egg strings from the field; the aquarium staff at Dunstaffnage Marine Laboratory for their support with the rearing experiments; Fiona Murray and Linda Key, Department of Zoology, University of Aberdeen, for assistance with data collection, and Ian Wilson of the Department of Mathematical Sciences, University of Aberdeen, for advice and guidance on numerical approaches to paternity analysis. This work was supported by the Natural Environment Research Council through the Developmental Ecology of Marine Animals thematic programme (DEMA project GST/02/1723). Additional support from the NERC was provided for related work on developmental errors and stress effects during incubation (GR9/04295).

References

Arnold, J.M. (1965) The inductive role of the yolk epithelium in the development of the squid, *Loligo pealeii* (Lesueur). *Biol. Bull.* **129**: 72–78.

Arnold, J.M. (1971) *Cephalopods. Experimental Embryology of Marine and Freshwater Invertebrates* (ed G. Reverberi). North Holland, Amsterdam.

Atkinson, B.G. (1973) Squid nidamental gland extract: isolation of a factor inhibiting ciliary activity. *J. Exp. Zool.* **184**: 335–340.

Baeg, G.H., Y Sakurai, Y. and Shimazaki, K. (1992) Embryonic stages of *Loligo bleekeri* Keferstein (Mollusca: Cephalopoda). *Veliger* **35**: 234–241.

Boletzky, S.V. (1987) Embryonic phase. In: *Cephalopod Life Cycles*, Vol. 2 (ed P.R. Boyle). Academic Press, London, pp. 5–31.

Boletzky, S.V. (1994) Embryonic development of cephalopods at low temperatures. *Antarctic Sci.* **6**: 139–142.

Boletzky, S.V. (1999) Recent studies on spawning, embryonic development and hatching in the Cephalopoda. *Adv. Mar. Biol.* **25**: 85–115.

Bower, J.R. and Sakurai, Y. (1996) Laboratory observations on *Todarodes pacificus* (Cephalopoda: Ommastrephidae). *Am. Malacolog. Bull.* **13**: 65–71.

Boyle, P.R. (ed) (1983) *Cephalopod Life Cycles. Vol. 1. Species Accounts.* Academic Press, London, p.475.

Boyle, P.R. (ed) (1987) *Cephalopod Life Cycles, Vol. 2. Comparative Reviews.* Academic Press, London, p.441.

Boyle, P.R. (1990) Cephalopod biology in the fisheries context. *Fish. Res.* **8**: 303–321.

Boyle, P.R. and Boletzky, S.V. (1996) Cephalopod populations: definition and dynamics. *Phil. Trans. R. Soc. Lond. B* **351**: 985–1002.

Boyle, P.R. and Ngoile, M.A.K. (1993). Population variation and growth in *Loligo forbesi* (Cephalopoda: Loliginidae) from Scottish waters. In: *Recent Advances in Cephalopod Fisheries Biology* (eds T. Okutani, R.K. O'Dor, and T. Kubodera). Tokai University Press, Tokyo, pp. 49–60.

Boyle, P.R. and Pierce, G.J. (eds) (1994a) Fishery biology of north east Atlantic squid. *Fish. Res.* **21**: 1–311.

Boyle, P.R. and Pierce, G.J. (1994b) Fishery potential of northeast Atlantic squid stocks: overview. *Fish. Res.* **21**: 1–15.

Boyle, P.R., Pierce, G.J and Hastie, L.C. (1995) Flexible reproductive strategies in the squid *Loligo forbesi. Mar. Biol.* **121**: 501–508.

Clarke, G.M. (1988) The genetic basis of developmental stability. IV. Individual and population asymmetry parameters. *Heredity* **80**: 562–567.

Collins, M.A., Pierce, G.J. and Boyle, P.R. (1997) Population indices of reproduction and recruitment in *Loligo forbesi* (Cephalopoda: Loliginidae) in Scottish and Irish waters. *J. Appl. Ecol.* **34**: 778–786.

Cushing, D.H. (1982) *Climate and Fisheries*. Academic Press, London.

Emery, A.M., Shaw, P.W., Greatorex, E.C., Boyle, P.R. and Noble, L.R. (2000) New microsatellite markers for assessment of paternity in the squid *Loligo forbesi* (Mollusca: Cephalopoda). *Mol. Ecol.* **9**: 110–112.

Fent, K. (1992) Embryotoxic effects of tributyltin on the minnow *Phoxinus phoxinus. Environ. Pollut.* **76**: 187–194.

Fent, K. and Meier, W. (1994) Effects of tributylintin on fish early-life stages. *Archi. Environ. Contam. Toxicol.* **27**: 224–231.

Fisher, B.R., Heredia, D.J. and Brown, K.M. (1995) Induction of hsp72 in heat-treated rat embryos – a tissue-specific response. *Teratology* **52**: 90–100.

Forsythe, J.W. (1993) A working hypothesis on how seasonal temperature change may impact on the field growth of young cephalopods. In: *Recent Advances in Cephalopod Fisheries Biology*. (eds T. Okutani, R.K. O'Dor, and T. Kubodera.). Tokai University Press, Tokyo, pp. 133–144.

Hanlon, R.T. (1990) Maintenance, rearing and culture of tenthoid and sepioid squids. In: *Squid as Experimental Animals* (eds D. Gilbert, H. Adelman and J.M. Arnold). Plenum Press, New York, pp. 35–62.

Hanlon, R.T. and Hixon R.F. (1983) Laboratory maintenance and culture of octopuses and Loliginid squids. In: *Culture of Marine Invertebrates* (ed C.J. Berg). Hutchinson Ross, Stroudsberg, PA, pp. 44–61.

Hanlon, R.T. and Messenger, J.B. (1996) *Cephalopod Behaviour*. Cambridge University Press, Cambridge, p. 232.

Mangold, K. (1963) Biologie des cephalopodes benthiguis et nectonigues de la mer catalane. *Vie Milieu Suppl.* **13**: 285.

Mangold, K. and Boletzky, S.V. (eds) (1985). *Biology and Distribution of Early Juvenile Cephalopods. Vie et Milieu.* Banyuls sur Mer, Laboratoire Arago.

Mangold, K., Boletzky, S.V. and Frosch, D. (1971) Reproductive biology and embryonic development of *Eledone cirrosa* (Cephalopoda: Octopoda) *Int. J. Life Ocean. Coast* **8**: 109–117.

McGowan, J.A. (1954) Observations on the sexual behaviour and spawning of the squid, *Loligo opalescens* at la Jolla, California. *Calif. Fish Game* **40**: 47–54.

McMahon, J.J. and Summers, W.C. (1971) Temperature effects on the developmental rate of squid (*Loligo pealei*) embryos. *Biol. Bull.* **141**: 561–567.

Moxon, E.R. and Wills, C. (1999) DNA Microsatellites: agents of evolution? *Sci. Am.* **280**: 94–99.

Naef, A. (1928) Die Cephalopoden, Embryologie. In: *Fauna e Flora del Golgo di Napoli*, Vol. 35. Vendita Presso R. Freidlander, Berlin, pp. 1–357.

O'Dor, R. K. and Balch, N. (1985) Properties of *Illex illecebrosus* egg masses potentially influencing larval oceanographic distribution. *NAFO Sci. Council Stud.* **9**: 69–76.

O'Dor, R.K., Balch, N. and Amaratunga, T. (1982a) Laboratory observations of midwater spawning by *Illex illecebrosus*. NAFO Report no. N493.

O'Dor, R.K., N. Balch, N., *et al.* (1982b) Embryonic development of the squid, *Illex illecebrosus*, and effect of temperature on development rates. *J. Northwest Atlantic Fish. Sci.* 3: 41–45.

Paulij, W.P. (1990) The impact of photoperiodicity on hatching of *Loligo vulgaris* and *Loligo forbesi*. *J. Mar. Biol. Assoc. UK* 70: 597–610.

Paulij, W.P., Herman, P.M.J., Roozen, M.E.F. and Denuce, J.M. (1991) The influence of photoperiodicity on hatching of *Sepia officinalis*. *J. Mar. Biol. Assoc. UK* 71: 665–678.

Pierce, G.J., Boyle, P.R., Hastie, L.C. and Key, L. (1994) The life history of *Loligo forbesi* in Scottish waters. *Fish. Res.* 21: 17–41.

Polifka, J.E., Rutledge, J.C., Kimmel, G.L., Dellarco, V. and Generoso, W.M. (1996) Exposure to ethylene-oxide during the early zygotic period induces skeletal anomalies in mouse fetuses. *Teratology* 53: 1–9.

Sakurai, Y., Young, R.E., Hirota, J., Mangold, K., Vecchione, M., Clarke, M.R. and Bower, J. (1995) Artificial fertilization and development through hatching in the oceanic squids *Ommastrephes bartramii* and *Sthenoteuthis oualaniensis* (Cephalopoda: Ommastrephidae). *Veliger* 38: 185–191.

Sakurai, Y., Bower, J.R., Nakamura, Y., Yamamoto, S. and Watanabe, K. (1996) Effect of temperature on development and survival of *Todarodes pacificus* embryos and paralarvae. *Am. Malacol. Bull.* 13: 89–95.

Sauer, W.H.H. and Smale, M.J. (1993) Spawning behaviour of *Loligo vulgaris reynaudii* in shallow coastal waters of the south-eastern Cape, South Africa. In: *Recent Advances in Cephalopod Fisheries Biology* (eds T. Okutani, R.K. O'Dor and T. Kubodera). Tokai University Press, Tokyo, pp. 489–498.

Segawa, S., Yang, W.T., Marthy, H.-J. and Hanlon, R.T. (1988) Illustrated embryonic stages of the eastern Atlantic squid *Loligo forbesi*. *Veliger* 30: 230–243.

Shaw, P.W. (1997) Polymorphic microsatellite markers in a cephalopod: the veined squid Loligo forbesi. *Mol. Ecol.* 6: 297–298.

Shaw, P.W. and Boyle, P.R. (1997) Multiple paternity within the brood of single females of *Loligo forbesi* (Cephalopoda) demonstrated with microsatellite markers. *Mar. Ecol. Prog. Ser.* 160: 279–282.

Somarakis, S., Kostikas, I. and Tsiminedes, N. (1997) Fluctuating asymmetry in the otoliths of larval fish as an indicator of condition: conceptual and methodological aspects. *J. Fish. Biol.* 51: 30–38.

Vecchione, M. (1988) *In-situ* observations on a large squid-spawning bed in the eastern gulf of Mexico. *Malacologia* 29: 135–141.

Villanueva, R. (1994) Decapod crab zoeae as food for rearing cephalopod paralarvae. *Aquaculture* 128: 143–152.

Villanueva, R. (1995) Experimental rearing and growth of *Octopus vulgaris* from hatching to settlement. *Can. J. Fish. Aquat. Sci.* 52: 2639–2650.

Weischer, M. L. and Marthy, H.J. (1983) Chemical and physiological properties of the natural tranquilliser in the cephalopod eggs. *Mar. Behav. Physiol.* 9: 131–138.

Willmer, P. (1990) Evidence from embryology and larvae. In: *Patterns in Animal Evolution.* Cambridge University Press, Cambridge, pp. 100–130.

Young, R. E. and Harman, R.F. (1988) Larva, paralarva and subadult in cephalopod terminology. *Malacologia* 29: 201–207.

14

Offspring size responses to maternal temperature in ectotherms

David Atkinson, Simon A. Morley, David Weetman and Roger N. Hughes

1. Introduction

Many environmental influences can cause a mother to alter the size of her offspring (eggs, neonates). These include maternal diet or food availability, quality of oviposition site, density of conspecifics, and predation risk (Fox and Czesak, 2000). Differences in offspring size within species are also commonly observed between seasons (e.g. in fish, Chambers, 1997; in arthropods, Kerfoot, 1974; Culver, 1980; Fox and Czesak, 2000) and along latitudinal and altitudinal gradients. In most cases, temperature differences between seasons, latitudes or altitudes are inversely correlated with offspring size, although there are exceptions (Chambers, 1997; Fox and Czesak, 2000).

At least three factors can make seasonal variation ecologically important. First, the variation can be substantial (e.g. mean carbon content of eggs of the marine copepod *Pseudodiaptomus marinus* can vary by up to 100% between seasons; Liang and Uye, 1997). Second, the trade-off between offspring size and number can explain much of the observed seasonal variation in fecundity, and hence affect population dynamics (e.g. in the sea slater, *Ligia oceanica*; Willows, 1987). Third, the size of offspring is often clearly linked to their chances of survival, e.g. large newborn of *Daphnia* spp. have greater resistance to starvation than do smaller ones (Tessier and Consolatti, 1989).

Negative correlations between offspring size and environmental temperature associated with season or geographical location have been used as evidence that ectothermic animals usually produce larger offspring at lower temperatures (Yampolsky and Scheiner, 1996). Yet in such studies several other factors will co-vary with temperature (e.g. food availability, predation, rainfall). Also, these correlations may reflect genetic differences, developmental plasticity or a combination of the two, and may be complicated further by differences during ontogeny (i.e. between

Environment and Animal Development: Genes, Life Histories and Plasticity, edited by D. Atkinson and M. Thorndyke.
© 2001 BIOS Scientific Publishers Ltd, Oxford.

successive clutches; Chapter 1). Thus, to identify thermal induction of differences in the size of offspring, controlled experiments are required.

Fox and Czesak (2000) provide examples of temperature-dependent differences in egg size in arthropods, which lend some support to the conclusion of Yampolsky and Scheiner (1996), and usefully attempt to distinguish between temperature experienced during rearing and that during oviposition. Here, we present a broader review containing more arthropod studies and additionally includes molluscs, fish, a bryozoan, an amphibian and a reptile. Of course the composition of offspring may also be affected by temperature, with important ecological consequences (Bernardo, 1996) but here we focus on total size.

The processes that affect offspring size will differ depending on whether temperature affects maternal investment or subsequent egg development. We therefore distinguish between effects of temperature on: (i) maternal resource allocation to offspring size; (ii) the development of eggs or embryos that are brooded by mothers prior to birth; and (iii) developing eggs or embryos of oviparous species after all parental investment in them has ceased. Our first aim is to assess how widespread an inverse relationship is between maternal temperature and offspring size in ectothermic animals (i.e. (i) and (ii) above).

An inverse relationship between rearing temperature and final adult size (or of other developmental events late in ontogeny such as pupation or final juvenile moult) is widespread. In a review of 109 studies on diverse taxa, this relationship was observed in 83% of cases (Atkinson, 1994). Offspring size is also often correlated with the size of the mother (e.g. McKee and Ebert, 1996; Sheader, 1996). Therefore, when analysing the results of controlled experiments we take care to separate, where possible, the direct effects of maternal temperature on allocation to offspring size from its effects due solely to thermally determined variation in maternal size.

Our second aim is to evaluate the range of hypotheses offered to account for thermal effects on offspring size, and discuss the empirical evidence for these.

2. Quality of data

Experiments that provide data on offspring size responses to temperature usually use too few values to determine the shape of the relationship. We therefore focus mainly on patterns of qualitative response to temperature (statistically significant increases vs reductions in offspring size; $p < 0.05$). These studies can also often lack information that is important for distinguishing between alternative causes. We therefore took the following precautions to reduce sources of error when interpreting thermal effects on offspring size. Data on the effects of temperature were not accepted if experiments did not control for variation in factors other than temperature or when experimental temperatures were considered to be extreme for the animals, being outside the range normally experienced in nature and associated with exceptionally high mortality. Also studies were excluded from the review if there was evidence that resource availability limited the rate at which offspring dry mass was produced (e.g. in studies that compared effects of high and low food rations, data from the 'low-food' treatment was ignored if it caused a reduced rate of production of offspring biomass). However, no study actually proved that resources were always non-limiting. For instance, even though food may have been provided *ad libitum*, oxygen availability was usually not mentioned. Therefore, in practice, studies were only accepted if resources provided to the animals appeared to be abundant. Measures of offspring size varied between studies (e.g. volume or dry weight of eggs; length or dry weight of newborn), and it was not possible to convert them all to a common unit. However, care was taken only to accept measurements taken at the same stage of development (e.g. egg 'stage I' in *Gammarus insensibilis*, Sheader, 1996). Where possible, differences in maternal size and age between temperature treatments were accounted for (e.g. by calculating relative offspring size = offspring size/maternal size).

3. A general response to maternal temperature?

3.1 *Variation in offspring size*

The effort spent seeking suitable studies was similar to that which yielded about 100 appropriate studies for the synthesis by Atkinson (1994). Despite this, only 32 suitable examples were discovered that examined effects of maternal temperature on offspring size (*Tables 1* and 2).

Table 1. Number of studies showing either a significant negative, positive or no significant effect of maternal temperature during egg production or rearing on size of egg or offspring. The three right-hand columns refer to effects of temperature after accounting for effects of a correlation with size response of mother (percentages in parentheses). The literature sources are listed in Table 2

Taxon	Relationship between temperature and propagule size			Relationship after controlling for maternal size effects		
	Negative	No significant effect	Positive	Negative	No single or significant effect	Positive
Chordata						
Pisces	3	0	1	2	0	1
Amphibia	1	0	0	1	0	0
Reptilia	1	0	0	1	0	0
Arthropoda						
Crustacea						
Cladocera	10	1	0	5	3	0
Copepoda	1	0	0	1	0	0
Amphipoda	1	0	0	1	0	0
Insecta						
Coleoptera	1	0	0	1	0	0
Diptera	3	0	0	2	0	0
Orthoptera	0	2	1	0	2	1
Collembola	0	1	1	0	1	1
Mollusca						
Bivalvia	1	1	0	1	1	0
Nematoda	1	0	0			
Bryozoa	0	1	0	0	1	0
Total (%)	22 (71)	6 (19)	3 (9)	15 (58)	8 (31)	3 (11)

Table 2. Nature of relationship between maternal temperature and size of egg or newborn

Taxon	Relationship (+ any change after accounting for maternal size effects)	Reference
CHORDATA		
Osteichthyes		
Cyprinodon nevadensis nevadensis	–ve	Shrode and Gerking (1977)
Engraulis japonica	–ve(?)	Imai and Tanaka (1987)
Etheostoma spectabile	–ve	Marsh (1984)
Gambusia affinis	+ve	Vondracek *et al.* (1988)
Amphibia		
Bombina orientalis	–ve	Kaplan (1987)
Reptilia		
Sceloporus jarrovi	–ve	Beuchat (1988)
ARTHROPODA		
Crustacea		
Cladocera		
Daphnia pulex		
(clone 1)	NS	Walls and Ventelä (1998)
(clone 2)	–ve	Walls and Ventelä (1998)
(clone 3)	–ve	Walls and Ventelä (1998)
D. pulex	–ve (?)	DuFresne and Hebert (1998)
(early season)	–ve (?)	Brambilla (1982)
(late season)	–ve	Brambilla (1982)
D. pulex	–ve(+/–)	Weetman (2000)
D. middendorffiana	–ve (?)	DuFresne and Hebert (1998)
D. magna	–ve	Sakwinska (1998)
D. magna	–ve*(?)	McKee and Ebert (1996)
Simocephalus vetulus	–ve	Perrin (1988)
Copepoda		
Sinocalanus tenellus	–ve	Kimoto *et al.* (1986)
Amphipoda		
Gammarus insensibilis	–ve	Sheader (1996)
Insecta		
Coleoptera		
Notiophilus biguttatus	–ve	Ernsting and Isaaks (1997, 2000)
Diptera		
Drosophila melanogaster	–ve (?)	Crill *et al.* (1996)
D. melanogaster	–ve	Imai (1934)
Scathophaga stercoraria	–ve	Blanckenhorn (2001)
Orthoptera		
Chorthippus brunneus	NS	Willott and Hassall (1998)
C. brunneus	+ve	Vosper (1998)
Omocestus viridulus	NS	Willott and Hassall (1998)
Collembola		
Folsomia candida		
('York' clone)	+ve	Stam *et al.* (1996)
('Brunoy' clone)	NS	Stam *et al.* (1996)

Table 2. continued

Taxon	Relationship (+ any change after accounting for maternal size effects)	Reference
MOLLUSCA		
Bivalvia		
Cerastoderma edule	−ve	Honkoop and van der Meer (1998)
Mytilus edulis	NS	Honkoop and van der Meer (1998)
NEMATODA		
Caenorhabditis elegans	−ve	van Voorhies (1996)
BRYOZOA		
Celleporella hyalina	NS	Morley *et al.* (in preparation)

'−ve' indicates a significant ($p<0.05$) negative relationship between propagule size and temperature over at least two temperature values; '+ve' indicates a significant positive relationship; 'NS' indicates no effect between any pair of temperatures; '(?)' indicates that when adult size is accounted for it is not known whether a significant relationship remains, so the study was excluded from that analysis; '(+/−)' indicates that, after accounting for adult size, there were significant positive and negative relationships, depending on temperature; *indicates result is derived from a recalculation using original data.

Significant inverse relationships between maternal temperature and offspring size occurred in 23 studies (72%), whereas a positive relationship occurred in only 3 examples (9%; *Table 1*). There is thus a suggestion that an inverse relationship may be widespread among ectotherms, although it is not universal. We next consider whether this trend is simply a consequence of thermal effects on maternal size, when this correlates with offspring size. Then, we examine several of the exceptions (positive relationship or no response) in Sections 4 and 5, in order to identify possible causes of exceptions to this proposed rule.

3.2 Effects of mother's size

A positive relationship between maternal and offspring size would be expected when offspring size is at an upper physical limit of the mother (e.g. if diameter of ovipositor or birth canal is limiting). Such a constraint is most likely when one or a few offspring are released at a time (see Sinervo and Licht, 1991a, for an experimental demonstration of this in lizards). An alternative, adaptive, explanation for a correlation between size of mother and offspring was suggested by Parker and Begon (1986). They predicted that, in species adapted to an environment in which offspring experienced high competition, large mothers would maximize their fitness by investing at least some of any extra capacity for egg production in producing larger offspring. This would lead to a positive correlation between maternal and egg or offspring size.

Whatever the cause of such a positive correlation, temperature could affect the size of the offspring if temperature experienced by the growing mother subsequently

affected her size. Any effects of temperature on maternal size were excluded by experimental designs in which temperature treatments were applied only after somatic
growth had ceased (e.g. the carabid beetle *Notiophilus biguttatus*, Ernsting and Isaaks,
1997, 2000). Thermal effects on maternal size were similarly irrelevant if there was no
relationship between maternal and offspring size (e.g. the field grasshopper
Chorthippus brunneus, Vosper, 1998; the amphipod *Gammarus insensibilis*, Sheader,
1996). In some studies insufficient information was available to deduce the effects of
temperature after accounting for effects on maternal size, so the total number of
examples was reduced from 32 to 26. All the studies that were excluded because of this
lack of information showed a negative effect of temperature on offspring size. (*Table
1*). Therefore, 15 (58%) still showed an inverse relationship, and 3 (11%) showed a
positive relationship. For simplicity, the magnitudes of temperature effects on
offspring size are not given in *Table 1*, but typically, when temperature is lowered by
between 5 and 10°C, the increase in offspring mass or volume is equivalent to about
1.8–2.4% per °C (Atkinson, in preparation).

In conclusion, the rather more specific suggestion that temperature has a widespread negative effect on maternal investment into offspring size may still be true to a
degree, but needs to be tentative because the number of examples is still quite small.

Even when offspring size is inversely proportional to temperature, the effect may
not be simple. For instance, in a clonal population of the cladoceran *Daphnia pulex*
studied by Weetman (2000), significant differences in mean offspring size were only
found in two of three comparisons between temperatures (*Figure 1a*).

Also, the size of newborns varied significantly between successive broods even
within temperatures and while absolute offspring size showed the positive relationship
with maternal size expected for *Daphnia* (Boersma, 1997), offspring size actually
declined relative to the weight of the mother (*Figure 1b*) – a trend also reported for
Daphnia magna (McKee and Ebert, 1996). Moreover, while relative offspring sizes
were initially inversely proportional to temperature, their decline with successive
clutches differed between thermal regimes, so that by the fifth clutch there was no
significant difference between 13 and 21°C (*Figure 1b*). This illustrates an increased
level of complexity in the offspring size–temperature relationship when ontogenetic
changes are considered.

From the analysis in *Table 1* we conclude that indirect effects of temperature on
offspring size, via its effects on maternal size, are clearly insufficient to account for
most of the inverse relationships between maternal temperature and offspring size.
Other explanations are therefore required. Two complementary types of explanation
are physiological mechanisms (Section 4), and hypotheses based on the adaptive
significance of the maternal investment in offspring size (Section 5).

4. Potential mechanisms

4.1 *Effects on maternal reproductive allocation*

It is useful first to identify stages of the mother's ontogeny when temperature acts to
alter offspring size. Blanckenhorn (2001) transferred adult female dung flies
Scathophaga stercoraria between temperatures and showed rapid and reversible
changes in offspring size between clutches, consistent with temperature acting during
vitellogenesis.

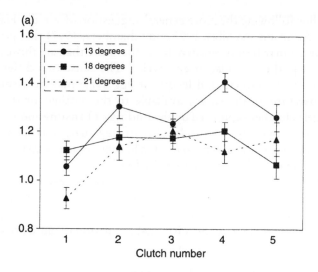

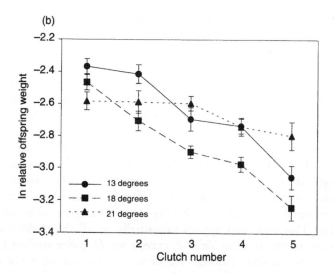

Figure 1. *Effect of temperature on: (a) weight (ln transformed), and (b) relative weight (i.e. weight divided by maternal weight, ln transformed) of newborn from a clone of* Daphnia pulex. *Means and 95% confidence intervals are shown. Offspring are larger at 13°C than at the higher temperatures. Relative weight decreases with age and on average is higher at both 13 and 21°C than at 18°C (from Weetman, 2000).*

In many Crustacea, the reproductive cycle is intimately linked to the moulting cycle. Sheader (1996) proposed that the larger size of eggs produced by *G. insensibilis* at lower temperatures was because moults, which force the termination of oocyte growth, are delayed more by low temperature than is the rate of yolk accumulation in oocytes. Thus with more time for yolk to accumulate, larger eggs are produced. A conceptually similar mechanism was proposed by Ernsting and Isaaks (1997, 2000) to explain the inverse relationship between maternal temperature and egg size in the carabid beetle, *Notiophilus biguttatus*. They argued that the accumulation of vitellogenin in oocytes was less sensitive to temperature than was the rate of production of new oocytes; thus following the more general suggestion of van der Have and de Jong (1996; see also Atkinson and Sibly, 1997 and Chambers, 1997) that growth processes (e.g. vitellogenesis) may be less sensitive to temperature than are those associated with differentiation (e.g. the follicular or egg cycle). The cause of a differential effect of temperature on oocyte growth and length of egg cycle may be associated with the effects of temperature on the energy available to the mother for the production of yolk. This model is further supported by the finding of Ernsting and Isaaks (2000) that an increase in the total rate of production of egg mass was associated with the production of smaller eggs, irrespective of whether total egg mass production was increased by temperature, by day length or by food.

The idea that egg production rate constrains egg size is consistent with the results of Sinervo and Licht (1991b), working on lizards. After performing endocrine and surgical manipulations of egg production, Sinervo and Licht (1991b) concluded that egg size is largely constrained by the hormonal regulation of clutch size.

4.2 Effect on embryo growth while in the brood pouch

For ectothermic species that brood their eggs in a pouch, maternal effects may continue for some time after completion of her investment of yolk in each egg, with a mother's body temperature also potentially affecting offspring size. Between 10 and 15°C, an inverse relationship was found between incubation temperature and egg volume at the last stage of development in an *in vitro* study of the embryonic development of *D. pulex* (Gulbrandsen and Johnsen, 1990). (Their results from incubation at 5°C appear to reinforce this trend, but are not truly comparable because eggs were removed from the brood pouch at a later developmental stage). Therefore, even though egg and newborn sizes are often positively correlated in *Daphnia* (e.g. Ebert, 1993), it is possible that at least some of the effect of maternal temperature on size of newborns occurred after the mother had completed her investment of yolk in the eggs. Larger size of newborns at lower incubation temperatures could arise if a lower proportion of the yolk is used in respiration by the developing embryo, allowing a greater amount to be converted into embryonic tissue. This argument applies potentially to all the cladoceran examples whose findings are summarised in *Table 1*, except for *D. pulex* studied by Brambilla (1982), in which eggs in early cleavage stages were removed from the mother and measured. Studies to investigate thermal effects on changes in embryo size *in situ* (i.e. within the brood pouch) are now required to test this idea further.

4.3 Effects on embryo growth after oviposition

Although the present study is primarily concerned with effects of maternal temperature, in oviparous species environmental temperature after oviposition can also affect embryonic growth, and hence the size of hatchlings. For example, Brittain *et al.* (1984) observed that the body lengths of hatchlings of the stonefly *Capnia atra* were significantly greater when eggs were incubated at 8°C than at 12 and 16°C (and also at the extreme temperature of 4°C), while the shortest nymphs hatched at 20 and 24°C. This inverse relationship with temperature is common in other taxa including fish (Blaxter, 1992). However, an exception could be found in the Orthoptera: at cooler egg incubation temperature (24°C) hatchlings of the locust *Schistocerca gregaria* had lower dry weights than did those from 31°C (Bernays, 1972).

As we suggested for embryos developing in brood pouches ((ii), above), larger hatchlings may be produced following embryonic development at cooler temperatures if the efficiency of conversion of yolk to body tissue is greater. This has been observed in a salmonid (Heming, 1982).

In summary, there are at least three ways in which lower environmental temperatures can lead to the production of larger offspring: (i) through its effect on adult size which sometimes correlates with offspring size; (ii) through the effect on weight-specific maternal investment in oocyte size, which is at least sometimes linked to total egg mass production; and (iii) through the subsequent efficiency of conversion of yolk into tissue of offspring. This latter mechanism can apply both to species that brood their embryos, and to those that oviposit and then leave their eggs to hatch.

5. Adaptive explanations

5.1 *The role of environmental predictability*

An alternative to asking '*how* does temperature affect offspring size?' is to ask '*why?*' (Chapter 1). Specifically, what, if anything, is the adaptive significance of the phenotypically plastic maternal adjustment of offspring size in response to temperature (Bernardo, 1996; Mousseau and Fox, 1998; Fox and Czesak, 2000)? A central assumption when applying this approach is that conditions experienced by eggs or offspring can be predicted to some extent by the thermal conditions experienced by the mother as she is allocating resources to her offspring (Bernardo, 1996; McGinley *et al.*, 1987). Factors increasing predictability are:

(i) short time periods between maternal investment in oocyte size and either hatching or birth. Conversely, the occurrence of egg diapause, which may take several months before it is broken, would tend to reduce predictability;

(ii) slow rate of change of environmental conditions in relation to the period between maternal investment in oocyte size and hatching or birth, or at least a very regular (predictable) change. Temperatures change relatively slowly in large water bodies such as seas or lakes, and also buried within sediments with high thermal inertia, compared with above-ground terrestrial habitats.

On the basis of the first argument, an adaptive change in offspring size in response to maternal temperature should not normally be expected in species with egg diapause. Both of the grasshopper species (Orthoptera) whose responses are summarized in *Table 1* exhibited egg diapause, and hatchlings in the field emerge several months after the eggs are laid (Marshall and Haes, 1988; Cherrill and Begon, 1991). Only one slight (5.6%) change in egg dry weight was observed in *Chorthippus brunneus* (Vosper, 1998); the other grasshopper studies revealed no effect, as expected (Willott and Hassall, 1998). We found no evidence that any other species in the review had egg diapause.

On the basis of the second argument, adaptive changes in offspring size might be expected more commonly in species from large aquatic or subterranean habitats than from above-ground terrestrial ones.

Because increased temperature generally shortens egg development time, conditions around the time of hatching are likely to be more predictable as temperatures increase. For mothers who experience conditions favourable for reproduction, we therefore propose the hypothesis that the reduced predictability at lower temperatures would favour larger offspring to cope with greater uncertainty (e.g. in food availability). In principle, this could provide a general explanation for inverse relationships between maternal temperature and offspring size, although we consider it more likely to be one of several possible contributory factors, outlined below.

5.2 *Advantages from shifts in offspring size*

Four hypotheses are now discussed which, individually or in combination, could explain thermally induced shifts in offspring size in those species in which conditions for developing embryos, hatchlings or neonates are predictable from maternal thermal conditions. We have made a start at investigating one of these – the oxygen limitation hypothesis (hypothesis iii, below) – but the rest still require testing.

It should be noted that, in principle, the thermal effect on offspring size may result from a fundamental physiological constraint (e.g. van der Have and de Jong, 1996) rather than an adaptation to different thermal environments. Blanckenhorn (2001), for instance, was unable to find fitness gains from producing smaller eggs at higher temperatures in the yellow dung fly, *Scathophaga stercoraria*.

Hypothesis i: adapting to the expected growth rate of newborn and hatchlings. When resources are abundant, the growth rate of newborn and hatchlings is usually slowed by cool temperature (Perrin, 1988). If this slow growth is predictable from conditions experienced by the mother during egg production, natural selection is predicted to favour reaction norms in which larger offspring are produced at lower temperatures (Sibly and Calow, 1983; Taylor and Williams, 1984; Perrin, 1988). This could provide a general adaptive explanation for the inverse relationship between maternal temperature and offspring size when conditions for newborn or hatchlings are predictable from conditions affecting the mother during the period of investment in offspring production.

Hypothesis ii: adapting to expected length of juvenile period. Another explanation that could apply generally was proposed by Yampolsky and Scheiner (1996). They argued that, since juvenile development is slowed by lower temperature, the benefits from a lowered instantaneous mortality rate caused by increased offspring size are magnified by the longer time spent as a juvenile. Yet the length of the juvenile period may not be readily predictable from maternal thermal conditions, especially in species with long generation times.

Thermal effects on juvenile mortality can also affect the optimal response of offspring size to temperature. This could oppose the predictions of hypotheses i and ii if daily juvenile mortality rate decreases with decreasing temperature. Indeed, Yampolsky and Scheiner note that the effect they describe can disappear altogether if daily mortality decreases faster than the increase in time to maturity. Sibly and Atkinson (1994) found that in all eight insect studies they analysed, daily survival rates were improved by (non-extreme) lowered temperature. Although these mortality data were from laboratory rather than field experiments, they highlight a potentially important confounding factor. These ideas should now be tested.

There are still several exceptions to the inverse association between temperature and offspring size that have not been dealt with by the mechanistic hypotheses and the two adaptationist arguments outlined so far. The following hypotheses are able to address some of these.

Hypothesis iii: adapting to expected oxygen supply and demand. For species in which oxygen supply to large developing embryos is potentially limiting, we predict that large embryos prior to hatch or birth would be more likely to suffer oxygen limitation at higher than at lower temperatures. This is because rates of respiration are generally able to accelerate faster under increased temperature than diffusion of oxygen (Woods, 1999). The relative paucity of oxygen at higher temperatures would be most acute for eggs or embryos in aquatic media (cf. air), because the concentration of dissolved oxygen in water surrounding the eggs or embryos decreases as temperature is raised (Davis, 1975). The difference in oxygen availability between conditions of high and low temperature may be increased still further for eggs or embryos that are surrounded by respiring or decaying organic material (e.g. in organic soil, or within non-photosynthesizing tissue of ectothermic species), which will deplete any available oxygen more quickly as temperature is raised. If limiting oxygen supply were predictable from thermal conditions experienced by the mother during the production of offspring, then it would be adaptive for the mother to avoid wasteful or potentially injurious investment in large offspring at increased temperatures. Instead, smaller size would be favoured because of a greater surface area:mass ratio (Atkinson and Sibly, 1996; Woods, 1999), an advantage magnified by the resulting increase in potential clutch size. Ernsting and Isaaks (1997) applied to the carabid beetle, *N. biguttatus*, the idea that the larger surface area:volume ratio of smaller eggs allows a higher metabolic rate (Bradford, 1990). They argued that large eggs would survive less well at high developmental temperatures, but observed no association between egg size and their survival. However, this result does not invalidate the 'adaptation to oxygen limitation' hypothesis. It is not clear, for example, whether the oxygen availability for developing eggs in the experiment was as un/limiting for metabolism as it is in natural oviposition sites to which the species is assumed to be adapted. Oxygen depletion at increased temperature would be greater in natural sites if, for instance, they were more organic-rich.

We have begun to investigate the role of oxygen limitation on the colonial modular bryozoan *Celleporella hyalina* (*Figure 2a*), which lives attached to seaweeds and other hard substrata in the littoral and shallow sub-littoral zones (S.A. Morley, R.N. Hughes and D. Atkinson, in preparation). The oxygen limitation hypothesis predicts, in the absence of perfect compensatory responses that would prevent oxygen limitation, that under normoxia maximum embryo size will be smaller at the higher temperatures, and that experimentally providing additional oxygen would counter this to some extent. However, maximum embryo size in *C. hyalina* was not reduced by temperature or by oxygen, but morphological changes that are consistent with adaptations to reduce oxygen limitation were indeed observed. Brood chambers (*Figure 2b*) containing individual embryos had both a greater density and total number of membrane-bound pores (through which gaseous exchange is thought to occur) at the higher temperature (10°C, mean±95% CI = 115±13 pores mm^{-3}; 18°C, 170±22 pores mm^{-3}; $F_{1,9}$=44.26, $p<0.01$), and were themselves smaller, hence reducing diffusion distance for oxygen (10°C, 0.0070±0.0007 mm^3; 18°C, 0.0062±0.0006 mm^3; $F_{1,8}$=13.06, $p<0.01$). Thus, if oxygen limitation were an important and widespread selective pressure on the temperature-induced alteration of propagule size, then the lack of response in this bryozoan (indicated in *Table 1*) may be because the adult instigates mechanisms to prevent oxygen limitation for the embryos.

Neither of these changes responded directly to increased oxygen tension (pore density $F_{1,9}$=0.05, $p=0.83$; ovicell volume $F_{1,8}$=4.09, $p=0.08$). We tentatively suggest

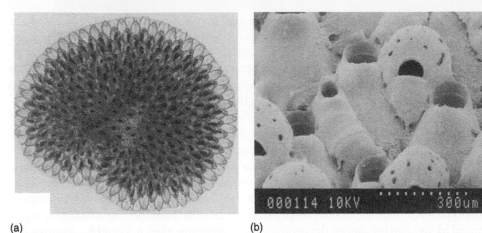

(a)   (b)

Figure 2. (a) *The colonial modular bryozoan* Celleporella hyalina *viewed from below through a clear acetate sheet. Individual feeding modules can be seen. The maximum diameter is ca 10 mm.* (b) *A scanning electronmicrograph showing the skeleton of part of a colony of* C. hyalina – *an oblique view from above. Near-spherical brood pouches (ovicells) are covered in small pores through which oxygen may pass to supply the individual developing embryo. The ovicells are distributed among two types of more elongated modules – feeding modules (autozooids) and smaller frontal male modules. Summer temperature (18°C) induces the production of smaller ovicells yet with more pores than at a spring/autumn temperature (10°C). This is consistent with morphological responses that would increase oxygen supply to the developing embryo (see text).*

that temperature rather than oxygen induces a morphological response during the construction of the ovicell because it is a better predictor of conditions throughout the ovicell's functional life. An ovicell provides a brood pouch for at least two successive embryos over several weeks, depending on temperature (personal observation). Because of the gradual seasonal change in sea temperature, the temperature at one time will correlate with that occurring several weeks into the future. By contrast, we expect any deviations from full oxygen saturation in seawater that is experienced by a colony in the field to be associated with eutrophic episodes and frequently changing balance of production to respiration within the seaweed habitat – and hence be unreliable predictors of conditions several weeks or months later. Hence, we expect temperature but not oxygen to be a reliable cue for conditions affecting the whole lifetime of the functioning ovicell. From this, natural selection would be expected to favour the use of temperature, rather than oxygen, to induce morphological responses that would reduce future and present oxygen limitation. Further investigations are planned, in which the total pore area is manipulated under different conditions of temperature and oxygen.

The only other aquatic species in our review for which there was a positive association between temperature and offspring size was the poeciliid fish, *Gambusia affinis* (Vondracek *et al.*, 1988). This differed from the three other fish species, which released eggs into the aquatic environment, by being ovoviviparous. Here, the oxygen limitation hypothesis could account for the difference in the phenotypic response by this species compared with the others if the lack of an egg membrane, and parental care by the brooding parent, prevented oxygen limitation for large embryos at increased temperatures. This suggestion of mechanisms to prevent oxygen limitation parallels that for *C. hyalina* (above). By contrast, eggs released into the surrounding habitat by the other species will lose the opportunity for parental care that could help compensate for effects of oxygen limitation for large embryos at high temperature.

Hypothesis iv: adapting to expected desiccation risk. Above 0°C, warming normally leads to increased risks of desiccation due to accelerated evaporation, especially in drier, terrestrial habitats. Large size can enhance survival in desiccating conditions, probably because of the smaller surface area:volume ratio for water loss (e.g. in hatchling grasshoppers, Cherrill, 1987). Large hatchlings or newborn may also be able to progress faster into a developmental stage, or disperse to a habitat, in which the animal is less vulnerable to desiccation. Hence, it is predicted that in environments that are potentially desiccating for offspring at increased temperatures, and if this can at least partly be predicted from maternal thermal conditions, then selection should favour the production of larger offspring at higher maternal temperatures. The only insect in our review without egg diapause that produced larger eggs at higher maternal temperature was the 'York' clone of the collembolan *F. candida*. Notably, Stam *et al.* (1996) observed that hatchlings of this species are naturally prone to desiccation. It is not known whether the 'Brunoy' clone, which showed no change in egg size, was adapted to an environment with a weaker correlation between temperature and desiccation risk.

5.3 *Correlation with adaptive shifts in clutch size*

An important point is that it may be difficult to predict the magnitude, or even direction, of a temperature effect on offspring size without considering effects on clutch size (Ernsting and Isaaks, 2000; Yampolsky and Scheiner, 1996). An increase in number of offspring per clutch or brood is normally at the expense of potential size of individual offspring. A negative relationship, or trade-off, between offspring size and number, when adult size was accounted for, was observed in several of the studies reviewed (e.g. McKee and Ebert, 1996; Weetman, 2000). It is therefore possible that thermally induced shifts in offspring size could have arisen as by-products of stronger selection on offspring number, for example, if increased temperature were associated with increased competition for limited resources. Such a relationship would favour smaller clutches at increased temperatures, but even this can result at least partly from selection for large offspring rather than for small clutch size per se (Parker and Begon, 1986). Moreover, this size response to temperature is the opposite of that actually observed in most species reviewed here.

The effect of temperature on optimal egg size may also depend on clutch size when embryos compete for oxygen, as is observed in aquatic egg masses (Lee and Strathmann, 1998; Woods, 1999). Here, the size of interstices between eggs can be important for oxygen supply, which will depend on both clutch size and egg size (Lee and Strathmann, 1998), and which should be at a greater premium at increased temperatures.

However, the inverse relationship between temperature and offspring size need not rely on changes in clutch size. Blanckenhorn (2001) observed the usual inverse relationship even though clutch size was unaffected by temperature.

6. Conclusion

We have reviewed the qualitative shift in the size of eggs and newborn produced by mothers who experience altered temperature (including during gametogenesis) in ectothermic animals. Studies were carefully screened and a greater taxonomic breadth was covered than in previous reviews. Our findings lend considerable weight to previous claims that temperature is usually (but not always) inversely associated with offspring size. This applies, but to a lesser extent, even when maternal size effects on offspring size are excluded. However, the suggestion that an inverse relationship is truly taxonomically widespread must remain tentative until more examples are accumulated. There are several lines of empirical evidence that show a reduction in egg size is associated with an increase in the total rate of production of egg mass. This association may be a physiological constraint on offspring size, or might represent a mechanism that will produce an adaptive shift in offspring size (e.g. as a correlate of conditions for offspring growth or development). We argue that the adaptationist hypotheses that relate to fitness consequences around the time of birth or hatching should not be applied to species or populations for which maternal temperature is a poor predictor of environmental conditions faced by offspring around that time (e.g. those with egg diapause). Some of our findings are consistent with the hypothesis that temperature may affect oxygen limitation for embryos prior to hatch or birth thereby altering optimal offspring size. This is most likely to apply in aquatic and organic-rich environments. Our data on responses by the bryozoan, *Celleporella hyalina*, show that maximum embryo size is not reduced by increased temperature and not limited by oxygen tension at normoxia, yet two potential adaptations (volume of brood chamber and density of pores in its skeleton) do respond to temperature in a manner that would increase oxygen supply to embryos. An association between maternal temperature and predictable desiccation risk for eggs, hatchlings or neonates in some terrestrial environments might account for positive relationships between temperature and offspring size, but like the other hypotheses presented here, it requires testing. Indeed, the fact that much of this chapter has discussed ideas rather than data indicates the need for more experimental work to understand an important biological relationship which, as we have shown here, is rather widely observed.

Acknowledgements

DW was supported by a NERC (UK) studentship. The bryozoan work was supported by NERC grant GR3/11355 to RNH and DA. We thank Wolf Blanckenhorn for allowing us to refer to his unpublished results prior to publication.

References

Atkinson, D. (1994) Temperature and organism size: a biological law for ectotherms? *Adv. Ecol. Res.* 25: 1–58.

Atkinson, D. and Sibly, R.M. (1996) On the solutions to a major life history puzzle. *Oikos* 77: 359–365.

Atkinson, D. and Sibly, R.M. (1997) Why are organisms bigger in colder environments? Making sense of a life history puzzle. *Trends Ecol. Evol.* 12: 235–239.

Bernardo, J. (1996) The particular maternal effect of propagule size, especially egg size: patterns, models quality of evidence and interpretations. *Am. Zool.* 36: 216–236.

Bernays, E.A. (1972) Some factors affecting size in first-instar larvae of *Schistocerca gregaria* (Forskal). *Acrida* 1: 189–195.

Beuchat, C.A. (1988) Temperature effects during gestation in a viviparous lizard. *J. Therm. Biol.* 13: 135–142.

Blanckenhorn, W.U. (2001) Temperature effects on egg size and their fitness consequences in the yellow dung fly *Scathophaga stercoraria. Evol. Ecol.* (in press).

Blaxter, J.H.S. (1992) The effect of temperature on larval fishes. *Neth. J. Zool.* 42: 336–357.

Boersma, M. (1997) Offspring size and parental fitness in *Daphnia magna. Evol. Ecol.* 11: 439–450.

Bradford, D.F. (1990) Incubation time and rate of embryonic development in amphibians: the influence of ovum size, temperature, and reproductive mode. *Physiol. Zool.* 63: 1157–1180.

Brambilla, D.J. (1982) Seasonal variation of egg size and number in a *Daphnia pulex* population. *Hydrobiologia* 97: 233–248.

Brittain, J.E., Lillehammer, A. and Saltveit, S.J. (1984) The effect of temperature on intraspecific variation in egg biology and nymphal size in the stonefly, *Capnia atra* (Plecoptera). *J. Anim. Ecol.* 53: 161–169.

Chambers, R.C. (1997) Environmental influences on egg and propagule sizes in marine fishes. In: *Early Life History and Recruitment in Fish Populations* (eds R.C. Chambers and E.A. Trippel). Chapman and Hall, London, pp. 63–102.

Cherrill, A.J. (1987) *The Development and Survival of the Eggs and Early Instars of the Grasshopper* Chorthippus brunneus *Thunberg in North West England*. Unpublished Ph.D. thesis, University of Liverpool.

Cherrill, A. and Begon, M. (1991) Oviposition date and pattern of embryogenesis in the grasshopper *Chorthippus brunneus* (Orthoptera, Acrididae). *Holarctic Ecol.* 14: 225–233.

Crill, W.D., Huey, R.B. and Gilchrist, G.W. (1996) Within- and between-generation effects of temperature on the morphology and physiology of *Drosophila melanogaster. Evolution* 50: 1205–1218.

Culver, D. (1980) Seasonal variation in the sizes at birth and first reproduction in Cladocera. In: *Evolution and Ecology of Zooplankton Communities* (ed W.C. Kerfoot). University Press of New England, Hanover, pp. 358–366.

Davis, J.C. (1975) Minimal dissolved oxygen requirements of aquatic life with emphasis on Canadian species: a review. *J. Fish. Res. Bd Can.* 32: 2295–2332.

DuFresne, F. and Hebert, P.D.N. (1998) Temperature-related differences in life-history characteristics between diploid and polyploid clones of the *Daphnia pulex* complex. *Ecoscience* 5: 433–437.

Ebert, D. (1993) The trade-off between offspring size and number in *Daphnia magna* – the influence of genetic, environmental and maternal effects. *Archiv Hydrobiol. Suppl.* 90: 453–473.

Ernsting, G. and Isaaks, A. (1997) Effects of temperature and season on egg size, hatchling size and adult size in *Notiophilus biguttatus. Ecol. Entomol.* 22: 32–40.

Ernsting, G. and Isaaks, A. (2000) Ectotherms, temperature and trade-offs: size and number of eggs in a carabid beetle. *Am. Nat* 155: 804–813.

Fox, C.W. and Czesak, M.E. (2000) Evolutionary ecology of progeny size in arthropods. *Annu. Rev. Entomol.* 45: 341–369.

Gulbrandsen, J. and Johnsen, G.H. (1990) Temperature-dependent development of parthenogenetic embryos in *Daphnia pulex* De Geer. *J. Plankton Res.* 12: 443–453.

Heming, T.A. (1982) Effects of temperature on utilization of yolk by Chinook salmon (*Oncorhyncus tshawytscha*) eggs and alevins. *Can. J. Fish. Aquat. Sci.* 39: 184–190.

Honkoop, P.J.C. and van der Meer, J. (1998) Experimentally induced effects of water temperature and immersion time on reproductive output of bivalves in the Wadden Sea. *J. Exp. Mar. Biol. Ecol.* 220: 227–246.

Imai, C. and Tanaka, S. (1987) Effect of sea water temperature on egg size of Japanese anchovy. *Nippon Suisan Gakkaishi* 53: 2169–2178.

Imai, T. (1934) The influence of temperature on egg size and variation in *Drosophila melanogaster. Wilhelm Roux Arch. Entwicklungsmech.* 132: 206–219.

Kaplan, R.H. (1987) Developmental plasticity and maternal effects of reproductive characteristics in the frog, *Bombina orientalis. Oecologia* 71: 273–279.

Kerfoot, W.C. (1974) Egg size cycle of a cladoceran. *Ecology* 55: 1259–1270.

Kimoto, K., Uye, S.-I. and Onbé, T. (1986) Egg production of a brackish-water calanoid copepod *Sinocalanus tenellus* in relation to food abundance and temperature. *Bull. Plankton Soc. Japan* 33: 133–145.

Lee, C.E. and Strathmann, R.R. (1998) Scaling of gelatinous clutches: effects of siblings' competition for oxygen on clutch size and parental investment per offspring. *Am. Nat* 151: 293–310.

Liang, D. and Uye, S. (1997) Seasonal reproductive biology of the egg-carrying calanoid copepod *Pseudodiaptomus marinus* in a eutrophic inlet of the Inland Sea of Japan. *Mar. Biol.* 128: 409–414.

Marsh, E. (1984) Egg size variation in central Texas populations of *Etheostoma spectabile* (Pisces: Percidae). *Copeia* 1984: 291–301.

Marshall, J.A. and Haes, E.C.M. (1988) *Grasshoppers and Allied Insects of Great Britain and Ireland.* Harley Books, Colchester.

McGinley, M.A., Temme, D.H. and Geber, M.A. (1987) Parental investment in offspring in variable environments – theoretical and empirical considerations. *Am. Nat* 130: 370–398.

McKee, D. and Ebert, D. (1996) The interactive effects of temperature, food level and maternal phenotype on offspring size in *Daphnia magna. Oecologia* 107: 189–196.

Mousseau, T.A. and Fox, C.W. (eds) (1998) *Maternal Effects as Adaptations.* Oxford University Press, Oxford.

Parker, G.A. and Begon, M. (1986) Optimal egg size and clutch size – effects of environment and maternal phenotype. *Am. Nat* 128: 573–592.

Perrin, N. (1988) Why are offspring born larger when it is colder? Phenotypic plasticity for offspring size in the cladoceran *Simocephalus vetulus* (Müller). *Func. Ecol.* 2: 283–288.

Sakwinska, O. (1998) Plasticity of *Daphnia magna* life history traits in response to temperature and information about a predator. *Freshwater Biol.* 39: 681–687.

Sheader, M. (1996) Factors influencing egg size in the gammarid amphipod *Gammarus insensibilis. Mar. Biol.* 124: 519–526.

Shrode, J.B. and Gerking, S.D. (1977) Effects of constant and fluctuating temperatures on reproductive performance of a desert pupfish *Cyprinodon n. nevadensis. Physiol. Zool.* 50: 1–10.

Sibly, R.M and Atkinson, D. (1994) How rearing temperature affects optimal adult size in ectotherms. *Func. Ecol.* 8: 486–493.

Sibly, R.M. and Calow, P. (1983) An integrated approach to life-cycle evolution using selective landscapes. *J. Theor. Biol.* 102: 527–547.

Sinervo, B. and Licht, P. (1991a) Proximate constraints on the evolution of egg size, number, and total clutch mass in lizards. *Science* 252: 1300–1302.

Sinervo, B. and Licht, P. (1991b). Hormonal and physiological control of clutch size, egg size and egg shape in side-blotched lizards (*Uta stansburiana*): constraints on the evolution of lizard life histories. *J. Exp. Zool.* 257: 252–264.

Stam, E.M., Vandeleemkule, M.A. and Ernsting, G. (1996) Trade-offs in the life-history and energy budget of the parthenogenetic collembolan *Folsomia candida* (Willem). *Oecologia* 107: 283–292.

Taylor, P.D. and Williams, G.C. (1984) Demographic parameters at evolutionary equilibrium. *Can. J. Zool.* 62: 2264–2271.

Tessier, A.J. and Consolatti, N.L. (1989) Variation in offspring size in *Daphnia* and consequences for individual fitness. *Oikos* 56: 269–276.

van der Have, T.M. and de Jong, G. (1996) Adult size in ectotherms – temperature effects on growth and differentiation. *J. Theor. Biol.* 183: 329–340.

van Voorhies, W.A. (1996) Bergmann size clines: a simple explanation for their occurrence in ectotherms. *Evolution* 50: 1259–1264.

Vondracek, B., Wurtsbaugh, W.A. and Cech J.J. Jr (1988) Growth and reproduction of the mosquitofish, *Gambusia affinis*, in relation to temperature and ration level: consequences for life history. *Environ. Biol. Fishes* 21: 45–57.

Vosper, K.A. (1998) *An Investigation into the Effects of Radiant Heat on Life History and Behaviour in the Field Grasshopper,* Chorthippus brunneus. Unpublished Ph.D. Thesis, University of Liverpool.

Walls, M. and Ventelä, A.-M. (1998) Life history variability in response to temperature and *Chaoborus* exposure in three *Daphnia pulex* clones. *Can. J. Fish. Aquat. Sci.* **55**: 1961–1970.

Weetman, D.W. (2000) *An Experimental Investigation of Temperature- and Predator-Induced Phenotypic Plasticity in* Daphnia *spp. and Guppies,* Poecilia reticulata. Unpublished Ph.D. thesis, University of Liverpool.

Willott, S.J. and Hassall, M. (1998) Life-history responses of British grasshoppers (Orthoptera: Acrididae) to temperature change. *Func. Ecol.* **12**: 232–241.

Willows, R.I. (1987) Intrapopulation variation in the reproductive characteristics of two populations of *Ligia oceanica* (Crustacea, Oniscoidea). *J. Anim. Ecol.* **56**: 331–340.

Woods, H.A. (1999) Egg-mass and cell size: effects of temperature on oxygen distribution. *Am. Zool.* **39**: 244–252.

Yampolsky, L.Y. and Scheiner, S.M. (1996) Why larger offspring at lower temperatures – a demographic approach. *Am. Nat* **147**: 86–100.

15

Growth strategies of ectothermic animals in temperate environments

Karl Gotthard

1. Introduction

Individual growth is a fundamental biological process and a typical component of juvenile development. The study of juvenile growth and development has long been of interest for a range of biological sub-disciplines including physiology, ecology and evolution. This chapter mainly concerns evolutionary aspects of juvenile growth, and in particular I will try to relate empirical findings to recent advances in life history theory. I will focus on how the evolution of growth strategies may be influenced by time constraints that are caused by seasonality. To keep this a manageable review I have restricted it to ectothermic animals, mainly because they share physiological features that are important in environments where temperature varies seasonally. I will first review the relevant theory and then discuss empirical results from three major groups of ectothermic animals (insects, fish, amphibians) in the light of the theoretical framework.

2. Theory

2.1 *Life history theory and growth*

Life history theory sees the scheduling of events such as growth, sexual maturation and reproduction as the result of strategic decisions over an organism's life (Roff, 1992; Stearns, 1992; McNamara and Houston, 1996). The theory attempts to define organisms by relatively few demographic traits, called life history traits, that show a close relationship with fitness (i.e. size at birth, growth patterns, age and size at maturity, age and size-specific investments in reproduction, number and quality of offspring). Life history theory tries to explain how natural selection has shaped these traits to produce life cycles that fit the requirements of the environment. Life history analysis is also a tool for investigating hypotheses of evolutionary causes and of adaptations in general (Stearns, 1986; McNamara and Houston, 1996; Nylin and Gotthard, 1998). This is because any heritable

Environment and Animal Development: Genes, Life Histories and Plasticity, edited by D. Atkinson and M. Thorndyke.
© 2001 BIOS Scientific Publishers Ltd, Oxford.

feature of an organism, be it behavioural, morphological or physiological, that affects the traits listed above will be under some type of selection.

Juvenile growth, age and size at maturity. The onset of reproduction is a key event in the life of most organisms, and in evolutionary terms it divides life into preparation and fulfilment (Stearns, 1992). Age and size at first reproduction are fundamental life history traits that often have a strong influence on fitness. A short juvenile period is beneficial because it reduces the risk of being killed before reproduction and entails a short generation time, while a large adult size is often correlated with high female fecundity and competitive ability in both males and females (Roff, 1992; Stearns, 1992). At some level a trade-off between these traits is expected since, all other things being equal, it will take longer to grow to a larger size. Nevertheless, the range of possible combinations of age and size at maturity for a particular individual is deter-mined by its juvenile growth trajectory (*Figure 1*). Consequently, variation in growth patterns will affect the age and size at maturity and the study of juvenile growth may be crucial for understanding life history evolution in general. Variation in growth trajec-tories may be due to inevitable variation in the quality of the environment but, as will be argued in this chapter, it may also be the result of an adaptive balancing of benefits and costs associated with different patterns of growth.

Growth rate is a pivotal aspect of growth trajectories and most life history models have assumed that selection should favour a maximization of juvenile growth rate. The rationale for this assumption is quite intuitive; individuals that maximize their juvenile growth rates will have the potential to reach the largest possible size in the shortest possible time. Growth rate will then be directly determined by the quality of the environment that in turn depends on factors such as food availability and ambient

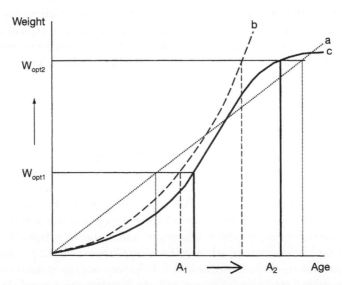

Figure 1. *Three hypothetical growth trajectories where growth rates are constant, increasing or decreasing at large sizes (i.e. they differ in how the slope of the growth trajectories change with size). When the optimal size of maturation for some reason changes from W_{opt1} to W_{opt2} the age at maturation increases in all three cases but the time sequence of maturation of the three growth trajectories is altered (from $a - b - c$ to $b - c - a$).*

temperature. Despite these arguments there is a growing body of evidence that individual organisms are often growing at a lower rate than they are physiologically capable of (Case, 1978; Arendt, 1997; Nylin and Gotthard, 1998). Moreover, genetic variation within and between populations has frequently been identified (Case, 1978; Arendt, 1997; Blanckenhorn, 1998a; Simons *et al.*, 1998). These observations suggest that the *optimal* growth rate of an individual organism in a particular environmental setting is not necessarily the same as the *maximal*, and that high growth rates might be associated with fitness cost. Consequently, much of the observed variation in growth rate may be the result of an adaptive balancing of costs and benefits associated with growth, which may result in different optima in different populations. Alternatively, individuals within a population may adjust their growth rates in relation to environmental conditions that influence the balance between costs and benefits of a particular growth pattern. In other words individual growth may be determined by a genetically imprinted *growth strategy* that can be modified by natural selection (Gotthard, 1999).

State-dependent life history theory. State-dependent life history theory is emerging as one of the major conceptual and theoretical tools for understanding the evolution of life histories (McNamara and Houston, 1996). The basic idea of this approach is that the outcome of life history decisions (i.e. when and how to grow, mature and reproduce) depends on the *state* of the organism. In any particular case the relevant individual state may include different aspects of the organism's physiological condition but also external circumstances. A life history strategy is seen as a genetically coded decision rule that specifies how an individual should respond to its internal state and its environment. These decision rules are the targets of natural selection (Houston and McNamara, 1992; McNamara and Houston, 1996). To some extent, state-based approaches have been used to model strategic decision making during development and growth of individuals (Rowe and Ludwig, 1991; Hutchinson *et al.*, 1997). Moreover, state-dependent decision making may serve as a conceptual model for finding mechanisms of strategic growth (Gotthard, 1998; Metcalfe *et al.*, 1998; Gotthard *et al.*, 1999, 2000).

2.2 *The costs of juvenile development*

The costs of growing fast. An important result of theoretical and empirical studies of adaptive growth adjustment is that high juvenile growth rates are likely to carry fitness costs, and in certain circumstances it is beneficial for individuals to avoid these costs by growing more slowly (Sibly and Calow, 1986; Werner, 1986; Lima and Dill, 1990; Stockhoff, 1991; Werner and Anholt, 1993; Gotthard *et al.*, 1994; Skelly, 1994; Anholt and Werner, 1995, 1998; Abrams *et al.*, 1996; Abrams and Rowe, 1996; Arendt, 1997; Blanckenhorn, 1998a; Nylin and Gotthard, 1998; Gotthard, 2000). The costs of high growth rates that have been proposed can roughly be divided into three categories: physiological, developmental and ecological costs.

Physiological costs. Arguments for the presence of physiological costs are based on the notion that high growth rates are associated with physiological conditions that entail a lower ability to endure adverse environmental conditions. This could, for instance, be periods of starvation or sub-optimal temperatures (Sibly and Calow, 1986; Conover and Present, 1990; Stockhoff, 1991; Gotthard et al., 1994; Arendt, 1997). A possible mechanism for this trade-off is that high growth rates are associated with high metabolic rates that would more quickly deplete stored resources during, for example, a period of starvation (Conover and Present, 1990; Stockhoff, 1991; Gotthard et al., 1994). Alternatively, fast growing individuals are allocating more resources to growth processes and relatively less to energy storage that could be used during periods of food shortage (Chippindale et al., 1996).

Developmental costs. It has been suggested that a high juvenile growth rate may be associated with a lower degree of developmental 'quality control' and thus with a higher probability of developmental error (Sibly and Calow, 1986; Arendt, 1997). This mechanism may explain observations of negative phenotypic and genetic correlations between growth or developmental rates and various measures of both juvenile and adult viability (see references in Arendt, 1997; and in Nylin and Gotthard, 1998).

Ecological costs. In animals a high growth rate typically necessitates high foraging activity, which increases the encounter rate with, and detection by, predators. This relationship leads to a fundamental trade-off between growth rate and predation risk (Lima and Dill, 1990; Werner and Anholt, 1993). It is clear that a wide range of animals lower their foraging activity in response to perceived increases in predation risks and that this lowers their growth rates. Moreover, there is direct observational and experimental evidence that high levels of activity and high growth rates increase the risk of predation (Bernays, 1997; Anholt and Werner, 1998; Gotthard, 2000).

The cost of a long juvenile period. Although slow juvenile growth may reduce *mortality rate* a slowly growing individual will typically have a long developmental period, which will increase the risk of being killed before sexual maturity *per se*. This is in fact the most commonly assumed cost of juvenile development in life history theory (Roff, 1992; Stearns, 1992). Hence, there are two fitness costs of juvenile growth that exert opposing selection pressures and it will be important to determine which of these costs are likely to be most important.

The chance of a juvenile surviving to reproduction (S_{juv}) can be expressed as a function of the daily survival chance (s) and the juvenile period in days (t):

$$S_{juv} = s^t \qquad (1)$$

Equation (1) shows that, in the simplest possible case, differences in development time (t) will typically be more important than differences in daily survival rate (s) and, consequently, the *costs of growing fast* might be less important than the *costs of growing for a long time*. However, in temperate areas the seasonal timing of life history events is of great importance and this is likely to influence the simple equation of juvenile survival presented above. In the next section I will discuss how seasonality might influence the relative importance of a high survival rate (s) and a short development time (t).

2.3 Seasonality – time horizons

Time horizons. Practically all environments are seasonal in that there is predictable variation in climatic variables over the year. In temperate regions a large proportion of the year is unsuitable for growth and reproduction for most organisms, and for ectotherms this is particularly true. The life cycles of temperate ectotherms must be completed within the limits set by seasonality, and the timing of life history events will have a strong influence on fitness. For example, reproductive success typically depends on the occurrence of resources that vary seasonally (i.e. food and temperatures that allow activity). Moreover, in sexually reproducing species it is indeed of crucial importance to start reproductive activities (mate searching, establishing mating territories, mating etc.) at a time of year when sexually mature potential mates are likely to be present. As a consequence, populations in temperate regions typically show a pattern of strong seasonal synchronization of generations. The amount of time that is available for growth and reproduction will be limited and therefore organisms will face *time horizons*. A time horizon can be defined as 'a period of time of predictable length' and they are typical components of temperate habitats.

In order to survive the winter most temperate ectotherms have to reach a certain developmental stage or a minimum body size before the onset of bad conditions. If an early maturing genotype has time enough to produce a whole new generation that can reach the overwintering stage/size before the onset of winter, a shortening of the juvenile period (smaller t in equation (1)) is likely to be beneficial. However, if the prevailing time horizon rules out an additional generation it will instead be important to maximize survival rate (s in equation (1)) until the next reproductive opportunity. Hence, when time horizons restrict potential shortening of the juvenile period (t) it might be beneficial to increase survival rate (s) by lowering juvenile growth rate, and thus, increase the total chance of juvenile survival (S_{juv}).

Consider a semelparous organism with an annual life cycle in a region where time horizons never permit more than one generation per year. The exact length of the time horizon will vary between years due to weather conditions and between individuals within years because of differences in microclimate and time of start. Within a population some individuals are likely to have more time available for growth and development than others, although there is never time enough to shorten the juvenile period. This type of life cycle is common in temperate insects but the logic can easily be extended to populations with either more or fewer than one generation per year, as well as populations where the number of generations varies among years. In fact, most organisms in temperate areas are likely to experience selection for synchronization of the life cycle *simultaneously* as they experience selection for utilizing the available time efficiently. Thus, we might expect to find adaptations that allow organisms to predict time horizons and to adjust their growth and development accordingly.

Predicting time horizons. In temperate areas variation in climatic variables allow organisms to predict time horizons and the most commonly used cue of seasonal change is the photoperiod, which gives relatively noise-free information about the time of year (*Figure 2a*). Photoperiodic sensitivity has been documented in a wide range of temperate organisms (Beck, 1980; Gwinner, 1981; Hoffman, 1981; Saunders, 1981; Tauber *et al.*, 1986; Danks, 1994). Temperature is another important seasonal cue, although a large proportion of the variation in temperatures is non-predictable (*Figure 2b*). Furthermore, the ambient temperature has large direct effects on practically all physiological processes and this is of course particularly important in ectothermic organisms. Hence, in respect to growth and development of temperate ectotherms the ambient temperature has a dual role: to some degree it predicts time

(a)

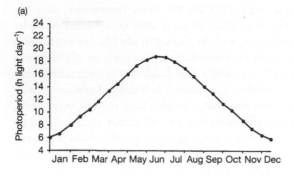

(b)

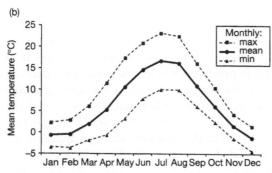

Figure 2. Example of yearly variation in climatic variables in central Sweden that predict time of year. (a) Photoperiod in the Stockholm area excluding twilight, and (b) mean monthly temperature in Målilla (in the region of Småland) flanked by the monthly maximal and minimal temperatures (values are means of daily recordings during the period 1988–1997; the data were provided by the Swedish Meteorological and Hydrological Institute). The difference between minimum and maximum temperatures represents variation that does not predict time of year. In contrast, photoperiod is a relatively 'noise-free' predictor of time of year.

horizons, while at the same time it is an important direct determinant of growth and developmental rates (Atkinson, 1994).

2.4 *Life history models*

Most models of optimal age and size at maturity have not incorporated the possibility that individuals adaptively adjust their growth by balancing it against juvenile mortality (reviewed in Roff, 1992; Stearns, 1992). Gilliam and Fraser (1987) presented a model including a trade-off between juvenile growth and mortality rate, which suggested that the optimal growth strategy of an individual is to choose habitats that minimize the ratio of mortality rate (μ) and growth rate (g) – the 'minimize μ/g' rule. However, seasonal time horizons were not incorporated into this model. In contrast, Ludwig and Rowe (1990) and Rowe and Ludwig (1991) analysed models of age and size at maturity/metamorphosis that included a growth/mortality trade-off and time horizons due to seasonality. The models assume that the trade-off is mediated by a choice between two habitats that differ in growth and mortality rates. Werner and Anholt (1993), furthermore, investigated adaptive responses in foraging activity (speed of foraging and time spent foraging) when individual growth rate and mortality risk are functions of activity, and time horizons are present. All these models indicate that when there are seasonal time constraints, the optimal age and size at maturity should typically vary with the time horizon for juvenile growth.

This last conclusion is also in line with the work of Abrams *et al.* (1996) and Abrams and Rowe (1996), which represents perhaps the most explicit life history modelling of adaptive growth strategies in a seasonal environment. In some contrast to the previous models Abrams *et al.* (1996) treats juvenile growth rate as an adaptively flexible trait that individuals can adjust depending on the situation. As in the other models there is a trade-off between growth and mortality rates, as well as a cost of maturing at a sub-optimal time of year. The model also assumes that individuals can predict time horizons by means of seasonal cues (i.e. photoperiod, temperature). The study then focuses on the optimal responses in final size, development time and growth rate when the amount of time available for growth varies. The results show that juveniles should adjust their growth according to information about prevailing time horizon. The model makes the intuitive prediction that with less time available for growth (shorter time horizon) juvenile development time should decrease. Such a shortening of the developmental period can in principle be achieved by maturing at a smaller size, or by growing faster, or by a combination of the two. In essence, the analysis shows that the optimal combination of juvenile growth rate and size at maturity depends on how costly it is to be a small adult, in relation to how costly it is to grow faster as a juvenile. Since these relationships are poorly known and the number of combinations that are theoretically possible is large, the range of predicted life history responses is also large. However, for both biological and theoretical reasons some relationships between final size and fitness and between growth rate and juvenile mortality are less probable than others (Abrams et al., 1996). The authors, therefore, argue that the most likely effect of decreasing the time horizon for growth is that growth rate increases (or stays constant), adult size decreases (or stays constant), as the development time becomes shorter. These life history responses can be viewed as reaction norms in response to some environmental cue of the time horizon such as the photoperiod, and this plasticity would represent an optimal *growth strategy*.

3. Empirical findings

The object of this part is to review empirical data that relate to various aspects of the theory presented above and I have chosen model cases where the development and growth can be clearly linked to the ecological setting of the animals. To show the generality of the phenomena discussed I concentrated on three major groups of ectothermic animals representing a large taxonomic range: insects, fish and amphibians. Rather than finding *all* relevant examples in the literature, I have focused on systems that illustrate some important aspects of the theory. Reviews by Case (1978), Arendt (1997) and Nylin and Gotthard (1998) provide other examples of adaptive growth adjustment.

3.1 *Insects*

Temperate insects survive the winter in a hormonally-controlled diapause that in the typical case can only take place in one species-specific life stage (i.e. egg, larva, pupa or adult). Furthermore, many insect populations produce several, consecutive, generations per year (bivoltine or multivoltine populations). This is possible because these insects have alternative developmental pathways, allowing direct development to reproduction early in a favourable season and development to winter diapause as conditions deteriorate (Tauber et al., 1986; Danks, 1994). Individuals that enter winter diapause then postpone reproduction to the next year. The pathway taken by an individual is typically determined by environmental cues of seasonal progression that are experienced during development (Tauber et al., 1986; Danks, 1994). The amount of time that is available for growth and reproduction is clearly an important resource for temperate insects, and selection is likely to favour an efficient usage of time (Gotthard, 1999). For example, the start of development for any given generation may vary between years due to varying weather, and between individuals due to the time of oviposition. Nevertheless, any insect larva must develop in such a way that it, or its offspring, will reach the diapausing stage before the onset of winter, independently of when in the favourable season it finds itself (Reavey and Lawton, 1991). Therefore, several researchers have argued that seasonality should favour phenotypic plasticity in the timing of growth and development within developmental pathways of insects (Masaki, 1978; Nylin et al., 1989; Leimar, 1996; Nylin and Gotthard, 1998).

This last prediction has been supported by experiments with several species of temperate butterflies that have been shown to adjust their growth in response to information about time horizons as given by photoperiod (Nylin et al., 1989, 1995, 1996; Nylin, 1992; Leimar, 1996; Gotthard, 1998; Gotthard et al., 1999, 2000). Most of the evidence comes from work on species from the tribe Pararginii (Satyrinae: Nymphalidae) in which larval growth decisions in response to photoperiod show adaptive variation among species (Nylin et al., 1996; Gotthard et al., 1999) and among populations within species (Nylin et al., 1989, 1995; Gotthard, 1998). There is also evidence that individual growth decisions in response to the photoperiod and ambient temperature depend on the seasonal state of growing larvae (Figures 3 and 4; Gotthard et al., 1999, 2000). This state-dependence of individual growth decisions appears to be

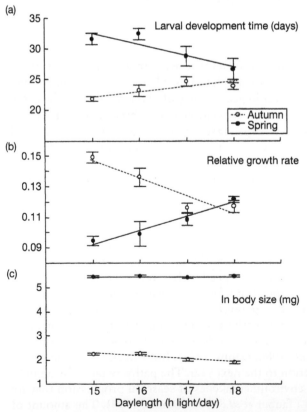

Figure 3. The effect of experimental manipulation of the photoperiod during larval growth on (a) larval development time, (b) growth rate, (c) ln final size of the butterfly Lasiommata maera (Reproduced from Gotthard et al., 1999, © Oikos, 1999). The same cohort of larvae was reared first in autumn, before entering winter diapause as half-grown larvae, and then in spring until pupation. Larvae used the photoperiod to adjust their growth patterns both in autumn and spring, but the relationship between photoperiod and time horizon differed between autumn and spring. This difference between the two growth periods seems to be adaptive since the information about time horizons that the photoperiod gives is qualitatively different in the two periods (in autumn the photoperiod is decreasing with time while in spring it is increasing with time). This experiment shows that growth decisions of individual larvae in response to information given by the photoperiod are state-dependent.

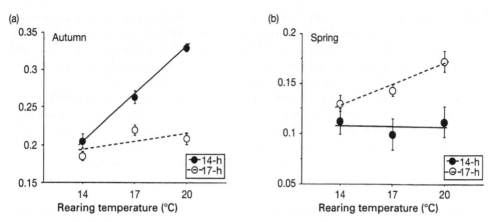

Figure 4. *The effect on relative larval growth rate of a combination of experimental variation in photoperiod and temperature in the butterfly* Lasiommata maera, *during (a) autumn and (b) spring (Reproduced from Gotthard et al., 2000, © 2000 The Royal Society). During both growing seasons larvae increased their growth rates in higher temperatures more if the photoperiod indicated short time horizon, and thus high time stress. In autumn a short photoperiod indicates high time stress, while in spring a relatively long photoperiod indicates a later date of the season.*

adaptive to allow larvae to adequately estimate time horizons from the photoperiod both before and after summer solstice (Gotthard *et al.*, 1999, 2000). Moreover, within the Pararginii there is experimental support that high larval growth rates may carry fitness costs both in terms of reduced starvation endurance (Gotthard *et al.*, 1994; Gotthard, 1999) and increased predation rate (Gotthard, 2000). As predicted by most of the theoretical models a shorter time horizon for larval growth in these species had the effect of reducing larval development time (*Figure 3*). This reduction in time was always associated with an increase in larval growth rate but only rarely was the final size of the developing larvae affected (*Figure 3*; Nylin *et al.*, 1989, 1995, 1996; Gotthard, 1998; Gotthard *et al.*, 1999, 2000). This implies that, when there is a shortage of time for growth in these species, the cost of increasing larval growth rate is typically lower than the cost of finishing at a smaller size. On the other hand when there is a surplus of time (that is not long enough to allow an additional generation), the benefit of growing slower is typically greater than the benefit of ending up at a larger size (Gotthard, 1999).

In a series of papers Blanckenhorn (1997a,b, 1998a,b) reports on both field and laboratory experiments investigating growth and developmental strategies of the yellow dung fly (*Scathophaga stercoraria*). The populations studied produce several overlapping generations each year. Female dung flies oviposit into cattle dung, which the developing larvae feed on and thereby deplete. In addition to seasonal time constraints, competition for dung imposes food and time constraints on development. In contrast to most cases of insects, temperature rather than photoperiod is the most important cue for the induction of diapause in *S. stercoraria* (Blanckenhorn, 1998b). Nevertheless, when the direct effects of temperature variation are controlled, individuals increase their growth rates as the time horizon for development prior to winter diapause becomes shorter. This increase in growth rate allowed late-hatching individuals in the field to maintain large final body sizes even though developmental periods in degree-days were shortened (Blanckenhorn, 1997b, 1998a). On the other hand, the increase in growth rate also increased larval mortality, indicating a trade-off between growth and mortality rates (Blanckenhorn, 1998a). The author relates his results to the model of Abrams *et al.* (1996) and concludes that the developmental strategy of *S. stercoraria* can only be fully predicted if larval growth rate varies adaptively and independently of developmental time (Blanckenhorn 1998a).

The relationship between the mountain birch (*Betula pubescens czerepanovii*) and its insect herbivores in northern Finland has been studied extensively over several decades (Haukioja *et al.*, 1978, 1988; Hanhimäki *et al.*, 1995; Tammaru, 1997; Kause *et al.*, 1999). The system is characterized by a seasonal decline in the nutritional quality of leaves for a range of insects feeding on the mountain birch. This decline is especially fast in early summer and during this period there is extensive within-canopy variation in leaf quality. Later in the season the quality of leaves is generally low and there is little variation in quality within the foliage of a tree. Early in the season the deterioration of food quality creates time horizons and in 'early season growers' such as the autumnal moth *Epiritta autumnata* selection apparently acts to maximize larval growth rate (Ayres and MacLean, 1987; Tammaru, 1997). Kause *et al.* (1999) showed how the nutritional quality of birch leaves affected larval growth and its physiological and behavioural determinants in six species of sawflies, which ranged from species growing early in the season to those growing later. All species were capable of maintaining relatively stable larval growth rates in a gradient of birch leaf quality (due to known differences among 20 tree genotypes), but the underlying behavioural and physiological mechanism of larval feeding varied among species. Early summer species that were given low-quality leaves typically increased growth effort by dispersing to new feeding sites, which increased the probability of finding young and nutritious leaves that are heterogeneously distributed in the tree canopy. In contrast, the late season species that typically meet little spatial variation in plant quality responded to poor leaf quality by increasing consumption rate. These results also suggest that a high growth effort carries some fitness cost in both seasonal categories since larvae growing on good-quality leaves could apparently have grown faster by altering their growth effort. In other studies, Kause (2000) investigated how the variation in diet-imposed time horizons among birch feeding insects (one butterfly species and six sawfly species) influenced the genetic architecture of developmental traits. The genetic correlation between development time and final size was positive only in species feeding on mature leaves of stable quality (mid-season species). However, in species that typically feed on rapidly growing or senescing leaves a longer development time did not lead to a greater final size and the genetic correlation was negative or absent. Hence, variation in time horizons among species seems to have resulted in community-wide patterns in the genetic architecture of growth and development (Kause, 2000).

3.2 *Fish*

Most fish may continue somatic growth after sexual maturation and fish growth is typically described as indeterminate and highly habitat-dependent (Sebens, 1987). Nevertheless, it is likely that juvenile growth and development are strongly correlated to fitness and this review concentrates on the juvenile growth trajectory.

The life history of the Atlantic salmon (*Salmo salar*) has been well studied because of its economic and conservation value (Metcalfe, 1998). Juvenile growth and development of Atlantic salmon occur in freshwater streams and after a developmental event which transforms juvenile salmon to smolts, most individuals migrate to the sea in late spring. The survival rate of migrants is positively correlated with body size at the time of seaward migration (Mangel, 1996; Thorpe and Metcalfe, 1998), and therefore there is selection against small body size at migration (Mangel, 1996). The age at migration to the sea varies between rivers and there is a strong negative correlation between the age at migration and length of the growth season (i.e. populations with a shorter time horizon for growth are on average older at migration; Metcalfe and Thorpe, 1990; Metcalfe, 1998). Moreover, the age at seaward migration varies between individuals, and within a single river there are often several age classes of smolt (Metcalfe and Thorpe, 1990; Metcalfe, 1998). The developmental decision of whether or not to migrate in any given year is made in autumn, some 9 months prior to the migratory event in spring, and the outcome is largely due to the environmental conditions experienced by individual fish. After the decision point in autumn a cohort is divided into those fish that will migrate at the next spring and those that will postpone it for at least another year (Thorpe, 1977; Nicieza et al., 1991; Metcalfe, 1998). During the months between the smolting decision and actual migration the growth rate of early migrants is much higher than that of delayed migrants of the same age (Metcalfe, 1998). This bimodal pattern of growth is most likely adaptive since early migrants will experience size-selective mortality in the sea and must grow fast to reach a large size, while the delayed migrants should be more prone to avoid the potential costs of high growth rates (Nicieza et al., 1994; Metcalfe, 1998; Metcalfe et al., 1998, 1999; Nicieza and Metcalfe, 1999). Moreover, Nicieza et al. (1994) showed that the smolting decision varied adaptively with the time horizon for growth: fish from a southern population (Spain) where post-decision growth is possible for a long period, initiated smolting at a lower threshold size than did a more northern population (Scotland) that has a shorter time horizon for growth after the smolting decision is made. This suggests that developmental decisions and growth patterns of these populations are locally adapted to the typical time horizons at their places of origin. Other studies show that growth decisions of individual juvenile salmon depend on their physiological state (i.e. early or late migrant, body size; Metcalfe et al., 1998), as well as their seasonal state (growing during winter or summer; Bull et al., 1996; Metcalfe, N.B., Bull, C.D. Mangel, M., unpublished results). These patterns are in line with adaptive predictions from models of state-dependent behaviour and life history (Metcalfe et al., 1998; Metcalfe, N.B., Bull, C.D., Mangel, M., unpublished results). Thus, there is evidence that the growth of juvenile salmon is controlled by a state-dependent decision process that appears to take into account both the costs and the benefits of high growth rates at several points in juvenile development.

The relationship between seasonal time horizons and growth has also been the subject for several studies in the Atlantic silverside (Menidia menidia; Conover and Present, 1990; Present and Conover, 1992; Schultz and Conover, 1997; Billerbeck et al., 2000). This species inhabits shallow waters along the east coast of North America from northern Florida (30°N) to the Gulf of St Lawrence (46°N) and throughout this geographical range the entire life cycle is completed in one year: all fish are sexually mature at age one and a very low proportion reaches a second breeding season. The length of the first growing season decreases by a factor of 2.5 going northwards along the latitudinal gradient, yet there is no decline in the final body size with increasing latitude. Body size at the end of the first growing season is related to fitness since larger fish survive the winter better than smaller fish (Conover and Present, 1990). Common garden experiments in the laboratory revealed that more northern populations of *M. menidia* had a greater capacity for growth, and that this latitudinal cline in growth rate had a genetic basis. Unexpectedly, fish from northern populations primarily grew faster than fish from southern populations in the high temperatures that southern populations are more often exposed to. The faster growth of high latitude fish was due to behavioural and physiological differences (Present and Conover, 1992; Billerbeck *et al.*, 2000). This latitudinal pattern is interpreted as a result of strong selection in northern populations that favours a rapid elevation in growth rate during the brief period of the year when temperatures are favourable (Conover and Present, 1990). The authors argue that the lower growth rates of the southern populations are most likely due to some costs of high growth rates, one of which may be a reduced swimming performance (Billerbeck *et al.*, 2000). Thus, it seems that the optimal growth strategy of these fish changes with the prevailing time horizons, which vary with latitude.

3.3 *Amphibians*

Most studies of growth and development of amphibians have analysed juvenile development to metamorphosis of aquatic larvae (reviewed by Wilbur, 1980; Werner, 1986; Newman, 1992), while relatively few have investigated entire juvenile development to sexual maturity (Berven, 1982b; Werner, 1986; Smith, 1987; Semlitsch *et al.*, 1988; Goater, 1994). However, these latter studies all suggest that phenotypic variation induced in the larval stage influenced adult fitness. Metamorphosing early and at a large size was typically positively associated with both juvenile and adult components of fitness (i.e. juvenile survival, age and size at maturity, reproductive output; Smith, 1987; Semlitsch *et al.*, 1988; Goater, 1994). While these studies indicate that high growth rates in larval amphibians may be positively associated with adult fitness, other studies have shown that the high activity needed to achieve a high larval growth rate is also costly in terms of higher predation risk (Werner and Anholt, 1993; Skelly, 1994, 1995; Anholt and Werner, 1995, 1998). Hence, the mode of growth and development of amphibian larvae is likely to affect important fitness components and many studies have focused on how variation in time horizons may influence the balance between larval growth- and survival-rates. For example, many amphibians breed and grow as larvae in temporary ponds that are occasionally filled by rain and then dry out (at a rate that depends on pond characteristics and weather conditions). As a pond dries out the larvae in it are exposed to a high mortality risk and in order to survive they must complete metamorphosis and enter the terrestrial phase of life (Newman, 1992). A large number of studies have investigated the relationship between pond duration and larval development, with the adaptive expectation that individuals in short-duration ponds should metamorphose earlier than individuals in ponds of longer duration (Semlitsch and Gibbons, 1985; Petranka and Sih, 1987; Newman, 1988a,b; Semlitsch and Wilbur, 1988; Semlitsch *et al.*, 1990; Semlitsch, 1993; Tejedo and Reques, 1994). This prediction is also often supported by field observations and experimental studies of species that inhabit temporary ponds, but not in species that inhabit more stable aquatic environments (Newman, 1992). The within-species variation in development time is typically associated with variation in growth and developmental rates and, to a lesser degree, with changes in the size at metamorphosis (Semlitsch and Gibbons, 1985; Petranka and Sih, 1987; Newman, 1988a,b; Semlitsch and Wilbur, 1988; Semlitsch *et al.*, 1990; Semlitsch, 1993; Tejedo and Reques, 1994).

The work of Berven and colleagues (Berven et al., 1979; Berven, 1982a,b) provides examples of a geographic variation in growth strategies of frog tadpoles that most likely represents adaptation to altitudinal variation in time horizons for growth. Wood frog larvae (*Rana sylvatica*) from high altitude populations grew and developed faster than lowland larvae in a range of rearing temperatures in the laboratory (Berven, 1982a). Consequently, in a common environment the mountain larvae completed metamorphosis both faster and at a larger size than their low-elevation conspecifics, and this difference had a clear genetic basis. The growing season is almost 50% longer in the lowlands compared to the mountain conditions (Berven, 1982b), and despite their lower genetic growth capacity the lowland populations metamorphose earlier in the field (Berven, 1982a). The higher growth capacity of mountain larvae is most likely due to the shorter time horizon for growth, while a different body size–fitness relationship may explain the relatively lower growth rates of the low-altitude larvae (Berven, 1982a,b). Moreover, recent evidence indicates that costs of fast larval growth may also be an important factor for explaining variation in growth strategies in wood frogs (Anholt and Werner, 1998).

4. Conclusions

Organisms in temperate environments are likely to experience selection for synchronization of the life cycle, as well as selection for utilizing the available time efficiently. Life history analysis of the optimal age and size at maturity predicts that growth and developmental strategies of ectotherms should covary with time horizons for growth (Rowe and Ludwig, 1991; Abrams et al., 1996; Abrams and Rowe, 1996). This correlation can be expected when there is variation in time horizons for growth among populations. Moreover, since time horizons for growth are bound to vary between individuals, we should expect that organisms estimate time horizons and adjust their growth and development accordingly. The predictions are supported by empirical evidence from a taxonomically large range of ectothermic animals, which emphasizes the generality of these phenomena. When time horizon for growth decreases the juvenile period typically becomes shorter and this is often associated with a higher juvenile growth rate and a smaller size at maturity. The optimal growth strategy will depend on how fitness is related to adult body size, juvenile development time and juvenile growth rate. The notion that high juvenile growth rates may carry fitness costs is a relatively new and important finding that has emerged from studies of adaptive growth strategies. State-dependent life history theory is emerging as a promising tool for analysing adaptive growth strategies. Within this framework a growth strategy can be viewed as a set of decision rules that specify how an individual should respond to its internal state and its environment. The decision rules have a genetic basis and this is the target for natural selection.

The presence and nature of growth strategies are of central importance for understanding the evolution of age and size at maturity. However, there is still a lack of basic empirical information such as quantitative estimates of how fitness depends on body size, development time and juvenile growth rate in a more natural setting (Abrams et al., 1996; Leimar, 1996; Tammaru et al., 1996; Tammaru, 1997; Blanckenhorn, 2000). Unfortunately, such estimates are often difficult to obtain and they are also likely to vary among organisms. Nevertheless, a combination of carefully designed field and laboratory experiments could produce adequate estimates. If so, it would add pieces of information that would increase the predictive power of models of the optimal age and size at maturity and thereby enhance our understanding of the evolution of developmental strategies and life histories.

Acknowledgements.

I thank David Atkinson for inviting me to the Annual Meeting of The Society for Experimental Biology. I also thank W. Blanckenhorn, N. Metcalfe and A. Kause for supplying me with their unpublished manuscripts.

References

Abrams, P.A., Leimar, O., Nylin, S. and Wiklund, C. (1996) The effect of flexible growth rates on optimal sizes and development times in a seasonal environment. *Am. Nat.* **147**: 381–395.

Abrams, P.A. and Rowe, L. (1996) The effects of predation on the age and size of maturity of prey. *Evolution* **50**: 1052–1061.

Anholt, B.R. and Werner, E.E. (1995) Interaction between food availability and predation mortality mediated by adaptive behavior. *Ecology* **76**: 2230–2234.

Anholt, B.R. and Werner, E.E. (1998) Predictable changes in predation mortality as a consequence of changes in food availability and predation risk. *Evol. Ecol.* **12**: 729–738.

Arendt, J.D. (1997) Adaptive intrinsic growth rates: An integration across taxa. *Q. Rev. Biol.* **72**: 149–177.

Atkinson, D. (1994) Temperature and organism size – a biological law for ectotherms? *Adv. Ecol. Res.* **25**: 1–58.

Ayres, M.P. and MacLean S.F. Jr (1987) Development of birch leaves and the growth energetics of *Epiritta autumnata* (Geometridae). *Ecology* **68**: 558–568.

Beck, S.D. (1980) *Insect Photoperiodism*. Academic Press, New York.

Bernays, E.A. (1997) Feeding by lepidopteran larvae is dangerous. *Ecol. Entomol.* **22**: 121–123.

Berven, K.A. (1982a) The genetic basis of altitudinal variation in the wood frog *Rana sylvatica* II. An experimental analysis of larval development. *Oecologia* **52**: 360–369.

Berven, K.A. (1982b) The genetic basis of altitudinal variation in the wood frog *Rana sylvatica* I. An experimental analysis of life history traits. *Evolution* **36**: 962–983.

Berven, K.A., Gill, D.E. and Smith-Gill, S.J. (1979) Countergradient selection in the green frog, *Rana clamitans*. *Evolution* **33**: 609–623.

Billerbeck, J.M., Schultz, E.T. and Conover, D.O. (2000) Adaptive variation in energy acquisition and allocation among latitudinal populations of the Atlantic silverside. *Oecologia* **122**: 210–219.

Blanckenhorn, W.U. (1997a) Altitudinal life history variation in the dung flies *Scathophaga stercoraria* and *Sepsis cynipsea*. *Oecologia* **109**: 342–352.

Blanckenhorn, W.U. (1997b) Effects of temperature on growth, development and diapause in the yellow dung fly – against all the rules? *Oecologia* **111**: 318–324.

Blanckenhorn, W.U. (1998a) Adaptive phenotypic plasticity in growth, development, and body size in the yellow dung fly. *Evolution* **52**: 1394–1407.

Blanckenhorn, W.U. (1998b) Altitudinal differentiation in the diapause response of two species of dung flies. *Ecol. Entomol.* **23**: 1–8.

Blanckenhorn, W.U. (2000) The evolution of body size: what keeps organisms small? *Q. Rev. Biol.* **75**: 385–407.

Bull, C.D., Metcalfe, N.B. and Mangel, M. (1996) Seasonal matching of foraging to anticipated energy requirements in anorexic juvenile salmon. *Proc. R. Soc. Lond. B* **263**: 13–18.

Case, T.J. (1978) On the evolution and adaptive significance of postnatal growth rates in the terrestrial vertebrates. *Q. Rev. Biol.* **53**: 243–282.

Chippindale, A.K., Chu, T.J.F. and Rose, M.R. (1996) Complex trade-offs and the evolution of starvation resistance in *Drosophila melanogaster. Evolution* 50: 753–766.

Conover, D.O. and Present, T.M.C. (1990) Countergradient variation in growth rate: compensation for length of the growing season among Atlantic silversides from different latitudes. *Oecologia* 83: 316–324.

Danks, H.V. (1994) Diversity and integration of life-cycle controls in insects. In: *Insect Life-cycle Polymorphisms* (ed H.V. Danks). Kluwer Academic, Dordrecht, pp. 5–40.

Gilliam, J.F. and Fraser, D.F. (1987) Habitat selection under predation hazard: a test of a model with foraging minnows. *Ecology* 68: 1856–1862.

Goater, C.P. (1994) Growth and survival of postmetamorphic toads: interactions among larval history, density, and parasitism. *Ecology* 75: 2264–2274.

Gotthard, K. (1998) Life history plasticity in the satyrine butterfly *Lasiommata petropolitana*: investigating an adaptive reaction norm. *J. Evol. Biol.* 11: 21–39.

Gotthard, K. (1999) Life history analysis of growth strategies in temperate butterflies. Ph.D. thesis, Stockholm University.

Gotthard, K. (2000) Increased risk of predation as a cost of high growth rate: an experimental test in a butterfly. *J. Anim. Ecol.* 69: 896–902

Gotthard, K., Nylin, S. and Wiklund, C. (1994) Adaptive variation in growth rate – life history costs and consequences in the speckled wood butterfly, *Pararge aegeria. Oecologia* 99: 281–289.

Gotthard, K., Nylin, S. and Wiklund, C. (1999) Seasonal plasticity in two satyrine butterflies: state-dependent decision making in relation to daylength. *Oikos* 84: 453–462.

Gotthard, K., Nylin, S. and Wiklund, C. (2000) Individual state controls temperature dependence in a butterfly (*Lasiommata maera*). *Proc. R. Soc. Lond. B* 267: 1–5.

Gwinner, E. (1981) Circannual Systems. In: *Handbook of Behavioral Neurobiology* (ed J. Aschoff). Plenum Press, New York, pp. 391–410.

Hanhimäki, S., Senn, J. and Haukioja, E. (1995) The convergence in growth of foliage-chewing insect species on individual mountain birch trees. *J. Anim. Ecol.* 64: 543–552.

Haukioja, E., Niemelä, P., Iso-Iivari, L., Ojala, H. and Aro, E.-M. (1978) Birch leaves as a resource for herbivores. I. Variation in the suitability of leaves. *Rep. Kevo Subarctic Res. Station* 14: 5–13.

Haukioja, E., Neuvonen, S., Hanhimäki, S. and Niemelä, P. (1988) The autumnal moth in Fennoscandia. In: *Dynamics of Forest Insect Populations. Patterns, Causes, Implications* (ed A. A. Berryman). Plenum Press, New York, pp. 163–178.

Hoffman, K. (1981) Photoperiodism in Vertebrates. In: *Handbook of Behavioral Neurobiology* (ed J. Aschoff). Plenum Press, New York, pp. 449–473.

Houston, A.I. and McNamara, J.M. (1992) Phenotypic plasticity as a state-dependent life-history decision. *Evol. Ecol.* 6: 243–253.

Hutchinson, J.M.C., McNamara, J., Houston, A.I. and Vollrath, F. (1997) Dyar's rule and the investment principle: optimal moulting strategies if feeding rate is size-dependent and growth is discontinuous. *Phil. Trans. R. Soc. Lond. B* 352: 113–138.

Kause, A. (2000) Environmental and genetic determination of phenotypes in insects feeding on mountain birch. Ph.D. thesis, University of Turku.

Kause, A., Haukioja, E. and Hanhimäki, S. (1999) Phenotypic plasticity in foraging behavior of sawfly larvae. *Ecology* 80: 1230–1241.

Leimar, O. (1996) Life history plasticity: Influence of photoperiod on growth and development in the common blue butterfly. *Oikos* 76: 228–234.

Lima, S. and Dill, L.M. (1990) Behavioral decision made under the risk of predation: a review and prospectus. *Can. J. Zool.* 68: 619–640.

Ludwig, D. and Rowe, L. (1990) Life history strategies for energy gain and predator avoidance under time constraints. *Am. Nat.* 135: 686–707.

Mangel, M. (1996) Computing expected reproductive success of female Atlantic salmon as a function of smolt size. *J. Fish Biol.* 49: 877–882.

Masaki, S. (1978) Seasonal and latitudinal adaptations in the life cycles of crickets. In *Evolution of Insect Migration and Diapause.* (ed H. Dingle). Springer, New York, pp. 72–100.

McNamara, J.M. and Houston, A.I. (1996) State-dependent life histories. *Nature* 380: 215–221.

Metcalfe, N.B. (1998) The interaction between behavior and physiology in determining life history patterns in Atlantic salmon (*Salmo salar*). *Can. J. Fish. Aquat. Sci.* 55 (Suppl. 1): 93–103.

Metcalfe, N.B. and Thorpe, J.E. (1990) Determinants of geographical variation in the age of seaward-migrating salmon, *Salmo salar. J. Anim. Ecol.* 59: 135–145.

Metcalfe, N.B., Fraser, N.H.C. and Burns, M.D. (1998) State-dependent shifts between nocturnal and diurnal activity in salmon. *Proc. R. Soc. Lond. B* 265: 1503–1507.

Metcalfe, N.B., Fraser, N.H.C. and Burns, M.D. (1999) Food availability and the nocturnal vs. diurnal foraging trade-off in juvenile salmon. *J. Anim. Ecol.* 68: 371–381.

Newman, R.A. (1988a) Adaptive plasticity in development of *Scaphiopus couchii* tadpoles in desert ponds. *Evolution* 42: 774–783.

Newman, R.A. (1988b) Genetic variation for larval anuran (*Scaphiopus couchii*) development time in an uncertain environment. *Evolution* 42: 763–773.

Newman, R.A. (1992) Adaptive plasticity in amphibian metamorphosis. *BioScience* 42: 671–678.

Nicieza, A.G. and Metcalfe, N.B. (1999) Costs of rapid growth: the risk of aggression is higher for faster growing salmon. *Funct. Ecol.* 13: 793–800.

Nicieza, A.G., Brana, F. and Toledo, M.M. (1991) Development of length-bimodality and smolting in wild stocks of Atlantic salmon, *Salmo salar* L., under different growth conditions. *J. Fish Biol.* 38: 509–523.

Nicieza, A.G., Reyes-Gavilan, F.G. and Brana, F. (1994) Differentiation in juvenile growth and bimodality patterns between northern and southern populations of Atlantic salmon (*Salmo salar* L.). *Can. J. Zool.* 72: 1603–1610.

Nylin, S. (1992) Seasonal plasticity in life history traits: growth and development in *Polyonia c-album* (Leipidoptera: Nymphalidae). *Biol. J. Linn. Soc.* 47: 301–323.

Nylin, S. and Gotthard, K. (1998) Plasticity in life history traits. *Annu. Rev. Entomol.* 43: 63–83.

Nylin, S., Wickman, P.-O. and Wiklund, C. (1989) Seasonal plasticity in growth and development of the speckled wood butterfly, *Pararge aegeria* (Satyrinae). *Biol. J. Linn. Soc.* 38: 155–171.

Nylin, S., Wickman, P. O. and Wiklund, C. (1995) Life-cycle regulation and life history plasticity in the speckled wood butterfly: are reaction norms predictable? *Biol. J. Linn. Soc.* 55: 143–157.

Nylin, S., Gotthard, K. and Wiklund, C. (1996) Reaction norms for age and size at maturity in Lasiommata butterflies: predictions and tests. *Evolution* 50: 1351–1358.

Petranka, J.W. and Sih, A. (1987) Habitat duration, length of larval period, and the evolution of a complex life cycle of a salamander, *Ambystoma texanum. Evolution* 41: 1347–1356.

Present, T.M.C. and Conover, D.O. (1992) Physiological basis of latitudinal growth differences in *Menidia menidia*: variation in consumption or efficiency? *Funct. Ecol.* 6: 23–31.

Reavey, D. and Lawton, J.H. (1991) Larval contribution to fitness in leaf-eating insects. In: *Reproductive Behaviour of Insects* (eds W. J. Bailey and J. Ridsdill-Smith). Chapman & Hall, London, pp. 293–329.

Roff, D.A. (1992) *The Evolution of Life Histories.* Chapman & Hall, New York.

Rowe, L. and Ludwig, D. (1991) Size and timing of metamorphosis in complex life cycles: time constraints and variation. *Ecology* 72: 413–427.

Saunders, D.S. (1981) Insect photoperiodism. In: *Handbook of Behavioral Neurobiology.* (ed J. Aschoff). Plenum Press, New York.

Schultz, E.T. and Conover, D.O. (1997) Latitudinal differences in somatic energy storage: adaptive responses to seasonality in an estuarine fish (Atherinidae: *Mendia mendia*). *Oecologia* 109: 516–529.

Sebens, K.P. (1987) The ecology of indeterminate growth in animals. *Annu. Rev. Ecol. Syst.* **18**: 371–407.

Semlitsch, R.D. (1993) Asymmetric competition in mixed populations of tadpoles of the hybridogenetic *Rana esculenta* complex. *Evolution* **47**: 510–519.

Semlitsch, R.D. and Gibbons, J.W. (1985) Phenotypic variation in metamorphosis and paedomorphosis in the salamander *Ambystoma talpoideum*. *Ecology* **66**: 1123–1130.

Semlitsch, R.D. and Wilbur, H.M. (1988) Effects of pond drying time on metamorphosis and survival in the salamander *Ambystoma talpoideum*. *Copeia* **1988**: 978–983.

Semlitsch, R.D., Scott, D.E. and Pechmann, J.H.K. (1988) Time and size at metamorphosis related to adult fitness in *Ambystoma talpoideum*. *Ecology* **69**: 184–192.

Semlitsch, R.D., Harris, R.N. and Wilbur, H.M. (1990) Paedomorphosis in *Ambystoma tapoideum*: maintenance of population variation and alternative life-history pathways. *Evolution* **44**: 1604–1613.

Sibly, R.M. and Calow, P. (1986) *Physiological Ecology of Animals: an Evolutionary Approach.* Blackwell Science, Oxford.

Simons, A.M., Carrière, Y. and Roff, D. (1998) The quantitative genetics of growth in a field cricket. *J. Evol. Biol.* **11**: 721–734.

Skelly, D.K. (1994) Activity level and the susceptibility level of anuran larvae to predation. *Anim. Behav.* **48**: 465–468.

Skelly, D.K. (1995) A behavioral trade-off and its consequences for the distribution of Pseudacris treefrog larvae. *Ecology* **76**: 150–164.

Smith, D.C. (1987) Adult recruitment in chorus frogs: effects of size and date at metamorphosis. *Ecology* **68**: 344–350.

Stearns, S.C. (1986) Natural selection and fitness, adaptation and constraint. In: *Patterns and Processes in the History of Life* (eds D. M. Raup and D. Jablonski). Springer, Heidelberg, pp. 23–44.

Stearns, S.C. (1992) *The Evolution of Life Histories.* Oxford University Press, Oxford.

Stockhoff, B.A. (1991) Starvation resistance of gypsy moth, *Lymantria dispar* (L.) (Leipidoptera: Lymantriidae): tradeoffs among growth, body size, and survival. *Oecologia* **88**: 422–429.

Tammaru, T. (1997) Size and time for a geometrid moth: evolutionary forces and consequences for population dynamics. Ph.D. thesis, University of Turku.

Tammaru, T., Ruohomaki, K. and Saikkonen, K. (1996) Components of male fitness in relation to body size in *Epirrita autumnata* (Lepidoptera, Geometridae). *Ecol. Entomol.* **21**: 185–192.

Tauber, M.J., Tauber, C.A. and Masaki, S. (1986) *Seasonal Adaptations of Insects.* Oxford University Press, Oxford.

Tejedo, M. and Reques, R. (1994) Plasticity in metamorphic traits of natterjack tadpoles: the interactive effects of density and pond duration. *Oikos* **71**: 295–304.

Thorpe, J.E. (1977) Bimodal distribution of length in juvenile Atlantic salmon (*Salmo salar* L) under artificial rearing conditions. *J. Fish Biol.* **11**: 175–184.

Thorpe, J.E. and Metcalfe, N.B. (1998) Is smolting a positive or a negative developmental decision? *Aquaculture* **168**: 95–103.

Werner, E.E. (1986) Amphibian metamorphosis: growth rates, predation risk and the optimal size at transformation. *Am. Nat.* **128**: 319–341.

Werner, E.E. and Anholt, B.R. (1993) Ecological consequences of the trade-off between growth and mortality rates mediated by foraging activity. *Am. Nat.* **142**: 242–272.

Wilbur, H.M. (1980) Complex life cycles. *Annu. Rev. Ecol. Syst.* **11**: 67–93.

16

Mechanisms and patterns of selection on performance curves: thermal sensitivity of caterpillar growth

Joel G. Kingsolver

1. Introduction

Body temperature impacts nearly every aspect of physiological and organismal performance in animals. For ectothermic organisms that do not regulate a constant body temperature, variation in environmental conditions will generate variation in body temperature and hence in rates of organismal performance. As a result, an individual animal's performance may vary continuously as a function of environmental temperatures. The relationship between temperature and organismal performance – sometimes termed a thermal performance curve (TPC; Huey and Stevenson, 1979) – often has a characteristic form in which performance increases linearly or geometrically at lower temperatures, reaches a maximum at some 'optimal' temperature, and declines rapidly for temperatures above the optimum (*Figure 1*; Huey and Kingsolver, 1989). How do we study selection and evolution of traits, like TPCs, that are continuous functions of the environment?

Here I will explore this question as it applies to thermal performance curves for short-term growth rates in ectotherms. There are physiological, environmental and population aspects to the problem. At the physiological level, growth rate is the result of a series of underlying physiological processes: ingestion, digestion, absorption, metabolism and synthesis. How do the thermal sensitivities of these underlying processes combine to determine the shape and position of the TPC for growth rate? Are different physiological processes rate-limiting for growth at different temperatures? At the environmental level, an individual animal may experience a substantial range of environmental temperatures during its lifetime as a result of diurnal and weather variation. How do we best characterize variation in the thermal environment of an individual? How do TPCs and the thermal environment combine to determine patterns of growth in the field? At the population level, there may be phenotypic and

Environment and Animal Development: Genes, Life Histories and Plasticity, edited by D. Atkinson and M. Thorndyke.
© 2001 BIOS Scientific Publishers Ltd, Oxford.

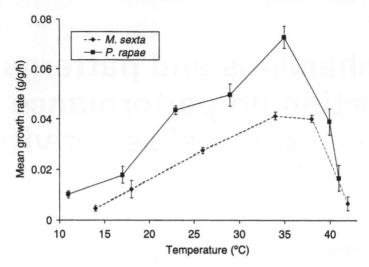

Figure 1. *Mean (± 1 SE) mass-specific growth rates (g/g/h) as functions of temperature for early fifth instar* Manduca sexta *(dashed line, diamonds) and* Pieris rapae *(solid line, squares) caterpillars. A total of 15–20 caterpillars of each species were measured at each temperature, using 4 h test periods in* M. sexta *and 6 h test periods in* P. rapae.

genetic variation among individuals in the shapes and positions of TPCs. How do we quantify patterns of (co)variation in traits, like TPCs, which are continuous functions? Does variation in TPCs generate variation in fitness, leading to selection on the shape or position of TPCs? How does selection on TPCs in a population relate to the patterns of temperature variation?

In this paper I will outline an approach towards addressing these questions, and illustrate this approach using our recent studies of TPCs of short-term growth rates in herbivorous caterpillars. This approach requires four main steps:

(i)   characterize the shape and position of the TPC for organismal performance and for underlying physiological processes in the lab;

(ii)   quantify the patterns of environmental temperatures experienced by populations in the field;

(iii) estimate the patterns of phenotypic and genetic covariation in TPCs within populations;

(iv)   estimate phenotypic selection on performance as a function of temperature, and relate selection to the patterns of environmental temperatures in the field.

Caterpillar growth rate is a useful model system for applying this approach for several reasons. First, caterpillars are eating and growth machines: under ideal conditions an individual caterpillar may increase in mass by 1000–10 000-fold in 2–3 weeks. Many aspects of the digestive physiology and nutritional ecology of caterpillars relevant to growth have been investigated (Slansky, 1993). Second, the effects of temperature on growth and development rates in caterpillars are well documented (Casey, 1993). Many caterpillars do not actively regulate their body temperature, and it is possible to characterize their thermal environment under natural outdoor conditions. Third, temperature and growth in caterpillars are directly relevant to important components of fitness, including larval survival, development time and time to pupation (generation time), pupal mass (adult body size), and fecundity. This may lead to natural selection on variation in TPCs for growth in the field. Finally, the capacity to combine lab studies of TPCs with field studies of fitness and of the thermal environment allows us to quantify patterns of population variation and selection in TPCs in caterpillars. Here we will describe such studies with two species of caterpillars: the tobacco hornworm *Manduca sexta*, and the small cabbage white *Pieris rapae*.

2. Thermal sensitivity of caterpillar growth rate

2.1 *Components of caterpillar growth*

Temperature has important effects on growth, consumption, and survival of caterpillars. Thermal performance curves for growth and consumption rates have a similar shape in many caterpillar species: at low temperatures, performance increases with increasing temperatures, reaches a maximum at intermediate temperatures, then declines rapidly with further increases in temperature (Casey, 1993; *Figure 1*). Most studies of temperature effects have used constant temperature conditions to estimate average growth rates over the duration of one or more larval instars (on the order of days to weeks; Reynolds and Nottingham, 1985). However, because thermal conditions change over much shorter time scales in the field, short-term measurements of short-term growth rates on the scale of hours are needed to explore selection on TPCs (Kingsolver and Woods, 1997).

For many herbivorous caterpillars, including *M. sexta* and *P. rapae*, the low concentrations of protein (nitrogen) in leaves are a major determinant of growth rates (Reynolds *et al.*, 1985; Scriber and Slansky, 1981; Slansky, 1993; Slansky and Feeny, 1977). It is useful to consider the transformation of leaf protein into insect tissue in terms of four basic physiological processes acting in series: consumption of protein; digestion of protein into amino acids; absorption of amino acids across the gut wall; and construction of tissue from absorbed amino acids (Woods and Kingsolver, 1999). In addition, rates of respiration will also impact growth rates. In our studies with *M. sexta* feeding on artificial diets (Kingsolver and Woods, 1997), we have measured the effects of temperature on rates of consumption, protein digestion, amino acid (methionine) uptake, and respiration, to explore how the thermal sensitivities (Q_{10}s) of these physiological components affect the thermal performance curve for growth rate (*Figure 2*). One striking result is that the decline in growth rates at high temperatures (>36°C) is strongly associated with a decline in consumption rates, but not with declines in absorption or digestion rates or accelerated respiration rates. This is counter to the common notion that thermal stability of rate-limiting enzymes in digestion and uptake determine upper temperatures limits for growth (Heinrich, 1977; Somero, 1978), and suggests that consumption rate is rate-limiting for growth at higher temperatures. Modelling and measurements of protein and amino acid concentrations along the gut in *M. sexta* suggest that, at intermediate temperatures, absorption rates are strongly limiting to growth rate (Woods and Kingsolver, 1999), at least for feeding on artificial, mechanically soft diets. Similarly, studies with the grasshopper *Melanoplus bivittatus* showed that growth rate at low temperatures was strongly limited by the rate of throughput in the digestive tract (Harrison and Fewell, 1995).

These studies suggest that for caterpillars and other herbivorous insects different physiological processes may limit growth rates at different temperatures (Casey and

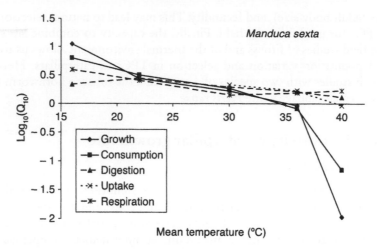

Figure 2. Q_{10} *values (on a log_{10} scale) of growth, consumption, protein digestion, methionine uptake, and respiration rates as functions of temperature for fifth instar* M. sexta *caterpillars. Modified from Kingsolver and Woods (1997).*

Knapp, 1987). As temperature conditions in the field change, different physiological processes may become more strongly limiting to overall growth. This suggests that selection on growth rate at different temperatures may affect different physiological processes.

2.2 *The thermal environment of caterpillars*

Many caterpillars, including *M. sexta* and *P. rapae*, are thermoconformers: they do not behaviourally orient to solar radiation or wind direction (except perhaps to avoid deleteriously high body temperatures >40°C), and do not use evaporative cooling or metabolic heat production to regulate body temperature (Casey, 1976, 1993). For example, *P. rapae* caterpillars are cryptic green, and typically rest on the shady undersides of leaves, generally moving to the leaf margins to feed (Jones, 1977; Wehling *et al.* unpublished data). We have used physical models of caterpillars, designed to mimic the body temperatures of early fifth instar *P. rapae* caterpillars, as a means of assessing the thermal environment of caterpillars in the field (Kingsolver, 2000).

Measurements of operative (model) caterpillar temperatures in an experimental collard garden in Seattle, WA clearly illustrate the short-term variability of thermal environments experienced by individual caterpillars (*Figure 3*). For example, diurnal variation in caterpillar temperatures over the course of 12 h can span most of the thermal range in which substantial growth may occur (*Figure 1*). Over this 1 week period, approximately the duration of the fifth larval instar under these conditions, operative temperatures ranged from 7 to 33°C. To characterize this variation, we can compute the frequency distribution of operative temperatures, $f(T)$, over this time period (*Figure 4a*). The mean of this distribution (for 25–30 July 1997, the time period shown in *Figure 3*) is near 19–20°C, but such 'average' temperatures are encountered infrequently by caterpillars. The distribution has major modes near 28 and 15°C, reflecting common midday and night-time temperatures during this time period. We have also plotted frequency distributions of operative temperatures for several other

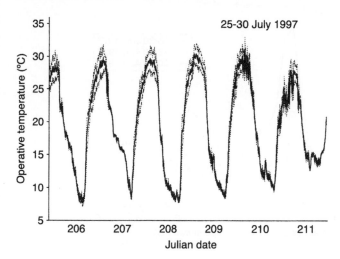

Figure 3. *Mean (± 1 SD) operative temperature of 10 models of fifth instar P. rapae caterpillars as a function of time, for models under collard leaves in an experimental collard garden in Seattle, WA, 25–30 July 1997. Reproduced from Kingsolver (2000). Physiol. Biochem. Zool. 73: 621–628, with permission of University of Chigaco Press.*

time periods during July–August 1997 (*Figure 4a*), to illustrate the substantial variation in thermal conditions within a single summer season (Kingsolver, 2000).

Thus, even in the mild summer conditions of the Pacific Northwest, caterpillars routinely face a wide range of thermal conditions at the time of a single larval instar. How does this variability impact overall growth in the field? One way to address this question is to combine our estimates of the TPC for growth rate, $g(T)$, and the frequency distribution of caterpillar temperatures, $f(T)$, (Kingsolver, 2000). Their product, $g(T) \times f(T)$, represents a weighted average growth rate indicating the relative contribution to overall growth rate made at each temperature, T, (Huey and Slatkin, 1976). (Note that this calculation ignores possible allometric changes in growth rate with size, and therefore cannot be used to make quantitative predictions of total growth over one or more instar.) Computing $g(T) \times f(T)$ as a function of temperature (*Figure 4b*) clearly shows how growth rates at higher temperatures can contribute disproportionately to overall mean growth rate, even when higher temperatures are relatively infrequent. For example, growth at temperature >28°C contributed 14–50% of the overall growth rate for the time periods represented in *Figure 4*, even though such temperatures represented only 9–28% of the total time (Kingsolver, 2000).

These results illustrate that higher temperatures can have a major impact on growth in the field, even when such temperatures are infrequent. In addition, if overall growth rate contributes to fitness, for example by decreasing generation time and larval mortality, selection on growth rate at higher temperatures might typically be stronger than on growth rate at lower temperatures (Gilchrist, 1995). To address this possibility, we need to assess patterns of phenotypic and genetic variation in growth rate, and how these relate to temperature (see Section 3).

3. Variation in thermal performance curves

How do thermal performance curves for growth rate vary among individuals and among genotypes? One obvious but fundamental point is that the quantitative trait of interest here, growth rate, is a continuous and non-linear function of an individual's environmental state, in this case temperature. This has two important implications for

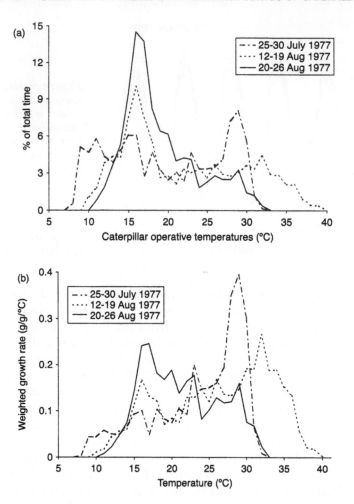

Figure 4. (a) *Frequency distributions (percentage of total time) of mean operative temperatures of 10 model P.* rapae *caterpillars in a collard garden in Seattle, WA, for three time periods in 1997: 25–30 July, 12–19 August, and 20–26 August. From Kingsolver (2000). (b) Weighted mean mass-specific growth rate (g/g/ °C) as a function of temperature for fifth instar P.* rapae *caterpillars in a collard garden in Seattle, WA, for three time periods in 1997: 25–30 July, 12–19 August, and 20–26 August. Reproduced from Kingsolver (2000).* Physiol. Biochem. Zool. **73**: 621–628, *with permission of University of Chigaco Press. See text.*

assessing patterns of variation in TPCs. First, growth rate at temperature T_1 cannot be independent of growth rate at temperature T_2, if the two temperatures are similar: as $T_2 - T_1 \rightarrow 0$, the growth rates at the two temperatures must converge. This basic fact places important contraints on possible patterns of variation in TPCs and other continuous reaction norms, which must be incorporated into our methods for analysing such variation (Gomulkiewicz and Kirkpatrick, 1992; Kirkpatrick and Lofsvold, 1992; Kirkpatrick *et al.*, 1990). Second, the asymmetric and non-linear shape of TPCs suggest that methods using simple parametric models for reaction norms will not adequately characterize variation in the shapes of TPCs (Lynch and Gabriel, 1987; Scheiner, 1993; Scheiner *et al.*, 1991; Scheiner and Lyman, 1991; Taylor, 1981). Instead, we will consider analytic methods that make essentially no assumptions about the shapes of TPCs (Gomulkiewicz and Kirkpatrick, 1992; Kirkpatrick and Heckman, 1989; Kirkpatrick *et al.*, 1990).

To pose the discussion of TPC variation in a more physiological context, imagine a hypothetical population of individuals or genotypes that vary in their TPCs in three alternative ways or axes (*Figure 5*; Huey and Kingsolver, 1989, 1993). One axis of variation (hotter–colder) might be variation in the position of the TPC, such that some individuals or genotypes have higher growth rates at hotter temperatures, whereas other individuals or genotypes have higher growth rates at colder temperatures (*Figure 5*, top). A second axis (faster–slower) might be variation in the overall height of the TPC, such that some individuals or genotypes have higher growth rates at all temperatures than other individuals or genotypes (*Figure 5*, middle). A third axis (generalist–specialist) might be variation in the shape of the TPC, such that individuals or genotypes with higher growth rates at intermediate temperatures have lower growth rates at low and high temperatures (specialists), whereas individuals or genotypes with lower growth rates at intermediate temperatures have higher growth rates at low and high temperatures (generalists; *Figure 5*, bottom). These different modes of variation imply different population-level patterns of covariation in growth rate across temperatures (*Figure 6a*). The hotter–colder mode implies that growth rate at lower temperatures is negatively correlated with growth at higher temperatures (and vice versa); the faster–slower mode implies that growth rate is positively correlated across all temperatures; and the generalist–specialist mode implies that growth rate at intermediate temperatures is negatively correlated with growth at both low and high temperatures (*Figure 6a*). In this sense, analyses of phenotypic and genetic variation can help us understand underlying constraints on axes of variation in TPCs (Gomulkiewicz and Kirkpatrick, 1992; Kirkpatrick and Lofsvold, 1992).

Of course, real populations will have more complex patterns of variation in TPCs, but we can still estimate the relative contributions of these different modes to total variation in a population. In fact, the patterns of covariation or loadings described in *Figure 6a* are the eigenfunctions of the phenotypic or genetic variance–covariance functions, which may be estimated from repeated measurements of short-term growth rate across a set of temperatures for a set of individuals within a particular breeding design (for technical details, see Gomulkiewicz and Kirkpatrick, 1992; Kirkpatrick and Heckman, 1989; Kirkpatrick *et al.*, 1990).

To illustrate the approach, we have applied these methods to our data on TPCs for growth rate in *P. rapae* caterpillars from Seattle, WA (Kingsolver, 2000). In these

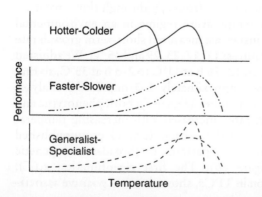

Figure 5. Hypothetical patterns of variation among individuals or genotypes in the shapes or positions of thermal performance curves.

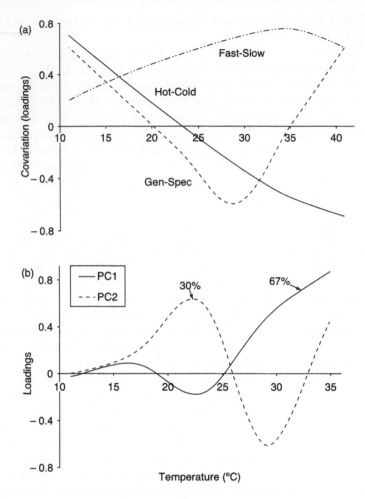

Figure 6. *(a) Patterns of covariation in performance as a function of temperature – that is, the principal eigenfunctions – that result from the hypothetical patterns of TPC variation in Figure 5. (b) Estimates of the first (PC1) and second (PC2) eigenfunctions of the (broad-sense) genetic variance – covariance function for short-term growth rate as a function of temperature, estimated from 19 full-sib families of fourth instar P. rapae caterpillars (see Figure 7).*

studies, we raised 21 full-sib families of caterpillars from egg through third instar on collard leaves in a fluctuating (10–30°C) temperature regime in an environmental chamber. Following moult into the fourth instar, we measured short-term growth rate of each caterpillar at a series of five temperatures: 11, 17, 23, 29 and 35 °C. We adjusted the duration of the measurement period from 12–16 h at 11°C to 2–3 h at 35°C, so that the mean growth at each temperature was approximately 1–3 mg. Initial analyses showed substantial variation in TPCs among families (*Figure 7*); both phenotypic and (broad-sense) genetic variances in growth rates increased with increasing temperatures. A preliminary eigenfunction analysis of these data (Kingsolver, unpublished results) showed several interesting results about patterns of (broad-sense) genetic variation in TPCs for growth rate (*Figure 6b*). The first eigenfunction $e_1(T)$, explaining 67% of the total genetic variation in TPCs, showed strong positive associa-

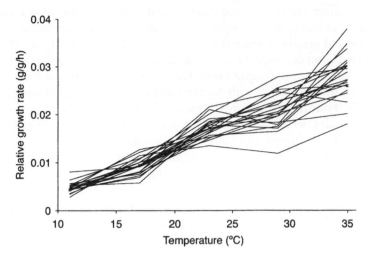

Figure 7. *Mean relative growth rates (g/g/h) as a function of temperature for 19 full-sib families of fourth instar P. rapae caterpillars.*

tions between growth rates from ~25 to 35°C, but growth rate at these temperatures was weakly correlated or uncorrelated with growth at lower temperatures. This indicates that most genetic variation in growth TPCs is simply related to faster or slower growth rates – the faster–slower mode – but only at higher (25–35°C) temperatures. The second eigenfunction, $e_2(T)$, explaining 30% of the total genetic variation, suggested strong negative correlations between growth rates around 29°C with growth at lower (17–23°C) or at higher (35°C) temperatures. This suggests that some genetic variation in TPCs is associated with negative correlations between the breadth and height of the TPC between 17 and 35°C, suggesting a possible generalist-specialist mode of variation in this temperature range. These preliminary results are based on a small number of full-sib families and any definite conclusions are premature, but they illustrate the utility of the approach and the sorts of information obtained from such analyses.

Regardless of the detailed structure of genetic variation in TPCs for growth rate in these and other caterpillars, our results do indicate that there is substantial phenotypic and genetic variation in thermal performance curves for short-term growth rate within populations of *P. rapae* caterpillars. This variation is primarily at higher temperatures at which growth rates are highest, and thus may have a major impact on variation in total growth. If growth rate affects components of fitness in field conditions, this variation provides the potential for selection on thermal performance curves to occur.

4. Selection on TPC variation

4.1 *Temperature, growth and fitness*

Temperature effects relevant to growth can influence fitness in four major ways. First, very low or high temperatures can directly reduce larval survival. For example, in *P. rapae* caterpillars, even short-term (12–24 h) exposures to temperatures above ~38°C increase mortality (Kingsolver, 2000). Second, increased growth and development rates at higher temperatures reduce the time to pupation. For multivoltine populations, including *P. rapae* in Seattle, reducing time to pupation can increase the effective number of generations per year and greatly increase the intrinsic rate of increase (Taylor, 1981). Third, increased growth and developmental rates decrease the time caterpillars are exposed to enemies. For example, for *P. rapae* caterpillars feeding on cultivated *Brassica oleracea*, slower growth was correlated with higher mortality due to parasitoids (Benrey and Denno, 1997). Fourth, in many insect species higher temperatures result in smaller pupal and adult size, which is frequently associated with reduced egg production and lifetime fecundity (Atkinson and Sibly, 1997; Berrigan, 1991; Wiklund and Karlsson, 1988). For example, for Vancouver (BC) populations of *P. rapae*, pupal mass decreases continuously with increasing temperature in the lab, and pupal mass is linearly related to lifetime fecundity (Jones *et al.*, 1982). As a consequence, we can approach the study of selection here by measuring TPCs for caterpillars in the laboratory and combining these with estimates of larval survival, time to pupation and pupal mass.

4.2 *Selection on TPCs*

One standard way to quantify the strength of directional selection on a quantitative trait, z, is in terms of the linear selection gradient, β, which relates variation in the trait (in units of standard deviation of the trait) to variation in relative fitness (where mean fitness in a population or sample is defined to equal 1; Lande, 1979). For a set of traits, $z = z_1, z_2 ... z_n$, the selection gradient is a vector $\beta = \beta_1, \beta_2 ... \beta_n$, that represents the strength of direct directional selection on each trait (Lande and Arnold, 1983). There are several attractive features of selection gradients (Arnold and Wade, 1984a,b; Lande and Arnold, 1983). First, the βs represent the direct strength of selection on each trait, adjusting for the phenotypic correlations among the traits. Second, selection gradients are relevant to models for the evolution of quantitative traits: in particular, β is directly proportional to the evolutionary change in population mean trait value resulting from selection. Third, β can be readily estimated with partial regression analyses, in studies where both trait values and fitness components are measured for a set of individuals in a population.

How do we apply this approach to the selection on thermal performance curves, in which the trait of interest is a function? Kirkpatrick and collaborators (Gomulkiewicz and Kirkpatrick, 1992; Kirkpatrick and Heckman, 1989; Kirkpatrick and Lofsvold, 1992; Kirkpatrick *et al.*, 1990) have extended this basic analytic framework to apply to reaction norms, age trajectories and other traits that are continuous functions: here the selection gradient is also a continuous function rather than a vector. In the case of TPCs, the selection gradient becomes a function of temperature, $\beta(T)$, indicating the strength of selection on performance at different temperatures.

To illustrate the approach, we have applied these methods to estimate selection gradient functions for TPCs of growth rate in *P. rapae* caterpillars. As described above, we raised ~300 caterpillars from egg through to third instar on collard leaves in a fluctuating (10–30°C) temperature regime in an environmental chamber. Following moult into the fourth instar, we measured short-term growth rate (mass gain) of each caterpillar at 11, 17, 23, 29 and 35°C. We estimated relative growth rate (RGR = ln(final mass–initial mass)/time) for each caterpillar at each temperature from these data. We then placed the caterpillars (now late fourth instars) on individual collard plants in an experimental collard garden in Seattle in July 1999, and monitored survival and time to pupation and pupal mass for each caterpillar. The collard garden was covered with coarse-mesh bridal veil netting to exclude social wasps and other large predators. We simultaneously monitored operative caterpillar temperatures in the garden using 25 model 'caterpillars', as described above.

We can use these data to estimate selection gradient functions, β(T), that relate variation in growth rate as a function of temperature, as measured in the lab, to variation in survival, time to pupation and pupal mass measured in the field *(Figure 8a)*. Our analyses indicated no significant relationship between the TPCs for growth rate and larval survival or development time, but detected significant directional selection on

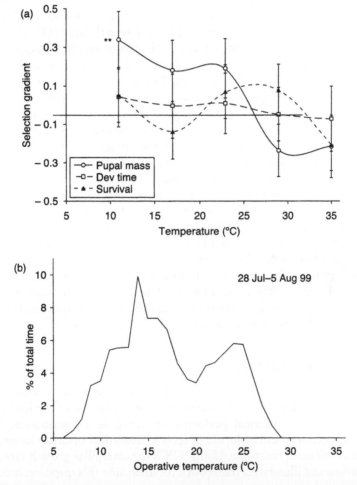

Figure 8. (a) Selection gradient functions β(T) for selection on TPCs for growth rate in fifth instar P. rapae caterpillars in a collard garden in Seattle, WA, 28 July–5 August 1999. Selection via three fitness components are indicated: survival to pupation (triangles); developmental time to pupation (squares); and pupal mass (circles). The gradient function β(T) for selection via pupal mass was significantly different from zero: the individual estimate of β (**) for selection via pupal mass on growth rate at 11°C was significantly different from zero. (b) Frequency distributions (percentage of total time) of mean operative temperatures of 24 model P. rapae caterpillars in a collard garden in Seattle, WA, for the period of the selection episode represented in (a), 28 July–5 August 1999.

TPCs for growth via effects on pupal mass, with positive selection at lower temperatures and negative selection at higher temperatures. If we consider temperatures individually, there was significant selection via pupal mass for increased growth rate at 11°C but not at any other temperature. These results suggest that caterpillars with relatively higher growth rates at low temperatures had relatively greater pupal masses, whereas caterpillars with relatively higher growth rates at higher (29–35°C) temperatures had similar or relatively small pupal masses.

Of course, one would expect that the patterns of intensity of selection on TPC to depend on the temperature conditions experienced during the period in which selection is occurring. In fact, theoretical models for reaction norms imply that the selection gradient function should depend directly on the distribution of environmental states occurring during the selection episode (Gomulkiewicz and Kirkpatrick, 1992); in the present case, selection on the TPC for growth rate, $\beta(T)$, depends on the distribution of caterpillar temperatures, $f(T)$. The frequency distribution of operative caterpillar temperatures during selection above (*Figure 8b*) showed that the modal temperature was only 15°C, and more than 75% of the period was at temperatures below 23°C; operative temperatures never exceeded 29°C during the time period. Interestingly, our estimates of $\beta(T)$ (*Figure 8a*) indicated significant directional selection for increased growth rates at lower temperatures (11–23°C), but not at higher temperatures, as one would expect if the strength of selection on TPCs is directly related to the frequency distribution of temperatures.

One final point is that the frequency distribution of temperatures experienced by a population may vary substantially within seasons, seasonally and annually with the vagaries of weather and climate. At the time scale of the fifth instar of *P. rapae* (~1 week), during which more than 70% of all larval growth occurs, the distribution of caterpillar temperatures may vary dramatically, especially at higher temperatures, for example compare *Figures 4 and 8b*). This suggests that patterns of selection on TPCs will fluctuate within and between seasons, in both predictable and unpredictable ways (Kingsolver and Huey, 1998); the analyses developed here provide a quantitative framework for interpreting how variation in the thermal environment translates into selection on thermal performance curves.

5. Summary and prospects

In this paper we have explored the question, how do we study selection and evolution of traits such as thermal performance curves that are continuous functions of the environment? We view the thermal performance curve as a continuous, non-linear, reversible reaction norm, and make use of recent theoretical advances in modelling the evolution of continuous functions. Using TPCs for caterpillar growth rate as a model system, I outline and illustrate a framework for analysing this question involving four main steps.

First, the shape and position of the TPC for organismal performance, $g(T)$, of many individuals (*Figure 1*) must be characterized . For many aspects of organismal performance including growth rate, TPCs often have a characteristic unimodal and asymmetric shape (Huey and Kingsolver, 1989). However, the thermal sensitivities of underlying physiological processes that determine organismal performance may vary (*Figure 2*); as a result, different physiological processes may limit performance at different temperatures. This is key to understanding how variation and selection on performance at different temperatures may involve different physiological processes, and may lead to evolutionary changes in the thermal sensitivity of those processes.

Second, the patterns of temperature variation experienced by individuals and populations in the field must be determined. We have quantified thermal variation in terms of the frequency distribution of operative caterpillar temperatures, $f(T)$ (*Figures 4 and 8*). These frequency distributions indicate that 'mean' and 'median temperatures provide little information about thermal conditions in natural environments: temperature distributions are often bimodal and/or strongly asymmetric, with a long and variable 'tail' towards higher temperatures. In addition, $f(T)$ may vary markedly within and between generations and years. These patterns of thermal variation provide the environmental context in which selection on TPCs may occur.

Third, the patterns of phenotypic and genetic covariation in TPCs within populations must be estimated. The fact that TPCs are continuous functions places important constraints on the possible modes and magnitudes of variation in TPCs (Gomulkiewicz and Kirkpatrick, 1992; Kirkpatrick and Lofsvold, 1992). We have quantified TPC variation in terms of the eigenfunctions $e_i(T)$ of the phenotypic and genetic variance-covariance functions (*Figure 6*), and related the shapes of such eigenfunctions in terms of common ecological hypotheses about covariation in TPCs (*Figure 5*). Our preliminary analyses of growth rate in *P. rapae* caterpillars suggest that most genetic variation in growth rate is at higher temperatures, and that growth rate is strongly positively correlated across temperatures from 25 to 35°C (*Figure 6b*). This implies that there may be important constraints on independent variation in growth rate at higher temperatures (but see Bennett *et al.*, 1990, 1992).

Fourth, phenotypic selection on performance as a function of temperature in the field must be determined. We have quantified directional selection on TPCs in terms of a selection gradient function, $\beta(T)$, which reflects how variation in performance at different temperatures determines variation in relative fitness; this selection function depends directly on the frequency distribution of temperatures $f(T)$ experienced by individuals in the population during the episode of selection. Our preliminary analyses from one field study of *P. rapae* caterpillars suggest directional selection for increasing growth rates at lower temperatures, but not at higher temperatures; and this pattern reflected the predominance of cooler environmental temperatures during this selection episode (*Figure 8*). How patterns of selection, as assessed by $\beta(T)$, change as $f(T)$ varies within and between seasons remains unknown.

The data and analyses presented here on TPCs for growth rate of *P. rapae* caterpillars are preliminary, and any firm conclusions about patterns of variation and selection in this study system are premature. However, these analyses do illustrate the potential of the approach for understanding selection and microevolution of physiological reaction norms in general, and for the evolution of temperature effects on growth in particular.

Acknowledgements

The studies of growth and feeding with *Manduca* summarized here were done in collaboration with Art Woods. I thank Kristina Williams, Yvette Maylett, Gwen Schlichta and Martha Wehling for help with the laboratory and field experiments, and Doug Ewing for his green thumb with collards. I thank Mark Kirkpatrick for providing the *Mathematica* software for estimating variance–covariance functions and eigenfunctions, and he and Richard Gomulkiewicz for introducing me to these outwardly exotic but basically friendly mathematical beasts. Ray Huey and the members of KingHuey laboratory group at the University of Washington provided valuable discussion on many issues about temperature and growth. David Atkinson, Pat Carter, Richard Gomulkiewicz and Ray Huey provided helpful suggestions on the manuscript. This research was supported by NSF grants IBN 9419850 and IBN 9818431 to the author.

References

Arnold, S.J. and Wade, M.J. (1984a) On the measurement of natural and sexual selection: applications. *Evolution* 38: 720–734.

Arnold, S.J. and Wade, M.J. (1984b) On the measurement of natural and sexual selection: theory. *Evolution* 38: 709–719.

Atkinson, D. and Sibly, R.M. (1997) Why are organisms usually bigger in colder environments? Making sense of a life history puzzle. *Trends Ecol. Evol.* 12: 235–239.

Bennett, A.F., Dao, K.M. and Lenski, R.E. (1990) Rapid evolution in response to high-temperature selection. *Nature* 346: 79–81.

Bennett, A.F., Lenski, R.E. and Mittler, J.E. (1992) Evolutionary adaptation to temperature. I. Fitness responses of *Escherichia coli* to changes in its thermal environment. *Evolution* 46: 16–30.

Benrey, B. and Denno, R.F. (1997) The slow-growth-high-mortality hypothesis: a test using the cabbage butterfly. *Ecology* 78: 987–999.

Berrigan, D. (1991) The allometry of egg size and number in insects. *Oikos* 60: 313–321.

Casey, T.M. (1976) Activity patterns, body temperature and thermal ecology in two desert caterpillars (Lepidoptera: Sphingidae). *Ecology* 57: 485–497.

Casey, T.M. (1993) Effects of temperature on foraging of caterpillars. In: *Caterpillars: Ecological and Evolutionary Constraints on Foraging* (eds N.E. Stamp and T.M. Casey). Chapman & Hall, New York, pp. 5–28.

Casey, T.M. and Knapp, R. (1987) Caterpillar thermal adaptation: behavioral differences reflect thermal sensitivities. *Comp. Biochem. and Physiol.* 86A: 679–682.

Gilchrist, G.W. (1995) Specialists and generalists in changing environments. I. Fitness landscapes of thermal sensitivity. *Am. Nat.* 146: 252–270.

Gomulkiewicz, R. and Kirkpatrick, M. (1992) Quantitative genetics and the evolution of reaction norms. *Evolution* 46: 390–411.

Harrison, J.F. and Fewell, J.H. (1995) Thermal effects on feeding behavior and net energy intake in a grasshopper experiencing large diurnal fluctuations in body temperature. *Physiol. Zool.* 68: 453–473.

Heinrich, B. (1977) Why have some animals evolved to regulate a high body temperature? *Am. Nat.* 111: 623–640.

Huey, R.B. and Kingsolver, J.G. (1989) Evolution of thermal sensitivity of ectotherm performance. *Trends Ecol. Evol.* 4: 131–135.

Huey, R.B. and Kingsolver, J. G. (1993) Evolutionary responses to extreme temperatures in ectotherms. *Am. Nat.* 143: S21–S46.

Huey, R.B. and Slatkin, M. (1976) Cost and benefits of lizard thermoregulation. *Q. Rev. Biol.* 51: 363–384.

Huey, R.B. and Stevenson, R.D. (1979) Integrating thermal physiology and ecology of ectotherms: a discussion of approaches. *Am. Zool.* 19: 357–366.

Jones, R.E. (1977) Search behaviour: a study of three caterpillar species. *Behaviour* 60: 237–259.

Jones, R.E., Hart, J.R. and Bull, G.D. (1982) Temperature, size and egg production in the cabbage butterfly, *Pieris rapae* L. *Aust. J. Zool.* 30: 223–232.

Kingsolver, J.G. (2000) Feeding, growth and the thermal environment of cabbage white caterpillars, *Pieris rapae* L. *Physiol. Biochem. Zool.* 73: 621–628.

Kingsolver, J.G. and Huey, R.B. (1998) Selection and evolution of morphological and physiological plasticity in thermally varying environments. *Am. Zool.* **38**: 545–560.

Kingsolver, J.G. and Woods, H.A. (1997) Thermal sensitivity of feeding and digestion in *Manduca* caterpillars. *Physiol. Zool.* **70**: 631–638.

Kirkpatrick, M. and Heckman, N. (1989) A quantitative genetic model for growth, shape, reaction norms, and other infinite-dimensional characters. *J. Math. Biol.* **27**: 429–450.

Kirkpatrick, M. and Lofsvold, D. (1992) Measuring selection and constraint in the evolution of growth. *Evolution* **46**: 954–971.

Kirkpatrick, M., Lofsvold, D. and Bulmer, M. (1990) Analysis of the inheritance, selection and evolution of growth trajectories. *Genetics* **124**: 979–993.

Lande, R. (1979) Quantitative genetic analysis of multivariate evolution, applied to brain: body size allometry. *Evolution* **33**: 402–416.

Lande, R. and Arnold, S.J. (1983) The measurement of selection on correlated characters. *Evolution* **37**: 1210–1226.

Lynch, M. and Gabriel, W. (1987) Environmental tolerance. *Am. Nat.* **129**: 283–303.

Reynolds, S.E. and Nottingham, S.F. (1985) Effects of temperature on growth and efficiency of food utilization in fifth instar caterpillars of the tobacco hornworm, *Manduca sexta*. *J. Insect Physiol.* **31**: 129–134.

Reynolds, S.E., Nottingham, S.F. and Stephens, A.E. (1985) Food and water economy and its relation to growth in the fifth-instar larvae of the tobacco hornworm, *Manduca sexta*. *J. Insect Physiol.* **31**: 119–127.

Scheiner, S.M. (1993) Genetics and evolution of phenotypic plasticity. *Annu. Rev. Ecol. System.* **24**: 25–68.

Scheiner, S.M. and Lyman, R.F. (1991) The genetics of phenotypic plasticity II. Response to selection. *J. Evol. Biol.* **4**: 23–50.

Scheiner, S.M., Caplan, R.L. and Lyman, R.F. (1991) The genetics of phenotypic plasticity III. Genetic correlations and fluctuating asymmetries. *J. Evol. Biol.* **4**: 51–68.

Scriber, J.M. and Slansky, F. Jr (1981) The nutritional ecology of immature insects. *Annu. Rev. Entomol.* **26**: 183–211.

Slansky, F.J. (1993) Nutritional ecology: the fundamental quest for nutrients. In: *Caterpillars. Ecological and Evolutionary Constraints on Foraging* (eds N.E. Stamp and T.M. Casey). Chapman & Hall, New York, pp. 29–91.

Slansky, F.J. and Feeny, P. (1977) Stabilization of the rate of nitrogen accumulation by larvae of the cabbage butterfly on wild and cultivated food plants. *Ecol. Monogr.* **47**: 209–228.

Somero, G.N. (1978) Temperature adaptation of enzymes: biological optimization through structure-function compromises. *Annu. Rev. Ecol. System.* **9**: 1–29.

Taylor, F. (1981) Ecology and evolution of physiological time in insects. *Am. Nat.* **117**: 1–23.

Wiklund, C. and Karlsson, B. (1988) Sexual size dimorphism in relation to fecundity in some Swedish satyrid butterflies. *Am. Nat.* **131**: 132–138.

Woods, H.A. and Kingsolver, J.G. (1999) Feeding rate and the structure of protein digestion and absorption in Lepidopteran midguts. *Arch. Insect Biochem. Physiol.* **42**: 74–87.

17

Fitness consequences of seasonal reproduction: experiments on the polychaete *Nereis virens* Sars

P.J.W. Olive, C. Lewis, V. Beardall, K. Last and M.G. Bentley

1. Introduction

1.1 *Synchronous reproduction in a seasonally variable environment*

A substantial component of the marine invertebrate fauna exhibits highly seasonal and synchronized reproduction (Babcock *et al.*, 1992; Babcock *et al.*, 1986; Giese and Knatani, 1987; Harrison *et al.*, 1984; Watson *et al.*, 2000). In temperate waters this reproductive strategy delivers offspring into specific conditions because of the inherent seasonality of the environment. The environmental conditions that vary seasonally are very complex; they include relatively simple signals such as the seasonal change in the duration of scotophase and photophase (the photoperiodic signal) and the more derived annual temperature signal, both of which approximate to a sine wave form. The annual temperature cycle, however, has a much lesser degree of predictability due to the greater component of between season and spatial variability or 'noise'. The seasonal signals also include more complex physical factors that affect the movements, mechanical energy and hydrodynamics of the water masses and which may have important consequences for the dispersal of pelagic offspring (Bhaud and Cha, 1992; Denny and Shibata, 1989). Seasonally variable environmental signals also include a host of secondary biotic factors such as food availability, the presence and abundance of competitors, predators and the presence and concentration of biologically active substances. Biotics include toxins, such as those produced by dinoflagellates (Southgate *et al.*, 1984; Dutz, 1998; Ianora *et al.*, 1999) and diatoms (Ianora *et al.*, 1996; Buttino *et al.*, 1999; Miralto *et al.*, 1999), the excretory products of metabolically active organisms and pheromones released to modify the behaviour of con-specifics, as demonstrated for a number of Nereidae (Hardege and Bartels-Hardege, 1995;

Environment and Animal Development: Genes, Life Histories and Plasticity, edited by D. Atkinson and M. Thorndyke.

Hardege *et al.*, 1994, 1996, 1998; Zeeck *et al.*, 1990). In this sense, the seasonally changing environment is both influenced by, and partially constructed by, the inter-acting organisms. Many of the seasonally variable physical environmental signals increase in amplitude with latitude and are strongest in polar regions. Seasonal envi-ronmental fluctuations become attenuated with depth such that the deep sea envi-ronment is typically thought of as a quasi-aseasonal environment at least in its physical components (Tyler, 1988; Tyler *et al.*, 1982; Tyler and Young, 1992). Contrary to earlier concepts and predictions, however, seasonal reproduction is not restricted to the coastal and shelf regions of temperate and polar regions and is observed in such envi-ronments.

1.2 *Degrees of seasonality*

Seasonal reproduction may be defined as 'a pattern of reproduction in which the production of offspring is restricted to a period that is less than the entire solar year and where there is also a fixed phase relationship between the periods of offspring production and the solar cycle'. Synchronized seasonal reproduction may be defined as 'a pattern of seasonal reproduction in which offspring production is during much less than the solar year'. This requires that most members of any population respond to their environment in a similar way so that they achieve sexual maturity and breeding takes place during a period that is much less than the entire year. Following the termi-nology of Watson *et al.* (2000), reproduction that involves the near simultaneous spawning of virtually all members of a local population of a single species is referred to as 'epidemic' spawning while simultaneous spawning by several species at a given locale is referred to as 'mass' spawning. In extreme cases of both epidemic and mass spawning events virtually all offspring production occurs during only a few hours of one day in the year (Hay, 1997). In addition, highly synchronized seasonal reproduction (epidemic spawning) may be 'pulsed' so that a series of batches of offspring are released from the population with a secondary periodicity that is frequently related to the moon phase (Caspers, 1984; Hauenschild, 1960). In both epidemic and mass spawning events, as above, physiological responses to the environment (i.e. reaction norms after Berrigan and Koella, 1994) must have evolved that cause gametogenic events and spawning to have a precise phase relationship to geophysical cycles such as the solar year, solar day, lunar and tidal cycles.

To predict the consequences of change in the physical environment on the repro-ductive and developmental biology of organisms, it is necessary to understand not only those reaction norms and the environmental signals that are transduced, but also the selective pressures ('ultimate factors') that determine the optimum time for breeding. These selective pressures however are largely unknown (see Olive, 1992; Olive *et al.*, 2000 for discussion).

1.3 *Developmental consequences of seasonal reproduction*

Synchronized seasonal (epidemic) reproduction in a seasonal environment delivers offspring into environmental conditions that are less variable than those experienced by the parents during the entire year and by the offspring of organisms with aseasonal reproduction (Olive, 1970; Olive and Clark, 1978). A commonly accepted paradigm is that the conditions into which the offspring are released have been optimized by selection so that 'Seasonal reproduction places offspring in favourable environmental conditions of temperature, light and food, thereby enhancing their chances of survival' (Giese and Knatani, 1987). We regard this as a hypothesis that is largely untested (Olive, 1992, 1995) and some observations have been made to suggest that the paradigm is at best only partly true (Sewell and Young, 1999).

The environmental conditions that will be experienced by the offspring of an organism with synchronized seasonal reproduction are, to some degree but not entirely, pre-determined by the parental reaction norms (physiological responses). These determine the period of sexual maturation and also trigger the spawning process. The reaction norms controlling the timing of gametogenic processes have, for instance, been determined for a variety of polychaetes (Bentley and Pacey, 1992; Clark, 1988; Olive,1980; Olive *et al.*, 1998; Watson *et al.*, 2000). Fine control of spawning may involve interactions with environmental temperature (Goerke, 1984; Watson *et al.*, 2000), lunar or tidal signals (Franke, 1986; Hauenschild, 1960) or chemical biotic signals, e.g. those caused by phytoplankton blooms (Starr *et al.*, 1990, 1992). Despite these responses to environmental signals, the conditions experienced by offspring are only partly pre-determined, because of inter-annual and small-scale local variation and long-term drift resulting from climate change.

In order to predict the consequences of such long-term drift, it is clearly necessary to understand the effects of small variations in environmental conditions on developmental processes and how synchronized seasonal reproduction impacts on fitness as a whole.

Here we use the polychaete *Nereis virens* Sars as a model to identify and test a number of hypotheses about the fitness components that arise from synchronized seasonal reproduction and report on experiments to quantify their relative importance.

1.4 *The life cycle and time of breeding in* Nereis virens *Sars*

The polychaete *N. virens* has a number of features that make it ideal for the experimental analysis of fitness components associated with seasonal reproduction.

Like all Nereidae *N. virens* is strictly semelparous but age at maturity is variable. Age at maturity can be manipulated by changing food availability, temperature and photoperiod (Olive *et al.*, 1997). Reproduction, however, is always seasonal and highly synchronized (epidemic) under ambient photoperiods whatever the age at maturity. The time of breeding is normally during the early spring in the UK (Brafield and Chapman, 1967) but, as in some other Nereidae (Fong and Pearse, 1992a,b), can be artificially manipulated by varying photoperiodic and temperature conditions (Olive *et al.*, 1998; Rees and Olive, 1999). In colder environments this species breeds later, for example after ice melt (May) in the St Lawrence estuary (Desrosiers *et al.*, 1985, 1994; Olivier *et al.*, 1992) and late spring on the north east seaboard of the USA (Creaser and Clifford, 1982). Spawning in the White Sea is reported to be in June (T. Andreeva and C.E. Cook, personal communication). These observations raise a number of interesting questions: have different reactions to photoperiodic inputs evolved to cause larval release at those different times of the year, and does the observed timing cause larvae to be released into optimum conditions for their development at these different locations?

We describe below experiments to test the hypothesis that the temporal organization of the life cycle causes spawning to occur at times that are optimal for fertilization and/or delivers larvae into environments that are optimal for their development. To do this we were able to use techniques for the cryopreservation of *N. virens* larvae (Olive and Wang, 1997; Wang and Olive, 1996) and photoperiodic manipulation so that larvae were made available for experiments throughout the year.

The spawning behaviour of *N. virens* appears to vary between populations. Males have been reported to be epitokous (undergoing somatic metamorphosis) and to swarm at the sea surface at the time of breeding in some UK populations (Brafield and Chapman, 1967). At Burntisland (Fife, east Scotland), however, males have been observed spawning on the sediment surface in shallow pools at low water of spring tides (Williams and Bentley, in press). Female *N. virens* undergo a much lesser degree of epitoky and do not leave the substratum prior to gamete release. Fertilization therefore takes place in the vicinity of the female parental burrow system. Sperm competition will occur as sperm are dispersed in the water column prior to coming into contact with the eggs. High rates of fertilization success are achieved *in situ* (Williams, 1998).

The pelagic larval phase in *N. virens* is relatively short. When reared at 12°C larvae hatch from the jelly coat as trochophore larvae and within 5 days the larvae reach the nechtochaete stage and exhibit a combination of swimming and searching behaviour as they enter the adult substrate. Larval dispersal is therefore expected to be limited and larvae are likely to recruit substantially to the parental habitat. Female *N. virens* have a very high maximum fecundity as approximately 70% of the available energy is allocated to germinal tissues (unpublished data, see also Gremare and Olive (1986) for similar data for *N. diversicolor*) and the fecundity of a 5 g female is about 500 000. Under these conditions dense patchiness in the spatial distribution of larvae is likely and competitive exclusion of late arriving larvae may occur. Experiments have been carried out to determine the importance of this in selection for the optimum time of breeding.

The life cycle of *N. virens* can be characterized as a semelparous life cycle with mixed maturation strategy (MMS; Heino *et al.*, 1997) since members of any offspring cohort may become sexually mature at any age between 1 and 7 or more years. In mild temperate regions the life cycle is normally completed in 2 or 3 years but in colder climatic conditions the potential life span is substantially longer (Olive *et al.*, 1997). Although age at maturity is variable, breeding in natural populations remains strictly modular and occurs only on very few days in any year.

1.5 Adult and larval components of fitness

Charnov (1997) suggested that the optimum life history strategy of an organism can be explored through the trade off relationships expressed by:

$$R_0 = s(\alpha).b.E(\alpha) = s(\alpha).V(\alpha) \tag{1}$$

Where, using Charnov's symbols, R_0 is the net reproductive rate, $s(\alpha)$ is survival rate to age at first reproduction, b is average adult fecundity, $E(\alpha)$ is average adult survival rate and $V(\alpha)$ is the reproductive value of an average individual at age α. For a strictly semelparous organism such as *N. virens* this is equivalent to defining fitness as 'the mean number of surviving offspring produced by an average adult' (Peck and Waxman, 2000). Since the larvae have no direct contact with adults and since egg size or quality is not thought to vary with age at maturity, $s(\alpha)$ is assumed to be independent of age at maturity. $V(\alpha)$ is a measure of the future expectation of offspring at age α and is useful for investigation of the partitioning of mortality schedules on adult components of fitness (Goodman, 1982).

We constructed the hypothesis that the observed timing of reproduction may enhance fitness through impacts on either $s(\alpha)$ or $V(\alpha)$. In order to test this hypothesis, we have developed a model of the life cycle of the semelparous organism *N. virens* (Olive et al., 2000) and designed experiments to test the predictions of the model in relation to fitness impacts of seasonal changes in the environment. The model components are summarized below.

We suppose that a population of *N. virens* consists of a set of age classes. For any one age class the mortality schedule has two components, mortality due to breeding and non-breeding mortality.

For an individual in an obligately semelparous population, the breeding mortality y has the value $y = 1$ or $y = 0$. For a year class at any age x the breeding mortality y_x is the proportion B_x of animals in the population N_x at age x for which $y = 1$.

Non-breeding mortality z_x is defined as the proportion of individuals in the population N_x that die in the time interval between age x and $x+1$ as a result of mortality other than that associated with breeding y_x. Reproductive value V_x at any age x is the lifetime expectation of future offspring by an average member of the population at age x. Age at first possible reproduction in a semelparous organism with a seasonal pattern of reproduction is 1. We therefore compute V_1 as the fitness component $V(\alpha)$ as the sum of the products of survival to all future ages and fecundity at each age.

For an average animal in a near steady state population V_1 is defined by equation 2 as:

$$V_1 = \sum_{x=1}^{x=\infty} \frac{N_x y_x}{N_1} \times f_x \qquad (2)$$

Where f_x is the average fecundity of an individual at age x, N_x is the number of individuals surviving to age x, N_1 the number of individuals at age 1 and y_x is the proportion of animals breeding at each age x.

We have used published data summarized by Olive et al. (1997) to parametize this model and have calculated V_1 for a variety of simulated conditions using the terms defined below:

α age at first reproduction in this model $=1$

ω age at last reproduction

a growth rate constant

b proportion of body weight at maturity allocated to germinal tissues $(=0.7)$

B_x proportion of individuals breeding in age class X

m_x average fecundity of an individual aged x

G_{max} the maximum observed growth rate under optimized conditions

E the number of feeding events per hour (E_{max} is set at 300 for animals exhibiting maximum growth based on unpublished actograph observations; see Last et al., 1999)

K the growth constant in the von Bertalanffy growth curve used to estimate different growth rates with values of: fast, $K=3$; medium, $K=1.0$; slow, $K=0.5$

l_x the probability of surviving from birth to age class X

t time

V_x reproductive value at age x

V_1 reproductive value at age 1.

W_x weight at age x

W_∞ weight at age infinity (the asymptotic age for the von Bertanlanffy growth curve) and here given values of: 15 g for fast growth ($K=3$); 40 g for medium growth ($K=1$); and 70 g for slow growth ($K=0.5$)

x age

X age class = age interval x to $x+1$

y_x the mortality due to reproduction at age x ($B_x.y$) where y is the mortality that is incurred by an individual on breeding where, for a strictly semelparous organism, $y=1$

z_x the probability of dying between x and $x+1$ (excluding y_x)

$z(f)$ the mortality associated with a single feeding event, given values 10^{-5}, 10^{-6} and 10^{-7}.

1.6 *Testing the 'ultimate factors' that determine the time of breeding*

In order to predict changes associated with enhanced rates of climate change it is necessary to understand the selective pressures that determine the optimum time for breeding. To address this we have combined an experimental and modelling approach to determine the extent to which the 'juvenile' and 'adult' components of fitness, that is $s(\alpha)$ and $V(\alpha)$ (following the terminology of Charnov, 1997) may be affected by changes in climatic factors that influence the timing of seasonal reproduction.

2. Materials and methods

2.1 *Numerical studies*

Calculation of $V(\alpha)$. The minimum age at maturity was assumed to be 1. Schedules for V_1 were calculated by iterative procedures to calculate V_1 in equation (2) following the methods of Olive *et al.* (2000). Initial parameters were derived where possible from published sources.

Simulation of growth. The weight of an individual at age x, W_x, is a function of time elapsed t since age zero when weight was assumed to have been zero. The weight would more precisely be modelled as egg weight at zero time (= 2.1×10^{-5} g from unpublished data) which may be discounted as effectively zero. The growth rate was modelled in various ways – as a constant rate that may itself be a function of external modifiers such as food, temperature or photoperiod when:

$$W \cong c + at \text{ and } \therefore \frac{dw}{dt} \cong a \tag{3}$$

or by a form of the von Bertalanffy growth curve where

$$W = W_\infty \times (1 - e^{-Kt})^3 \tag{4}$$

and where the asymptotic size W_∞ and the growth parameter K were given values that mimic the growth and age at maturity derived from published data for three different populations of *Nereis virens* (Olive *et al.*, 1997). The three growth rate schedules were:

(i)   fast (based on a cultured population) where $W_\infty = 15$ g, $K=3$;

(ii)   medium (based on Thames estuary population) where $W_\infty = 40$ g, $K=1$;

(iii)   slow (based on New Brunswick population) where $W_\infty = 70$ g, $K=0.5$.

The maximum growth rate G_{max} that we have observed under optimized laboratory conditions is 0.05 g per worm per day. These growth rates are referred to as fast, medium, slow and max.

Simulation of fecundity. In Nereidae the proportion of the total body weight allocated to reproductive tissues has been shown to be approximately 0.7. We assume that egg size is not changed by age at maturity hence:

$$m_x = \frac{b \times W_x}{\text{Egg weight}} \tag{5}$$

where b has the value 0.7 for modelled semelparity. Egg weight is derived from the data of Olive *et al.* (1997) and has the value of 2.1×10^{-5} g per egg.

2.2 *Fertilization*

Fertilizations were carried out using oocytes extracted from females in which the modal oocyte diameter >180 μm and in which 100% of the eggs gave a positive fertilization reaction. Eggs from three females were pooled and subjected to a preconditioning exposure of 2–3 h in seawater at 5, 6, 8, 12, 14, 18 and 22°C. Samples of 10^4 pre-conditioned eggs were placed in petri dishes in 10 ml of 0.2 μm filtered seawater at the pre-conditioning temperatures and exposed to samples of spermatozoa at 10^6 sperm per ml and a sperm:egg ratio of 1000:1. The eggs were washed after 10 min with temperature controlled seawater and left for 24 h prior to fixation in 4% formaldehyde. Fertilization success was determined by examination of the eggs, after Hoescht staining, using a Leitz Dialux 20 fluorescence microscope with a HBO 50 W/AC mercury short arc lamp and filter block A exciting at 340–380 nm and suppressing at 430 nm, and NPL FLUOTAR objectives of magnification 16/0.45 and 40/0.70. Evidence of first cleavage was taken as an indicator of successful fertilization.

2.3 *Growth and segment proliferation rate*

Post-nechtochaete larvae derived from artificial fertilizations were transferred to the laboratory at an age of 15 days and maintained at moderate density (1300 individuals m^{-2}) and high density (3300 larvae m^{-2}) under two fixed photoperiods, light–dark 16:8 and 8:16, for 70 days at a temperature of 12°C. Weight and segment number were recorded at intervals of 70 days.

2.4 *Competition*

Fertilizations of *N. virens* were carried out at 30 day intervals as described above and larvae reared to an age of 14 days at 12°C when they had reached a post-nechtochaete 3 setiger stage of development. Plastic boxes measuring 21 × 15 cm and containing a thin layer of sand were used as test chambers. Larvae were introduced at day zero of the experiment in two boxes at each of the densities equivalent to 10 000, 5000, 1000 and 500 larvae m^{-2}, based on data in Desrosiers *et al.* (1991), and maintained at 16°C under simulated ambient summer photoperiod (light–dark 16:8). A commercial polychaete food (Seabait Ltd; *N. virens* larval food) was introduced in minimal quantities to maintain growth but prevent fouling. After 30 days the number of animals surviving and the numbers of segments were counted. In one of the boxes at each density a further sample of larvae at the same density was added as a 'second late recruitment'. The boxes were examined again at day 60 and the survival rate and numbers of segments again measured. Because of the clear difference in size of the initial and second batches of larvae a distinction between the batches was possible.

2.5 *Age at maturity*

Individuals from the same genetic stock held by the company Seabait Ltd were fertilized at different times using the procedures to modify the time of maturity outlined by Olive *et al.* (1998) and Rees and Olive (1999). Specimens were stocked at a density of 500 individuals m^{-2} and reared in seawater derived from a power station output at a temperature of 18 ± 1°C and fed daily with pelletized food (Seabait Ltd, NVF1) at a ration of 2% of calculated body weight per day. These conditions were such that sexual maturation could be completed at the first possible age of maturity – 1 year. Batches of animals with the same birth date were kept under ambient photo-periods (natural light outside) or were kept under an artificial photoperiod using flourescent strip lights controlled by interval timers at light–dark 16:8 until November and 8:16 thereafter.

From July onwards 25 individuals were collected from each treatment at monthly intervals and a sample of the coelomic fluid taken using a hypodermic syringe. The coelomic biopsies were examined microscopically and assigned to the following cate-gories: immature (no discernible oocytes or spermatocyte cells); female – coelomic oocytes less than 80 μm in diameter, coelomic oocytes 80–120 μm in diameter, coelomic oocytes 120–160 μm in diameter, and coelomic oocytes greater than 170 μm in diameter. Males were recorded as immature or as being with tetrads and/or disso-ciated sperm.

Repeating the sampling at approximately monthly intervals monitored the progression of sexual maturation with time.

3. Results

3.1 *Impacts of seasonality on the 'juvenile' component of fitness s(α)*

Fertilization, development of* Nereis *larvae and growth of juveniles. *N. virens* breeds in northeast England and western Scotland during February (Williams and Bentley, in press) at a time when the seawater temperature is near to the annual minimum and the duration of daylight is less than at the spring equinox. The relationship between fertil-ization rate under standardized conditions and temperature (*Figure 1*) shows that the highest levels of fertilization success are achieved at temperatures substantially above those that are recorded in the natural habitat at the normal time of spawning. We conclude that spawning takes place in the UK at times when the environmental temperature is well below the optimum for this process.

Growth in juvenile Nereidae involves segment proliferation and segment enlargement. Growth and development therefore can be conveniently measured by the rate of segment proliferation and the rate of weight gain. Both growth components are influenced by photoperiod. The larvae reared under two static light–dark cycles, 16:8 and 8:16, at constant temperature were found to have significant differences in

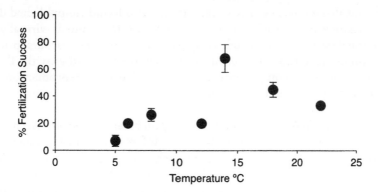

Figure 1. *The success rate for fertilizations of* Nereis virens *oocytes under standard conditions in a range of temperatures. The temperature of the seawater at the natural time of breeding is around 5°C. Fertilization success is defined as the numbers of eggs entering cleavage within 24 h of sperm addition. Vertical bars represent the standard error of the mean.*

segment number after 70 days both at intermediate densities (*Figure 2*, Kruskal–Wallace H = 4.93; $p<0.05$) and high densities (Kruskal–Wallace H = 31.43; p < 0.001). Similarly the rate of weight gain was influenced by photoperiod. The growth of individuals reared for 270 days at constant temperature can be expressed by the linear relationship for y = weight (g) and x = age (days):

$$y = 0.034x + 0.462 \ (r^2 \ 0.989) \text{ under } 16:8$$

$$y = 0.025x + 0.33 \ (r^2 \ 0.991) \text{ under } 8:16$$

The differences in slope are significant (p <0.001).

The rate of segment proliferation during regeneration is also clearly influenced by the photoperiod (Last and Olive, 1999) being substantially greater under 16:8.

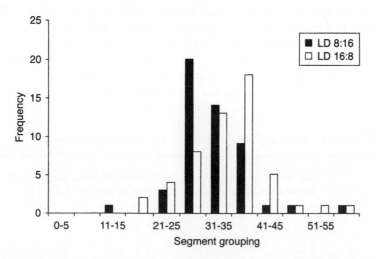

Figure 2. *The frequency of* Nereis virens *larvae grouped by number of segments when grown under standard conditions for 70 days under light–dark (LD) 8:16 and 16:8 at 18°C.*

Temperature also influences growth rate, it has been found (unpublished data) that growth is maximized by rearing at temperatures of 18°C and is virtually zero at temperatures below 4°C. We conclude that spawning of *N. virens* occurs 'prior' to the onset of optimum conditions for fertilization, development and growth. There must therefore be some other components of selection that drive reproduction to occur prior to the onset of optimum conditions.

Competitive interactions between larvae in a patchy environment. We hypothesize that intraspecific competition between larvae of different ages may be one component that would confer a selective advantage on early breeding. We suppose that, if two batches of larvae were to enter the same habitat under otherwise similar conditions at densities near to the carrying capacity of the environment, those larvae that entered the habitat patch first would have a competitive advantage. We have used our ability to manipulate the birth date of *N. virens* to test this hypothesis.

Table 1 shows the survival to age 60 days of two cohorts of larvae introduced at a range of numerical densities to the same substrate patch on day 1 and day 30. The survival of the second cohort was significantly lower than those of the first cohort. Furthermore an examination of the number of segments in the surviving larvae at day 60 (*Table 2*) suggests that late arriving larvae acted as a food resource for the larvae that entered the substrate patch 30 days earlier.

3.2 *Impacts of reproductive season on adult components of fitness V(α)*

Age at maturity in **Nereis virens** *and fitness.* Age at maturity in natural populations of *Nereis virens* varies between 2 and 7 or more years, but under conditions of intensive culture sexual maturation can be achieved within 1 year (Olive, 1999; Olive *et al.*, 1997).

We used the model above to conduct numerical studies to explore the potential relationship between fitness and age at maturity in *N. virens*. We assume that the survival rate of offspring is independent of the age of the parents. It has been demonstrated using this approach that the optimum age at maturity then depends on the external non-breeding mortality rate under a wide variety of life cycle conditions (Olive *et al.*, 2000). In a natural population the proportion of animals that breed in each year class was found to increase with age (Olive *et al.*, 1997) from data in Creaser *et al.* (1983). *Figure 3* shows the relationship between reproductive value at age of first maturity V_1 and external mortality rate z_x, in a simulation where where the proportion of animals breeding B_x was doubled each year (in short-lived populations the proportion B_x

Table 1. The survival of larvae of Nereis virens *in competition experiment*

Density (no. m^{-2})	Survival of original larvae after 30 days (%)	Survival of larvae introduced after 30 days at 60 days (%)
10 000	23.5	3.7
5000	15.0	5.5
1000	87.5	22.5
500	85.0	10.0

Table 2. *The segment growth of* Nereis virens *larvae in competion experiments*

Density (no. m^{-2})	Mean number of segments for original larvae + SEM 'without'	Mean number of segments for original larvae + SEM 'with'
10 000	32.4 ± 0.99	40.0 ± 1.07
5000	35.9 ± 1.97	48.6 ± 2.06
1000	39.8 ± 1.52	62.6 ± 2.75
500	51.6 ± 2.19	47.2 ± 1.55

increases steeply with age, while in long-lived populations this proportion increases more slowly; this illustration represents an intermediate condition). The curves represent model outputs for a range of breeding mortality rates ($y_x = yB_x$; $y = 1$) showing V_1 as a function of non-breeding mortality z_x for initial values y_x from 0.1 to 1. In this simulation the average age at maturity of the modelled population is defined by y_x such that, when the breeding mortality rate $y_x = 1$, all animals die at the first age of maturity and the average age at maturity increases as y_x declines. This numerical study demonstrates that the average age at maturity that confers maximum fitness (maximum V_1) varies with the external mortality rate z_x. There is a critical crossover point (marked by the arrow in *Figure 3*). When the external mortality rate exceeds this critical value (0.36 in *Figure 3*) maximum fitness occurs when reproduction takes place at the youngest possible age, that is when $y_x = 1$. Under these conditions there would be a fitness cost associated with delayed reproduction. This result is robust for a wide variety of simulations using a variety of schedules for B_x (Olive and Beardall, unpublished findings, see also Olive *et al.*, 2000).

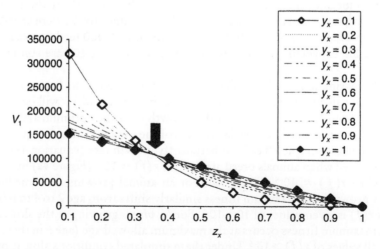

Figure 3. *Sensitivity analysis for reproductive value at the age one V_1 in relation to external mortality rate z_x. In this simulation the maximum age for breeding is assumed to be 6 and all animals die before age 7. The proportion of each year class that breeds and then dies at each age 1–6 y_x doubles at each age. Calculated outputs are presented for initial values of y_x from 0.1 to 1. The arrow points to the critical external mortality rate. When z_x exceeds this value maximum fitness arises from reproduction at the earliest achievable age.*

In the numerical study illustrated in *Figure 3*, environmental conditions were assumed to be constant. However, food availability and predation pressure are both likely to vary seasonally and Olive *et al.* (2000) show that, in an environment that has seasonal variation in food availability and/or foraging risk, the fitness component $V(\alpha)$ is influenced not only by age at maturity but also by the time of breeding. Specifically fitness will be maximized when the period of reduced foraging associated with egg maturation in *N. virens* coincides with a period of reduced food availability and/or increased predation risk.

Growth rate and age at maturity. Rapid growth of *N. virens* has been found to lead to earlier reproduction at a smaller average size than is observed under conditions that lead to lower growth rates (Olive *et al.*, 1997). We here explore this relationship by conducting numerical studies to investigate the impact of growth rate on optimum age at maturity when there is a direct link between growth rate a, which may be temperature dependent, and non-breeding mortality z_x.

A variety of growth and maturity schedules were simulated using a modified von Bertalanffy growth model to mimic the growth of *N. virens*. In addition we assume that an individual is exposed to a finite risk of predation $z(f)$ each time it feeds. This is reasonable because *N. virens* is a burrow dwelling animal that partially emerges from the burrow to feed, when it is at risk of predation from birds or fish. Foraging activity is normally restricted to the scotophase (Last *et al.*, 1999), but can be induced in 'hungry' specimens during the photophase by the offer of food. The number of feeding events E was given a maximum value of 300 events h^{-1} based on video footage of individual worms without disturbance under conditions to maximize growth rate under light–dark 16:8 (Last, unpublished). We assume that the benefit of feeding (energy consumed per feeding event) is, on average, constant. Mortality is then proportional to the rate of growth. The relative scale of this effect was modelled using the foraging mortality quotient $z(f)$ to determine the strength of the relationship between growth rate a and non-breeding mortality z_x. Values for V_1 were calculated for simulations in which $z(f) = 10^{-5}$, 10^{-6} and 10^{-7} which, with 300 feeding excursions h^{-1} give realistic life expectancies. The outputs for $z(f) = 10^{-5}$, 10^{-6} are summarized in *Figure 4*.

Under the maximum growth rate all individuals must breed in the first year, because if they do not all will suffer predation prior to reaching age 2 and V_1 will be zero for all values of $z(f)$ from 10^{-5} to 10^{-7}. For all other growth rates the optimum age at maturity (maximum V_1) is influenced by the relative value of the foraging mortality quotient $z(f)$.

For an animal growing at a 'fast' von Bertalanffy growth rate (equation 4) maximum fitness is achieved when animals breed at age 1 for $z(f) = 10^{-5}$ (*Figure 4a*) but is shifted to age 2 when $z(f) = 10^{-6}$ (*Figure 4b*). For an animal growing at a 'medium' von Bertalanffy growth rate, maximum fitness similarly shifts from age 2 to 4 to 6 (data not shown) as $z(f)$ is given values of 10^{-5}–10^{-7}. For *N. virens* growing at the slowest simulated rate, maximum fitness occurs at the maximum allowed age (age 6 in these simulations) for all values of $z(f)$ >10^{-5}. Under these simulated conditions slow growth and low risk favour breeding late but rapid growth and high risk favour breeding at an earlier age.

There is therefore an implied fitness cost if breeding occurs at an age greater than the optimal age. In the following section we show that birth date has a direct effect on the minimum age at maturity.

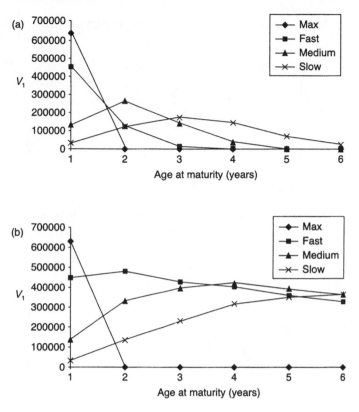

Figure 4. Sensitivity analysis for reproductive value at the age one V_1 in relation to age at maturity when all individuals in a population breed at some age x between 1 and 6 years when the external mortality rate z_x is a function of growth rate. This is reasonable for Nereis virens, which must emerge from its burrow to feed and during each such emergence there is an enhanced risk of predation. The mortality risk per foraging event $z(f)$ has the value 10^{-5} in (a) and 10^{-6} in (b). Simulations are shown for the maximum growth rate under captive conditions (Max) and for three simulated growth rates fast, medium and slow, following data in Olive et al. (1997).

Experimental investigation of the influence of birth date on age at maturity. If, as suggested by our simulations of the life cycle of *N. virens*, there is an optimum age at maturity for any schedule of external mortality, growth rate and fecundity, it follows that there will be a fitness cost associated with delayed reproduction and a benefit associated with early reproduction. To test possible effects of birth date on age at maturity we have manipulated the environmental conditions experienced by members of a single year class of *N. virens* to achieve different effective birth dates. Batches of juvenile *N. virens* produced by *in vitro* fertilizations of females and males drawn from the same genetic stock having birth dates in January and March were produced. The larvae were reared at 18°C and with unlimited food to maximize the growth rate. Under these conditions a proportion of the animals are expected to reach maturity within one year. The batches of larvae were reared under ambient and an artificially controlled light regime to extend the effective 'summer' period. The sexual maturity of the females was estimated by examining the oocyte diameters in biopsies of coelomic fluid taken at intervals from July/August onwards. The progressive maturation of these animals is summarized in *Figure 5*. None of the larvae whose birth date was in

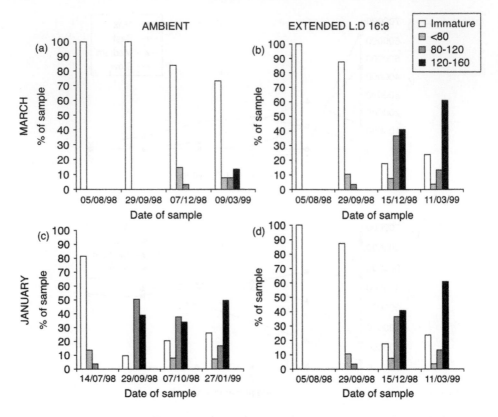

Figure 5. *Progression of sexual maturation in female* Nereis virens *reared under standard conditions under ambient photoperiod and an extended light–dark 16:8 photoperiod regime. Data are shown for animals of the same stock with a birth date in March and in January. The histograms show the percentage of each sample in the maturity class based on oocyte diameter (μm).*

March, reared under ambient light regimes, became mature at an age of 1 year in the March following their birth (*Figure 5a*). Substantial numbers of similar larvae reared under an artificial 16:8 light–dark regime until November and then transferred to 8:16 did, however, become mature by the following March (*Figure 5b*). In contrast to this, substantial numbers of the larvae with a birth date in January did become mature at age 1 under both ambient conditions (*Figure 5c*) and conditions of an artificially extended summer (*Figure 5d*). Under optimum conditions for growth and natural photoperiods the proportion of animals reaching sexual maturity at age 1 is clearly dependent on the birth date (*Table 3*).

Reproduction in *N. virens* is strictly modular under ambient conditions and breeding can only occur at yearly intervals, at ages that are some integer multiple of 1. We conclude that early birth date under ambient conditions can increase the proportion of animals able to become mature in any one year and therefore reduce the mean generation time. Our simulations (Olive *et al.*, 2000, and *Figures 3* and *4* above) show that early reproduction will therefore increase V_α and increase fitness when the external non-breeding mortality rate is greater than the critical value (0.36 in *Figure 3*). Similarly delayed reproduction due to late birth date will impose a fitness cost by causing an increase of at least 1 year in the average age at maturity.

Table 3. Effect of photoperiod and birth date on sexual maturation at age 1 in Nereis virens *cultured under conditions to maximize growth rate (χ^2 tests based on arc sin transformed data)*

Birth date	Photoperiod	Animals mature at age = 1	χ^2 test on effect of photoperiod
January	Ambient	50%	$\chi^2 = 0.117$
January	Extended light–dark 16:8	54%	$p > 0.500$
March	Ambient	14%	$\chi 2 = 45.97$
March	Extended light–dark 16:8	58%	$p < 0.001$

4. Discussion

By focusing on the developmental ecology of *N. virens* in the context of the overall lifetime fitness, we have been able to show that the selective pressures on the time of reproduction in this species are highly complex. We would not claim that we have as yet examined all the fitness parameters that may be influenced by 'time of breeding'. Further experiments, to quantify the possible size advantage accruing from early reproduction (Conover, 1992) and to more completely characterize the interaction between the effects of environmental temperature on developmental processes and effects of timing on competitive interactions, are underway (Lewis, unpublished).

Our studies of *N. virens* exemplify the interaction between the time of breeding (seasonality) and reproductive value at the age of maturity $V(\alpha)$. This species exhibits capital resource breeding (Bonnet *et al.*, 1998) since during the final stages of sexual maturation resources that have been stored in the somatic tissues are re-deployed in germinal tissues. In female *N. virens* this re-deployment of energy reserves occurs during a rapid phase of egg growth (Fischer and Hoeger, 1993; Fischer *et al.*, 1991) the onset of which is triggered in September by the critical photoperiodic transition (Last *et al.*, 1999; Olive *et al.*, 1998). Following this critical transition, the lipoprotein vitellin, lipid yolk and complex carbohydrate precursors of the jelly coat are progressively accumulated in the oocyte cytoplasm. During this process approximately 70% of the energy resources accumulated during the entire prior lifetime are transferred to the germ cells. Energy accumulated after the onset of rapid oocyte growth is unlikely to impact directly on female fitness by increasing fecundity in the immediate breeding season and furthermore, since the animals are strictly semelparous, cannot contribute to fecundity during any future breeding season. The control of the temporal organization of this switch between somatic and sexual development appears to be one of the major roles of the nereid endocrine system. It ensures that regenerative growth (addition of lost caudal segments for instance) only occurs when newly acquired resources can contribute to future fecundity (Golding, 1985; Golding and Olive, 1978; Golding and Yuwono, 1994).

Last and Olive (1999) observed that, after the transition through the threshold photoperiod, all members of a mixed age population, regardless of whether they would breed during the following spring or not, exhibit reduced foraging behaviour and suppressed growth rate. This may indicate anticipation of lower food availability during the winter, or may indicate suppression of foraging activity in anticipation of breeding. The decision to enter the final stage of sexual maturation that culminates in spawning and death can be interpreted as a 'state-dependent' decision (McNamara and Houston, 1996) that can only be made following the environmental signal associated with transition through the critical photophase in the autumn. Animals that are in a suitable physiological state at that time become destined to breed during the following spring breeding season and show permanently suppressed foraging behaviour. These animals complete sexual and somatic maturation, show permanent changes in endocrine activity (Golding and Yuwono, 1994), suppress foraging activity and partially resorb the alimentary system. Such behaviour we interpret as minimizing non-breeding mortality in the last few months prior to breeding. Those individuals that have not reached an appropriate physiological state will not breed in the following breeding season ($y = 0$). Such animals spontaneously revert to high levels of foraging behaviour 3 months after the time of the threshold photoperiodic transition (Last and Olive, unpublished results).

It has been suggested as a general rule that early reproduction confers a fitness advantage (Sibly and Calow, 1986). Our model of the life cycle of *N. virens* conforms to this prediction when non-breeding mortality z_x is greater than critical values (see *Figure 3* and Olive *et al.*, 2000). We have shown that, under standardized conditions that favour rapid growth, the proportion of a year class that enters the critical physiological state in the autumn following their birth depends on the birth date. The proportion of such individuals, with a birth date in March, was significantly lower than the proportion of individuals of the same genetic stock whose birth date was in January. This difference due to birth date was suppressed if the animals were reared under an artificial lighting regime that delayed the autumn photoperiodic transition by 2 months.

Our experiments also show that there is a competitive advantage in a patchy environment that may be gained by early reproduction. The larvae that recruit to any environmental patch before potential competitors have the capacity to exclude late arriving larvae. Reproduction prior to the onset of 'optimum' conditions for development may therefore enhance larval survival $S\alpha$.

Since both elements of the equation $R_0 = s(\alpha).V(\alpha)$, may be increased by early reproduction, it is essential to address the question what limits the advantage that accrues from early reproduction? One possibility is that the mortality rate of larvae that are exposed to sub-optimal conditions for growth is eventually sufficient to offset the advantages that accrue from early reproduction. This trade-off will tend to set an effective limit to the earliest time of reproduction and is the subject of current investigations.

There is also a possible trade-off affecting the adult component of fitness $V(\alpha)$. Foraging should be restricted to those times when food is abundant and/or risk is low and foraging suppressed at other times. If, in a semelparous organism such as *N. virens*, some finite period of time is required during which acquired resources are reallocated to germinal tissues, a mismatch between times when conditions are optimal for feeding (high benefit/low risk) and times when foraging is suppressed during resource reallocation, will lead to reduced fitness. This is because a smaller proportion of the members of any year class will then attain the physiological state which enables them to breed during the following spring and, as we have demonstrated, this will increase the mean age at maturity and therefore reduce fitness.

We believe, therefore, that the observed time of breeding for a population of *N. virens* represents a dynamic equilibrium between selective processes that push the time of breeding forwards and selective processes that push the time of breeding back. This concept is visualized in *Figure 6*, where we suggest that the timing of spawning may be the result of a trade-off between a complex of factors that impact on either larval

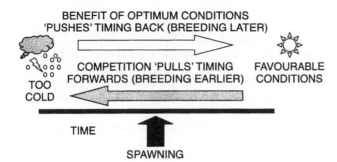

Figure 6. *Representation of a dynamic equilibrium model for the time of spawning in* Nereis virens *that may be applicable to the timing of epidemic breeding for a variety of marine invertebrates in seasonal environments.*

survival $s(\alpha)$ or reproductive value $(V(\alpha))$ and which tend to select for earlier or later reproduction.

What then would be the consequences of an overall increase in environmental temperature? If the timing of reproduction delivered larvae into optimal conditions as in the Giese Stevenson hypothesis, any change in temperature at the time of breeding would represent an increased stress leading to a loss of fitness. In our hypothesis, however, such a change may lead to increased competitiveness due to enhanced larval growth rate as environmental conditions tend more towards the optimal at the time of breeding. It may, however, reintroduce selection for earlier reproduction.

It has been suggested (Reise et al., 1999) that *N. virens* may be a relatively recent introduction to the North Sea fauna having arrived within the last 300 years from a site of origin in the eastern seaboard of the USA or the St Lawrence estuary. Populations of *N. virens* in the St Lawrence estuary (Desrosiers et al., 1985, 1991, 1994) and in New Brunswick (Snow and Marsden, 1974) are described as having a greater mean age at maturity than populations in the UK (Brafield and Chapman, 1967) and the commercially exploited populations further south on the American eastern seaboard in Maine (Creaser and Clifford, 1982). The time of breeding in the St Lawrence populations is constrained by the time of ice melt and does not normally occur until May. This factor together with the relatively short growth period in the St Lawrence may impose a constraint on the minimum age at maturity. A reasonable hypothetical construct from these observations together with our numerical studies and experimental observations is that, on introduction to the warmer conditions of the North Sea, a major constraint on the time of reproduction of *N. virens* was removed. This allowed a forwards drift in the time of reproduction because of the selective pressures we have identified limited by one of the opposing selection pressures outlined above.

Acknowledgements

This work was carried out with the support of NERC (grant no. GST/02/2164) as a contribution to the thematic programme DEMA. We would also like to acknowledge support received from the Department of Marine Sciences and Coastal Management, University of Newcastle upon Tyne enabling VB to contribute to the programme and to acknowledge the support given by Seabait Ltd through the supply of materials and for access to unpublished data. Procedures for manipulation of the time of breeding of *Nereis virens* are subject to patent. Applications to use these procedures under license should be addressed to Seabait Ltd, Woodhorn Village, Ashington, Northumberland, NE63 9NW, UK.

References

Babcock, R.C., Bull, G.D., Harrison, P.L., Heyward, A.J., Oliver, J.K., Wallace, C.C. and Willis, B.L. (1986) Synchronous spawnings of 105 scleractinian coral species on the Great Barrier Reef. *Mar. Biol.* **90**: 379–394.

Babcock, R., Mundy, C., Keesing, J. and Oliver, J. (1992) Predictable and unpredictable spawning events – in-situ behavioral data from free-spawning coral-reef invertebrates. *Invertebr. Reprod. Dev.* **22**: 213–228.

Bentley, M.G. and Pacey, A.A. (1992) Physiological and environmental control of reproduction in polychaetes. *Oceanogr. Mar. Biol. Ann. Rev.* **30**: 443–481.

Berrigan, D. and Koella, J.C. (1994) The evolution of reaction norms – simple-models for age and size at maturity. *J. Evol. Biol.* **7**: 549–566.

Bhaud, M. and Cha, J.H. (1992) Sources of fluctuations et de stabilité dans le déroulement du cycle de vie de *Eupolymnia nebulosa* en Mediterranée. *Ann. Inst. Oceanogr.* **68**: 25–35.

Bonnet, X., Bradshaw, D. and Shine, R. (1998) Capital versus income breeding: an ectothermic perspective. *Oikos* **83**: 333–342.

Brafield, A.E. and Chapman, G. (1967) Gametogenesis and breeding in a natural population of *Nereis virens. J. Mar. Biol. Assoc. UK.* **47**: 619–627.

Buttino, I., Miralto, A., Ianora, A., Romano, G. and Poulet, S.A. (1999) Water-soluble extracts of the diatom *Thalassiosiro rotula* induce aberrations in embryonic tubulin organisation of the sea urchin *Paracentrotus lividus. Mar. Biol.* **134**: 147–154.

Caspers, H. (1984) Spawning periodicity and habitat of the palolo worm *Eunice viridis* (Polychaeta: Eunicidae) in the Samoan islands. *Mar. Biol.* **79**: 229–236.

Charnov, E.L. (1997) Trade-off – invariant rules for evolutionarily stable life histories. *Nature* **387**: 393–394.

Clark, S. (1988) A two phase photoperiodic response controlling the annual gametogenic cycle in *Harmothoe imbricata* (L). *Invert. Reprod. Devl.* **14**: 245–266.

Conover, D.O. (1992) Seasonality and the scheduling of life history at different latitudes. *J. Fish Biol.* **41** (Suppl. B): 161–178.

Creaser, E.P. and Clifford, D.A. (1982) Life history studies of the sandworm *Nereis virens* in the Sheepscote estuary, Maine. *Fish. Bull.* **80**: 735–743.

Creaser, E.P., Clifford, D.A., Hogan, M.J. and Sampson, D.B. (1983) A commercial sampling programme for sandworms *Nereis virens* Sars and bloodworms *Glycera dibranchiata* Ehlers, harvested along the Maine coast. NOAA technical report, NMFS SSRF-767, pp. 1–56.

Denny, M.W. and Shibata, M.F. (1989) Consequences of surf-zone turbulence for settlement and external fertilization. *Am. Nat.* **134**: 859–889.

Desrosiers, G., Caron, A., Olivier, M., and Miron, G. (1985) Cycle de developpement d'une population de *Nereis virens* (Polychaeta: Nereidae) sur la rive sud de l'estuaire maritime du Saint Laurent. *Oceanol. Acta* **17**: 178–190.

Desrosiers, G., Olivier, M. and Vincent, B. (1991) Variations in density and growth-rate of *Nereis virens* (Sars) recruits (Annelida, Polechaeta) in the intertidal zone. *Can. J. Zool.* **69**: 560–566.

Desrosiers, G., Caron, A., Olivier, M. and Miron, G. (1994) Life history of the Polychaete *Nereis virens* (Sars) in an Intertidal flat of the lower St-Lawrence estuary. *Oceanol. Acta* **17**: 683–695.

Dutz, J. (1998) Repression of fecundity in the neritic copepod *Acartia clausi* exposed to the toxic dinoflagellate *Alexandrium lusitanicum*: relationship between feeding and egg production. *Mar. Ecol. Progr. Ser.* **175**: 97–107.

Fischer, A. and Hoeger, U. (1993) Metabolic links between somatic sexual maturation and oogenesis in nereid annelids- a brief review. *Invert. Reprod. Devl.* **23**: 131–138.

Fischer, A., Rabien, H. and Heacox, A.E. (1991) Specific, concentration-dependent uptake of vitellin by the oocytes of *Nereis virens* (Annelida, Polychaeta) in-vitro. *J. Exp. Zool.* **260**: 106–115.

Fong, P.P. and Pearse, J.S. (1992a) Photoperiodic regulation of parturition in the self fertilising viviparous polychaete *Neanthes limnicola* from central California. *Mar. Biol.* 112: 81–89.

Fong, P.P. and Pearse, J.S. (1992b) Evidence for a programmed circannual life cycle modulated by increasing daylengths in *Neanthes limnicola* (Polychaeta: Nereidae) from central California. *Biol. Bull.* 182: 289–297.

Franke, H.D. (1986) The role of light and other endogenous factors in the timing of the reproductive cycle of *Typosyllis prolifera* and some other polychaetes. *Am. Zool.* 26: 433–445.

Giese, A.C. and Kanatani, H. (1987) Maturation and spawning. In: *Reproduction in Marine Invertebrates*, Vol. 9 (eds A.C. Giese and J.S. Pearse). Academic Press, Los Angeles, CA, pp. 251–329.

Goerke, H. (1984) Temperature dependence of swarming in North Sea Nereidae. In: *Polychaete Reproduction* (eds A. Fischer and H.D. Pfannensteil). Gustav Fischer, Stuttgart, pp. 39–51.

Golding, D.W. (1985) Brain body interactions in *Nereis* – reactivation of the cerebral neuroendocrine system by experimental manipulation. *Int. J. Invert. Reprod. Devl.* 8: 51–59.

Golding, D.W. and Olive, P.J.W. (1978) Patterns of regenerative growth, reproductive strategy, and endocrine control in polychaete annelids. In *Comparative Endocrinology* (eds P.J. Gaillard and H.H. Boer). Elsevier/North Holland, Amsterdam, pp. 117–120.

Golding, D.W. and Yuwono, E. (1994) Latent capacities for gametogenic cycling in the semelparous invertebrate *Nereis*. *Proc. Nat. Acad. Sci. USA* 91: 11777–11781.

Goodman, D. (1982) Optimal life histories, optimal notation, and the value of reproductive value. *Am. Nat.* 119: 803–823.

Gremare, A. and Olive, P.J.W. (1986) A preliminary study of fecundity and reproductive effort in two polychaetous annelids with contrasting reproductive strategies. *Int. J. Invert. Reprod. Devl.* 9: 1–16.

Hardege, J.D. and Bartels-Hardege, H.D. (1995) Spawning behavior and development of *Perinereis nuntia* Var *brevicirrus* (Annelida, Polychaeta). *Invert. Biol.* 114: 39–45.

Hardege, J.D., Bartels-Hardege, H.D., Yang, Y., Wu, B.L., Zhu, M.Y. and Zeeck, E. (1994) Environmental control of reproduction In: *Perinereis nuntia* Var *Brevicirrus*. *J. Mar. Biol. Assoc. UK* 74: 903–918.

Hardege, J.D., Bentley, M.G., Beckmann, M. and Muller, C. (1996) Sex-pheromones in marine polychaetes – volatile organic-substances (vos) isolated from *Arenicola marina*. *Mar. Ecol. Prog. Ser.* 139: 157–166.

Hardege, J.D., Muller, C.T., Beckman, M., Bartels-Hardege, H.D. and Bentley, M.G. (1998) Timing of reproduction in marine Polychaetes: the role of sex pheromones. *Ecoscience* 5: 395–404.

Harrison, P.H., Babcock, R.C., Bull, G.D., Oliver, J.K., Wallace, C.E. and Willis, B.L. (1984) Mass spawning in tropical reef corals. *Science* 223: 1186–1189.

Hauenschild, C. (1960) Lunar periodicity. *Cold Spring Harbor Symp. Quant. Biol.* 25: 491–497.

Hay, M. (1997) Synchronous spawning: when timing is everything. *Science* 275: 1116–1118.

Heino, M., Metz, J.A.J. and Kaitala, V. (1997) Evolution of mixed maturation strategies in semelparous life histories: the crucial role of dimensionality of feedback environment. *Phil. Trans. R. Soc. Lond. Ser. B Biol. Sci.* 352: 1647–1655.

Ianora, A., Poulet, S.A., Miralto, A. and Grottoli, R. (1996) The diatom *Thalassiosira rotula* affects reproductive success in the copepod *Acartia clausi*. *Mar. Biol.* 125: 279–286.

Ianora, A., Miralto, A., Buttino, I., Romano, G. and Poulet, S.A. (1999) First evidence of some dinoflagellates reducing male copepod fertilisation capacity. *Limnol. Oceanogr.* 44: 147–153.

Last, K. and Olive, P.J.W. (1999) Photoperiodic control of growth and segment proliferation by *Nereis* (*Neanthes*) *virens* Sars in relation to real time and state of maturity. *Mar. Biol.* 134: 191–200.

Last, K., Olive, P.J.W. and Edwards, A. (1999) An actographic study of diel activity in the semelparous polychaete *Nereis* (*Neanthes*) *virens* Sars in relation to the annual cycle of growth. *Invert. Reprod. Devl.* 35: 141–145.

McNamara, J.M. and Houston, A.I. (1996) State-dependent life histories. *Nature* 380: 215–221

Miralto, A., Baronne, G., Romano, G., Poulet, S.A., Lanora, A., Russo, G.L., Buttion, I., Mazzarella, G., Laabir, M., Cabrini, M. and Giacobbe, M.G. (1999) The insidious effect of diatoms on copepod reproduction. *Nature* 402: 173–176.

Olive, P.J.W. (1970) Reproduction in a Northumberland population of the polychaete *Cirratulus cirratus*. *Mar. Biol.* 5: 259–273.

Olive, P.J.W. (1980) Control of the reproductive cycle in female *Eulalia viridis* (Polychaeta, Phyllodocidae). *J. Mar. Biol. Assoc. UK.* 61: 941–958.

Olive, P.J.W. (1992) The adaptive significance of seasonal reproduction in marine- invertebrates – the importance of distinguishing between models. *Invert. Reprod. Devl.* 22: 165–174.

Olive, P.J.W. (1995) Annual breeding cycles in marine invertebrates and environmental temperature: – probing the proximate and ultimate causes of reproductive synchrony. *J. Thermal Biol.* 20: 79–90.

Olive, P.J.W. (1999) Polychaete aquaculture and polychaete science: a mutual synergism. *Hydrobiologia* 402: 177–186.

Olive, P.J.W. and Wang, W.B. (1997) Cryopreservation of *Nereis virens* (Polychaeta, Annelida) larvae: the mechanism of preservation of a differentiated metazoan. *Cryobiology* 34: 284–294.

Olive, P.J.W. and Clark, R.B. 1978 Physiology of reproduction. In: *Physiology of the Annelids* (ed. P.J. Mill). Academic Press, London, pp. 271–368.

Olive, P.J.W., Fletcher, J., Rees, S. and Desrosiers, G. (1997) Interactions of environmental temperature with photoperiod in determining age at maturity in a semelparous Polychaete *Nereis (Neanthes) virens* Sars. *J. Thermal Biol.* 22: 489–497.

Olive, P.J.W., Rees, S.W. and Djunaedi, A. (1998) The influence of photoperiod and temperature on oocyte growth in the semelparous polychaete *Nereis (Neanthes) virens* Sars. *Mar. Ecol. Prog. Ser.* 172: 169 – 183.

Olive, P.J.W., Lewis, C. and Beardall, V. (2000) Fitness components of seasonal reproduction: an analysis using *Nereis virens* as a life history model. *Oceanol. Acta* 23: 377–389 .

Olivier, M., Desrosiers, G. and Lechapt, J.P. (1992) Description of the developmental stages of recruits of *Nereis virens* (Sars) (Annelida, Polychaeta). *Cah. Biol. Mar.* 33: 319–329.

Peck, J.R. and Waxman, D. (2000) Mutation and sex in a competitive world. *Nature* 406: 399–404.

Rees, S.W. and Olive, P.J.W. (1999) A new *in vitro* fluorescence assay indicates that photoperiod and temperature influence vitellin incorporation by *Nereis* oocytes. *Comp. Biochem. Physiol. A* 123: 213–220.

Reise, K., Gollasch, S. and Wolff, W.J. (1999) Introduced marine species of the North Sea coasts. *Helg. Meeresunsters.* 52: 219–234.

Sewell, M.A. and Young, C.M. (1999) Temperature limits to fertilisaton and early development in the tropical sea urchin *Echinometra lacunter*. *J. Exp. Mar. Biol. Ecol.* 236: 291–305.

Sibly, R. and Calow, P. (1986) Why breeding earlier is always worthwhile. *J. Theor. Biol.* 123: 311–319.

Snow, D.R. and Marsden, J.R. (1974) Life cycle, weight and possible age distribution in a population of *Nereis virens* from New Brunswick. *J. Nat. Hist.* 8: 513–527.

Southgate, T., Wilson, K., Cross, T.F. and Myers, A.A. (1984) Recolonisation of a rocky shore in S.W. Ireland following a toxic bloom of the dinoflagellate, *Gyrodinium auroleum*. *J. Mar. Biol. Assoc. UK* 64: 485–492

Starr, M., Himellmann, J.H. and Theriault, J.C. (1990) Direct coupling of marine invertebrate spawning and phytoplankton blooms. *Science* 247: 1071–1074.

Starr, M., Himellmann, J.H. and Theriault, J.C. (1992) Isolation and properties of a substance from the diatom *Phaeodactylum tricornatum* which induces spawning in the echinoderm *Strongylocentrotus droebachiensis*. *Mar. Ecol. Progr. Ser.* 79: 275–287.

Tyler, P.A. (1988) Seasonality in the deep-sea. *Oceanogr. Mar. Biol. Annu. Rev.* 26: 227–258.

Tyler, P.A. and Young, C.M. (1992) Reproduction in marine invertebrates in "stable" environments: the deep sea model. *Invert. Reprod. Devl.* 22: 185–192.

Tyler, P.A., Grant, A., Pain, S.L. and Gage, J.D. (1982) Is annual reproduction in the deep sea echinoderms a response to variability in their environment? *Nature* 300: 747–749.

Wang, W. and Olive, P.J.W. (1996) Confocal and cryomicroscope study of fractional volume in the cryopreservation of a differentiated metazoan. *Cryobiology* 33: 636.

Watson, G.J., Williams, M.E. and Bentley, M.G. (2000) Can synchronous spawning be predicted from environmental parameters? A case study of the lugworm *Arenicola marina*. *Mar. Biol.* 136: 1003–1017.

Williams, M.E. (1998) Fertilisation success of marine invertebrates. Ph.D. thesis, University of St Andrews.

Zeeck, E., Hardege, J.D. and Bartels-Hardege, J.D. (1990) Sex pheromones and reproductive isolation in two nereid species, *Nereis succinea* and *Platynereis dumerilii*. *Mar. Ecol. Progr. Ser.* 67: 183–188.

18

Overwintering biology as a guide to the establishment potential of non-native arthropods in the UK

J.S. Bale and K.F.A. Walters

1. Introduction

Temperature is widely regarded as the most important environmental variable which acts through physiological processes (development, reproduction, survival) to determine ecological outcomes including species distributions, abundance and range margins. These linkages between abiotic factors, physiology and ecology provide a framework by which understand the ways in which temperature can influence the establishment of species transported between different climatic zones. Introductions of alien species into the UK are mainly 'accidental', involving phytophagous insects associated with plants and plant products, which can subsequently become established as pest species, for example the Western flower thrips *Frankliniella occidentalis* on various glasshouse crops. Other introductions are 'intentional', the most obvious examples being novel biological control agents, natural enemies of pest species such as the predatory mirid *Macrolophus caliginosus* against the glasshouse whitefly *Trialeurodes vaporariorum*.

The discovery of a reliable predictive relationship between aspects of the thermal biology of alien species and the probability of establishment in a new country would represent a significant development in the strategic planning and management of such events in a number of key areas: identifying environments where the species could survive (glasshouses, open fields); indicating the risk of spread to neighbouring countries (e.g. within Europe); and providing a longer-term perspective, such as the impact of climate warming. This review outlines the risks associated with alien introductions, summarizes the theoretical relationships of insect thermal biology which affect establishment in new environments, and describes a series of experiments on closely related native and non-native phytophagous species designed to evaluate the use of thermal data as part of a risk analysis protocol for alien species in the UK.

Environment and Animal Development: Genes, Life Histories and Plasticity, edited by D. Atkinson and M. Thorndyke.
© 2001 BIOS Scientific Publishers Ltd, Oxford.

2. Risks associated with alien introductions

The risks associated with the introduction of non-native species differ between plant feeding and natural enemy species. Herbivorous insects may become established as crop pests in the new environment, as is anticipated with the potential invasion into the UK of the Colorado beetle, *Leptinotarsa decemlineata*, from continental Europe, or utilize rare plant species as a food resource. Most introductions of novel biological control agents are made for release into glasshouses, such as the use of the parasitoid wasp, *Encarsia formosa*, against the glasshouse whitefly, *T. vaporariorum*, with the assumption that any escaping individuals will be unable to survive at the lower temperatures encountered in outdoor situations. Whilst there is the risk that the introduction of an additional natural enemy species to a glasshouse may disrupt the balance in existing control schemes, arguably the risks are greater if the species is able to establish in the wider environment and then parasitize or predate non-target species, some of which may be rare or endangered. Risk analysis following the introduction of both plant feeding and natural enemy species is therefore a two-stage process involving both abiotic and biotic factors: (i) is the new climatic environment favourable for survival, development and reproduction; (ii) are there suitable food resources available (plants or prey), and are any of these species of conservation importance.

2.1 *Abiotic factors affecting establishment*

Temperature exerts a dominant effect on two physiological processes which affect the establishment of both plant feeding and natural enemy species in new climatic environments. First, there must be an adequate thermal budget (number of day degrees) above the developmental threshold to allow the species to develop from egg to adult and then to reproduce. Second, the species must be able to produce an overwintering stage, capable of surviving at low temperature, often in a starved condition. These two requirements therefore form the basis of the experimental programme used to assess the establishment potential of alien species in the UK. A number of UK-resident species overwinter successfully without entering a diapause stage, suggesting that such obligatory or facultative dormancy, whilst advantageous, is not essential for winter survival.

2.2 *Biotic factors affecting establishment*

The most likely biotic interactions affecting non-native herbivorous insects are the availability of one or more suitable host plants, competition for such resources, and the activity of any natural enemy species. The analogous factors for novel biological control agents include availability of prey, competition for hosts (parasitoids) or prey (predators), and the activities of their own natural enemies such as hyperparasitoids. For both types of introduction, availability of a food supply and the effects of competition are more likely to be important in the short term since the transportation of alien species rarely includes the simultaneous movement of their natural enemy fauna.

3. Theoretical relationships affecting establishment

In the temperate climate of the UK the two processes that are most affected by temperature are development and winter survival. For both of these ecophysiological relationships there is a theoretical framework within which to place experimental data.

3.1 *Effects of temperature on development*

For any species of insect there is a temperature below which there is no development, commonly known as the developmental threshold. Although it is usual to refer to this threshold as a feature of the species, it is important to note that there will be variability in this value between individuals and between different stages of development (Hart *et al.*, 1997). The developmental threshold is used to calculate the thermal budget (number of day degrees above the threshold required to complete one generation), and in turn the voltinism (number of generations per year).

When samples of insects drawn from the same population are reared at a range of constant temperatures (e.g. 5–35°C), the rate of development usually shows a characteristic pattern (*Figure 1*). The relationship is approximately linear at intermediate temperatures (zone B in *Figure 1*), with an asymptotic decrease at lower temperatures (zone A) and a rapid decrease in rate at temperatures above an optimum (zone C Campbell *et al.*, 1974). The relationship between the rate of development and temperature in zone B can be approximated to a straight line (expressed as $y = a + bt$); extrapolation of this line to the temperature (x) axis provides an estimate of the developmental threshold, calculated as $-a/b$.

The thermal budget (day degrees per generation) is derived from the developmental threshold by the equation $D \times (R{-}T)$, where R is the rearing temperature, T the threshold and D the number of days required to complete development at temperature R. The thermal budget is therefore a constant. As an example, if an insect with a threshold of 5°C is reared at 10°C and takes 60 days to complete development from egg to adult, the thermal budget is $60 \times (10{-}5)$ or 300 day degrees. If the calculation was based on data obtained at 20°C in the knowledge that the threshold was 5°C and the thermal budget was 300 day degrees, it would be predicted that development would be completed in 20 days, that is $20 \times (20{-}5)$. A generalized estimate of the thermal budget can be obtained as the reciprocal of the gradient of the straight line, $1/b$ (Eckenrode and Chapman, 1972; Campbell *et al.*, 1974; Lamb, 1992).

Difficulties can arise in achieving an accurate estimate of the development threshold (and hence the thermal budget and voltinism), because at temperatures close to the

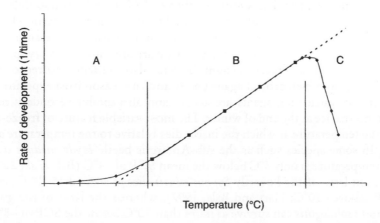

Figure 1. *Generalized relationship between rate of development and temperature (modified after Campbell et al., 1974).*

threshold, the rate of development is not linearly related to temperature (zone A of *Figure 1*). Various analytical methods are available to overcome these problems, as discussed in Hart *et al.* (1997).

With an accurate assessment of the developmental threshold and a consistent value for the thermal budget across a range of temperatures, the day degree requirements can be compared with meteorological records for different microhabitats (soil, land surface, aerial locations) and different climatic regions of a country, to obtain an estimate of the likely number of generations that could be completed in an annual cycle (Hart *et al.*, 1997; McDonald *et al.*, 1998).

3.2 *Effects of temperature on survival*

In the temperate climate of the UK the most likely deleterious effect of temperature is the impact on survival in winter. The low temperature mortality of insects is a function of both the temperature and the duration of exposure; within the potentially lethal range, the lower the temperature the more rapidly the insect will die. In ecological terms, it is the interaction between the severity of the cold stress experienced in winter and the cold tolerance of individual insects that determines the level of post-winter survival and hence the size of the base population for subsequent development and reproduction (Bale, 1987). The effects of low temperature in winter on the spring abundance and outbreaks of the autumnal moth, *Epirrita autumnata*, in Scandinavia (Nilssen and Tenow, 1990; Tenow and Nilssen, 1990) and anholocyclic clones of the aphids *Myzus persicae* and *Sitobion avenae* in the UK (Harrington *et al.*, 1991) exemplify the economic importance of these inter-relationships in pest species.

It has been a common perception that freezing is a major threat to overwintering insects, leading to the view that there are two main strategies of cold hardiness: freeze tolerance and freeze avoidance. It is certainly the case that some insects can survive at temperatures below that at which they freeze (the so-called supercooling point or SCP) whilst others survive at very low sub-zero temperatures and die only if they freeze. Also, there is now a clear understanding of the seasonal changes in the physiology and biochemistry of insects that enhance their winter cold hardiness. In most freeze-tolerant insects, freezing is initiated above –10°C in 'safe' extracellular areas of the body by ice nucleating agents (proteins). Polyols such as glycerol which accumulate from shifts in metabolic pathways associated with the diapause state and increase further in concentration at low temperature are thought to cryoprotect the partially frozen tissues. Freeze-tolerant insects also contain antifreeze (thermal hysteresis) proteins which protect against early and late season frost exposures before and after the ice nucleating agents are active, and also inhibit secondary recrystallization of internal ice at the end of winter. The most variable feature of freeze-tolerant species is the temperature at which the insect dies relative to the temperature at which it freezes. In some species such as the sub-Antarctic beetle *Hydromedion sparsutum* the lethal temperature is only 5°C below the mean SCP of –4°C (Bale *et al.*, 2000). The freeze tolerant larva of the hoverfly *Syphus ribesii* dies at around 15–20°C below the mean SCP (above –10°C; Hart and Bale, 1997), whereas the larva of the goldenrod gallfly *Eurosta solidaginis* can survive at more than 30°C below the SCP of –8°C (Bale *et al.*, 1989). In all these species, the difference between the freezing and lethal temperatures has been assessed in brief exposures under laboratory conditions; in natural environments with prolonged and severe frost exposure, the lethal temperature may be higher than that observed in the laboratory.

As the description implies, freeze-intolerant insects are killed by freezing and must therefore avoid this lethal event to survive. These 'freeze-avoiding' species contain the same polyols and antifreeze proteins found in freeze-tolerant insects, but with a different function. The polyols act as antifreeze agents, depressing the 'whole body' supercooling point of the insect by their colligative properties, that is, there is an inverse relationship between the increasing concentration of glycerol and other polyols from summer to winter and the progressive lowering of the SCP. In mid-winter, most freeze-avoiding larvae, pupae and adult insects freeze between –20 and –30°C and overwintering eggs down to –40°C. In these species, winter cold hardiness is further enhanced by antifreeze proteins that 'stabilize the supercooled state' by binding to embryonic ice crystals, hence preventing the formation of larger crystals with sufficient critical mass to initiate spontaneous nucleation in the insect.

The problem with the freeze-tolerance–freeze-avoidance two-strategy concept is that on a worldwide basis there are very few insects that are freeze-tolerant and perhaps even fewer that are killed only when they freeze. Most insects die at temperatures above the laboratory-measured SCP with mortality increasing at lower temperatures and in longer exposures. Bale (1996) proposed five categories of cold tolerance that represented insects from the most to the least cold hardy: freeze tolerance, freeze avoidance, chill tolerant, chill susceptible and opportunistic survival. Within each of these categories, there is also some gradation of cold tolerance, but also one or more features that are common to all species in the group. Thus, all freeze-tolerant species can tolerate ice formation in their body tissues and fluids, although there is variation between species in the SCP, lower lethal temperature, activity of ice nucleating agents, site of nucleation and ability to survive intracellular freezing. The gallfly, *Eurosta solidaginis*, is a typical example with an SCP of –8°C initiated by an ice nucleating protein and a lower lethal temperature below –40°C (Bale *et al.*, 1989).

Apart from freeze-tolerant species, the vast majority of insects have low SCP in winter, often below –20°C. However, from the many species that have now been studied from different climatic zones, very few seem able to survive through prolonged periods of low sub-zero temperatures without the occurrence of some 'non-freezing' mortality above the SCP. One such species is the autumnal moth, *Epirrita autumnata*, in northern Scandinavia. The overwintering eggs of this moth freeze at around –36°C; in outbreak years, the areas of most damage to the birch forest by defoliating caterpillars in spring correlate well with locations where the minimum winter temperature was above –36°C, whereas in other well-defined sites where the minimum was below the SCP, few if any eggs survived and there was no damage to the trees. This suggests that eggs of *E. autumnata* can survive at very low temperatures close to the SCP for various periods of time without significant mortality (Nilssen and Tenow, 1990; Tenow and Nilssen, 1990).

In the majority of overwintering insects, pre-freeze mortality appears to be the most common cause of death at low temperature. In some species, termed 'chill tolerant', individuals can survive for long periods of time at sub-zero temperatures, although a proportion of the population die above the mean SCP; the Antarctic mite, *Alaskozetes antarcticus*, shows this pattern with a winter SCP of –30°C and 67% survival after 100 days at –20°C (Cannon, 1987). In the less severe climate of the UK, the beech leaf mining weevil, *Rhynchaenus fagi*, freezes at –25°C in winter, but there is substantial mortality in populations maintained at –10° and –15°C (Bale, 1991).

The main distinguishing feature between chill-tolerant and chill-susceptible species is the effect of the duration of exposure on mortality. Whereas chill-tolerant insects can survive for weeks and usually months at sub-zero temperatures, chill-susceptible species die within minutes or a few hours. As examples, anholocyclic clones of the aphids *Myzus persicae* and *Sitobion avenae* freeze at –25°C in winter but in exposures of only 1 min, 50% are killed above –10°C and there are few survivors below –15°C (Clough *et al.*, 1990).

The least cold hardy group are species that are unable to survive below their developmental threshold for any length of time and therefore must opportunistically seek out highly protected environments to ensure survival. In the UK, pupae of the housefly, *Musca domestica*, freeeze at –15°C but over 90% die after 5 days at 0°C (Coulson and Bale, 1990).

It is evident that across the spectrum from *E. solidaginis* to *M. domestica* there is a continuum of decreasing cold hardiness, where in most species survival depends on both the temperature and duration of exposure. In natural overwintering populations, such time-dependent mortality will be more common than extreme exposures close to the SCP, highlighting the need to combine a range of laboratory and field experiments when assessing the cold hardiness and winter survival of insects.

4. Establishment of alien phytophagous insects – a comparative experimental approach

Most accidental introductions of alien phytophagous species into the UK originate from countries where the climate is either similar or, more often, warmer. For many of these likely introductions there are related species that are resident and well established in the UK. With this knowledge it is possible to select pairs of resident and non-resident species with similar biologies and design experiments to compare various aspects of their ecophysiology. In general, where the resident and non-resident species have similar developmental thresholds, rates of development and cold tolerance, the risk of establishment will be greater than where, for instance, the non-resident species has a higher threshold, slower rate of development and is less cold tolerant. The aim therefore is to identify one or more 'thermal indices' which differ between resident and non-resident species and could be used as a reliable predictor of establishment potential in the UK.

The validity of this approach has been tested with two leaf-chewing lepidopteran pests, the large yellow underwing moth, *Noctua pronuba* (resident) and the dark sword grass moth, *Agrotis ipsilon* (non-resident); and two sap-sucking pests, the cabbage whitefly, *Aleyrodes proletella* (resident) and the tobacco whitefly, *Bemisia tabaci* (non-resident).

4.1 *Comparative thermal indices*

The thermal indices assessed in these experiments comprise:

(i)   *developmental threshold* – the temperature below which there is no development;
(ii) *thermal budget* – the number of day degrees above the developmental threshold required to complete one generation or other specified phase of development;

(iii) *freezing temperature (or supercooling point)* – the temperature at which freezing (spontaneous nucleation) occurs when an organism is cooled at a standard rate (e.g. 1°C min⁻¹);

(iv) *lethal temperature* – the temperature required to kill a given proportion of a population (e.g. 10, 50, 90%) in a constant time exposure (e.g. 1 min);

(v) *lethal time* – the time required to kill a given proportion of a population in a constant temperature exposure (e.g. –5°C);

(vi) *field survival* – the level and duration of survival of populations of the subject species under winter field conditions.

4.2 Noctua pronuba *and* Agrotis ipsilon

A summary of the thermal data obtained on *Noctua pronuba* and *Agrotis ipsilon* is shown in *Table 1*. The developmental thresholds of the two species are similar (6.1 and 5.6°C respectively); also, *A. ipsilon* requires fewer day degrees (DD) above the threshold to develop from egg to adult. It would appear therefore that temperatures in the UK are more than adequate to sustain the development of *A. ipsilon* through its univoltine life cycle and that the failure of the species to establish cannot be attributed to this factor.

In contrast, the data on the cold tolerance and survival of the overwintering larvae reveal clear and consistent differences between the species. Although larvae of *N. pronuba* and *A. ipsilon* both freeze at around –12°C, when placed in an acclimation regime, the SCP of the former is lowered to –15.4°C, but there is no response in *A. ipsilon*. The lethal temperature which kills 50% (LTemp$_{50}$) of acclimated *N. pronuba* (–15.8°C) is similar to the mean SCP (–15.4°C), indicating that in these brief 1 min exposures, larvae do not die until they freeze. With *A. ipsilon*, the LTemp$_{50}$ (–8.9°C) is over 3°C above the mean SCP, suggesting that some larvae are effectively dead before

Table 1. Comparative thermal data for a resident and a non-resident species of noctuid moth

Thermal index	Noctua pronuba (resident)	Agrotis ipsilon (non-resident)
Developmental threshold ± 95% confidence limits (°C)	6.1 4.2 – 8.4	5.6 3.8 – 7.6
Thermal budget ± 95% confidence limits (DD)	1086 986 – 1166	697 655 – 746
SCP non-acclimated ± SE (°C)	–12.1 ± 0.3	–12.4 ± 0.5
SCP acclimated ± SE (°C)	–15.4 ± 0.5	–12.6 ± 0.5
LTemp$_{50}$ acclimated ± 95% fiducial limits (°C)	–15.8 –15.0 to –16.5	–8.9 –7.4 to –10.1
LTime$_{50}$ acclimated at –5°C ± 95% fiducial limits (days)	9.0 6.2 – 11.2	5.8 5.2 – 6.2
Field survival	>6 months	6 weeks

DD, day degrees

they reach their freezing temperature. A further indication of this difference is gained from the lethal time data where, in a constant exposure at −5°C, the LTime$_{50}$ is markedly longer for *N. pronuba*.

When placed in the field at the beginning of winter with a food supply (there is no diapause in either species and larvae feed on 'warm' days in winter), approximately 40% of *N. pronuba* survived through to spring, pupated and emerged, whereas all the *A. ipsilon* larvae were dead after 6 weeks.

It is evident that overwintering larvae of *N. pronuba* are more cold hardy than *A. ipsilon*. The general lack of winter cold hardiness in overwintering larvae of *A. ipsilon* is therefore a likely explanation for the failure of this species to become established in the UK, despite the known occurrence of adult moths in summer.

4.3 Aleyrodes proletella *and* Bemisia tabaci

The cabbage whitefly, *Aleyrodes proletella*, is commonly found on various brassica species in winter, whereas *Bemisia tabaci* is regarded as a glasshouse pest in western Europe and might therefore be expected to be less well adapted to low temperatures. The data in *Table 2* support this view.

There is a substantial difference of over 3°C in the developmental thresholds of the two whitefly species. The non-linearity of the development data prevented a calculation of the day degree requirement per generation, but at temperatures above the respective thresholds, the development rate was faster in *A. proletella* than in *B. tabaci* (e.g. 20 and 24 days, respectively, from egg to adult at 28°C).

The freezing temperatures of non-acclimated and acclimated *A. proletella* and *B. tabaci* were similar, and all below −20°C. These values (which may be significantly different due to the small level of variation in the data) are typical of sap-feeding Hemiptera. However, as with a number of well-studied aphid species (Bale *et al.*,

Table 2. Comparative thermal data for a resident and a non-resident species of whitefly

Thermal index	Aleyrodes proletella (resident)	Bemisia tabaci (non-resident)
Developmental threshold ± 95% confidence limits (°C)	9.5 7.9 – 11.5	13.3 10.6 – 17.0
Thermal budget (egg to adult at 28° C in days)[a]	20	24
SCP non-acclimated ± SE (°C)	−23.4 ± 0.2	−25.5 ± 0.2
SCP acclimated ± SE (°C)	−23.4 ± 0.2	−23.9 ± 0.3
LTemp$_{50}$ acclimated ± 95% fiducial limits (°C)	−21.2 −20.4 to −21.7	−5.8 −4.3 to −6.9
LTime$_{50}$ acclimated ± 95% fiducial limits at −5°C (days)	8.1 6.4 – 9.2	1.2 0.6 – 1.7
Field survival	>6 months	3 weeks

[a] Due to the non-linearity of the data, development times from egg to adult at 20°C are shown.

1988), the supercooling point data is unlikely to be directly relevant to their cold hardiness, but provides the baseline temperature above which to investigate pre-freeze mortality and sub-lethal effects. In fact, the LTemp$_{50}$ of acclimated *A. proletella* (−21.2°C) is close to the mean SCP, indicating that, in such short exposures, there is little non-freezing death; in contrast, 50% of *B. tabaci* are killed in 1 min exposures over 15°C above the mean SCP. The LTime data also show this striking difference in cold tolerance between the resident and non-resident species in which 50% of *B. tabaci* are killed after about a day at −5°C. Similar differences were observed in the field exposure where the *B. tabaci* population died out after only 3 weeks, but *A. proletella* survived to the end of winter.

Collectively, the results from the whitefly experiments indicate that whilst *B. tabaci* has the potential to become a major glasshouse pest in the UK, any escaping individuals would die rapidly with the onset of winter.

5. Comparisons between different taxonomic groups

When viewed across the four species, a clear pattern emerges in which, as might be expected, the two resident species survive through an entire UK winter whereas the two non-resident species die out, but after different periods of time. The most useful risk assessment system would be one that did not involve, as a matter of routine, winter-long field experiments. Given that clear differences were apparent in winter survivorship, the maximum field survival times of the four species were plotted against different laboratory indices of their cold hardiness. A strong relationship was observed between time taken to kill 50% of each species at −5°C and their field survival (log$_{10}$ days; *Figure 2*). An independent validation of this relationship was obtained with the western flower thrips, *Frankliniella occidentalis*. This thrip species was first recorded in the UK glasshouses in 1986 and, despite attempts at eradication, has since become naturalized in Britain. With an LTime$_{50}$ of 7 days at −5°C (McDonald *et al.*, 1997), it would be predicted that overwintering populations of *F. occidentalis* would survive in the field for about 9–10 weeks, and this was found to be the case (*Figure 2*).

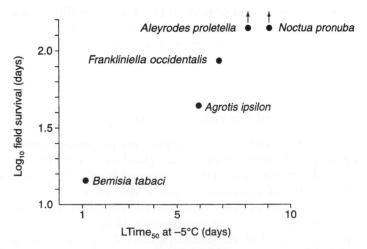

Figure 2. Relationship between LTime$_{50}$ at −5°C and field survival in winter of resident and non-resident herbivorous insects (arrows indicate survival beyond the end of the observation period).

6. Discussion

The establishment of a species in a new environment is a two-stage process determined by a combination of abiotic and biotic factors. If the species is unable to develop and reproduce, or overwinter successfully, establishment will not occur, even if suitable host plants are available and natural enemy pressure is low. Conversely, although an alien species may be ecophysiologically adapted to its new environment, the lack of food resources may prevent establishment.

This study focused on the influence of thermal factors on the establishment potential of two non-resident phytophagous insects. For most invading species, information is available on the host plant range and preferences from studies carried out in the native country, which can be matched against the occurrence of the same or related plants in the new environment. In general however, there is much less data on the thermal biology of the insects, especially on low-temperature survival.

It is now widely recognized that, for most insects, the supercooling point, in isolation, is not a reliable indicator of cold tolerance; the SCP simply indicates the lower limit of the supercooling process in very brief exposures at increasingly lower sub-zero temperatures, which in most species is biologically unrelated to the temperature at which the insect dies. For the same reason, data on the lethal temperature in similarly short exposures can also be misleading; although 'LTemp' data provide a rapid indication of mortality above the SCP, the absence of a 'time factor' masks the scale of mortality that may occur at higher sub-zero temperatures in extended exposures.

As an indication of long-term survival under natural conditions it might be expected therefore that laboratory experiments which include different temperature–time combinations would be most informative as an indicator of field survival. In these experiments, insects were exposed at constant temperatures (0, –5 and –10°C), whereas in nature temperatures fluctuate diurnally and over longer periods of time. However, the purpose of these laboratory experiments was not to mimic natural temperature patterns (which is virtually impossible), but to identify a quantifiable laboratory stress regime which correlated with the duration of survival of all the tested species under field conditions. On the basis of the results obtained with related resident and non-resident species of noctuid moth and whitefly, subsequently 'validated' with a species from a different taxonomic group, exposure at –5°C appears to be a good predictor of field survival in winter. The strength of this relationship now requires data from other resident and non-resident phytophagous insects, preferably from a range of different taxa.

7. Wider application of the risk assessment protocol

A number of alien insects and related arthropods are purposely introduced into the UK as biological control agents, mainly against glasshouse pests. Two such introductions, the parasitoid *E. formosa* against the glasshouse whitefly *T. vaporariorum* and the predatory mite *Phytoseiulus persimilis* against the glasshouse red spider mite *Tetranychus urticae* have proved highly successful and pose no threat to other potential prey species outside of the glasshouse environment. With more recent introductions, such as the predatory mirid *M. caliginosus* (against *T. vaporariorum*) and the mite *Amblyseius californicus* (against *T. urticae*), there have been reports of both species occurring outside of glasshouses in winter. Occurrence outside of glasshouses is not in itself evidence of establishment, and the virtual year-round production in glasshouses means that escaping individuals can recolonize favourable environments more or less immediately and do not necessarily have to overwinter in unprotected sites for many months. However, recent laboratory and winter field experiments with *M. caliginosus* and a predatory ladybird *Delphastus catalinae* (both of which are under consideration as novel controls for *T. vaporariorum*) have revealed clear differences in their cold tolerance and indicated that *M. caliginosus* has the potential to survive outdoors throughout a UK winter (Bale, Hart, Tullett and Walters, unpublished data).

In summary, an emerging relationship has been identified in which exposure of taxonomically related resident and non-resident phytophagous insects at low and sub-zero temperatures in the laboratory can be used to reliably predict their winter survival in the field, and hence their longer term establishment potential. Whilst initially developed as a risk assessment procedure for alien pest species, the system could be usefully applied to novel biological control agents introduced into glasshouse environments.

Acknowledgements

We are grateful to Arthur Austin, Jamie McDonald, Andrew Hart, Andrew Tullett and Barbara Russell for carrying out the experimental programmes and to MAFF and DETR for funding the research.

References

Bale, J.S. (1987) Insect cold hardiness: freezing and supercooling – an ecophysiological perspective. *J. Insect. Physiol.* 33: 899–908.

Bale, J.S. (1991) Insects at low temperature: a predictable relationship? *Func. Ecol.* 5: 291–298.

Bale, J.S. (1996) Insect cold hardiness: a matter of life and death. *Eur. J. Entomol.* 93: 369–382.

Bale, J.S., Harrington, R. and Clough, M.S. (1988) Low temperature mortality of the peach potato aphid *Myzus persicae. Ecol. Entomol.* 13: 121–129.

Bale, J.S., Hansen, T.N. and Baust, J.G. (1989) Nucleators and sites of nucleation in the freeze tolerant larvae of the gallfly *Eurosta solidaginis* (Fitch). *J. Insect Physiol.* 35: 291–298.

Bale, J.S., Block, W. and Worland, M.R. (2000) Thermal tolerance and acclimation response of the sub-Antarctic beetle *Hydromedion sparsutum. Polar Biol.* 23: 77–84.

Campbell, A., Frazer, B.D., Gilbert, N., Gutierrez, A.P. and Mackauer, M. (1974) Temperature requirements of some aphids and their parasites. *J. Appl. Ecol.* 11: 431–438.

Cannon, R.J.C. (1987) Effects of low temperature acclimation on the survival and cold tolerance of an Antarctic mite. *J. Insect Physiol.* 33: 509–521.

Clough, M.S., Bale, J.S. and Harrington, R. (1990) Differential cold hardiness in adults and nymphs of the peach potato aphid *Myzus persicae. Ann. Appl. Biol.* 116: 1–9.

Coulson, S.J. and Bale, J.S. (1990) Characterisation and limitations of the rapid cold hardening response in the housefly *Musca domestica. J. Insect Physiol.* 36: 207–211.

Eckenrode, C.J. and Chapman, R.K. (1972) Seasonal adult cabbage maggot populations in the field in relation to thermal unit accumulations. *Ann. Entomol. Soc. Am.* 65: 151–156.

Harrington, R., Howling, G.G., Bale, J.S. and Clark, S. (1991) A new approach to the use of meteorological and suction trap data in forecasting aphid problems. *EPPO Bull.* 21: 499–505.

Hart, A.J. and Bale, J.S. (1997) Evidence for the first strongly freeze tolerant insect found in the UK. *Ecol. Entomol.* 22: 242–245.

Hart, A.J., Bale, J.S. and Fenlon, J. (1997) Developmental threshold, day degree requirements and voltinism of the aphid predator *Episyrphus balteatus* (Diptera: Syrphidae). *Ann. Appl. Biol.* 130: 427–437.

Lamb, R.J. (1992) Developmental rate of *Acyrthosiphum pisum* (Homoptera: Aphididae) at low temperatures: implications for estimating rate parameters for insects. *Environ. Entomol.* **21**: 10–19.

McDonald, J.R., Bale, J.S. and Walters, K.F.A. (1997) Low temperature mortality and overwintering of the western flower thrips (*Frankliniella occidentalis*). *Bull. Entomol. Res.* **87**: 497–505.

McDonald, J.R., Bale, J.S. and Walters, K.F.A. (1998) The effect of temperature on the development of the western flower thrips (*Frankliniella occidentalis*). *Eur. J. Entomol.* **95**: 301–306.

Nilssen, A. and Tenow, O. (1990) Diapause, embryo growth and supercooling capacity of *Epirrita autumnata* eggs from northern Fennoscandia. *Entomol. Exp. Applic.* **57**: 39–55.

Tenow, O. and Nilssen, A. (1990) Egg cold hardiness and topoclimatic limitations to the outbreaks of *Epirrita autumnata* in northern Fennoscandia. *J. Appl. Ecol.* **27**: 723–734.

Index of italicized names